Organic Synthesis

The Science behind the Art

Organic Synthesis

The Science behind the Art

W. A. Smit
Zelinsky Institute of Organic Chemistry, Moscow, Russia

A. F. Bochkov
Institute of Biochemical Physics, Moscow, Russia

R. Caple
University of Minnesota, Duluth, Minnesota, USA

Front cover illustration taken from an original idea by Boris Gorovoy

ISBN 0-85404-544-9

A catalogue record for this book is available from the British Library.

Published by The Royal Society of Chemistry,
Thomas Graham House, Science Park, Milton Road, Cambridge CB4 4WF, UK

For further information see our web site at www.rsc.org

Typeset by Paston Press Ltd., Loddon, Norfolk
Printed by Redwood Books Ltd., Trowbridge, Wiltshire

Preface

At the beginning of its history organic chemistry was perceived as a branch of natural science dealing with a specific type of compounds, namely, those isolated from organisms, living or fossils. But pretty soon our great predecessors, who laid the foundations of organic chemistry, found themselves engaged in a feverish drive aimed at the synthesis of hundreds and hundreds of compounds which never before existed on this planet and have no resemblance to natural compounds. At that time, it came as a startling observation that this newly-born science may serve not only as an instrument for the discovery and study of natural phenomena, but that it is also capable of creating a wide variety of unnatural compounds, an entirely new object of exploration and practical utilization. Since then, owing to cumulative activity of several generations of chemists, more than a dozen million new compounds have been prepared and, as a result, at the end of this century we live in a world which is composed, at least to a significant extent, of artificially created substances and materials.

As a science in its own right, organic synthesis emerged at the beginning of this century, when chemists started to master the skills of manipulating compounds in a controlled and predictable fashion which eventually elaborated an arsenal of tools required for the preparation of various target products from simple starting materials. The spectacular progress achieved from this (especially over the last few decades in the development of synthetic methods), complemented by the discovery of new approaches to the analysis of synthetic problems, changed the very image of organic synthesis dramatically. The complexity of tasks increased tremendously and by now one may safely claim that almost any compound, isolated from natural sources or conceived in the chemist's mind, can be synthesized with a reasonable amount of time and effort. Modern organic synthesis, with its spirit of the most daring endeavor, coupled with the craftsmanship of the design and assemblage of diverse molecular structures of formidable complexity, may serve as a convincing illustration to the prophetic claims of M. Berthelot (1860) about the intrinsic capacity for creation as a distinctive feature of the science of chemistry. It also seems obvious that the outstanding synthetic achievements of this century should be listed properly among the top intellectual accomplishments of human genius.

Hundreds of research papers devoted to the problems of total synthesis are published annually. A formal and non-personal style of presentation, generally adopted for scientific publications, at times looks like as if it is specifically designed to hide all emotional and creative aspects of the underlying research stories most carefully. But nonetheless, quite often one cannot help but feel a sort of excitement mixed with admiration upon reading such matter-of-fact presentations which describe a successful synthesis of some molecular ensemble, incredibly sophisticated and truly marvellous for the chemist's eye. These feelings are not caused only by a spectacular manifestation of the predictive power and logical rigor of the scientific approach of modern synthesis, but also because of the aesthetic appeal of the synthetic goals and elegance of the elaborated problem solutions. It is this alloy of science and art that prompted the title of this book and, in fact, also determined its specific genre.

A lucky chance at the dawn of 'perestroika' and 'glasnost' brought all three of us together on a canoe trip in the spectacular region of Karelia in northern Russia. During this trip, over a campfire at white nights of this latitude, we spent many hours sharing our experiences and views about various aspects of our professional activity in organic chemistry. We also discussed a book previously authored by A.F. Bochkov and W.A. Smit. This text, published in Russia in 1989 (Nauka Publishers) and titled *Organic Synthesis* was actually an effort to provide an overview of the role of organic synthesis in chemistry and, in general, in science. The book turned out to be popular in Russia among both organic chemistry professionals and students, as well as those who used to have a rather peripheral contact with this area of organic chemistry. The success of this publication prompted us with the idea of writing jointly an updated, more detailed and elaborated English version of the book, based essentially upon the concepts of the Russian prototype.

We were well aware that a number of excellent monographs and textbooks had been already published that described both the synthetic methods and strategy of contemporary organic synthesis, which are still of exceptional value for teaching synthetic craftsmanship. Yet it was our feeling that almost no attempts had been made, on the whole, to highlight this amazing and flourishing area of intellectual activity from a historical viewpoint in conjunction with the analysis of its modern achievements, problems and major trends.

We fully understood, of course, that it was both an impossible and unnecessary task to be exhaustive and all-encompassing in such a text. As we saw it, our main objective was to present the aesthetics and ideology of pursuits in the area of organic synthesis, the evolution of the methodology specifically designed for the solution of tactical and strategic problems, and to discuss the main principles of molecular design as a truly challenging and most promising trend of current synthetic endeavors. In short, we strived to concentrate on those aspects which actually constitute the scientific background of the art of organic synthesis.

It is our hope that this book will prove to be stimulating reading to the young chemists wishing to pursue a career in this field, perhaps as a supplementary text to an advanced course in organic chemistry. The Russian forerunner of this book was used successfully in exactly this role. We furthermore hope that it might also be of interest to all of those who have already been touched, directly or indirectly, by this beautiful and highly creative area of modern science and who would like to learn more about its appeal and promise.

Acknowledgements. The very way in which this book was written required its careful reading by a number of experts, and their support and encouragement was most valuable to us. We are especially indebted to Profs. Roald Hoffmann of Cornell University, Fred Menger of Emory University, and Bob Carlson and Victor Zhdankin of the University of Minnesota-Duluth who took the trouble of reading the manuscript and made many of the most valuable comments. Our special thanks go to the contribution made by Prof. Becky Hoye of Macalester College, whose energetic and at times very critical comments were truly instrumental to improving the initially created text. We are particularly indebted to Susanne Sharpe of Macalester College for her invaluable and time-consuming assistance in the preparation and editing of the entire manuscript and for her most friendly support of all our efforts. Editorial comments suggested by Elizabeth Icks of the College of St-Scholastica are also highly appreciated. We are most thankful to our Russian colleges, Profs. Oleg Chizhov, Eduard Serebryakov, Nicolai Zefirov, Yuri Ustyniuk, Genrikh Tolstikov and Andrei Simolin, whose comments on the previously published Russian version of the text turned out to be extremely useful for us in making the present book.

We would like to thank the Fullbright and Soros exchange programs, which provided us with the opportunities to visit the respective institutions in Russia and America and thus enabled the drive to complete preparation of the manuscript. During these shuttle visits we enjoyed the hospitality of the Departments of Chemistry of the University of Minnesota-Duluth and Macalester College in the USA and the Zelinsky Institute of Organic Chemistry in Russia. The generous help and valuable support from the faculty and staff of these institutions are most gratefully acknowledged.

Contents

Introduction

> 'There is excitement, adventure, and challenge, and there can be great art in organic synthesis. These alone should be enough, and organic chemistry will be sadder when none of its partitioners are responsive to these stimuli.'
>
> R. B. Woodward, 1956

The term 'organic synthesis' means literally that its major goal is the construction of organic molecules. What for? From what? How? These are questions that face both newcomers to this field as well as experienced professionals.

The answer to the question 'from what?' seems more or less obvious — from simpler molecules. 'From simpler' usually means 'from more available'. Available natural sources of organic compounds include carbon dioxide, raw organic material from fossil sources (petroleum, gas, coal), and living organisms. Their composition ultimately delineates the spectrum of compounds which can be used as starting products for an organic synthesis. For example, a well known material of our century, polyethylene, can be produced in multiton quantities because its synthesis is easily achieved by the polymerization of a simple and available raw product, ethylene. An enormously large area of industrial and laboratory chemistry, dealing with aromatic compounds (polymers, dyes, explosives, medical drugs, *etc.*), is actually based upon the wide occurrence of the common basic element of their structures, the benzene nucleus, in the large number of aromatic hydrocarbons which are isolated during the regular processing of coal and petroleum. Viscose and acetate fibers, nitrocellulose materials and gun powder, and glucose also became industrial products because they are obtained by simple chemical reactions from polysaccharides, the most abundant class of organic compounds on Earth.

In the molecule of polyethylene or, for example, phenol, it is trivial to recognize the structural elements corresponding to available natural precursors and hence to elaborate a logistically simple scheme for the preparation of the target products. However, in the majority of cases the well-trained eye of the professional is required in order to identify the basic fragment(s) present in the complex target molecule which can be derived from a suitable precursor(s). This skill rests primarily in the ability to refer easily to the rich arsenal of synthetic methods, *i.e.* one should be able to answer the question 'how?'. In considering the latter question, however, it becomes clear that by no means can the problem be reduced only to the availability of possible starting materials. For example, it would be tempting to obtain acetic acid from the readily available gases methane and carbon dioxide:

$$CH_4 + CO_2 \rightarrow CH_3COOH$$

On paper, this route seems to be quite reasonable, inasmuch as it involves formally the simple combination of two molecules. In reality this preparation cannot be achieved as indicated in such a straightforward way. Yet, as we will see shortly, it is possible to elaborate indirect routes which will ultimately lead to such a conversion. In fact, the power of modern organic synthesis has reached the level when an organic chemist is able to prepare, at least in principle, 'whatever you need from whatever you choose'. However, this power is by no means a magic wand to be employed arbitrarily at one's will. The might of organic synthesis is based on the knowledge of rigorously established and rather strict laws governing the course of chemical reactions which comprise the set of the basic tools for doing a synthetic job. In every reaction there are formed and/or cleaved some 'specific' bonds between 'specific' atoms. It is this very 'specificity' of the chosen transformations that enables chemists to predict and control the overall results of synthetic operations. Thus the right choice among the set of available reactions is of paramount importance in order to solve the main tactical problem of organic synthesis: how to achieve a selective creation or rupture of the required bond(s) in the assembled structure?

The 'assemblage' of complex molecules from simple precursors most usually involves a step-by-step protocol and thus the entire process is broken into several separate steps, each one aimed at the creation of a particular bond(s) present in the final molecule or, more often, in an intermediate precursor to be employed at a later step(s) of the whole sequence. Only in special cases do these sequential steps turn out to be of the same type, and thus the final goal can be achieved as a result of a single operation (as is the case in the polymerization of ethylene into polyethylene). More usually the pathway of a complex synthesis includes a series of entirely different synthetic steps and realization of each step may represent an independent chemical problem. Furthermore, as a rule, more than one route might be envisaged for the preparation of a target compound and each of the alternative pathways may include different reaction sequences and starting materials. Therefore, in addition to the selection of suitable precursors and reactions for the creation of the chosen bonds in the target molecule, the synthetic chemist has to address a more general and often rather troublesome strategic task, namely the elaboration of an optimal plan for the entire synthesis.

In the rational planning of a synthesis, it is expedient to perform a mental 'disconnection' of the target molecule in order to arrive at the structure of the nearest precursor(s) which can be converted into the required structure with the help of known methods. Theoretically, one may start this disconnection procedure from any site of the target structure and then proceed retrosynthetically by applying the same procedure with any of the emerging precursors, thus arriving eventually at readily available starting materials. Obviously it would not be productive to undertake such an exhaustive search; the selection of a few rational options among a multitude of thus generated alternative pathways might be too formidable a task. In the elaboration of a synthetic strategy, one should also never forget that even the well-established procedures may fail when applied in a specific structural context and thus an otherwise chemically sound synthetic plan may prove to be unworkable. If such a 'misfire' occurs at the initial steps of the synthetic sequence, at most only a few days or weeks are lost. However, if it happens at the concluding step of a lengthy, for example, 40-step synthesis, it might cost an entire year of work, as this failure would never be found until the previous 39 steps were completed. Hence synthetic plans should have the maximum flexibility, with the most risky synthetic operations shifted to the earliest possible step of the entire sequence.

A number of criteria must be considered when making the final selection between the options that emerge for the total synthesis of a given compound. Among the most important are the length of the scheme (the fewer the steps, the better) and anticipated yield at each step; the availability and price of starting compounds and other materials, including solvents, catalysts and adsorbents; the complexity of the equipment needed, *etc*. In order to make an adequate assessment of all these, sometimes contradictory, requirements, one must have both an in-depth knowledge of a rich arsenal of available synthetic methods and a clear understanding of the ultimate goal of the whole endeavor. Here it is of the utmost importance to address the question 'why should this synthesis be undertaken?'. In fact, a synthetic plan designed for industrial application may appear nearly ideal from a purely chemical viewpoint, but nevertheless it might be turned down as absolutely unacceptable owing to cost considerations or the necessity of employing toxic or explosive materials or due to the problem associated with hazardous wastes. On the other hand, application of a reaction that requires an additional and rather meticulous elaboration of optimal conditions (say a heterogeneous catalysis process) can hardly be recommended as a procedure of choice for a laboratory synthesis. Yet this reaction might be extremely promising for the chemical industry because the laborious preliminary investigation may pay off once the procedure has been finely tuned and elaborated into a profitable large-scale process.

The question about the goals of organic synthesis reflects not only narrow professional interests, but in fact ascends to a more global and important problem regarding the destination and usefulness of pure science. Is it really imperative to spend time and money pursuing the goals of pure science which are not likely to bring immediate benefits to humanity in the foreseeable future? The history of modern civilization is ripe with conclusive evidence attesting to the pragmatic utility of even the most esoteric pursuits of scientific endeavor and we need not repeat here the well-known reasoning underlying this assertion. Nevertheless, the issue is never closed completely and the same questions keep emerging with reference to this or that particular area of science. This apparent 'lack of understanding' might be boring or even annoying for the scientists, who always tend to believe in the intrinsic merits and unquestionable values of their own pursuits. Yet, in our opinion, no researcher may feel free of the responsibility to answer these legitimate doubts of the layman.

People that directly or indirectly provide financial support for the development of science have the right to learn why we are so persistent in pursuing our goals. Thus it should not be surprising that from time to time they will question the expediency of some academic investigation that may appear as if conceived with one sole purpose, namely to satisfy the scientist's curiosity at the taxpayer's expense. When the discussion refers to synthetic studies directed at the manufacturing of an artificial foodstuff, then such efforts are likely to get approval almost without question ('There is nothing more indisputable than bread!' — so says the Great Inquisitor in 'The Karamazov Brothers' by Dostoevsky). However, when professionals assess the ingenious synthesis of chlorophyll (Woodward, 1960) as one of the benchmark achievements of organic chemistry, the non-specialists may view this undertaking as, politely speaking, dubious, since any green plant is capable of synthesizing chlorophyll every summer in abundant amounts and without our assistance. Such a perplexity is understandable and it should be clarified. Therefore we start our book with the question 'why?' in regards to the goals of organic synthesis.

This book refers almost exclusively to the laboratory and not industrial organic syntheses. The former is much more diversified in its goals and methods, but the fundamentals of both, of course, are the same. In the final analysis, any industrial synthesis was conceived in the laboratory and differs from ordinary bench chemistry only due to the necessity to satisfy a certain set of economical and technical requirements.

This book is not aimed at the comprehensive coverage of the whole area of organic synthesis. Our goal is to present the ideology and general principles and approaches employed in this branch of organic chemistry. Therefore we had to face a rather difficult task of making the choice of representative examples from an almost innumerable multitude of synthetic studies. The selection of material inevitably bears also the imprinting of the personal scientific interests and experiences of the authors. Nevertheless, it seems to us that inasmuch as the principles of modern organic synthesis bear a universal character, one can almost arbitrarily choose illustrative examples from any area, whether it is the chemistry of aliphatic or aromatic compounds, carbohydrates, organometallics, acylic compounds, or polycyclics.

Organic synthesis is a rather peculiar area of intellectual activity, creative in all its major aspects. Its methodology is based on both logistic and purely heuristic (and not amenable to easy formalization) approaches. Likewise, the immediate result of a synthesis might be not only finding the way to prepare a natural compound, but also the creation of artificial objects which had never existed before in Nature and may fortuitously exhibit an absolutely unexpected set of properties. In this area are merged together such different qualities as a rigorous scientific analysis of natural phenomena with its exact predictions, a search for aesthetically appealing solutions, a deep knowledge of chemistry, and an adroit 'feeling' for compounds, almost an intuitive apprehension of their behavior. That is why outstanding achievements of modern synthesis are often perceived as marvellously created pieces of art, having their intrinsic beauty fused with the expediency and laconism.

We attempted in this book to show not only the basic problems which are dealt with by synthetic chemists, but also a meaning and creative function of their activity. We fully apprehend the futility of attempts to describe our subject in a way equally acceptable for all potential readers, from graduate students to professional synthetic chemists. Nevertheless, it is still our hope that the former will be able to grasp some insights about the appeals of our science while the latter will not chastise us for oversimplification unavoidable in the presentation of complicated problems within the limited volume of this book.

CHAPTER 1

Goals of an Organic Synthesis

The role of organic synthesis in science and in practice is not easily defined in an unambiguous way. To answer the question about the goals of an organic synthesis, one cannot simply refer directly to the application or usefulness of the target compound, even if the term 'usefulness' is understood in the broadest sense. Nevertheless, we would like to start this chapter with just this obvious case — the synthesis of unquestionably useful organic compounds.

1.1 GOAL UNAMBIGUOUS AND UNQUESTIONABLE

From ancient times, mankind was enchanted by the marvelous colors arising from the treatment of cloth with the natural dyes extracted from various animals or plants. As early as the 13th century B.C., Phoenicians knew how to manufacture indigoid dyes (Tyrian purple) from the secretions of certain Mediterranean Sea mollusks. To produce 1 gram of the dye, 10 000 animals were required for a lengthy and laborious procedure. Its price was up to 10–20 times its weight in gold.

In ancient Rome, the skill of producing this dye became one of the most closely guarded state secrets. By Nero's decree, the right to wear garments dyed in purple was granted exclusively to the emperor himself (Royal Purple).[1a] This romantic aura persisted up to the second half of the 19th century, when a rationalistic approach in an emerging science, organic chemistry, mercilessly removed the curtain of mystery and identified the individual components responsible for the dying properties of the natural material (indigo **1** and 6,6′-dibromoindigo **2**, Scheme 1.1).

Shortly thereafter, an inexpensive procedure for the industrial production of **1** from readily available starting materials was elaborated (Bayer, 1878).[1b] In related efforts, chemists identified another compound, alizarine **3**, which was isolated from a certain species of plants (*Rubia tinctoria*). It was used for centuries as a natural dye. Originally very expensive, it soon became an inexpensive product owing to the ease of its synthesis from the aromatic hydrocarbon anthracene, present in coal tar (Grebe and Lieberman, 1868).[2]

These were truly triumphal achievements and they produced a deep impression, not only on chemists, but on the general public as well. It was convincing proof of the power and promise of this rapidly blossoming and daring newborn infant, organic synthesis.

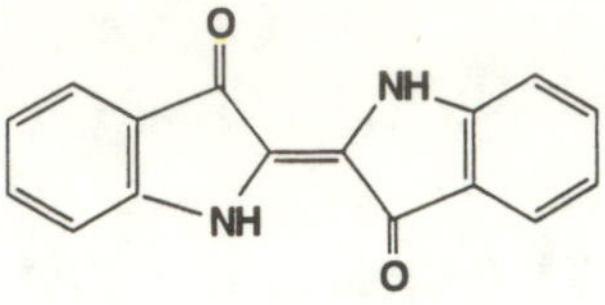

Scheme 1.1

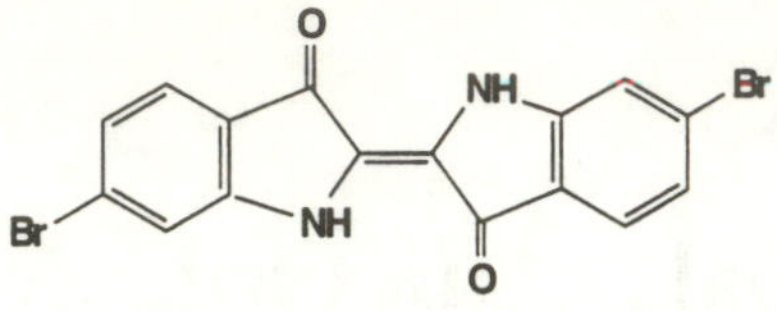

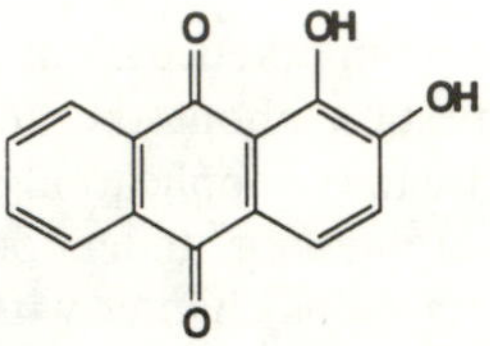

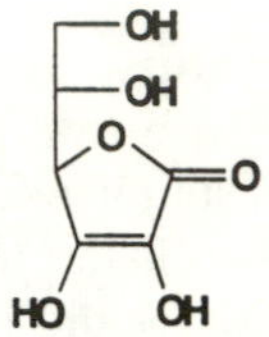

The thread of life, DNA, codes hereditary information for all living creatures. The well-known double helix structure of this molecule was proposed by Watson and Crick in 1953. As Khorana acknowledged later, 'Synthetic work related to this structure immediately began to be my ambition'.[3] The accomplishment of this dream required nearly two decades of intense work by a large group, but culminated in a brilliant success (and a Nobel Prize). Khorana's total synthesis of a biologically active gene, a fragment of DNA, coding the biosynthesis of tyrosine messenger RNA was a benchmark achievement. Its synthesis confirmed the fundamental principles of molecular genetics and provided a tremendous impact on the development of genetic engineering.

Ascorbic acid **4** is one of a set of essential vitamins. The consequences of a deficiency of this simple (but then unknown) ingredient in the diet were first encountered in the era of great geographical discoveries. Deaths among sailors, caused by the mysterious illness scurvy, were heavier than those by all other natural disasters taken together. Elucidation of the structure of ascorbic acid in 1928, followed by its laboratory synthesis (Rechstein, 1934)[4] and shortly thereafter by its industrial synthesis from D-glucose, forever eliminated this threat. According to Pauling, it provided us as well with reliable protection against a number of other diseases, including the common cold.

Prostaglandins (PGs) such as PGE_1, **5** (Scheme 1.2), first identified in the 1950s, were immediately recognized as extremely important bioactive substances. These regulators,[5a] present in nearly all tissues and fluids of mammals, powerfully affect the functioning of their respiratory, digestive, reproductive, and cardiovascular systems. PGs are produced in minute amounts (the human organism produces as little as 1 milligram per day), and there are no natural sources available for the isolation of PGs in substantial amounts. Additional complications in the study and collection of prostaglandins arise because of the high lability of these compounds.

Both the progress gained in the in-depth understanding of the mechanism of their action, and the achievements in the practical application of prostaglandins (in medicine and veterinary science), were made possible only by the success of synthetic chemists in developing efficient routes for the total synthesis of these compounds and their numerous analogs.[5b] Because of the exceptional activity of PGs and some of their more stable synthetic analogs, their production on a laboratory scale (hundred milligram quantities to several kilograms per year) is sufficient to satisfy the demands of an entire country. As a result, a synthetic program initially aimed at purely fundamental goals led directly to the development of a synthetic protocol useful for applied purposes.

'Is a tree worth a life?' — an article under this headline was published in *Newsweek* (August 5, 1991). 'Tree' refers to the evergreen Pacific yew tree, *Taxus brevifolia*, which grows in the forests of the western USA and Canada. A peculiar and rather fateful feature of the yew tree is its unique ability to produce the complicated molecule taxol **6** (Scheme 1.2), a significantly efficient anti-cancer drug.[6a,b] This drug passed phase III clinical trials and became one of the most promising medicines for the treatment of ovarian and breast cancer, especially those cases incurable by other forms of treatment.

Every year, breast cancer will kill about 45 000 women in the USA while an additional 12 000 will be victims of ovarian cancer. Treatment for one cancer patient requires the sacrifice of three 100-year-old trees to obtain 60 pounds of bark to produce a few grams of **6**. The Bristol-Myers pharmaceutical company alone needs 25 kilograms of pure taxol to broaden their clinical studies — a harvest of about 38 000 trees.[6a] With the survival of the Pacific yew at risk, the expression of great concern among the environmentalists is not surprising: 'Is a tree worth a life?' Fortunately it need not be a 'your money or your life' dilemma. Several options are in fact available which can save life without unacceptable sacrifices of the environment. Not surprisingly, the search for more abundant and renewable natural sources of taxol are carried out with extreme vigor. Efforts spent on the total synthesis of taxol and related compounds have been no less. The unique pattern of the carbon framework coupled with the extensive functionalization made the total synthesis of **6** a truly challenging goal. The first two total syntheses, reported independently in 1994 by Holton's[6c] and Nicolaou's teams,[6d] were properly acclaimed as brilliant successes of modern synthetic chemistry. Both preparations are rather lengthy and may seem to be of purely academic interest. Yet these and related studies pave the way for further exploration of structure–activity relationships aimed at elucidating more available and active taxol analogs of practical value.[6b,e]

The fascinating success of transplantation surgery is among the most spectacular achievements of modern medicine. Undoubtedly the development of ingenious surgery skills and carefully refined techniques was a necessary prerequisite for these achievements. Of no less importance was the discovery of drugs capable of modifying and controlling the reactions of a patient's immune system to prevent the rejection of grafted organs.[7a] One of the most efficient immunosuppressant agents, FK-506 **7**, was isolated from the fermentation broth of the microorganism *Streptomyces tsukubaensis* in 1987.[7b]

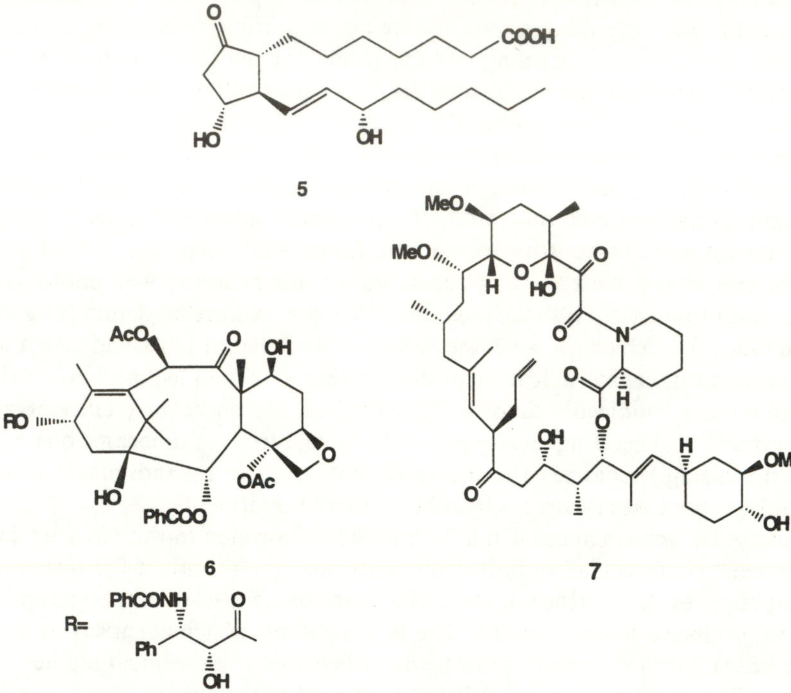

5

6

7

R=

Scheme 1.2

Despite its formidable complexity, the total synthesis of this compound was completed in less than two years by Shinkai's group at Merck Sharpe & Dohm.[7c] Certainly this synthesis is not suggested as an alternative route to the rather inexpensive microbiological process. However, elaboration of synthetic approaches toward **7** opened an entry into the preparation of its derivatives bearing isotope labels[7d] as well as analogs of **7**[7e,f] required for the study of the interactions of the drug with its receptors, an extremely difficult problem. These studies are of utmost importance and very promising since they may lead ultimately to the rational design of immune modulators which are simpler and more useful than the original compound **7**.

It is easy to put together a long list summarizing the achievements of organic synthesis that supply virtually every field of science and touch all aspects of our everyday life. The complexity of these syntheses, their scale (from a fraction of a milligram to a million tons) and the methods used vary tremendously. They differ as well in their ultimate significance for mankind. Whatever the targets are, however, synthetic rubbers and fibers, drugs and dyes, high octane gasoline and detergents, vitamins and hormones, or numerous reagents, they share one thing in common: in all cases the target possessed a set of useful properties warranting its synthesis. This direction of synthetic studies seems to be of unquestionable value and it corresponds exactly to the wishes of the taxpayers who want to see a quick reward for the investment of their money.

1.2 GOAL UNAMBIGUOUS BUT QUESTIONABLE

The importance of science, however, cannot be directly assessed using the criteria of immediate 'usefulness'. As organic chemistry has evolved, synthetic chemists have striven to synthesize any compound that could be isolated from natural sources, especially from living organisms, often without any obvious relevance to the possible utilitarian value of these compounds. Some of these syntheses took decades to accomplish. At present, the gap between the discovery of a new natural compound (and such discoveries are made, in the true sense of the word, every day) and its synthesis is reduced to a very few years, or even months. However, why spend so much effort for the preparation of a compound already synthesized in nature?

It is true that quite often the challenging complexity of the target *per se* serves as a powerful driving force to exercise the synthetic chemist's skill. Yet, the principal motivation stems from the perception that Mother Nature does nothing in vain. Everything she makes serves the essential needs of living organisms and, consequently, is of vital importance for mankind. This confidence continually finds credibility in both general as well as specific aspects in the course of the evolution of knowledge. Consider the following examples.

Among the multitude of natural compounds there exists a large class known as the isoprenoids (or terpenoids), composed of thousands of structurally diverse compounds related by a common biosynthetic origin. Some of these compounds, such as vitamins A and D or the steroidal hormones, have been known for a long time to be essential regulators for the normal functioning of a mammalian organism. In addition, there are compounds that, without doubt, are of practical value (camphor, natural rubber, menthol, carotene, *etc.*). Until the early 1950s the prevailing view was that the majority of isoprenoids were superfluous, devoid of both biological activity and applied usefulness. The reasons why living organisms took the trouble and consumed the energy to make these complicated structures remained obscure. It was commonly accepted that they were inert materials (secondary metabolites), their only destination being the removal of the end products of metabolism. It might have appeared that merely professional pedantry and the lack of any imagination compelled chemists to pursue endless and time consuming studies in the search to isolate ever-increasing numbers of natural isoprenoids from all imaginable sources, to establish their structures, and then to synthesize them. For decades, the only observable results of these studies were additional, yet seemingly meaningless, contributions to the inventory of the products created and stored in nature for unknown purposes.

In the 1960s, however, these views underwent truly dramatic changes. Doubts regarding the usefulness of terpenoids, both for the producing organisms and for us as customers, had to be abandoned. For instance, it became obvious that not only mammals, but insects as well, widely use various isoprenoids as hormones. Thus, one of the most amazing biological phenomena, insect metamorphosis (the emergence of an adult from a larval stage *via* periodic molting), is controlled by a carefully tuned interplay of a set of hormones released by several glands. A small gland known as the *corpora allata* releases a juvenile hormone (JH) **8** (Scheme 1.3), which is essential for the development of the larvae. At a certain moment the release of **8** is stopped. Molting into an adult then occurs, induced by a secretion of another hormone, ecdyson **9**, by the prothoracic gland.

This aging process can be completely stopped if, at this very moment, a fresh amount of **8** is introduced. As a result, a giant but not viable larva appears. Both **8** and **9** (Scheme 1.3) are modified terpenoids.[8a] The richest natural source of JH (adult male abdomens of the silk moth *Hyalphora cecropia*) gives no more than a couple of micrograms of **8** per insect. Nevertheless, **8** became a relatively available compound due to the tremendous efforts spent upon its total synthesis.[8b] Elucidation of the role and successful synthesis of JH triggered an avalanche of studies aimed at the creation of simpler and convenient analogs of this compound. These efforts ultimately led to the appearance of a new generation of pest-control chemicals.

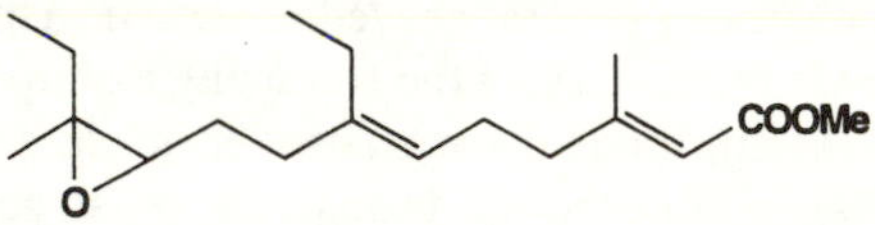

COOMe

O

8

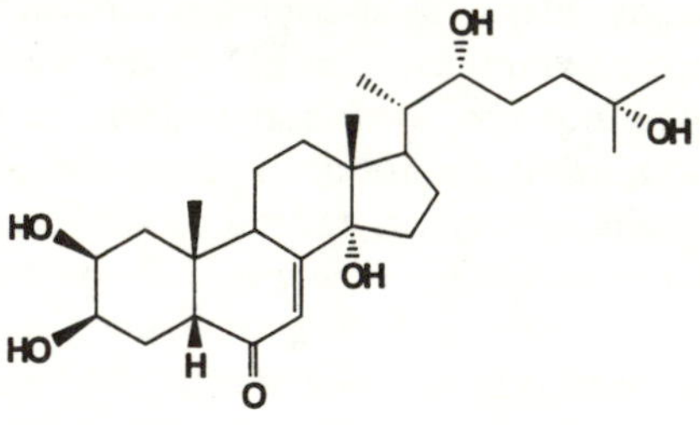

OH

OH

HO

HO

H

O

OH

9

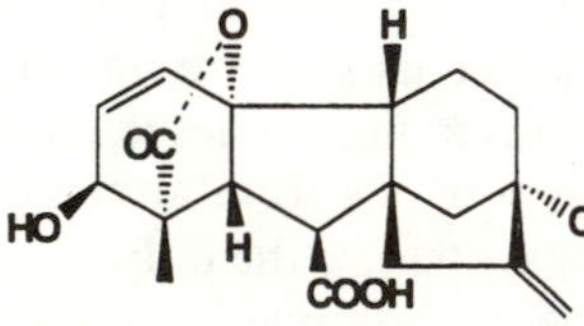

O

H

OC

HO

H

OH

COOH

10

OH

O

COOH

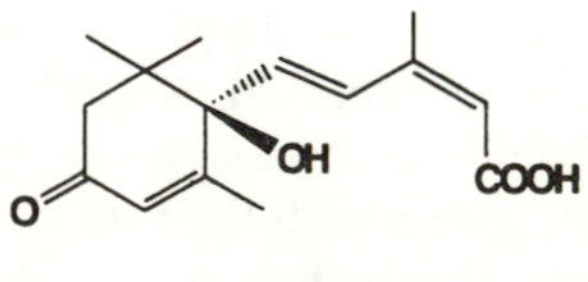

11

Scheme 1.3

In plants, some terpenoids are produced as vitally important hormones involved in the regulation of growth and development. Thus, the diterpenoid gibberellic acid **10**,[9a] widely distributed in the plant kingdom, is known to exercise numerous physiological functions. This compound was first identified as a metabolite of the fungus *Gibberella fujikuroi*, a fungus shown to cause abnormal growth and eventual death of afflicted rice seedlings. A later study discovered that **10** and its numerous analogs are produced by various plants as endogenous growth regulators. Synthesis of this compound by Corey's group[9b] stands as one of the top achievements that attests to the power of modern organic chemistry.

Another terpenoid, abcisin **11**,[10] which was isolated from a variety of plants, functions as a sort of antagonist to **10**. In fact, **11** was shown to be responsible for the inhibition of the growth of seedlings and induction of the formation of resting buds. Thus the changes from the state of active growth during long-day conditions to the dormancy period under short-day conditions are controlled by the balance in the production of these hormones.

Microorganisms and fungi are an especially rich source of isoprenoids of the most diverse structures. Among these products one may find powerful toxins, compounds with antitumor and anti-inflammatory activity or antibiotics. Very little is known about their role in the host organisms. However, the broad spectrum of the observed biological activity could be taken as at least circumstantial evidence to indicate the existence of some function mediated by these products and essential to their producers.

Nowhere in Nature can an individual live isolated, not participating in the intricate interactions between other members of the biological community (biocenosis). Therefore, a truly comprehensive understanding of the functions of natural compounds requires an in-depth investigation of their possible involvement as mediators in the interactions between organisms belonging to the same, or even entirely different, species. As a community we are at the very beginning of the studies of these aspects of chemical ecology. At the same time, numerous facts have already been accumulated which attest to the generality and vital importance of chemical communication channels at all levels of biological organization.

A special term, pheromones (exohormones, or more generally semiochemicals), was coined for compounds which fulfill the role of chemical signals transferring information from one organism to another. The isoprenoids described earlier are not the only group of compounds specifically designated to serve as chemical signals. Nature has no special preference in its choice of a particular group of organic chemicals for these purposes. It can choose a suitable compound to fulfill the required function from a broad array of products without any obvious limitation to the gross structure, complexity, or functionality of the candidates. The following examples exemplify the diversity of functions and chemical structures of semiochemicals used by various species.

It appears that insects have achieved the most spectacular results in elaborating an extremely intricate and ingenious system of chemical communication. With the help of pheromones they can pass information about species of the same type (recognition and classification signals), about the location of male or female species (sex attractants), about the closeness of an enemy (alarm pheromones) or the shortest route to the food source (route indicators), and many others.

The efficiency of long-distant interactions between individual insects mediated by pheromones is truly remarkable. A female Chinese silkworm moth, *Bombix mori*, is able to produce an attractant, bombicol, which can elicit a response from males over an incredibly long distance (for an insect) of approximately 10 kilometers. The challenge to isolate this compound was taken by Butenandt and his co-workers. After many years of laborious work they were able to isolate 3 milligrams of bombicol from more than 31 000 pheromone glands dissected from female *B. mori*.[11a] This compound was shown to possess the simple structure of (10*E*,12*Z*)-hexadeca-10,12-dien-1-ol **12** (Scheme 1.4), and its synthesis was realized in a matter of a few months.[11b]

Sex attractants of many other *Lepidopterans* (butterflies and moths) were shown to have rather primitive structures, such as long-chain unbranched primary alcohols (12–16 carbon atoms) containing one or more double bonds. Other insects have elaborated quite different kinds of chemicals for the same purposes, *e.g.* **13–15**.[12]

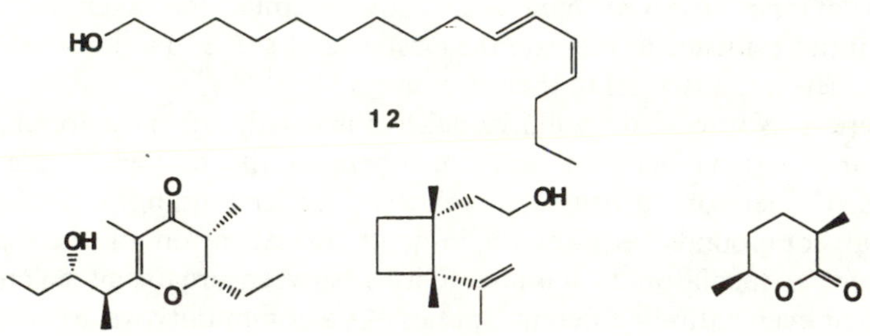

HO
12
OH O
13
(*Cigarette beetle*)
OH
14
(*Boll weevil*)
O O
15
(*Carpenter bee*)

Scheme 1.4

A mass attack of insects can cause serious devastation to crops, forests, food storehouses, *etc*. These invasions are usually triggered by the release of a set of pheromones. For example, upon landing on a ponderosa pine tree, the female western pine beetle, *Dendroctonous brevicomis*, releases *exo*-brevicomin **16** (Scheme 1.5) to attract males. Shortly after mating, the 'pioneers' start to release a mixture of compounds, **17–19**, which carries a sort of 'you are welcome' message to their kinsfolk. The flow of incomers increases a hundred-fold and as a result the tree is overwhelmed and killed.[12]

The chemical information channel is especially vital for social insects. The nearly ideal social order in a beehive is perpetuated as long as the honeybee's 'queen' maintains the ability to produce a very simple aliphatic compound, (*E*)-9-oxo-2-decenoic acid **20** ('queen substance').[13a,b] This multipurpose compound

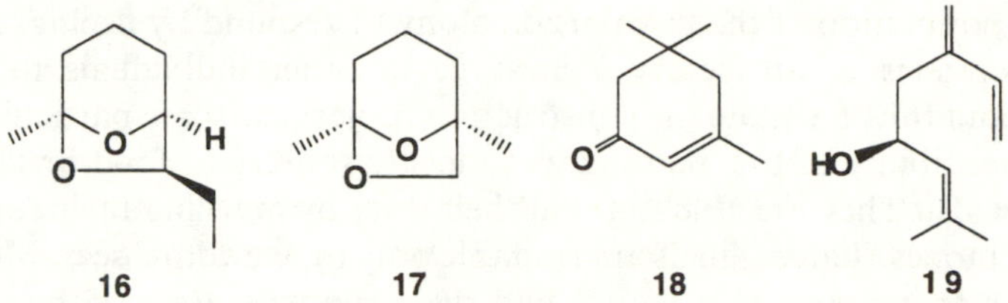

16 17 18 19

Scheme 1.5

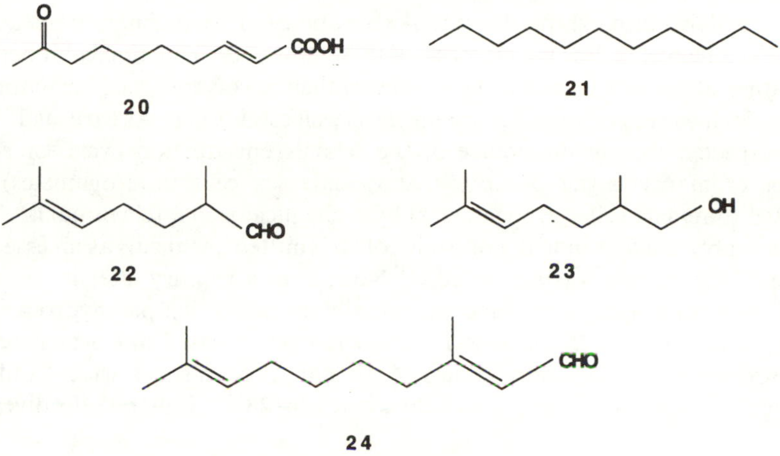

O COOH

20

21

CHO OH

22 23

CHO

24

Scheme 1.6

serves first as a powerful sex attractant for males. At the same time, it has a special, appealing taste to the other members of the beehive family. As a result of its consumption, the reproductive capacity of females is suppressed, as is their instinct to build the enlarged cells required for breeding. Thus, no competitors to the acting 'queen' appear and her power stays unchallenged to the end of her active life (when the queen substance is no longer emitted).

The strict hierarchy of ant colonies gives a well-defined and absolutely rigid specialization to every member. Their remarkable ability to 'organize' collective efforts such as building anthills, conducting army raids for food or laborers, or cultivating fungus or mealybugs ('domesticated animals'), are all features to be admired. Unambiguous evidence attests to the importance of chemical communication as one of the most significant links between individual insects. The chemicals serve as mediators controlling both 'horizontal' and 'vertical' relationships. For example, chemical signals ('caste pheromones') released by a queen determine whether a given ant will develop into a regular worker or become a soldier. The latter is specifically endowed with the ability to emit an alarm pheromone in the case of any threat. Once emitted, this lone signal is immediately amplified by other ants and in a few seconds the colony is ready to fight. In other circumstances the alarm pheromone can provoke panic, causing the entire population of the threatened colony to respond by fleeing. Aggregation pheromones of ants carry a message to other individuals to join the 'builders' and thus facilitate the construction projects at some particular site. It is the responsibility of the 'scout' ants to locate sources of food or targets for 'military raids'. They are able to fulfill their duty by marking their routes with trail pheromones. These signals are readable only by the addressees. More often than not, rather simple compounds and their mixtures are used by insects as 'carriers' of various messages. For example, alarm pheromones of certain ant species may contain, as the active components, the set of compounds **21–24** (Scheme 1.6) in variable amounts.[14]

Lower plants also exhibit the remarkable ability of producing a tremendous variety of chemicals. For the majority of these products the biological functions are rather obscure. In some cases, however, their involvement as mediators in interactions between individual organisms is well established. As early as 1854 it was suspected that in the course of the sexual reproduction cycle for some species of marine algae, *Fucus serratus*, male sex cells (androgametes) are directed to female cells (gynogametes) by a chemical signal (chemotaxis). This was probably the first time the possible role of emitted chemicals as an essential mediator for mating was suggested.[15a] Now, over a century later it has been established that this compound is actually a rather simple hydrocarbon, fucoserratene, C_8H_{12} **25** (Scheme 1.7), which when released into water, serves as a sex attractant for androgamets. A similar mechanism is widely utilized among other species of lower plants. Structures **26–28**[15a,b] illustrate the diversity

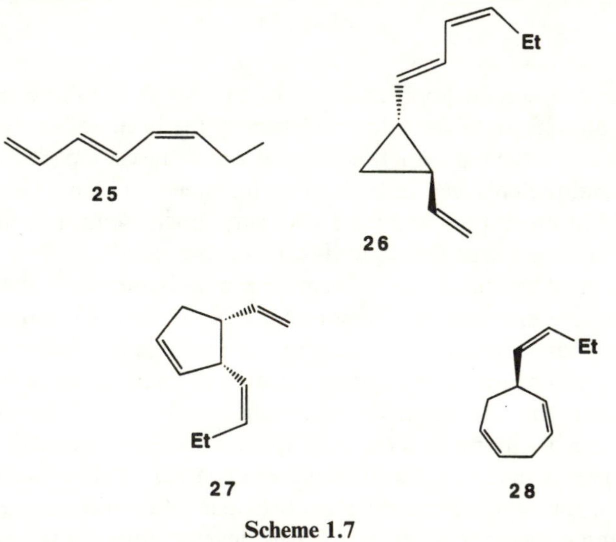

Et

25

26

Et

Et

27

28

Scheme 1.7

of hydrocarbons which may be produced for this purpose. These products are virtually insoluble in water, making the efficiency of these signals truly extraordinary. The threshold concentration sufficient for response can be as low as

6.5×10^{-12} mol L^{-1}. Calculations have shown that only 1 to 10 molecules of pheromone *per* gamete is sufficient to elicit the alluring response.[15a]

Female gametes of the water mold *Allomyces* prefer to use the sesquiterpene sirenin **29**[16] as an attractant for the opposite sex cells. The sea weed *Achlia bisexualis* mutually coordinates and controls the processes of the formation and growth of both male and female gametes by a sequential release of a set of exohormones into water. One of the most important participants during this interlude to mating is the steroid anteridiol **30**.[17]

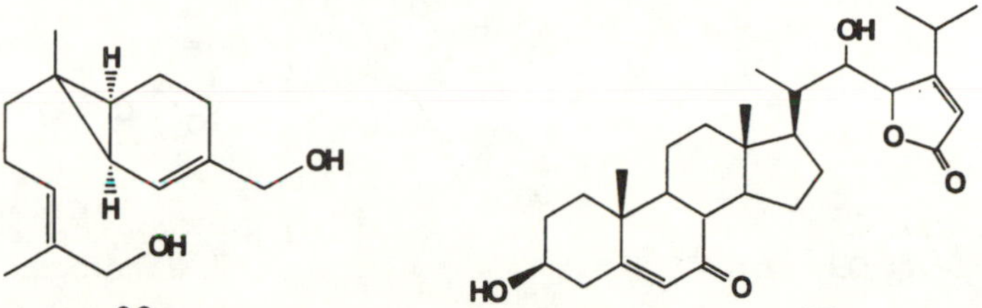

Scheme 1.8

Since ancient times, an abundance of data showing the rather complicated, and at times mysterious, pattern of relationships between different species of a given biocenosis has been accumulated. In many cases it was suspected that these interactions were monitored and controlled with the help of chemicals excreted into the surroundings. Later experimental studies provided convincing data attesting to the universal occurrence of this chemical channel and the efficiency of its operation for informational exchange in biological communities. Here are some examples.

Seeds of the parasitic plant *Striga asiatica*, 'witchweed', which may stay dormant and viable in the soil for up to 20 years, immediately start to germinate at the moment roots of the 'host plant' appear in their vicinity. This phenomenon is by far more interesting than merely another example of biological curiosity because its results may affect the lives of millions of people around the world. The 'host plant' refers to corn, sorghum, and other grasses and the growth of the parasite may cause severe damage to these crops. How do seeds of the parasite learn about the presence of the host roots and choose the right direction for their own growth to attach to the host? A series of meticulous studies showed that the growth of the parasite is triggered chemically by the release of a germination factor from the growing roots of the host plants.

One of the identified stimulators was the terpenoid strigol **31** (Scheme 1.9).[18a] In attempts to develop an efficient tool to eradicate the witchweed (by artificially provoking its growth prior to the growth of corn), numerous efforts were dedicated to the synthesis of **31** and its analogs.[18b] Later studies disclosed the presence of another active compound in the exudate of *Sorghum*, the substituted hydroquinone **32**.[18c] As is typical for hydroquinone derivatives, **32** was found to be quite amenable to oxidation to quinone **33**, which occurs readily in the soil.

Because of the ease of this process, the stimulation of the parasite seed growth by **32** may occur only within the short distance through which **32** can diffuse before being oxidatively deactivated into **33**. From the point of view of the parasite this seems to be a clever mechanism for stimulation since its seeds cannot travel in the soil and they produce roots of no more than 3 mm in length. Hence, it would seem unreasonable for them to become activated by host roots at larger distances. Why and how the host plant acquired such a self-destructing ability to stimulate its parasite is one of the mysteries of evolution.

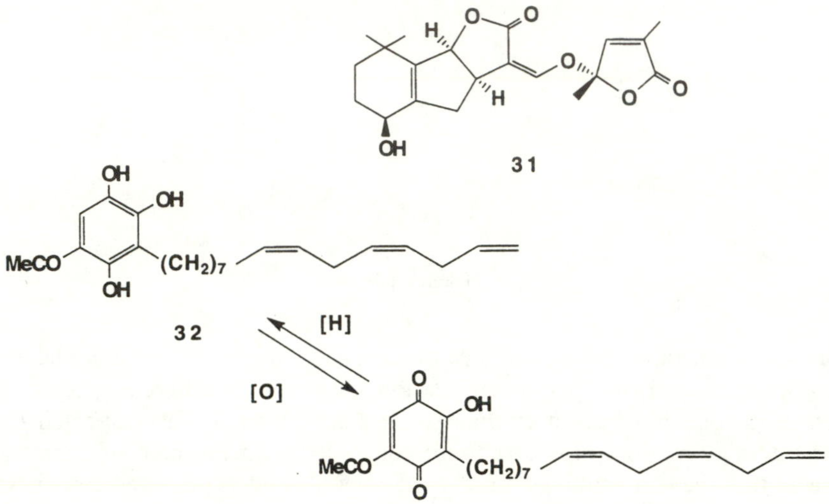

Scheme 1.9

A unique role is played by chemical communication in the interactions between plants and insects.[19] About half a million insect species feed on plants. The process of reproduction in many plant species is critically dependent upon pollination by insects. It is not surprising, then, to find among the numerous natural products of plants both attractants for 'useful' insects and repellents or even insecticides for plant-eating insects.[20] The remarkable diversity of the these compounds (the list includes acyclic and polycyclic compounds, isoprenoids, aromatic derivatives, heterocyclic compounds, *etc.*) illustrates the non-selectivity in the structure of the chemical mediators for biological applications. The intimate mechanism of their action is, unfortunately, still insufficiently understood.

One of the most instructive examples related to the mystery of chemical-mediated relations between plants and animals was described by Meinwald.[21] Alkaloids (natural compounds traditionally referred to as 'secondary', *i.e.* non-important, metabolites) are amply produced by various plants. In many cases, alkaloids have been shown to be part of a defense system against various herbivores. However, some plant-eating insects learned to break through this defense and, further, elaborate a clever approach to utilize the consumed alkaloids to serve their own interests. For example, larvae of the moth *Utetheisa ornatrix* are able to feed on *Crotolaria* plants, thereby ingesting large amounts of pyrrolizidine alkaloids such as monocrotonaline **34** (Scheme 1.10). As a result of this diet the adult moth acquires protection against the attack of predators such as birds or spiders. In the adult *U. ornatrix*, **34** undergoes transformation into another alkaloid, hydroxydanaidal **35**, an essential component of male pheromones. The higher the amount of **35** secreted by a male, the better are his chances of finding a mating partner. The reasons for the female's preference are rather straightforward. If the male's semen contains a lot of **35**, a substantial amount is transmitted to the eggs. The eggs then inherit from their 'dad' an efficient repellent for predators such as ladybugs (ladybirds). The evolutionary benefits of this mode of caring for the offspring are obvious and serve as an illustration of the multifaceted aspects of functions performed by semiochemicals.

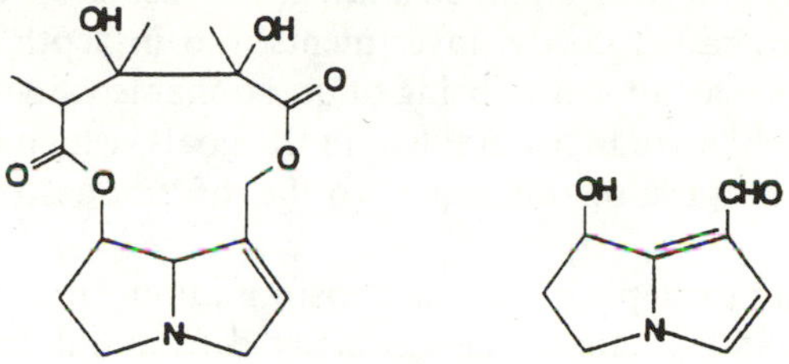

34 **35**

Scheme 1.10

These examples refer to rather simple cases where semiochemicals serve mainly as mediators in direct and well-defined interactions within a single organism or among a few individuals. In fact, relations among various biological partners form a very complicated network of short- and long-range biological interactions. The stability of integrated biological systems as a whole and the efficiency of their functioning depends primarily upon the control of the interactions between the individual parts. Chemical communication is actually one of the most essential and well-elaborated parts of this control system, although it is probably the least understood.

It would seem obvious that knowledge in the language of chemical communication, written in the alphabet of molecular structures, is in fact a mandatory condition to establish meaningful and mutually beneficial interactions between mankind and the environment. Only with the accumulation of this knowledge will we eventually be able to stop our endless wars with Nature in futile attempts to exterminate 'harmful' species, to communicate with them, and to exert reasonable control over their activity. Clearly this ultimate goal is still a remote prospect, but it will not be realized without the help of synthetic chemists.

There is no end to the stories about ingenious chemical tricks elaborated by various species in their struggle for survival, but why have we paid so much attention to these facts in a book devoted specifically to the problems of organic synthesis? Primarily because assessments regarding the usefulness of a given natural compound based solely upon data referring to its activity toward its producer are not legitimate. The assessment must be based upon a compound's involvement in mediating relationships between various species of the entire biological community.

Further, the isolation of a natural compound of novel structure, apparently devoid of any property useful for its host, must be taken as an incentive to look for other aspects of its potential activity. Unsuspected and important biological functions may be discovered. In a way, chemical studies in the area of natural compounds should be considered an essential part of life science aimed at the elucidation of factors that might control the vitality of biological creatures.

Organic synthesis is unique in its capability to prepare structurally diverse natural compounds and their modified analogs in reasonable quantities. These initial, and at times rather costly, investments into in-depth studies of novel natural compounds may appear as being of questionable value. In the long run, however, the dividends are indisputable and the goals, which might have been initially considered questionable, turn to be of unquestionable synthetic significance.

Here is one recent example of such a transformation. In 1968, the sesquiterpene (−)-ovalicin **36** (Scheme 1.11) was isolated from cultures of the fungus *Pseudorotium ovalis* Stolk. This compound exhibited a wide range of antibiotic activity, but otherwise did not appear to be very promising from the point of view of medicinal utilization. As is quite common for fungal metabolites, no data were available about the role of **36** for the organism-producer. Yet because

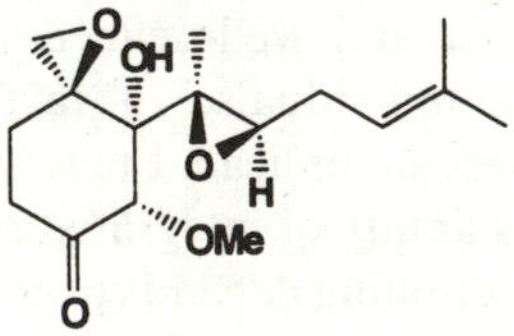

O OH O H OMe O

36

Scheme 1.11

of its structural peculiarity, ovalicin represented a challenging synthetic goal, a sort of testing ground to exercise the skills of organic chemists. Corey's synthesis of racemic **36** in 1985[22a] was undertaken merely as part of fundamental research aimed at investigating new synthetic methodologies. Rather unexpectedly, and nearly a decade later, interest in this compound was renewed owing to the accidental finding that related fungal metabolites are active inhibitors of new blood vessel growth (angiogenesis) and thus might be useful as anti-tumor agents. These findings prompted Corey to submit the natural enantiomer

(−)-**36** for bioassay, which revealed that this compound shows outstanding promise as an angiogenesis inhibitor and is less toxic and more stable than other agents. Hence, the elaboration of a practical route in synthetic methods ensued. As a result, the previously reported synthetic sequence was successfully modified and adjusted for the scaled-up preparation of the needed (−)-ovalicin.[22b]

1.3 SYNTHESIS AS A SEARCH (GOAL AMBIGUOUS BUT UNQUESTIONABLE)

Natural product synthesis is probably the most obvious goal for an organic synthesis, but it is only one of many facets. The depth of experience in organic chemistry shows, beyond any doubt, that natural product synthesis is far from being the only route available for obtaining compounds with useful properties. Useful properties were often discovered as a result of synthetic studies that had no relevance whatsoever to the chemistry of natural compounds. Here, however, one must face a very serious problem. What criteria can be used to choose a specific target and how can one predict properties of an unknown compound? Actually, quite a number of properties of an organic compound can be predicted in advance based exclusively upon its structural formula, even though the compound does not exist in Nature or in the laboratory. Thus it is not difficult to specify initially the type of structural elements which should be present in the compound in order to endow it with the property of dye, drug, perfume, insecticide, adhesive, plasticizer, *etc*. This goal can be achieved by simple structural analogy or by way of serious theoretical analysis. The most common intrinsic property of such solutions, however, turns out to be their ambiguity. Here are some examples.

The chemistry of organic dyes was one of the first areas dedicated to the creation of new compounds with predetermined properties. The correlations between the color of a compound and its structure were established empirically and subsequently confirmed by theory. Correlations were based upon the notion of chromophores — groups of atoms responsible for the characteristic property of a molecule to absorb light of a given wavelength. One of the most widely encountered chromophores in synthetic dyes is the diarylazo group, as is present in the structure of azobenzene **37** (Scheme 1.12). This color can be changed into intense bright red if additional substituents like the *para*-positioned dimethylamino and nitro groups are introduced, as in compounds of the general formula **38**. For an entire series of compounds with varying alkyl groups (R^1 and R^2), one can predict the very bright red color. With the color, as well, one can safely predict weakly basic properties (associated with the amino group) for the whole series of these compounds and a color dependence upon pH changes in the medium.

If our task were to synthesize a bright red azo dye with the properties of a base, theory would bring us to structure **38**, but would say nothing about the nature of the alkyl groups. The investigator must make a selection from a

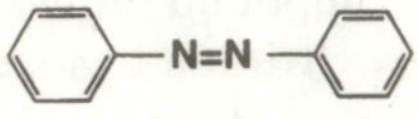

37 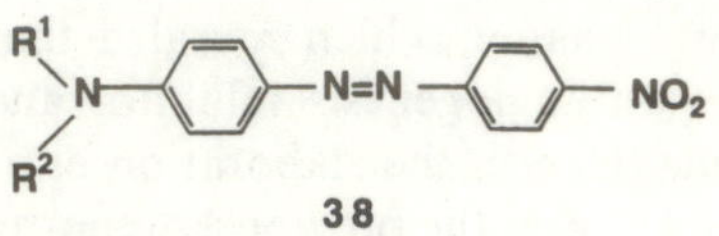 38

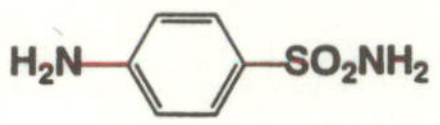

39 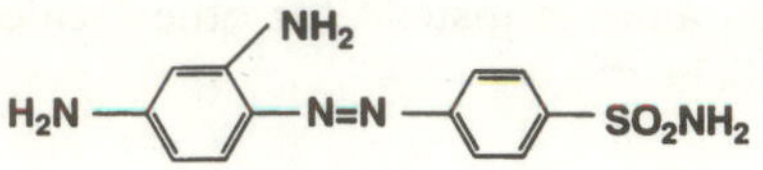 39a

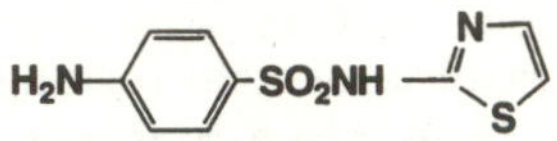

40 Sulfathiazole 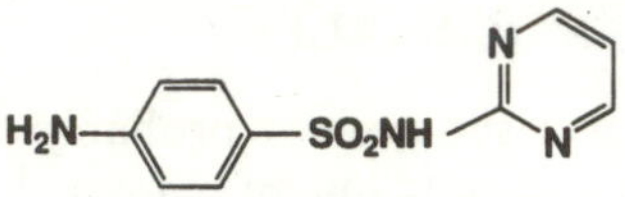 41 Sulfadiazine

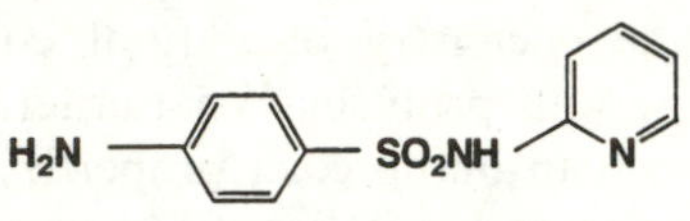

42 Sulfapyridine

Scheme 1.12

multitude of closely related, but different, compounds. Fortunately, the number of candidates is reduced by the need to meet additional requirements.

The nature of the alkyl groups will determine the actual basicity of the dye, its solubility in water and organic solvents, its melting point, the strength of its binding with the material to be dyed and, to some extent, its light and heat sensitivity. These important practical features are predicted much less accurately *a priori* than spectral characteristics. As a result, even after a very thoughtful analysis, several almost equivalent target structures usually remain. The choice, then, is to synthesize the whole set. Only after experimental studies of the essential properties of these products will one be able to select the one which possesses the optimal combination of required properties.

The empirical selection of promising candidates from many related compounds has been especially common for studies in the preparation of new medicinal drugs and biologically active materials. No general theory predicts their action. Rather, some empirical guidelines suggest that compounds containing a particular structural fragment or combination of some fragments will possibly possess a certain pattern of biological activity.

Besides the required pattern of activity, there are many other critically important characteristics of a potential drug such as its toxicity, possible short- and long-term side effects, its ability to accumulate in or to be excreted from an organism, its stability for storing and sterilization purposes, its compatibility with other medicines, *etc.* Therefore, the discovery of promising biological activity in a compound, artificially made or isolated from natural sources, is typically followed with a series of intensive studies aimed at the synthesis of a wide array of its structural analogs.

A classical example is the history of the creation of sulfanilamide drugs.[23] Initially, investigations in this area were stimulated by noting that the presence of the sulfanilamide group in the molecule of azo dyes greatly increased their affinity toward wool fabric. It was thought at the time, although in error, that bacteria cell walls were composed of protein and that sulfamide dyes might actively bind to the bacteria and inhibit their growth. In the course of these studies it was discovered (by mere chance!) that the dye sulfanilamide **39a** (red prontosil) exhibited an amazing activity against streptococci infection in mice. At that time (1932) there was no active medication against bacterial infection. The most surprising aspect of the action of **39a** was that while it was very active in experiments *in vivo*, its activity dropped almost to zero when checked *in vitro*. The answer to this mystery was soon found. In the test animal (mice), **39a** underwent reduction to sulfanilamide **39**, an active bactericide.

A burst of synthetic activity in this area ensued. By 1947, over 5000 sulfanilamides were synthesized and tested as medicinal drugs against a number of diseases. As a result of this wide screening, more than 100 candidates possessing the desired biological activity were selected. Less than a dozen from this list turned out to have the set of additional properties required for the drug. They were widely used over the next several decades. Some, for example the 'sulfa drugs' **39–42**, are still used today.

When it is possible to identify a definite structural feature responsible for biological activity (*e.g.* the sulfanilamide moiety in **39–42**), the search for an efficient drug is expedited. Usually, however, the situation is far from being that simple. Structure–activity correlations are not that easy to establish. As an example, let us briefly consider the typical patterns of biological activity in steroids.

The first steroid, cholesterol **43** (Scheme 1.13), was isolated in the 18th century from human gall stones. No specific biological activity was found for this compound. Since its discovery, hundreds of other steroids were isolated from a variety of natural sources. In addition to these, several thousands more were synthesized. You will ask, no doubt, why was this done? Consider, then, the structures of several bioactive natural compounds, **43–50**, of this class. It is easy to see the close structural relationship among these compounds. They all contain the perhydrocyclopentanophenanthrene system **51**. In spite of that basic relationship, their functions in a living organism differ in the most spectacular fashion. Cholesterol **43** is abundantly present in all normal animal tissues (an organism of 80 kg contains up to 0.23 kg of cholesterol) as an essential component of lipid membranes. It is also notorious in its role in the development of coronary heart disease. Estrone **44** and testosterone **45** are the female and male sex hormones respectively of mammals. Cortisone **46** is a hormone produced by the adrenal cortex with powerful anti-inflammatory (anti-arthritic) activity. Aldosterone **47** is a hormone that regulates salt metabolism and digitoxygenin **48** is a component of digitalis, a traditional

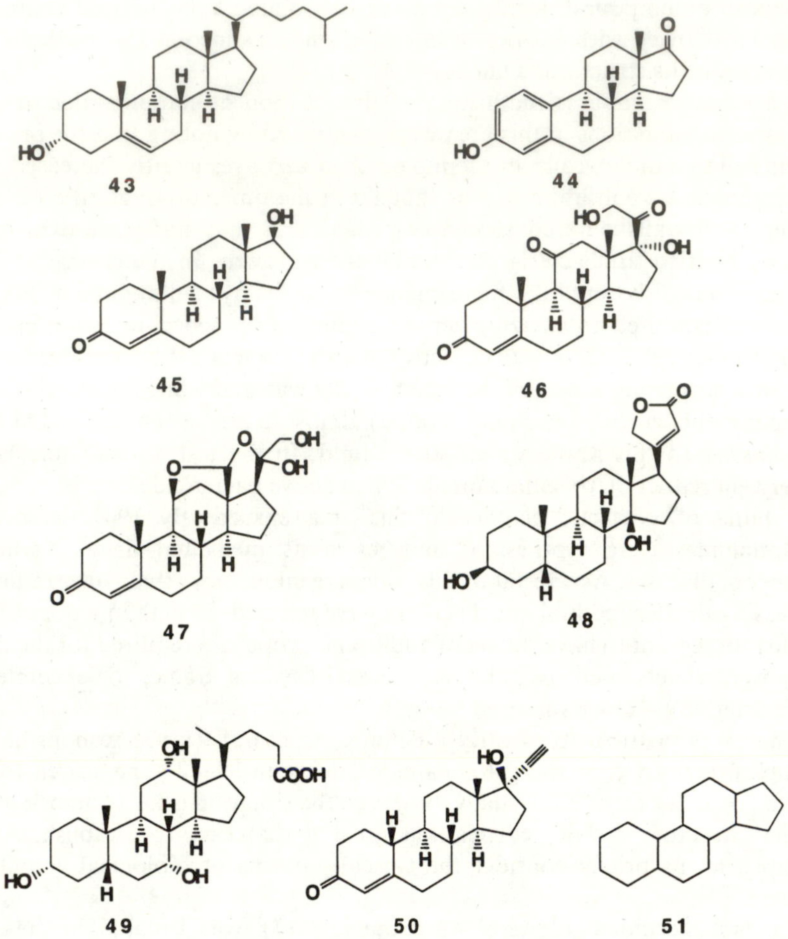

43 44 45 46 47 48 49 50 51

Scheme 1.13

drug from herbs used in heart therapy. One may add to this list the previously mentioned hormone of molting, ecdyson **9**, and the sex attractant of algae, anteridiol **30**.

Steroidal hormones are usually polyfunctional in their biological role. In medicine, more often than not, one needs a drug that possesses a rather focused set of specific pharmacological properties with as few side effects as possible. Yet even within a limited set such as compounds **43–48**, having nearly identical carbon skeletons, it is virtually impossible to establish any straightforward correlation between biological activity and a particular structural feature. A hydroxyl group at the same C-3 position is present in **9**, **30**, **43**, **44**, and **48**, whereas **45–47** contain a carbonyl group at that center. Additional aliphatic side chains of various structures are positioned at the same C-17 center for **9**, **30**, **43**, and **46–48**.

The only realistic way to create steroidal drugs with a narrower spectrum of properties is to synthesize numerous analogs in a steroid series and then test their biological activity. With human health at stake, tremendous amounts of resources have been directed to the synthesis of natural steroid hormones and their artificial analogs. Analogs were prepared bearing nearly all possible modifications on the carbon atoms of the original backbone. Among the outstanding achievements in this area was undoubtedly the partial synthesis of cortisone **46** from the readily available cholic acid **49**. This conversion initially required 37 steps and yielded the target compound in a negligible total yield (Sarett, 1946).[24a] One could hardly have expected this route to be of any practical viability. In less than three years, however, the yield was increased by a factor of more than two orders of magnitude. In a pilot plant, about 1 kilogram of **46** was prepared, an amount sufficient to do the clinical testing necessary for this promising drug.[24b]

For at least two decades, steroids remained as one of the most challenging synthetic targets.[24c] The intense pursuit of these studies[23c] made available a wide array of hormonal drugs. At the same time, considerable progress was gained in the elucidation of the intricate mechanism of steroidal hormonal action. One of the more spectacular results of this study was the elaboration of an approach to family planning. The creation of oral contraceptives, ('the Pill' **50**, currently used by at least 50 million women worldwide) resulted in tremendous social consequences worldwide.

A similar story can be told with respect to prostaglandins. Again, the vital importance of synthetic studies in this field was unquestionable from the very beginning owing to the unique biological functions of these regulators. The natural source for the isolation of these compounds was extremely limited, however, making the total synthesis of natural PGs (like PGE_1, **5**, Scheme 1.2) a worthy goal. It took enormous time and effort to synthesize hundreds of analogs and to screen their biological effects in order to identify only a few structures that showed promise for practical application.[5]

Consider one more example among more recent achievements in synthesis. The history of antimalaria drugs dates back to the 16th century when Jesuit missionaries in Peru learned from the natives that chewing cinchona bark was the most efficient treatment for malaria. In 1834, Pelletier isolated the active constituent, the alkaloid quinine **52** (Scheme 1.14). This compound was widely used as an antimalarial medication long before the basic principles of chemotherapy were formulated. The first attempt to synthesize quinine dates back to 1856. Then, 18-year-old William Perkin (destined to become a famous chemist only a few years later) undertook the daring assault to prepare this compound by oxidation of a mixture of toluidine and allyl bromide. He started the project without the slightest idea of the real structure of quinine and, expectedly, failed (although he prepared instead the first artificial dyestuff, mauveine). The quinine structure was elucidated only 50 years later. Numerous synthetic efforts finally culminated in Woodward's brilliant total synthesis of quinine in 1948. Starting in the 1930s a number of empirically found synthetic compounds (almost structurally unrelated to quinine) such as chloroquine **53**, mepacrine **54**, or proguanil **55** were offered as efficient drugs in the treatment of malaria.[25a]

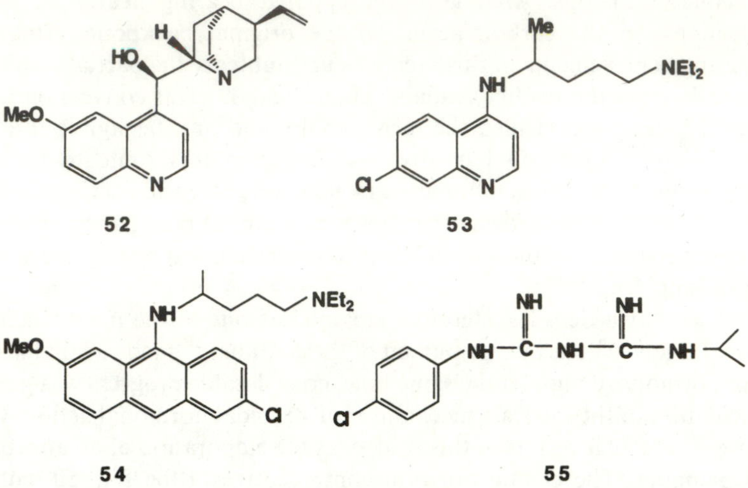

Scheme 1.14

It appeared as if the problem of curing malaria was finally solved. Yet *Plasmodium falciparum*, the parasite that causes malaria, exhibited resistance toward these drugs in new strains of the parasite. Renewed interest in the creation of novel antimalarial drugs resulted. Rather paradoxically, the key leads for their search were again found in ancient folk medicine, this time in Asia. In his *Compendium of Materia Medica of 1596*, the famous Chinese herbalist Li Shizhen described the use of preparations from qing hao (*Artemisia annua*, sweet wormwood) to combat the chills and fevers of malaria. In the late 1970s an active component of this treatment, the terpenoid artemisinin **56** (Scheme 1.15), was isolated from this plant.[25b] Artemisinin, which contains a very rare structural fragment of an endoperoxide, revealed a long-sought-after pattern of activity against chloroquine-resistant strains of *P. falciparum*. It yielded a cure rate of nearly 99%! Of special importance is its activity against cerebral malaria, an advanced and terminal form of the disease. The practical application of this promising drug, however, is hampered because of its insolubility in water, instability, and relatively high price. Both the synthesis of artemisinin and the search for simpler analogs are therefore unquestionably worthwhile goals. While several successful syntheses of artemisinin have been already described,[25c] none of them appears practical. Intensive studies in this field, however, elaborated a set of various structures with the endoperoxide fragment and made them available for structure–activity studies. Some of them, like **57** and **58**,[25d] showed activity in *in vitro* tests even higher than that of artemisinin itself.

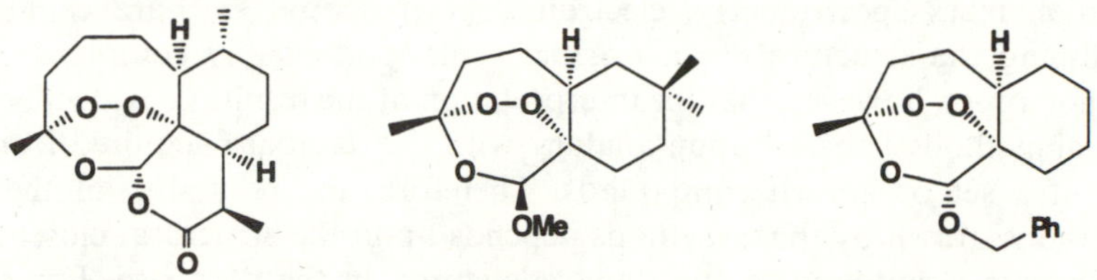

Scheme 1.15

These types of investigations are usually referred to as the 'syntheses of potentially useful compounds'. In all tasks with an ambiguous answer, a chemist is trained to think not only in terms of a single target structure, but rather in terms of a series of related compounds. A synthetic plan should be designed, then, with special attention paid to the elaboration of fairly general methods, adaptable for the preparation of a broad set of compounds with strategic variations in structural parameters. This type of plan is actually a prerequisite to carrying out further analog preparations in a rational manner. It may happen that a whole series composed of 1000 compounds should be prepared in order to arrive at one particular representative of unquestionable practical value. Even so, one cannot claim that the remaining 999 apparently useless products were made in vain.* It is less than likely that the final success in finding the desired compound could have ever been achieved in the absence of the massive database obtained by the syntheses and subsequent evaluation of the whole set of compounds.

1.4 SYNTHESIS AS AN INSTRUMENT OF EXPLORATION

The goals discussed above refer to the preparative function of synthesis to supply essential materials. The actual route used to acquire the product is immaterial: the final results are the same whether or not target compounds are isolated from natural sources or obtained through biochemical processes. (In the majority of cases, though, a chemical synthesis turns out to be a more common and reliable approach.)

There is, however, an area where synthesis does more than fulfill an auxiliary role. It represents the essence of the task. The most obvious examples are related to the problems of structural proof for novel compounds obtained as the result of previously unknown reactions or those isolated from natural sources. In such cases, final and unambiguous proof of the structure (as deduced from analytical data) is achieved by an independent synthesis of the same compound followed by direct comparison with an authentic sample. This claim might now appear obsolete. In fact, it may seem that modern powerful physical methods for structural analyses, such as nuclear magnetic resonance spectroscopy, high-

* These numbers are not merely speculative estimates. The ratio 1:1000 is typical and considered reasonable for many studies oriented toward the creation of biologically active compounds.

resolution mass spectrometry, electronic spectroscopy, *etc*., are capable of establishing the structure of an organic compound of any complexity. One must not forget, however, that the interpretation of the results provided by these physical methods is based upon analogy with the database acquired from the study of a set of known compounds. Therefore, the reliability of the final conclusions drawn by these methods depends upon the structural closeness of an unknown compound to the standards stored in the database. For a new compound with an unprecedented structure, the interpretation of spectral data may lead to inconclusive or even erroneous results.

Free from these limitations is an X-ray analysis. This method is considered the 'absolute' method for structural determination. Its universal applicability is limited, unfortunately, by the necessity to prepare monocrystalline samples, a task not always easy to accomplish. One classic method for establishing structure is based upon a stepwise degradation of an unknown molecule into smaller and known components. This degradation is followed with a logical restructuring of the starting material based upon the identity of the fragments. This route, while being generally applicable, is time consuming and requires a substantial amount of the compound to be degraded. In addition, in the course of the degradation, some structural information, especially that related to the stereochemistry of the broken bonds, will inevitably be lost.

An independent synthesis is devoid of these shortcomings. The structure confirmed by this route can be considered to be an absolute truth no less than if established by a trustworthy Supreme Court.

In many cases an independent synthesis turns out to be the only practical method to choose correctly between several tentative structures of an unknown compound. For cases in which the studied compound is available in a very small amount (less than a milligram or even a microgram), degradation methods are not feasible. Even instrumental methods in such cases may prove to be inconclusive in structural analysis. The exact structure of such natural products as the previously mentioned JH **8** or bombicol **12** were conclusively elucidated only after their total synthesis. Some more recent examples follow.

From 1976 to 1978, tremendous efforts were spent on the isolation of a major component (in a pure form) of the sex pheromone of the female cockroach, *Periplaneta americana*. The potency of the compound isolated, periplanone B, was truly amazing (threshold limits of 10^{-7} μg). It was considered to be a promising agent to control this pest. However, only trace amounts were isolated from natural sources (a total of 0.2 mg isolated from 75 000 specimens). Even the use of modern methods of instrumental analysis did not lead to the elucidation of its complete structure. With the help of these methods it was possible only to ascertain the basic structural features, as shown in formula **59a** (Scheme 1.16). The problem of its stereochemistry remained unanswered. This most difficult part of the problem was finally solved only after a total synthesis of three of four possible geometrical isomers of periplanone B.[26a] Comparison of the spectral parameters of synthetic samples with those of the natural periplanone B determined the stereochemistry as shown in structure **59b**. A minor component of the same pheromone, periplanone A, was available in even lesser amounts than periplanone B. Spectral data led to the erroneous determination of its structure to be as shown in **60a**. The correct structure **60b** was established only from the result of its total synthesis.[26b] Intensive synthetic efforts in this area coupled with structure–activity studies resulted in the preparation of bioactive analogs of **59a** (such as **61**) by a short synthetic route from readily available materials.[26c]

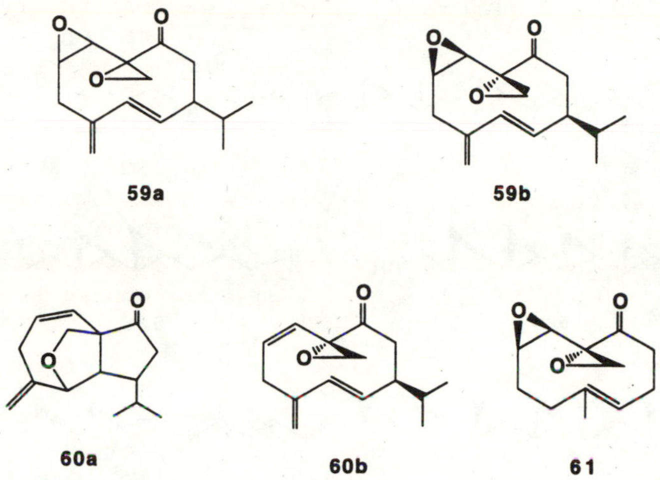

59a **59b**

60a **60b** **61**

Scheme 1.16

In 1990, Japanese chemists isolated from the fermentation broth of *Streptoverticillium fervens* a novel anti-fungal agent designated as FR-900848 and determined its gross structure as that shown in formula **62a** (Scheme 1.17).[27] This agent exhibited a promising pattern of activity toward a set of filamentous fungi, an especially dangerous infection for patients having AIDS or diabetes. For a while, the problem of determining the stereochemistry of this compound stayed unresolved, as even the most sophisticated versions of NMR methods failed to determine unambiguously the stereochemical relationships within the unique polycyclopropane carbon framework which bears 10 chiral centers as shown in **62a**. Rather obviously, the total synthesis of the 2^{10} possible stereoisomers also could not be considered as an appealing pathway to resolve this problem. However, the consideration of plausible biosynthetic pathways involved in the biosynthesis of this substance provided a good basis to suggest an all-*trans* configuration for the substituents at the cyclopropane rings.[27a] Partial degradation of the natural product yielded a symmetrical quatercyclopropane derivative which could only have been represented as either of the diastereomers **62b** or **62c**. Both diastereoisomers were prepared by stereospecific synthesis. Now the stereochemistry of the degradation product could be determined unambiguosly as all-*trans*, all-*syn* by direct comparison with an authentic sample of **62b**. This result, combined with other relevent data, led to the determination of the full structure of FR-900848 as shown in **62**[27b]. It is also worthwhile to comment in passing that the next representative of the poly-cyclopropane family was isolated recently from microbial fermentation of *Streptomyces* spp. This compound, which exhibited significant acitivity as a potential inhibitor of lipid deposition in arterial walls, differs from **62** by the presence of an additional cyclopropane ring and is likely to belong to the same stereochemical series.[27c]

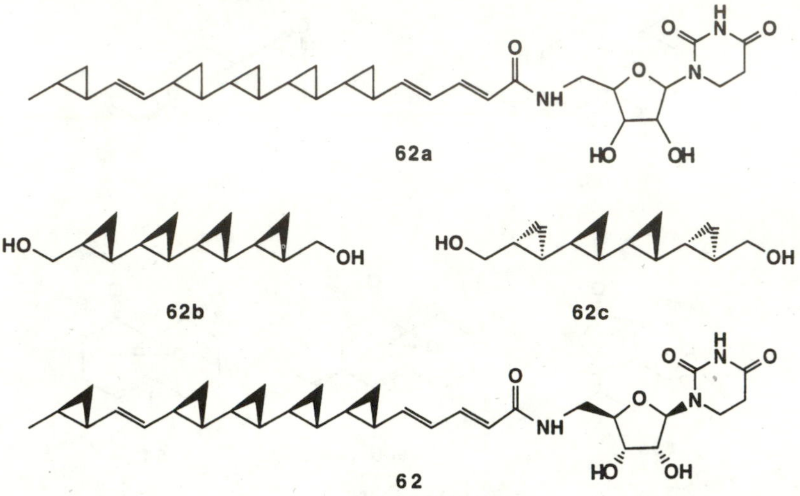

62a

62b **62c**

62

Scheme 1.17

Additional problems in structure elucidation can arise owing to the extreme lability of some natural compounds. In 1969 a series of thorough biomedical studies indicated that blood platelets produce (albeit in minute amounts) an extremely potent vasoconstrictor and platelet aggregating factor. There were good reasons to suggest that an overproduction of the substance, thromboxane A2 (TxA_2), is responsible for the development of heart attacks or strokes. Isolation of TxA_2 was achieved in 1975 but the unique lability of this compound (a half-life in aqueous solution at 37 °C of only 34 s!) actually precluded the possibility of its identification by the standard set of modern instrumental methods. Nevertheless, the established structure **63a** (Scheme 1.18) of the inactive decomposition product of TxA_2, combined with a thorough analysis of biosynthetic routes (leading to both TxA_2 and related prostaglandins), allowed chemists to propose the tentative structure **63** for TxA_2.[28a] This structure was so unusual for a natural compound that for the next 10 years its plausibility was hotly debated. In 1985, however, the structure of TxA_2 was substantiated beyond any doubt by a synthesis carried out by Stille.[28b] Curiously, the only criterion applicable in this case to ascertain finally the identity of the synthesized material with respect to natural **63** was a comparison of the biological properties of these two samples in a set of bioassays. The exceptional instability of **63** severely hampered experimental studies of its important physiological functions and therefore its chemically stable analogs were badly needed. Of the vast number of analogs synthesized, **64** proved to be useful as a substitute for **63**.[28c]

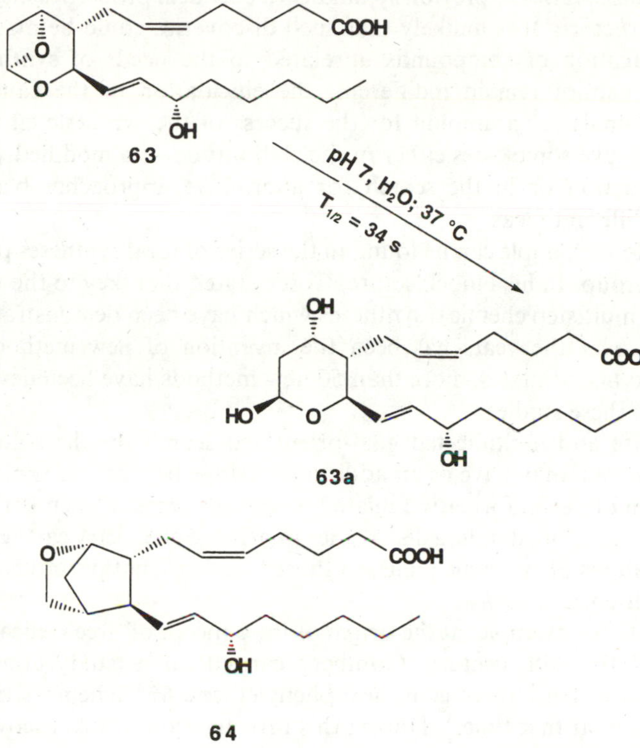

COOH
O
O
OH
63
pH 7, H_2O; 37 °C
$T_{1/2}$ = 34 s
OH
COOH
HO
O
OH
63a
O
COOH
OH
64

Scheme 1.18

Along with its role as a tool for structure elucidation, synthesis fulfills a different, less obvious, and perhaps more important function. The studies aimed at a total synthesis represent one of the most powerful routes to delve into the chemistry of the target compounds. An understanding of the chemical peculiarities of an organic compound is a necessary condition for the success of its synthesis. Yet the factual data necessary for such an understanding are most effectively accumulated in the process of synthesizing the targeted compound. Why?

The initial plan of any synthesis is usually based upon a set of well-known synthetic methods. In principle, these methods are suitable for solving a given task and should not introduce ambiguous results. If everything goes as planned, we then obtain direct experimental proof of the correctness of our theoretical prediction. This important point was recognized by one of the greatest synthetic chemists of our time, Nobel Laureate R.B. Woodward, who wrote: 'One can hardly deny that the successful result of a synthesis, consisting of more than 30 steps, appears as a stern test of the ability of science to predict and thus verify its strength of knowledge in the sphere of the research objectives'.[29]

It often happens that upon exploration into a new area of 'research objectives' these well defined methods do not work. Such 'misfires' are, in themselves, discoveries of previously unknown chemical properties intrinsic to a particular structure. It is unlikely that such discoveries could be made merely upon investigation of compounds unrelated to the needs of synthesis. The investigator cannot remain indifferent; the elucidation of the cause of the observed anomaly is paramount for the success of the synthetic efforts. Such analysis may give some clues either in the elaboration of a modified version of the same reaction or in the search for alternative approaches based upon completely different ideas.

An illustrious example can be found in the series of total syntheses performed by Corey's group. In his Nobel lecture, Corey stated that 'key to the success of many of the multistep chemical syntheses which have been demonstrated in our laboratories over the years has been the invention of new methodology'.[30] According to his estimates, more than 50 new methods have been developed in the course of these studies.

This intense and multifaceted goal-orientated search for the solution of a synthetic problem may have as an additional 'fringe benefit', an explosion-like accumulation of new information related to various areas of chemistry. In fact, the origin of many outstanding discoveries in organic chemistry can be traced to the initial failures of synthetic plans, to those 'misfires' that prompted chemists to revise their current views.

Let us look, for example, at the origin of the concept of 'free radicals'. In the beginning of the 20th century, Gomberg carried out a study aimed at the synthesis of a rather elusive goal, hexaphenylethane **65** (Scheme 1.19), a very exotic structure at that time.[31] During this investigation it was discovered that the standard route to the creation of a central carbon–carbon bond (condensation of halogen derivatives in the presence of a metal) did not lead to the desired result but gave products of a different type. A careful analysis of the dependence of the product structure upon the reaction conditions forced the investigator to conclude that the presumably formed hexaphenylethane was unstable and spontaneously falls apart with cleavage of the central carbon–carbon bond and

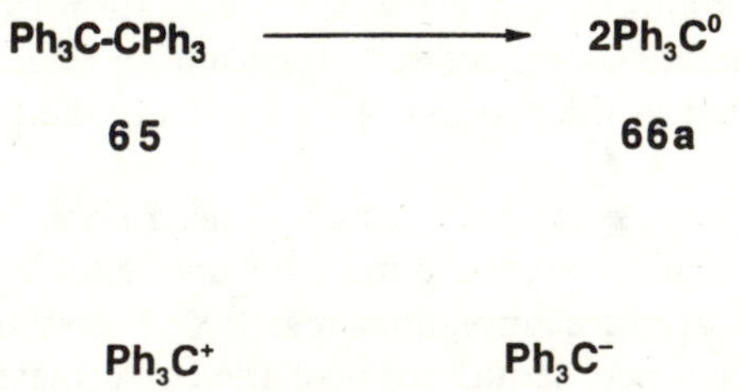

$$Ph_3C\text{-}CPh_3 \longrightarrow 2Ph_3C^{\bullet}$$

65 **66a**

Ph_3C^+ Ph_3C^-

66b **66c**

Scheme 1.19

formation of two triphenylmethyl radicals **66a**. The very possibility of the existence of a stable species containing a trivalent carbon atom was considered at that time to be absolutely unlikely and Gomberg's finding was met with disbelief and sharp criticism.

Both the results and the reasoning, however, were rigorously confirmed shortly thereafter. Eventually it became clear that Gomberg's synthetic 'failure' led to the discovery of stable radicals, a new class of organic species, derivatives of trivalent carbon. Further studies led to the understanding that the triarylmethyl system is extremely versatile and gives rise to the preparation of other types of derivatives of trivalent carbon, *e.g.* salts of the triphenylmethyl cation **66b** and the triphenylmethyl anion **66c**.

Preparation of the first representatives of stable organic cations, anions, and radicals, elucidation of the factors determining their stability, and peculiarities of their reactivity, inspired hypotheses regarding the generality of these species' involvement as intermediates in numerous organic reactions. This concept played a key role in the formulation of main ideas in classical theory dealing with the reactivity of organic compounds. These important ramifications all originated from the unexpected results obtained in Gomberg's studies, studies targeted at purely synthetic goals.

A more recent illustration attesting to the indisputable role of synthesis in the development of theoretical concepts can be found in the field of partial and total synthesis of steroidal hormones and their analogs. Investigators working in this area in the 1930s–1940s met with a series of unexpected problems both in the construction of the required tetracyclic skeleton and in accomplishing some otherwise trivial conversions, such as addition to C=C or C=O double bonds, oxirane ring openings, or even the transformation of an alcohol into a halide derivative. The requirements of the synthesis forced chemists to develop an abundance of alternative experimental procedures to bypass these problems, thus greatly enriching the arsenal of preparative methods. At the same time, the reasons for the abnormalities observed in the reactivity pattern were carefully studied.

Studies of these abnormalities ultimately led to the formulation of the main concepts of modern conformational analysis. As early as 1890 it was suggested by Sachse that cyclohexane is not flat and can actually possess a non-planar configuration.[32a] If one accepts this assumption then, as Sachse states in the concluding paragraph of his paper, 'all monosubstituted derivatives of cyclohexane can exist in at least two modifications'. No direct experimental evidence in favor of this suggestion was available at that time and it stayed nearly forgotten for more than 50 years, although its plausibility was implicated by a number of theoretical and physical organic studies. The full importance of this suggestion was appreciated only after the short publication 'Conformation of the Steroid Nucleus' in *Experientia* in August 1950.[32b] Its author, Barton, summarized numerous and seemingly contradictory experimental data about the reactivity of variously substituted steroids. He provided convincing evidence that the observed discrepancies are due to the dramatic differences in the properties of substituents, positioned in equatorial or axial positions. Eventually, the whole concept of conformational analysis was created[32c] with the Nobel prize awarded to Barton for his eye-opening study.

Today, conformational theory stands as one of the basic foundations of organic chemistry. It is difficult to comprehend that, for all practical purposes, this concept was non-existent for more than a century; its birth was largely stimulated by the urgent needs of organic synthesis.

Enormous efforts have been and still are spent on the synthesis of structures of ever-increasing complexity, compounds isolated from natural sources as well as those sought strictly by synthetic design. This increase in complexity is correspondingly accompanied by the appearance of unexpected and difficult problems due to the novelty of the molecular constructions. The necessity to resolve these problems not only stimulates the development of new synthetic methods but also serves as a challenge for the creation of new concepts in theoretical organic chemistry.

1.5 'CHEMISTRY CREATES ITS OWN SUBJECT ...'

As early as 1860, an outstanding chemist of the 19th century, Berthelot, claimed: 'Chemistry creates its own subject. This creative ability, similar to an art, is the main feature that distinguishes chemistry from the natural and humanitarian sciences'.[33] Let us try to analyse the basis of this perception of chemistry's select posture in the spectrum of the sciences.

From its early days, science has been studying Nature in the search for interrelations between various phenomena and the laws describing these phenomena. Nature, to the scientist, exists as a primordial object to be investigated. A biologist investigates living nature in the context of how it was molded by conditions on Earth. An astronomer studies the infinite number of celestial bodies and the universe as a whole. The organic chemist investigates organic compounds, their properties, reactions, and the principles of their behavior. Initially, organic chemistry began with the investigation of natural products. This area, as we have shown, still constitutes an important part of chemistry. In contrast to other natural sciences, however, the appearance of organic chemistry as a science became possible only after the artificial creation of its subject. An organic chemist had to be a Demiurge, creating their own material world filled with artificial materials never before existing in Nature. Here lies the principal distinction between organic chemistry and all other natural sciences. From the very beginning, organic chemistry has appeared as a nearly closed system that secures for itself both the objects for research and the problems to be solved. It evolves as a science in accordance with its own innate laws. An analogy to such an ability for self-development can be found only in mathematics.

This is how it was in the time of Berthelot. This is the way it is now, although the uniqueness of organic chemistry, in the sense of creating its own research objectives, might now be arguable (in view of the appearance of other artificially created areas of science, such as solid state physics, non-linear optics, genetic engineering, *etc.*). Nevertheless, the basic meaning of Berthelot's statement remains valid today: organic chemistry is permanently and vigorously engaged in creation and investigation into its own subject, a type of artificial nature. The properties of this 'newly created nature' turn out to be just as diverse, unexpected, and inexhaustible as is true for real nature. This is the main difference between newly synthesized organic compounds and the numerous classes of other man-made utilitarian items.

For example, various mechanical, electronic, or other technical devices created by man can be highly complex and totally unrelated to any natural objects. Yet these items share a common feature: they were designed with a definite goal. Qualitatively speaking, their basic properties were also known in advance. If we design and build an airplane, we know it will never serendipitously turn out to be a tape recorder or a meat chopper; it will be only a good airplane or a bad one.

In contrast, if we plan to synthesize a new compound, striving to obtain a medical drug, it might well turn out that this compound would exhibit instead the properties of a toxin, a defoliant, or a photosensitizer, to list just a few of the possible options. It is also likely that in carrying out 'the plan' one could obtain something having a totally different structure with correspondingly unexpected activity.

In the mid 1880s, a young Russian chemist, Zelinsky, a postgraduate student in Mayer's laboratory in Germany, took part in a project aimed at the elaboration of a new synthesis of tetrahydrothiophene **67** from 2-chloroethanol **68**, *via* an apparently simple route as shown in Scheme 1.20. However, Zelinsky's research progress was abruptly stopped shortly after he succeeded in preparing the advanced intermediate product β,β-dichlorodiethyl sulfide **69**. Instead of proceeding further with his experiments, Zelinsky had to spend several weeks in a hospital owing to the severe burns caused by this simple and innocent looking (on paper!) compound. Later, **69** became known as the infamous mustard gas.

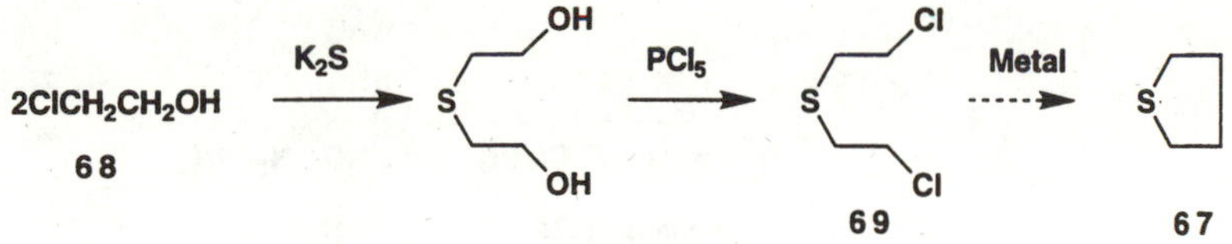

$2ClCH_2CH_2OH$ (**68**) $\xrightarrow{K_2S}$ $S(CH_2CH_2OH)_2$ $\xrightarrow{PCl_5}$ $S(CH_2CH_2Cl)_2$ (**69**) $\dashrightarrow$ (Metal) tetrahydrothiophene (**67**)

Scheme 1.20

The discovery of the physiological activity of this substance did not bring only disaster to mankind. A detailed investigation into the mechanisms of its action led to the emergence of a new direction in the chemotherapy of malignant tumors. These studies were initially based upon the use of structural analogs of mustard gas.

Such stories in the history of organic chemistry abound. Surely the study of man-made organic compounds is as rich a field for making unexpected discoveries as is any other area of natural science dealing with the research objects supplied by Mother Nature herself.

As is generally known, the uniqueness of carbon as a chemical element is due to the combination of two properties, its tetravalency and its ability to form stable bonds not only with atoms of many other elements but with other carbon atoms as well. These two factors are responsible for the fantastic variety in structural fragments that are found in organic molecules. Therefore, it is easy to imagine how large the number of possible organic compounds might be. In fact, this number is actually infinite. How can we formulate the mathematical notion of infinity for an infinite number of organic compounds? To any given, however complex, compound one may add one more carbon unit, thereby obtaining a new compound with which one can carry out the same operation, *etc.* Modification of a molecule can be achieved in a variety of routes and in every single case a novel entity is generated. The multiplicity of ways of increasing the complexity of a molecule and the number of variations grows roughly in proportion to the number of atoms already existing in the molecule.

An analogy to the multitude of possible structures of organic compounds can be found in a tree-like structure with an infinite number of branches. The coefficient of branching in this tree tends to increase with the increase in the number of previous divergency points. The total number of objects of this multitude should grow almost factorially ($n!$, where n is the number of carbons in a molecule). Even for rather small organic molecules, quite a number of permutations are possible and thus the total number of generated structures might be truly astonomical.

Consider, for example, compounds containing only 60 carbon atoms, a saturated aliphatic acids of general formula **70** in which the variable groups R^1 and R^2 can be located at any possible position along the chain. The number of such structures, arising from the variation of the nature and position of just one

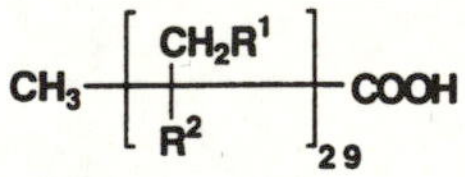

$$CH_3-\left[\begin{array}{c} CH_2R^1 \\ | \\ -C- \\ | \\ R^2 \end{array}\right]_{29}-COOH$$

70

R^1, R^2 = H, OH, F, Cl, Br, I, NH_2, NO_2, N_3, SH

Scheme 1.21

group R^1 (or only R^2), is 10^{29}. If both groups are varied, the total number of possible structures will be $10^{29} \times 10^{29} = 10^{58}$, a number greater, by a factor of 10^7, than the number of atoms on our planet! There are not enough carbon atoms present in our galaxy to prepare all these structures, even in milligram quantities. Moreover, owing to the presence of 29 chiral centers, there are 2^{29} or 5.4×10^8 stereoisomers possible for each positional isomer of **70**. Thus the total number of structures corresponding to the general formula **70** additionally increases to approximately 5.4×10^{66}. The total number of nucleons in the entire observable universe would not suffice for the preparation of all these compounds in 1 mg amounts. Thus conversion of an abstract notion about the infinite number of organic compounds into real numbers beautifully illustrates the inexhaustible multitude of possible organic structures.

What can be the source of this multitude? How can an abstract, enormously huge, and pale scheme of theoretically possible structures be painted with the plethora of vivid colors of actually existing materials? There are two sources: natural (living organisms and mined organic raw materials) and artificial (organic synthesis).

Organic chemistry began as a science dealing with natural compounds isolated from living organisms. Natural compounds, however, while being extremely diverse, fill in patches of the total area of possible organic compounds in a very peculiar and, from the viewpoint of organic systematics, sporadic fashion. If one rationally classifies organic compounds according to functional groups and fills in such a table with only natural compounds, a very strange picture would appear. We would see a system containing isolated clusters, overpopulated with some structures combined with areas populated only with dispersed and isolated representatives. Equally surprising is the realization that huge tracts of possible and at times even trivial structures would be left unoccupied.

For example, unbranched aliphatic acids with an even number of carbon atoms will be generously represented, while one finds a nearly complete absence of odd-numbered and branched acids. Nature produces an incredible diversity of the most ingeniously constructed cyclic products containing cycloaliphatic, aromatic, or heterocyclic moieties, but such derivatives like aniline or thiophenol, as well as plethora of other simple representatives of these classes, are not in the list of naturally occurring substances. Such important types as alkyl halides, nitro compounds, and diazo compounds would be sparsely represented by very rare (if any) examples. Even the simplest compounds like formaldehyde, chloroform, diethyl ether, dioxane, *etc.*, which are trivial to organic chemists, turn to be rather exotic for Nature. In the list of items provided by Nature one will notice the almost complete absence of various organometallic compounds, as well as many other classes of structures of immense scientific and practical significance.

Despite the richness and variety of compounds provided by Nature, the science of organic chemistry could not have been created with only such a limited and an absolutely inadequate set of materials. From the very beginning, the evolution of organic chemistry quickly refuted its initial and narrow definition as a science dealing exclusively with natural products. Tremendous efforts by several generations of chemists were spent in the preparation of thousands of previously unknown compounds in order to create a sound factual basis for this new science. Thorough investigation of the properties of these newly created objects led to the formulation of the main problems to be dealt with by organic chemistry, elucidation of its fundamental laws, and to the development of the necessary tools and methods to explore the phenomena of organic chemistry. As the result of this activity, there appeared an absolutely novel area of science which opened the routes for the creation of a new multibillion dollars worth branch of industry, based on chemical technologies. Our living conditions and even lifestyle have changed dramatically in this century and these changes were brought to us, by and large, mainly due to the achievements of organic chemistry. It can be said that in a way an artificial nature was created. This artificial nature supplies us with a multitude of products which are now used to dress, protect, decorate, cure, and even feed the present day population. Furthermore, this 'man-made nature' is becoming an unprecedented biogeochemical factor of global importance.*

Let us have a look at the major trends in the rich history of organic chemistry in which synthesis was (and still is!) playing a leading role.

1.5.1 Elucidation of the Functional Dependence between Properties and the Structure of Organic Compounds

In the chemical sciences, function refers to the chemical, physical, or biological properties of materials. The most fundamental variable for determining function is the structure of the molecule. A functional dependence of this type, in principle, cannot be established on a single example. In order to study, or even to discover, the existence of functional dependency, one needs to change variables systematically. One must evaluate a series of compounds. Systematic changes in structure can be achieved only as discrete steps and these steps must be well planned and well engineered, as changes that are too dramatic can cause indecipherable effect.

Every organic compound possesses a unique chemical individuality. Its properties, as a function of structure, depend upon several subtly interdependent structural parameters. Because of this peculiarity, 'structure–activity correlations' more often than not are amenable only to 'qualitative' rather than 'quantitative' formulations. Even so, it is unavoidable to search for a target material with a given set of properties on a somewhat empirical basis. These investigations include physico- and biochemical correlations within a series of related compounds. This crucial series of compounds are supplied by organic synthesis.

1.5.2 Creation of Unique Structures Especially Designed to Serve as Models for Investigation

Throughout the history of organic chemistry, urgent theoretical problems have emerged which required the study of structures which, at the time, were considered to be esoteric. It was thus necessary to synthesize the required

* Environmental pollution is only one but probably the most obvious aspect of its action. Of no less importance is the sudden appearance of the plethora of man-made compounds on our planet which may affect in an absolutely unpredictable way the course of natural evolution of the biosphere. An accelerated rate of formation of new varieties of microorganisms caused by the wide utilization of antibiotics might be considered as an instructive manifestation of the unpredictable and far-fetched effects caused by the operation of novel anthropogenic factors of natural selection.

compound in order to ascertain first the feasibility of its existence and subsequently to check the validity of predictions about its properties. A few examples given below will illustrate this point.

Isolated examples of isomeric organic compounds were known as early as the 1830s. This phenomenon turned to be a touchstone for the structural theory of organic chemistry formulated in the 1850s owing to the efforts of Kekulé, Cooper, and Butlerov. It was Butlerov who not only advanced the consistent explanation of isomerism in terms of this theory but also predicted the existence of all theoretically possible isomers for a number of simple C_4 and C_5 derivatives (1864). Subsequent synthesis of these isomers served as a convincing proof of the validity of the structural theory.

A further example concerns the verification of the theory of aromaticity. Synthesis of the eight-membered analog of benzene, cyclooctatetraene **71** (Scheme 1.22), accomplished in 1911 by Willstadter, was an outstanding achievement in its own right. Even more important was elucidation of the non-aromatic character of this molecule. This finding provided a powerful impact on the studies aimed at the elaboration of a more consistent explanation of aromaticity phenomena.

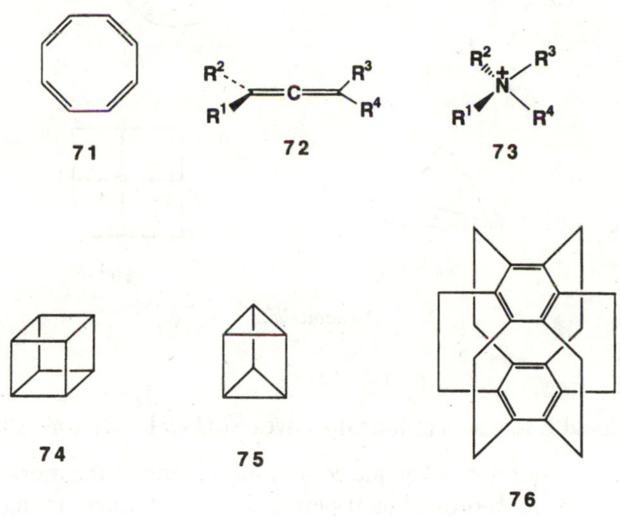

Scheme 1.22

It should also be mentioned that the predictions based on the van't Hoff–LeBel theory of stereochemistry were tested and ultimately substantiated by a plethora of experimental evidence, as, for example the preparation of optically active derivatives of allenes like **72** and quaternary ammonium salts of the type **73**.

Questions as to whether or not compounds of highly strained molecular frameworks could exist and, if so, what specific properties could be expected, prompted organic chemists to undertake the synthesis of such irresistibly charming structures as cubane **74**, prismane **75**, and superphane **76**, as well as many others that will be shown in the following chapters.

The synthesis of rotaxanes, compounds in which a cyclic molecule is fixed on a 'rope' (owing to the presence of bulky groups at both ends of the 'rope', as in **77**, Scheme 1.23), and catenanes, composed of two interlocking rings (like links in a chain in **78**), was prompted by a thorough theoretical analysis which revealed the possibility of the existence of these entirely new structural types and led to formulation of the structural requirements for their preparation. While this particular investigation could have been considered to be too esoteric to have any ramifications of practical importance, it was not long after that the catenane structure was found for some DNA molecules. In Chapter 4 we will present some recent and most promising developments in the area of catenane- and rotaxane-like compounds. Likewise in Chapter 4 we will highlight the challenges which emerged owing to the formulation of such exotic frameworks as shown for hypothetical molecules such as tetrahedrane **79** and fenestrane **80**.

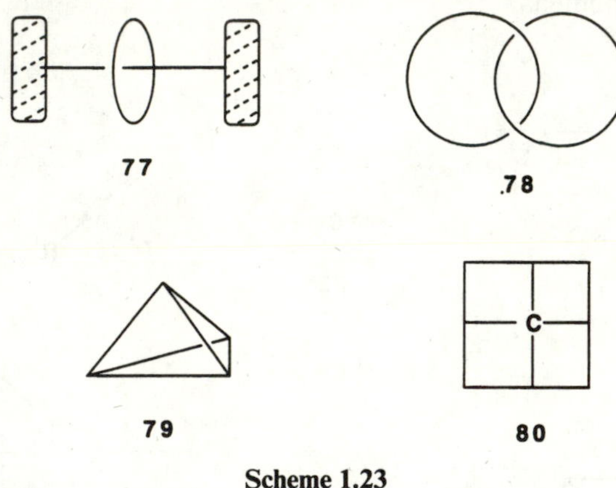

Scheme 1.23

1.5.3 Continual Expansion of the Objectives Studied by Organic Chemistry

Increasing the set of known organic compounds is one of the more unassuming but necessary jobs for a professional synthetic chemist; unassuming because the majority of such syntheses can have a very commonplace character and thus be rather unimpressive, even though they lead to the creation of tens or even hundreds of new compounds.

This route of self-development was a natural and unavoidable occupation for the immature science of organic chemistry. Now, when millions of organic compounds have already been described and their basic classes studied rather thoroughly, such a 'synthesis for the sake of synthesis' may seem extravagant. It is a most difficult dilemma to ascertain whether it is actually worthwhile to appropriate resources from a goal-orientated investigation to the synthesis of yet another million new compounds*, not knowing in advance if they will be useful.

As is the case with every fundamental science, organic chemistry probes the unknown. It is impossible to predict whether or not a significant discovery in some particular area may or may not happen. Even less can be said about the practical applications of potential discoveries. With certainty, however, one can predict that if organic chemistry is arrested in its evolution, both the discoveries in unexplored areas and their further applications will never occur.

Chemists who synthesized cholesterol benzoate about 100 years ago, a routine synthesis of a derivative of a known compound, had no way of knowing that they had opened a route to the creation of innumerable and various devices in which liquid crystals are used. This new state of a material was unexpectedly discovered in the course of studies which were narrowly focused at the preparation of various derivatives from the readily available natural compound cholesterol. Similarly, the epoch of modern chemotherapy originated with the discovery of 'sulfa drugs', which happened as an absolutely unexpected consequence in a broad investigation aimed at the synthesis of hundreds of most diversified derivatives of aromatic compounds, potentially useful as components of azo dyes.

The chemistry of fluoro organic compounds may serve as a typical example of the creation of a novel exploration solely for investigation purposes. This pursuit emerged from a purely academic interest, akin to childish curiosity. How would organic compounds 'look' if more and more atoms of hydrogen were replaced with fluorine? In the 1920s–1930s this was a truly difficult area for exploration, even on a laboratory scale. At that time, no one would dare to predict that these compounds might ever become the subjects of any interest outside a rather narrow sphere of interest of 'pure science'. However, investigations in this area not only enriched theoretical organic chemistry but led, again rather unexpectedly, to the discovery of new materials with unique physicochemical properties. These are found in fluoro plastics (PTFE),[34] polymers with an exceptional set of useful properties not even remotely duplicated by any known natural or synthetic materials; chlorofluorocarbons, the basic products of modern refrigerants and aerosol technologies; or highly fluorinated liquid compounds as, for example, perfluorotetrahydrofuran, which has been shown to possess an unsuspected ability for dissolving oxygen. To this list should be added the creation of fluorine-containing analogs of natural metabolites with promising patterns of physiological activity. While these fluorinated analogs do not differ in basic structure from the corresponding natural compound and thus may serve as false substrates for the respective enzyme systems, the presence of fluorine may inhibit the biochemical functions of the natural compound. Scores and scores of these sorts of derivatives are currently synthesized and checked in studies related to biochemistry and medicine.[35] Among the most known examples is 5-fluorouracil, a fluoro analog of one of the natural components of

*In total, about 15 million organic compounds have been prepared so far. This list keeps expanding at the astounding average rate of approximately 500 new componds per day!

nucleic acid. At present, 5-fluorouracil is used as a highly effective anti-tumor drug.

Artificially created organic compounds also provide an inexhaustible source for discoveries in seemingly unrelated, but important, areas of science. A spectacular illustration can be found in studies aimed at the creation of organic conductors and superconductors. The inability of organic compounds to conduct electricity has been known for centuries. The insulating property of polymers was, in fact, a key factor in bringing about their widespread use as coating materials. In the last two decades, however, it was found that certain types of polymers may have some of the properties of conductors. Thus, polymers of the general formula —$(CH{=}CH)_n$—, prepared in the 1960s by a Ziegler–Natta-type polymerization of acetylene, acquire metal-like conductivity upon 'doping' (partial oxidation with mild reagents like iodine). The conductivity of doped polyacetylene may be as high as 10^4 S cm^{-1} compared with 10^6 S cm^{-1} for silver and 10^{-18} S cm^{-1} for PTFE, the nearly perfect insulator. This was a landmark discovery. For the first time it was found that organic chemistry was able to create conductive materials, a skill once limited to inorganic chemistry. Not surprisingly, this discovery was immediately followed by an explosion of activity in the search for other organic compounds with metal-like properties.[36] Besides polyacetylene, polymers containing an extended polyene system like polyphenylene, polypyrrole, or polyaniline were shown to exhibit conducting properties under a variety of conditions.[37] (Polyaniline was prepared as early as 1862 but, until recently, no one suspected it to exhibit conductivity up to 200 S cm^{-1}.)

In parallel studies, an even more interesting physical property, the superconductivity of organic materials, has been discovered. For example, 1:1 charge-transfer complexes of tetrathiafulvalene **81** (Scheme 1.24) and tetracyanoquinodimethane **82** were shown to display not only metal-like conductivity at ambient temperature but also properties of superconductors at low temperatures. Numerous compounds of this and other types were synthesized and tested. Especially promising results were obtained with charge-transfer salts of bis(ethylenedithio)tetrathiafulvalene **83** with simple inorganic anions. Some of

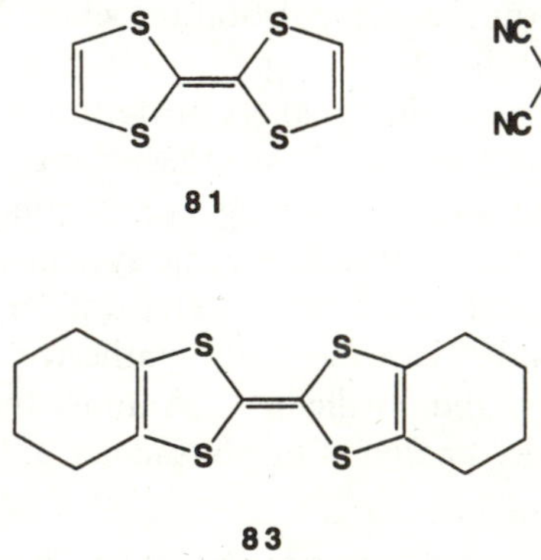

S S S S

81

NC CN NC CN

82

S S S S

83

Scheme 1.24

these complexes revealed superconducting properties at temperatures as high as 10.4 K.

Not surprisingly, the creation of new types of materials with the properties of high-temperature superconductors attracted special attention. Goal-oriented synthetic studies aimed at the preparation of various compounds with structural parameters required for possible candidates in the role of 'organic metals' added an entirely new dimension to organic synthesis. In this area the organic chemist has to apply his/her skills to finding the optimal solution for physical problems, a task previously assigned to general and inorganic chemistry. This new area for the study of physical phenomena emerged entirely owing to decades-long studies motivated by a purely academic interest toward preparation of novel polyconjugated systems, various heterocyclic compounds, and charge-transfer complexes.

The inherent creativity of this area of science manifests itself not only in its outstanding achievements but, in fact, with all the routine syntheses in the preparation of novel compounds. The synthesis of a new compound, regardless of whether or not it had the anticipated properties, represents a small but meaningful contribution to the treasury of knowledge. Someday it might even lead to a stunning discovery!

Only the most general and rather traditional goals of organic syntheis were briefly outlined in this chapter. A collection of essays exposing the underlying motivation, excitement, and adventure of organic synthesis, as well as its creative functions, applied utility, and potential threats can be found in Hoffmann's book *The Same and Not the Same.*[38] Interested readers are also referred to Seebach's insightful review, entitled 'Organic Synthesis — Where Now?'.[39] This review not only summarizes the most significant acheivements of organic synthesis during the last 25 years, but also highlights the main trends of its current development as a central discipline among natural and technical sciences.

REFERENCES

[1] (a) A detailed account of chemical history of royal purple is given in: McGovern, P. E.; Michel, R. H. *Acc. Chem. Res.*, **1990**, *23*, 152; see also: Hoffmann, R. *Am. Sci.*, **1990**, *78*, 308; (b) Bayer, A. *Ber.*, **1878**, *11*, 2128.

[2] Grebe, C.; Lieberman, C. *Ber.*, **1869**, *2*, 332.

[3] Khorana, H. G. in *Frontiers in Bioorganic Chemistry and Molecular Biology*, Ovchinnikov, Y. A.; Kolosov, M. N., Eds., Elsevier/North-Holland Biomedical Press, Amsterdam, **1979**, ch. 10, p. 191.

[4] Reichstein, T. A.; Gussner, A. *Helv. Chim. Acta*, **1935**, *18*, 608.

[5] (a) Bergstrøm, S. *Science*, **1967**, *157*, 382; (b) review: Curtis-Prior, P. B., ed., *Prostaglandins: Biology and Chemistry of Prostaglandins and Related Eicosanoids*, Churchill Livingstone, Edinburgh, **1988**.

[6] (a) A lively account of this story is given by Bowman, S. in *Chem. Eng. News*, **1991**, Sept. 2, 11; see also Freemantle, M. *ibid.*, **1994**, Aug. 1, 35; (b) reviews: Guenard, D.; Gueritte-Voegelein, F.; Potier, P. *Acc. Chem. Res.*, **1993**, *26*, 160; Nicolaou, K. C.; Guy, R. K.; Dai, W. M. *Angew. Chem., Int. Ed. Engl.*, **1994**, *33*, 15; (c) Holton, R. A.;

Kim, H. B.; Somoza, C.; Liang, F.; Biediger, R. J.; Boatman, P. D.; Shindo, M.; Smith, C. C.; Kim, S.; Nadizadeh, H.; Suzuki, Y.; Tao, C.; Vu, P.; Tang, S.; Zhang, P.; Murthi, K. K.; Gentile, L. N.; Liu, J. H. *J. Am. Chem. Soc.*, **1994**, *116*, 1599; (d) Nicolaou, K. C.; Claiborne, C. F.; Nantermet, P. G.; Couladorous, E. A.; Sorensen, E. J. *J. Am. Chem. Soc.*, **1994**, *116*, 1599; (e) see, for example: Blechert, S.; Kleine-Klausing, A. *Angew. Chem., Int. Ed. Engl.*, **1991**, *30*, 412.

[7] (a) The problem of immunosuppresive agents is highlighted in: Kessler, H.; Mierke, D. F.; Donald, D.; Furber, M. *Angew. Chem., Int. Ed. Engl.*, **1991**, *30*, 954; (b) for the outline of the story, see: Stinson, S. *Chem. Eng. News*, **1989**, Feb. 6, 30; (c) Jones, T. K.; Mills, S. G.; Reamer, R. A.; Askin, D.; Desmond, R.; Ryan, K. M.; Volante, R. P.; Shinkai, I. *J. Am. Chem. Soc.*, **1989**, *111*, 1157; (d) Nakatsuka, M.; Ragan, J. A.; Sammakia, T.; Smith, D. B.; Uehling, D. E.; Schreiber, S. L. *J. Am. Chem. Soc.*, **1990**, *112*, 5583; (e) Goulet, M. T.; Hodkey, D. W. *Tetrahedron Lett.*, **1991**, *32*, 4627; (f) Andrus, M. B.; Schreiber, S. L. *J. Am. Chem. Soc.*, **1993**, *115*, 10420.

[8] (a) Review: Siddal, J. B. in *Chemical Ecology*, Academic Press, New York, **1970**, ch. 11, p. 281; (b) Trost, B. M. *Acc. Chem. Res.*, **1970**, *3*, 120.

[9] (a) See: *Gibberellins and Plant Growth*, Krishnamurthy, H. N., Ed., Wiley, New York, **1975**; (b) Corey, E. J.; Danheiser, R. L.; Chandrasekaran, S.; Keck, G. E.; Gopalan, B.; Larsen, S. D.; Siret, P.; Gras, J. L. *J. Am. Chem. Soc.*, **1978**, *100*, 8034.

[10] Addicott, E. T.; Lyon, J. L.; Ohkuma, K.; Thiessen, W. E.; Carns, H. R.; Smith, O. E.; Cornforth, J. W.; Millborrow, B. V.; Ryback, G.; Wareing, P. F. *Science*, **1968**, *159*, 1493.

[11] (a) Butenandt, A.; Hecker, E.; Hopp, M.; Koch, W. *Justus Liebigs Ann. Chem.*, **1962**, *658*, 39; (b) Truscheit, E.; Eiter, K., *Justus Liebigs Ann. Chem.*, **1962**, *658*, 65.

[12] Review: Kelly, D. R. *Chem. Br.*, **1990**, *26*, 124.

[13] (a) Isolation: Butler, C. G.; Callow, R. K.; Johnston, N. C. *Nature*, **1959**, *184*, 1871; (b) synthesis: Bellassoued, M.; Majidi, A. *Tetrahedron Lett.*, **1991**, *32*, 7253 and references cited therein.

[14] Review: *Pheromones*, Birch, M. C., Ed., North-Holland, Amsterdam, **1977**; see also popular articles: *Slave-Making Ants*, Topoff, H. *Am. Sci.*, **1990**, *78*, 520; *Empire of the Ants*, Wilson, E. D. *Discoverer*, **1990**, #3, 44.

[15] (a) For a history and mechanism of action, see the review: Jaenicke, L.; Boland, W. *Angew. Chem., Int. Ed. Engl.*, **1982**, *21*, 643; (b) synthesis: Abraham, W. D.; Cohen, T. *J. Am. Chem. Soc.*, **1991**, *113*, 2313; Crouse, G. D.; Paquette, L. A. *J. Org. Chem.*, **1981**, *46*, 4272 and references cited therein.

[16] Synthesis: Corey, E. J.; Achiwa, K.; Katzenellenbogen, J. A. *J. Am. Chem. Soc.*, **1969**, *91*, 4318.

[17] MacMorris, T. C.; Barksdale, A.W. *Nature*, **1967**, *215*, 320; Arsenault, G. P.; Biemann, K.; Barksdale, A. W.; MacMorris, T. C. *J. Am. Chem. Soc.*, **1968**, *90*, 5635.

[18] Isolation: (a) Cook, C. E.; Whichard, L. P.; Turner, B.; Wall, M. E.; Egley, E. H. *Science*, **1966**, *154*, 1189; synthesis: (b) Johnson, A. W.; Govda, G.; Hassanali, A.; Knox, J.; Monako, S.; Razavi, Z.; Rosebery, G. *J. Chem. Soc., Perkin Trans I*, **1981**, 1734; (c) Chang, M.; Netzly, D. H.; Butler, L. G.; Lynn, D. G. *J. Am. Chem. Soc.*, **1986**, *108*, 7858.

[19] Reviews: Barbie, M. *Introduction à l'Ecologie Chimique*, Masson, Paris, **1976**; Levinson, G. *Naturwissenschaften*, **1972**, *59*, 477; Gerout, V. *Prog. Phytochem.*, **1970**, *2*, 143.

[20] For a review on insect antifeedants and their potential importance for crop protection, see: Ley, S. V.; Toogood, P. L. *Chem. Br.*, **1990**, *26*, 31.

[21] Eisner T.; Meinwald, J. in *Pheromone Biochemistry*, Prestwich, G. D.; Bloomquist, G. J., Eds., Academic Press, Orlando, **1987**, ch. 8, p. 251; a lively exposure of the topic is given in: Meinwald, J. *Eng. Sci.*, **1986**, *49*, #5, 14.

[22] (a) Corey, E. J.; Dittami, J. P. *J. Am. Chem. Soc.*, **1985**, *107*, 256; (b) Corey, E. J.; Guzman-Perez, A.; Noe, M. C. *J. Am. Chem. Soc.*, **1994**, *116*, 12109.

[23] For a history of the discovery, see: Roberts, R. M. *Serendipity. Accidental Discovery in Science*, Wiley, New York, 1989, ch. 24, p. 159.

[24] (a) Sarrett, L. H. *J. Biol. Chem.*, **1946**, *162*, 591; (b) for the concise history of the creation of steroid drugs, see: Hirschman, R. *Angew. Chem., Int. Ed. Engl.*, **1991**, *30*, 1278; (c) for a review, see: Shoppee, C. W. in *Perspectives in Organic Chemistry*, Todd, A. Ed., Interscience, New York, **1956**, p. 315.

[25] (a) For a review, see: Walker, J. *Chemotherapy*, in ref. 24(c), p. 430; (b) Klayman, D. L. *Science*, **1985**, *228*, 1049; (c) see, for example: Avery, M. A.; Chong, W. K. M.; Jennings-White, C. *J. Am. Chem. Soc.*, **1992**, *114*, 974; see also: Ye, B.; Wu, Y.-L. *J. Chem. Soc., Chem. Commun.*, **1990**, 726; (d) Posner, G. H.; Oh, C. H.; Milhous, W. K. *Tetrahedron Lett.*, **1991**, *32*, 4235 and references cited therein.

[26] (a) Still, W. C. *J. Am. Chem. Soc.*, **1979**, *101*, 2493; (b) Hauptman, H.; Muhlbauer, G.; Sass, H. *Tetrahedron Lett.*, **1986**, *27*, 6189; (c) Mori, M.; Okada, K.; Shimazaki, K.; Chuman, T. *Tetrahedron Lett.*, **1990**, *31*, 4037.

[27] (a) Highlights of this story are given in: Stinson, S. *Chem. Eng. News*, **1995**, Apr. 17, 22; (b) Barrett, A. G. M.; Kasdorf, K.; Tustin, G. J.; Williams, D. J. *J. Chem. Soc., Chem. Commun.*, **1995**, 1143; (c) Kuo, M. S.; Zielinsky, R. J.; Cialdella, J. I.; Marschke, C. K.; Dupuis, M. J.; Li, G. P.; Kloosterman, D. A.; Spilman, C. H.; Marshall, V. P. *J. Am. Chem. Soc.*, **1995**, *117*, 10629.

[28] (a) For a history and leading references, see: Cross, P. E.; Dickinson, R. P. *Chem. Br.*, **1991**, *27*, 911; (b) Bhagwat, S. S.; Hamann, P. R.; Still, W. C.; Bunting, S.; Fitzpatrick, F. A. *Nature*, **1985**, *315*, 511; (c) Bundy, G. L. *Tetrahedron Lett.*, **1975**, 1957; see also: Nicolaou, K. C.; Magolda, R. L.; Smith, J. B.; Aharony, D.; Smith, E. F.; Lefer, A. M. *Proc. Natl. Acad. Sci. USA*, **1979**, *76*, 2566.

[29] Woodward R. B. in ref. 24(c), p. 155.

[30] Corey, E. J. *Angew. Chem., Int. Ed. Engl.*, **1991**, *30*, 455.

[31] Gomberg, M. *Ber.*, **1900**, *33*, 3150; for a review, see: Gomberg, M. *Chem. Rev.*, **1925**, *1*, 91.

[32] (a) Sachse, H. *Ber.*, **1890**, *23*, 1363; (b) Barton, D. H. R. *Experientia*, **1950**, *6*, 315; (c) Barton, D. H. R. in ref. 24(c), p. 68.

[33] Quoted from ref. 29, p. 176.

[34] An exciting story of PTFE discovery can be found in ref. 23, ch. 27, p. 187.

[35] Representative examples can be found in the reviews: Mann, J. *Chem. Soc. Rev.*, **1987**, *16*, 381; Welch, J. T. *Tetrahedron*, **1987**, *43*, 3123.

[36] A concise report describing the history and scope of the studies in the area of conducting polymers can be found in: Kanatzidis, M. G. *Chem. Eng. News*, **1990**, *68*, #49, 36.

[37] See, for a review: Williams, J. M.; Beno, M. A.; Wang, H. H.; Leung, P. C. W.; Emge, T. J.; Geiser, U.; Carlson, K. D. *Acc. Chem. Res.*, **1985**, *18*, 261.

[38] Hoffmann, R. *The Same and Not the Same*, Columbia University Press, New York, **1995**.

[39] Seebach, D. *Angew. Chem., Int. Ed. Engl.*, **1990**, *29*, 1320.

CHAPTER 2

Tactics of Synthesis

INTRODUCTORY REMARKS

The title of this chapter may lead some to the conclusion that we intend to present all, or the majority, of the methods employed in modern synthetic practice. In fact, we have neither the intention nor the opportunity within the scope of this text. Such a comprehensive review can be found in excellent monographs mentioned in the list of references. Our goals are entirely different and more modest. We are going to present merely a step-by-step exposure of the basic guidelines that govern the elaboration and utilization of synthetic methods as tools for the solution of organic synthesis problems.

At the beginning are offered background data which illustrate the typical features of organic reactions, factors determining the course of the reaction, its rate, and the very possibility of the reaction occurring. Special emphasis is given to the key role of reaction intermediates, short-living species formed in the course of many organic reactions. Next we consider the problem of converting an organic reaction into a synthetic method to be applied as an infallible tool of organic synthesis. The critical role in assembling organic structures is held by the C–C bond-forming reactions and therefore we find it mandatory to review, briefly, the major routes of creating the C–C bond and the most popular methods employed for this purpose.

Functional group transformations represent another extremely important component of the synthetic arsenal. In fact, this material constitutes the main content of every organic chemistry course. For this reason, we limit our discussion to the main features of those reactions which govern their application to the solution of organic synthesis problems.

For the otherwise 'efficient' synthetic transformations, the lack of selectivity represents a roadblock and so our next topic deals with the most general aspects of selectivity problems and the fairly diverse approaches elaborated to secure realiable control over the regio- and stereochemical course of the reaction.

Organic synthesis has undergone the transformation from a purely heuristic discipline into a solid branch of science with its own logistics and technology based on the principles of retrosynthetic analysis. An essential part of retrosynthetic analysis involves the sequential disconnection of the target molecule into synthons (virtual building blocks) and the subsequent identification of the corresponding reagents to be used in the synthesis. The essence of the synthon approach is discussed in Section 2.6.

The concepts mentioned above can equally well be applied to the synthesis of acyclic and cyclic compounds. Yet owing to the geometrical peculiarities of the cyclic structures, one may expect to find a number of additional, and at times extremely troublesome, challenges to be encountered in this area. For this reason, the problems inherent to the preparation of cyclic compounds of various ring sizes and a set of specific methods elaborated for this purpose are thoroughly discussed in separate sections.

More often than not, the assemblage of the carbon framework in the course of a total synthesis also leads to the appearance of additional functional groups in the synthesized structure. These groups served the purpose of mandatory auxiliaries in earlier steps of the synthesis. As these groups must be removed, destructive methods (to cleave specific C–C bonds) are also presented as important tools in the arsenal of synthetic methods.

Finally, we consider reactions that lead to rearrangements of the carbon framework, as these reactions offer entirely new opportunities in a total synthesis. Their seminal importance for the synthesis of otherwise barely accessible and exotic polycyclic structures will be highlighted in the concluding paragraphs of this chapter.

As one might have noticed, this chapter deals actually with the general aspects of the tactics of an organic synthesis. Here we would like to provide an answer to the ultimate question: 'How can a specific bond or set of bonds at a given site of the final structure be efficiently created?'

PART I HOW TO ACHIEVE THE DESIRED TRANSFORMATION

2.1 GENERAL CONSIDERATIONS OF TRANSFORMATION OPTIONS

We discussed, in the Introduction, an attractive but nevertheless totally unrealistic route to the synthesis of acetic acid from methane and carbon dioxide:

$$CH_4 + CO_2 \not\rightarrow CH_3CO_2H$$

Is this route actually unrealistic? Both the elemental composition and the structure of the final compound suggest that this is the shortest and most direct synthetic pathway. Yet if you mix methane and carbon dioxide you will see that absolutely nothing happens, regardless of reaction conditions. The necessity to discard this simple and appealing route can be explained by at least two reasons.

The first reason is analogous to a simple observation: a sled goes downhill on its own volition but under no circumstances will it go uphill by itself. Acetic acid is richer in its energy content than a mixture of methane and carbon dioxide. Under certain conditions, acetic acid can be split into methane and carbon dioxide, but the reverse process cannot occur. To do so would require the expenditure of a large amount of free energy, just as it would to push a sled uphill.

Secondly, even if a reaction looks good on paper and even if thermodynamic factors are favorable (going downhill in free energy), there still must exist a realistic mechanism by which the reaction can occur. No such reaction channel is available for the direct conversion of methane and carbon dioxide into acetic acid. Yet there are ways of going around these apparently insurmountable obstacles. For example, in the present case the following stepwise detour can be proposed:

$$CH_4 + Br_2 \rightarrow CH_3Br + HBr \quad (1)$$

$$CH_3Br + Mg \rightarrow CH_3MgBr \quad (2)$$

$$CH_3MgBr + CO_2 \rightarrow CH_3COOMgBr \quad (3)$$

$$CH_3COOMgBr + HBr \rightarrow CH_3COOH + MgBr_2 \quad (4)$$

This four-step process seems cumbersome in contrast to the simplicity of the first scheme. It has, however, one unquestionable merit: it works! All four steps proceed smoothly and accomplish the desired conversion quite efficiently.

The question 'how?' immediately arises with respect to this success as it is obvious that the pathway (1)–(4) eventually leads to the same net result as the direct (uphill in terms of free energy) coupling of methane and carbon dioxide. First of all, this detour works because mechanisms exist for these sequential steps that permit the reactions to be carried out. Why, though, does the sled ultimately ascend the hill? What kind of force is responsible for driving the sled uphill?

If one summarizes the previous four reactions, the entire sequence can be reduced to the following two net conversions:

$$CH_4 + CO_2 \rightarrow CH_3CO_2H \quad (5)$$

$$Mg + Br_2 \rightarrow MgBr_2 \quad (6)$$

We can now easily account for the ultimate reason the detour works. The energy released by the highly exothermic reaction (6) more than compensates for the energy needed to effect the desired conversion (5).

A total gain of energy for the whole system by itself, however, does not guarantee a solution to the problem. The released energy produced by the reaction between magnesium and bromine could not be used directly to initiate the reaction between methane and carbon dioxide. The essence of the successful solution shown above lies in the fact that in the sequence of reactions (1)–(4) the energy of the exothermic counterpart of the process, reaction (6), is rendered into smaller fractions and consumed during key steps of the overall conversion

(5), rather than all at once. As a result, a substantial part of this energy is not wasted as mere heat, but instead is used to accomplish the chemical work that leads ultimately to the creation of new chemical bonds.

There is yet a third and independent complication which might seriously affect the efficiency of the transformation of organic compounds. In the majority of organic transformations, fairly different products can be obtained from the same set of starting materials. In fact, the diversity in available reaction pathways is a source of major problems and uncertainties arising in the course of planning an organic synthesis.

The four-step synthesis of acetic acid shown above is carried out in such a way that all the chosen reactions lead to the exclusive, or at least highly predominant, formation of one product. This is ensured both by the nature of the reagents and the reaction conditions used. In turn, the proper choice of the conditions optimal for every step is possible because the synthesis is carried out in a stepwise fashion. It would be an absolutely hopeless adventure even to attempt to elaborate conditions that would enable one to achieve the synthesis of acetic acid in one flask from a simple mixture all of the required components (methane, carbon dioxide, bromine, and magnesium).

This point can be further illustrated by the following mechanical analogy. Let us say we are faced with the task of fixing two plates together with a nut and bolt. It certainly would be strange to attempt to accomplish this task by placing all the components into a box and shaking the box vigorously until the desired result is achieved owing to the energy spent in the process of shaking. The rational approach, of course, is to split the overall process into separate operations: (1) place the plates on top of one another, (2) line up the holes, (3) place the bolt through the holes, and (4) screw on the nut. Each of these elementary operations is associated with its own peculiar procedures and utilization of the proper tools. Keep in mind it is also not possible to change the order of the operations and, certainly, there is no way to do them simultaneously. Though this mechanical model may seem to be by far too primitive, it well illustrates the general approach to the solution of organic synthesis problems.

In order to achieve the desirable overall result, however unrealistic it may appear, an organic chemist looks for an opportunity to break down the overall transformation into a sequence of chemically possible steps. 'Chemically possible' means that a well-defined chemical reaction can be utilized to achieve the required elementary transformation. Then it will be within the chemist's control, at least in principle, to select the proper reagents and reaction conditions ('tools' and 'procedures') for each step that would ensure the efficiency of every required transformation of the whole sequence and hence the viability of the conceived synthetic plan. This ideal scheme is rarely realized completely, but it is generally applicable within reasonable limits. In fact, identification of a set of simple and realistic steps in conjunction with a careful selection of the optimal reagents and reaction conditions for every single step is the underlying approach elaborated for controlled total synthesis.

An illustration of the factors controlling the selectivity of a reaction can be seen in the key step in the synthesis of acetic acid, the reaction of methylmagnesium bromide with carbon dioxide (3). Methylmagnesium bromide is a compound with a highly polarized **C–Mg** bond and a significant partial negative charge located on the carbon (indicated by the symbol $\delta-$ in the scheme). The **C–O** bond in carbon dioxide is also somewhat polarized, but this time with a partial positive charge ($\delta+$ on the carbon). While the mechanism of the reaction between methylmagnesium bromide and carbon dioxide is complex and in some aspects still obscure, the simplified schematic pathway given in Scheme 2.1 depicts the essence of the process and accounts for the observed pattern of its selectivity.

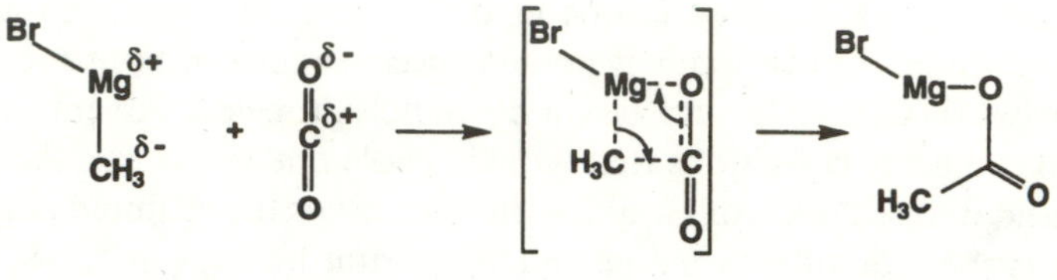

Scheme 2.1

Among the numerous variations in the mutual orientations for the colliding molecules of CO_2 and CH_3MgBr, there are two limiting cases which differ sharply in their consequences. In the case that the colliding partners have the wrong orientation, the repulsive coulombic forces between similar charges will drive the molecules apart (elastic collision) and no chemical reaction will occur. If, on the other hand, the molecules approach each other with a complementary orientation of opposite partial charges as shown in Scheme 2.1, an electrostatic attraction arises. As the distance between the molecules becomes shorter, the electrostatic interaction between the species leads to an increase in their intrinsic polarization. This increase of the partial charges leads, eventually, to the formation of a species that can be described as a provisional structure, a transition state **1**. This transition state corresponds to the partial dissociation of old bonds and the formation of new bonds. It is higher in energy than either of the starting compounds or the products. This higher energy state represents a potential barrier that can be overcome only if the colliding molecules possess sufficient energy.

Upon surmounting this barrier, a complete displacement of electron pairs is achieved with the creation of a **Mg–O** bond and a **C–C** bond. The overall process is energetically favorable because (i) it ultimately leads to the formation of a very stable covalent **C–C** bond and an ionic **Mg–O** bond and (ii) its course is governed by the intrinsic tendency of magnesium to lose electrons and of oxygen to acquire electrons.

The extreme importance of both the thermodynamic (free energy) and kinetic (mechanistic pathway) considerations makes it necessary to analyse these factors more thoroughly and from other perspectives.

2.2 THE THERMODYNAMIC ALLOWANCE OF THE PROCESS

Fossil raw materials, which serve as the ultimate source of starting compounds for organic synthesis, were formed as a result of bio- and geochemical processes over an extremely long period of time. During this time, an equilibrium (or near equilibrium) state was achieved. This means that the individual compounds isolated from fossils correspond to those of minimum free energy, at least under anaerobic conditions. Organic synthesis is generally targeted to produce compounds with a higher free energy content than the starting materials. If it were otherwise, the problems of organic synthesis would be extremely simple since the target compounds could be easily accessible *via* any route involving the equilibration of the starting materials (for example, upon heating). The increase in free energy in the target compound is accumulated through the formation of novel chemical bonds confined in a more ordered system. In order to achieve this non-equilibrium state, additional work must be done upon the initial system. Consequently, some external source of energy is required. This energy may be supplied in the form of heat, electrical, or light energy, but, most frequently, organic synthesis is based on the utilization of chemical energy.

Chemical energy is stored and supplied in the form of chemical reagents, which accumulated their energy during their preparation. Regardless of the refined preparative methods employed, the use of electrical energy is usually found at the initialization of the process. Thus in the example shown, magnesium and bromine serve as highly reactive agents, both having been produced by the electrolysis of their corresponding salts. This simple example makes it clear why the key role in many organic syntheses is played by reagents such as, for example, free halogens (F_2, Cl_2, and Br_2), metals such as Li, Na, K, Mg, and Zn, and hydrides like LiH, NaH, KH, $NaBH_4$, $LiAlH_4$, Et_3SiH, Bu_3SnH, B_2H_6, *etc.*

From a thermodynamic point of view, an organic synthesis can be likened to a complex and hazardous climb up a mountain, with many detours up and down but eventually leading to a certain point positioned at a higher level. The schematic profile of a hypothetical route from starting compound **A** to final product **P** is presented in Fig. 1. This profile graphically illustrates three important peculiarities which are generally featured in a typical synthetic

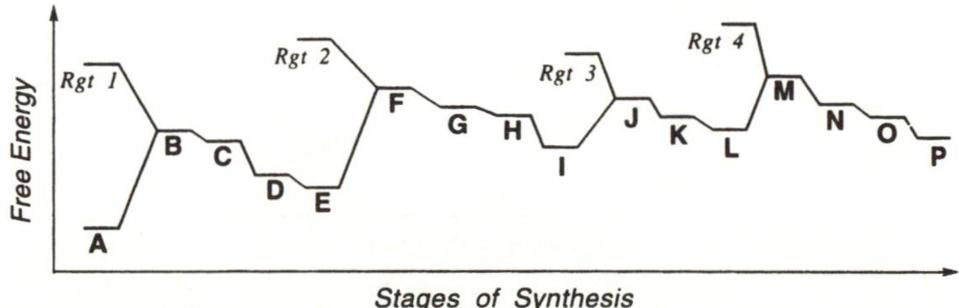

Figure 1 *Energy profile of multistep synthesis of product P from the starting compound A* via *intermediate products B, C,* etc.

sequence. First of all, successful movement along the route from **A** to **P** requires periodic introduction of free energy into the system. This is achieved through the help of additional key reagents (shown here as Rgt 1 to Rgt 4). Secondly, the acquired energy is expended gradually over a sequence of transformations, thereby achieving a measure of control over the reaction course, as was the case in the synthesis of acetic acid. There we saw that the energy stored in methylmagnesium bromide was used in successive reactions, first with CO_2 and then with HBr. Third, the intermediate products on the route from **A** to **P** possess, as a rule, an excess of free energy over that of their respective precursors. Therefore, the possibilty exists that these intermediates might alternatively slide down into a different 'potential well' and thus be converted into undesirable products. It is of utmost importance, then, to be able to 'channel' the energy accumulated in the reactive intermediates into the right direction, a goal not always easily accomplished.

2.3 THE AVAILABILITY OF A REACTION CHANNEL. KINETIC *VS.* THERMODYNAMIC CONTROL

To perform the thermodynamically allowed transformation X → Y, the reaction system X (where X refers to the whole set of reacting components) must, as a rule, surmount some energy barrier. This energy barrier results from the necessity to pass through a transition state of an energy higher than either the starting materials or the products. For a desirable reaction to occur, the colliding molecules must additionally have the necessary orientation. As a result, at any given moment only a small fraction of the molecules will be able to undergo the desired transformation. Organic reactions therefore do not occur instantly, but proceed with a measurable rate, ultimately dependent upon the height of the barrier (activation energy). If the barrier is low, the rate of the reaction is high, and *vice versa*. The availability of a suitable reaction channel implies the existence of a reasonably low-energy transition state, corresponding to a surmountable energy barrier (Fig. 2).

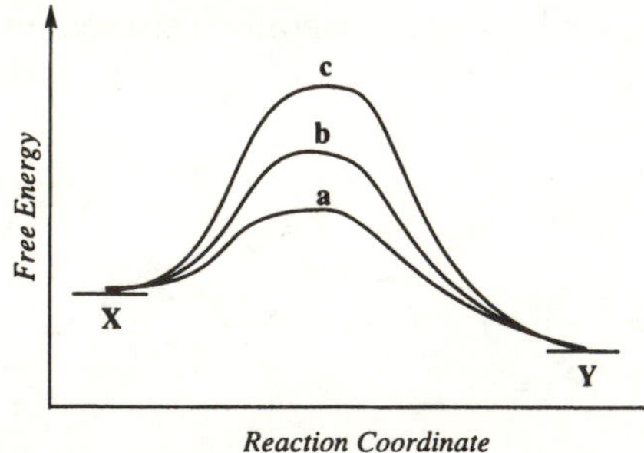

Figure 2 *Reaction potential barriers:* **a** *— very low barrier, reaction occurs almost instantaneously;* **b** *— medium barrier, measurable reaction rate;* **c** *— very high barrier, reaction rate is too slow to be measured*

The barrier between starting compounds and their products is never infinitely high. As a matter of fact, we always have the possibility, at least theoretically, of going over the highest possible barriers by breaking the starting materials into small fragments or even atoms and then reassembling them. Provided a transformation $X \rightarrow Y$ is thermodynamically allowable, this route may be theoretically possible, but realized only if no better allowable pathways exist for this system.

Thermodynamically driven high-temperature conversions have rather important ramifications. A notorious group of ecotoxins, the dioxins, are extremely stable thermodynamically. As a result, they are abundantly formed in the course of burning almost any type of halogen-containing organic compound. To ensure complete combustion of wastes containing chloroorganic materials [such as poly(vinyl chloride), PVC], special precautions need to be taken to prevent the thermodynamically driven formation of dioxins.

Let us consider next the same system, $X \rightarrow Y$, but with an additional component, the thermodynamically more stable by-product Z (Fig. 3). Starting system X may react in two directions, forming either the desired product Y or an unwanted by-product Z. Curve (a) portrays a favorable case for the generation of product Y. The barrier of the reaction $X \rightarrow Y$ is significantly lower than that of $X \rightarrow Z$ and hence the former reaction proceeds preferentially. Likewise, the high barrier for the reaction $Y \rightarrow Z$ prevents this unwanted reaction from occurring even though Z is a more stable product. In this case, Y is said to be kinetically stable and its synthesis from X is a viable process since the transformation of Y into the thermodynamically more stable product Z proceeds very slowly. In other words, Y still has the tendency of 'falling' into a lower-energy well but it is 'locked' at its comparatively higher energy level by virtue of the high barriers. It is precisely the existence of such relatively high energy barriers that account not only for the very existence of a large number of otherwise thermodynamically unstable organic compounds, but also for the opportunity for their synthesis along 'contra-thermodynamic' routes, as is shown by the uphill climb leading from **A** to **P** in Fig. 1.

Curves (b) and (c) in Fig. 3 describe unfavorable conditions for the synthesis of Y from X. In the first case, the side-reaction $X \rightarrow Z$ is highly predominant. In the second, the desired reaction leading to Y takes place initially, but Y turns out to be both thermodynamically and kinetically unstable and is rapidly depleted by the formation of Z.

The next two real examples serve to illustrate the role of kinetic and thermodynamic stability. The structure of benzene and its adequate representation was one of the most controversial topics for organic chemistry in the middle of the 19th century. Among various options, a hypothetical structure **2** for benzene was suggested by Dewar in 1867. This suggestion turned out to be basically wrong since experimental data unambiguously supported Kekulé's structure **3**. Almost a century later, Dewar's imaginative formulation served to inspire studies aimed specifically at the synthesis of compound **2**. The latter, named appropriately 'Dewar benzene', was finally prepared by van Tamelen.[1a] Not unexpectedly, it was found that Dewar benzene is considerably less stable in

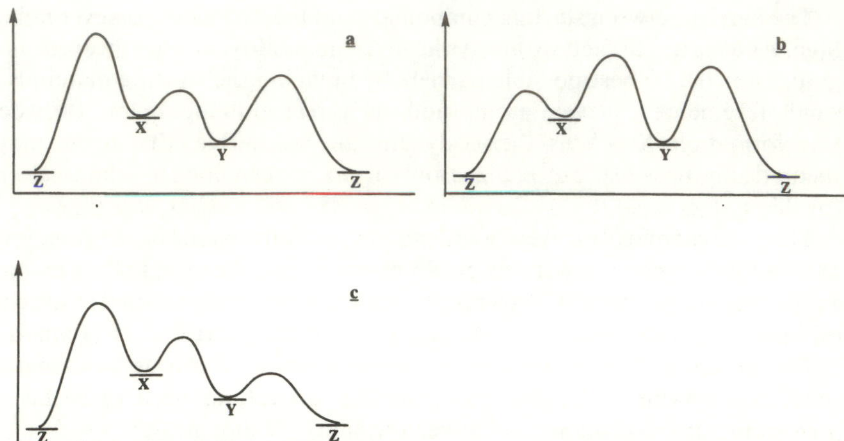

Figure 3 *(a) X → Y process is dominant, Y is kinetically stable; (b) X → Z process is dominant; (c) X → Y process is initially dominant but Y is kinetically unstable and undergoes fast conversion into a by-product Z*

the thermodynamic sense than 'normal' benzene. In fact, the conversion of Dewar benzene to benzene proceeds with a formidable energy release, approximately 60–70 kcal mol^{-1}! Nevertheless, Dewar benzene is still capable of existence at ambient temperature, with its conversion to benzene proceeding at a measurable rate (a half-life of approximately two days at 20 °C).

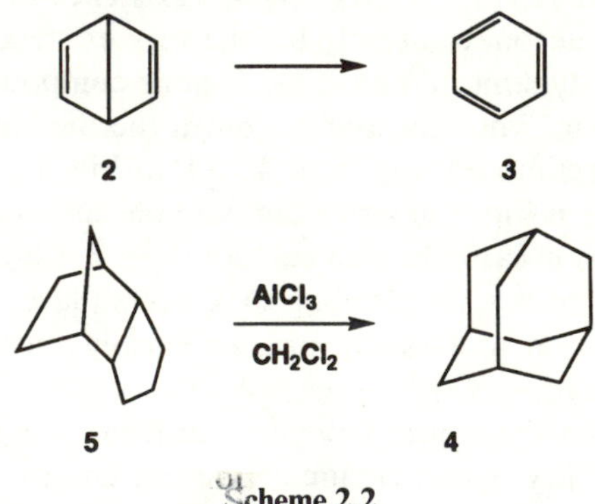

Scheme 2.2

These observations indicate the existence of high energy barriers that literally 'lock' the Dewar benzene inside the energy well and prevent its immediate conversion to benzene. Kinetic stability makes the synthesis of Dewar benzene feasible. The methods employed for its preparation are sufficiently mild so the opportunity for the concurrent reaction (the formation of the more stable isomer benzene **3**) was safely excluded.

The case of adamantane **4** may serve as an example of the reverse type. This hydrocarbon, $C_{10}H_{16}$, the skeleton of which represents a repeating unit of the crystal lattice of diamond, was considered as late as the 1940s to be an exotic compound, especially because its synthesis was a rather lengthy procedure. It was recognized later that, among all other possible structures for $C_{10}H_{16}$ isomers, the structure of adamantane **4** corresponds to the deepest minimum in free energy. In other words, adamantane is extraordinarily stable thermodynamically. These findings suggested an opportunity for the preparation of adamantane by a thermodynamically controlled isomerization of other $C_{10}H_{16}$ hydrocarbons. In fact it was discovered by von Schleyer that the readily available hydrocarbon **5** can be isomerized to adamantane **4**, albeit under rather drastic acidic conditions.[1b] This method was further developed for large scale preparations of adamantane. The fact that adamantane was found in petroleum, an abundant pool of thermodynamically equilibrated compounds, is partially accounted for by the observed ease of its formation from the less stable hydrocarbons.

It is necessary to emphasize that energy profile diagrams of the types depicted in Figs. 1–3 should not be considered as rigid descriptions, determined exclusively by the structure of the compounds involved. A significant role in the processes described by a given energy profile is played by the external conditions of the reactions. The latter include such factors as temperature, pressure, solvents, catalysts, light, *etc.* Their influence on the ease of a given pathway leading to a specific product is generally amenable to a rational interpretation. In many cases, with the proper choice of the reaction parameters, it is possible to channel a reaction along the lines dictated by the requirements of a particular synthetic task.

Let us illustrate this important claim with a simple model showing how the reactivity pattern of the substrate, toluene **6** (Scheme 2.3), can be controlled by variations in reaction partners and external conditions. There are two reactions of **6** that proceed with the same stoichiometry (equations 1 and 2) but yield isomeric products: benzyl bromide **7** and *p*-bromotoluene **8**. Under the appropriate conditions it is possible to carry out each reaction selectively with almost complete exclusion of the alternate process. In order to understand how this can be accomplished, it is necessary to analyse the mechanisms of these conversions. The concise description of reaction mechanisms requires the use of special symbols such as the curved pronged or half-pronged arrows shown below. These arrows indicate the movement of an electron pair or a single electron, respectively, within the dynamics of a reaction process.

The mechanism for the bromination of toluene to form benzyl bromide (equation 1) is represented in Scheme 2.4. The actual reagent attacking toluene is atomic bromine. This atomic bromine is formed from the reversible dissociation of the bromine molecule upon interaction with a photon. The bromine atom is a very reactive species, capable of abstracting an atom or radical from other molecules. Although there are eight hydrogen atoms in toluene, only the three located in the methyl group effectively react with atomic bromine, owing to the fact that removal of a methyl hydrogen leads

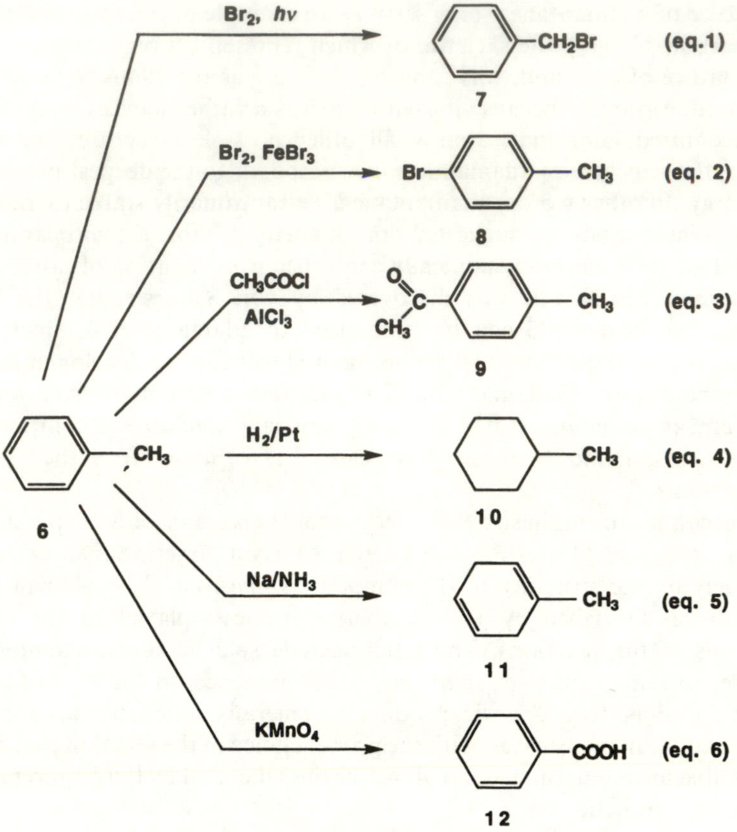

Br_2, $h\nu$ — CH_2Br **7** (eq.1)

Br_2, $FeBr_3$ — Br— —CH_3 **8** (eq. 2)

CH_3COCl, $AlCl_3$ — O=C(CH$_3$)— —CH_3 **9** (eq. 3)

6 CH_3 — H_2/Pt — CH_3 **10** (eq. 4)

Na/NH_3 — CH_3 **11** (eq. 5)

$KMnO_4$ — COOH **12** (eq. 6)

Scheme 2.3

Br—Br $\xrightarrow{h\nu}$ Br• + Br•

6 (H–C–H, H) + Br• $\xrightarrow{-\text{HBr}}$ •CH_2 **6a** + Br—Br

$\longrightarrow$ CH_2Br **7** + Br•

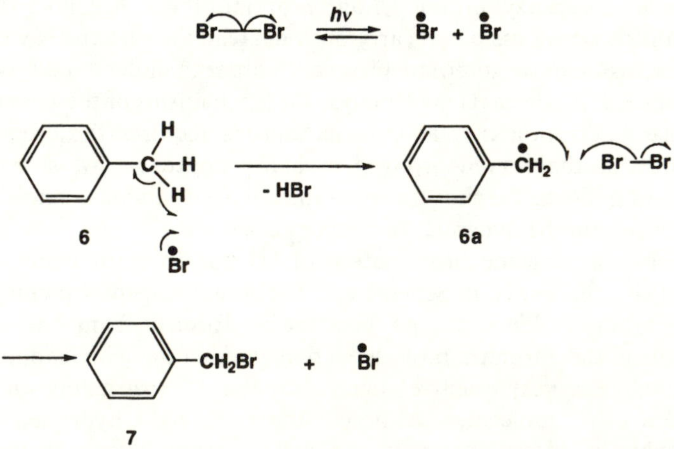

Scheme 2.4

to the relatively stable benzyl radical **6a**. Reaction of this radical with molecular bromine results in a formation of the final product, benzyl bromide **7**, accompanied by the regeneration of atomic bromine to serve further as a chain propagator.

The reaction of bromine with toluene in the presence of iron(III) bromide (equation 2) proceeds along an entirely different reaction pathway. Here, the actual reagent turns out to be neither molecular bromine nor atomic bromine. In this case, under the action of iron(III) bromide, molecular bromine is converted into a bromine cation and an anionic complex, $[FeBr_4]^-$, as is shown in Scheme 2.5. This is a reversible process and the stationary concentration of the bromine cation can be rather low (as in the former case of atomic bromine). It is, nevertheless, sufficient to make Br^+ a reactive partner.

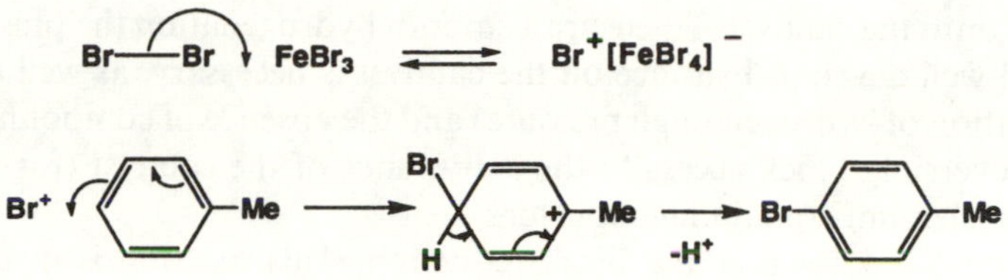

$$Br{-}Br + FeBr_3 \rightleftharpoons Br^+[FeBr_4]^-$$

Br⁺ + toluene (Me) → **13** (Br, H, +, Me) → (−H⁺) → Br–C₆H₄–Me **8**

Scheme 2.5

The methyl hydrogens of toluene are relatively inert to the attack of these charged particles. In contrast, the polarizable π-electron system of the aromatic nucleus can be easily perturbed by the approach of the bromine cation. In a simplified form, their interaction can be described in the sequence of the following steps. The attacking cation pulls an electron pair of the aromatic system towards itself to form a **C–Br** bond. A concerted shift of the next electron pair leads to the development of a positive charge on the methyl-bearing *para*-carbon atom. Loss of a proton from this σ-complex **13** leads to the formation of bromotoluene with restoration of the aromaticity in the system. Besides the *para* isomer of bromotoluene **8**, the corresponding *ortho* isomer is also formed, but in lesser amounts.

These mechanistic descriptions, in spite of their obvious oversimplification, may provide us with important clues about the conditions necessary to avoid completely the formation of 4-bromotoluene **8**, if our efforts are aimed at the preparation of benzyl bromide **7**, and *vice versa*. In order to prepare benzyl bromide **7** cleanly, toluene and bromine must not contain impurities capable of generating Br^+ (for example, it is not a good idea to use toluene stored in a rusty barrel without purification). The ionic bromination of toluene to form 4-bromotoluene **8** must be carried out in the dark to prevent the light-induced formation of the radical bromine species responsible for the formation of benzyl bromide.

Many other important reactions take place in much the same manner as the ionic bromination of toluene and are referred to as electrophilic aromatic substitutions. Among them is the reaction of acetyl chloride in the presence of aluminum chloride (equation 3, the Friedel–Crafts reaction). Here the attacking reagent is the MeCOCl–$AlCl_3$ complex which serves as a source of the acetyl cation, $MeCO^+$. The acetyl cation reacts with toluene to form a σ-complex analogous to the intermediate **13** formed during the preparation of bromotoluene.

Now let us compare two other reactions, both of which involve the addition of hydrogen (equations 4 and 5). Complete reduction of toluene under the action of hydrogen to form methylcyclohexane **10** proceeds easily in the presence of Group 8–10 metals, in particular platinum (equation 4). The process occurs on the surface of the metal catalyst by a rather complex mechanism. In essence, molecular hydrogen becomes reactive owing to its interaction with active centers on the surface of the metal catalyst. The activated hydrogen is able to add readily to the double bonds of the toluene, which is also adsorbed onto the catalyst. To ensure a smooth hydrogenation the presence of a clean and well developed surface on the catalyst is necessary, as well as a high concentration of hydrogen (high pressure) and the absence of compounds which might irreversibly block access to the active sites of the catalyst (for example, sulfur- or selenium-containing impurities).

Equation 5 is known as the Birch reduction. This reaction is mediated by Group 1 metals (usually sodium) in a liquid ammonia solution in the presence of a certain amount of alcohol. The net result is also the addition of hydrogen, but in this case only two hydrogen atoms are added. The reason for this peculiarity becomes apparent upon examination of the mechanism of this process. When a metal like sodium is dissolved in liquid ammonia, it dissociates into a sodium cation and a solvated electron (Scheme 2.6). The first step in the Birch reduction is the attack of the solvated electron on the aromatic system of toluene to

$$Na + NH_3 \longrightarrow Na^+ \cdot nNH_3 + e^- \cdot mNH_3$$

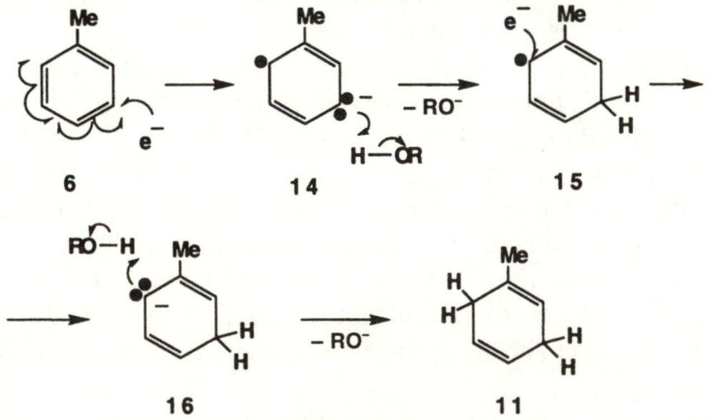

Me
e⁻
6
Me
14
H—OR
− RO⁻
e⁻
Me
H
H
15
RO—H
Me
H
H
16
− RO⁻
Me
H
H
H
H
11

Scheme 2.6

produce the radical-anion **14**. The radical-anion abstracts a proton from an alcohol and is transformed into the radical **15**. This radical in turn adds a second electron and is converted to a carbanion **16**. A final transfer of a proton from the medium to this anion gives the final product, the diene **11**.

Among possible alternative isomers, the preferential formation of diene **11**, with the indicated location of the double bonds, is determined by the structure of the initially formed, most stable intermediate, radical-anion **14**. Thus, the reduction of a single bond of toluene, as is represented in equation 5, requires the presence of an electron source (sodium), a solvent capable of electron solvation (liquid ammonia), and a proton donor (alcohol).

Toluene can also be oxidized, for example with permanganate to produce benzoic acid **12** (equation 6). The mechanism is somewhat analogous to radical bromination as it takes advantage of the susceptibility of the methyl hydrogens to attack by radicals. In this case a sequential replacement of all three hydrogens occurs, which results in the conversion of the methyl group into a carboxyl group. An interesting aspect of this conversion is that its course is determined primarily by the nature of the reagent and is minimally sensitive to the variations of the external conditions. Thus this conversion can be successful in a toluene medium or an aqueous solution, at room temperature or upon heating, *etc.* The predominant product, while it may be formed at different rates and in different yields, invariably turns out to be benzoic acid.

The reactions of toluene illustrate a very important peculiarity of many organic reactions: they proceed with the involvement of high-energy intermediates. The structure and reactivity of these species are actually the main factors that determine both the viability of the reactions and the pathways they would take. In the above examples, species like $Br^{\bullet}$, Br^{+}, CH_3CO^{+}, σ-complex **13**, radical-anion **14**, and carbanion **15** participated as intermediates. These intermediates, however, should never be confused with transition states even though they may often be similar in energy and structure. The differences can be best illustrated by means of the energy profile diagram presented in Fig. 4.

An intermediate, like **C**, corresponds to a fully defined chemical species which, at least in principle, can be isolated. It differs from 'normal' compounds

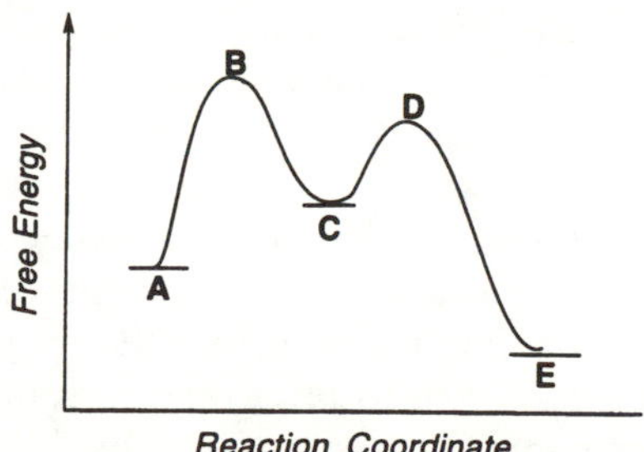

Figure 4 *Potential energy profile for a reaction pathway involving the formation of an intermediate.* **A** *— starting system;* **B, D** *— transition states;* **C** *— intermediate;* **E** *— final products*

like **A** or **E** only in a quantitative sense, owing to its higher energy content and the relatively low barriers that separate it from the more stable compounds. Both of these factors determine the high reactivity of typical intermediates. Their energy excess serves to promote reactions with otherwise unreactive compounds. Low barriers permit the reaction to occur rapidly.

A transition state (TS) is a completely different entity. It refers to a momentary state of a reacting system at a maximum of free energy. By definition, it cannot exist as a static structure. The energy profile diagram illustrates this in its own language: the absence of potential barriers at TS points **B** or **D**. Only a fleeting formation of the transition state is achieved; then the system must immediately slide down to a lower energy level like **A**, **C**, or **E**, where existing energy barriers maintain a static system.

Fig. 4 illustrates yet another important feature about reactions of this type. The barrier to be crossed on the route between the starting compounds and the intermediate is higher than the barrier between the intermediate and the final products. This means that the formation of the intermediate is the slowest process and will determine the overall rate of the reaction (since the conversion of **C** to products is extremely fast). The energy barrier for the transformation of the intermediate into products is low because the system has already accumulated a significant reserve of energy and has ascended almost to the summit of this barrier.

Two conclusions can be drawn from what has just been said. Typically, reaction intermediates are present only in low concentrations because they react faster than they are formed. Therefore, the isolation and structure elucidation of the intermediates can be a troublesome task. Secondly, if the structure of an intermediate has been completely ascertained and it can be independently prepared, then its use as the reagent can lead not only to the sharp acceleration of the reaction but also to an increase in its selectivity. This point is worthy of additional comments.

In order to form an intermediate from starting compounds it is frequently necessary to overcome a high energy barrier. Therefore, in order to achieve an acceptable rate of formation (and this is the rate-limiting step of the overall process!), it may be necessary to resort to various methods of forcing the reaction. These may include the use of high temperatures, irradiation, highly active reagents, or catalysts. If these 'forcing conditions' are applied directly to the system consisting of the initial compound and a reagent, then in addition to promoting the formation of the desired intermediates, these conditions can also facilitate various side-reactions of both the reactants and the final products. These complications can be largely alleviated if there is a way to carry out the overall reaction as a sequence of independent steps, the first step being the preparation of the intermediate from the reagent (in the absence of the substrate) and the second step to react the substrate with the pre-formed intermediate. Here tremendous promise is offered by the chance to prefabricate the required intermediate under the conditions required for its generation, with subsequent reaction of the intermediate with a given substrate under conditions optimal for that step. In fact, quite a number of revised and efficient versions of well-known classical methods have been developed recently owing to the utilization of this fairly general approach. We shall refer rather often to this point in the following sections, but here it is worthwhile to look at just one example related to the reactions of toluene mentioned earlier.

The Friedel–Crafts acetylation of toluene (equation 3, Scheme 2.3) takes place smoothly at room temperature, but aromatic systems containing electron-withdrawing groups, such as dichlorobenzene or nitrobenzene, react very sluggishly under these conditions. The use of more forcing conditions might be undesirable owing to the occurrence of side reactions.

As was already shown, the catalyst in this reaction serves to polarize the MeCO–Cl bond to form an intermediate tentatively ascribed the structure of the acetylium salt $[MeCO]^+[AlCl_4]^-$. It turns out that the acetylium ion can actually be prepared as a stable salt, for example as $[MeCO]^+[SbCl_6]^-$. Since the reactive species is already present in this salt, $[MeCO]^+[SbCl_6]^-$ serves as a mild and very active acetylating agent, even for unreactive aromatic compounds. Moreover, the reactions of this salt do not necessitate the presence of any acid catalyst. Hence, reagents of this type can also be utilized for unstable substrates.

We have seen, then, by the judicious selection of a reaction that is suitable for a given substrate and by the selection of proper reagents under well-defined conditions, that it is possible to regulate the reactivity of organic compounds and to direct their transformations as desired. The enormously rich potential of organic reactions in achieving most diversified transformations of organic compounds was well recognized long ago. At the same time it was also clear that not all of these reactions could be used as truly effective tools in organic synthesis. In fact, the main pathways for interconverting organic molecules had already been elucidated in the early 1930s. The synthetic potential of already discovered reactions was sufficiently diversified and the synthesis of almost any compound could have been considered as a possible enterprise, at least theoretically. Yet at that time the achievements in the area of total synthesis were rather modest. There was still a long way to go before the theoretically possible could be transformed into the practically viable.

In fact this situation is typical for almost any other area of science and technology. Thus, for example, the elucidation of the basic principles of aeronautics and successful design of the airplanes made possible the first intercontinental flights in the early 1930s. These spectacular achievements manifested an appearance of a new and most promising way of transportation. Yet it took several decades of research and development in order to make aviation reliable, efficient, and relatively cheap. Only owing to these efforts could a modern global system of air transportation have been created. Basic underlying principles stayed the same but their implementation had changed dramatically!

In much the same way, the transformation of the vast potential of organic chemistry of the 1930s and 1940s into the near omnipotence of modern organic synthesis was achieved not owing to the discovery of new *fundamental principles* of organic chemistry but rather to a truly *revolutionary developments of its tools and methodology*. This evolution required tremendous efforts aimed primarily at the elaboration of well-known reactions into reliable and practical synthetic working tools. It was mandatory to raise the status of standard organic transformations up to the rank of *synthetic method*.

We will now examine in greater detail what is required to consider an organic reaction to be a synthetic method.

2.4 ORGANIC REACTION *VS.* SYNTHETIC METHOD

There is no strict definition of the term 'synthetic method' but it is not difficult to describe the meaning of this notion. An ideal synthetic method can be likened to an operator in mathematics, to a 'black box' in cybernetics, or to *any device which can be applied to an object to achieve predictable changes*. In a similar way, the synthetic method must serve as a standard device to induce an unambiguous change over the structure of the treated compounds. This 'black box' for synthesis might contain a standard sequence of operations, including one or more chemical reactions with the required reagents, necessary solvents and catalysts, procedures of reaction monitoring and product isolation, *etc*.

As might be expected, the value of a given synthetic method is directly related to the nature of the transformation that can be accomplished with its help. This transformation must be focused and suitable for converting abundant starting materials into the less available compounds. For example, aromatic hydrocarbons are readily available starting materials, the products of coal and petroleum processing. Many of their important functional derivatives, however, need to be artificially created. Major pathways to achieve transformation of aromatic hydrocarbons into more valuable derivatives are based on electrophilic substitution such as Lewis acid catalysed bromination or acylation (see above for the examples). Therefore these reactions were thoroughly studied and thus eventually elaborated into reliable and highly efficient methods general in their scope of application.

The key transformation in a given method may imply the utilization of unstable intermediate compounds or reactants. However, if the net result of a sequence of two or three reactions can be tracked back to readily available materials, this sequence can also become the basis of a good synthetic method. For example, the reaction of organomagnesium compounds, the Grignard reagents, with carbon dioxide provides a smooth and fairly general entry into carboxylic acids. These reagents, however, are not very stable upon storage and only a limited number of them are commercially available. Fortunately, they are easily synthesized from readily available organic halides and magnesium and can be used immediately in their reaction with carbon dioxide. The sequence of three reactions shown below serves as an excellent method for the synthesis of a carboxylic acid from the corresponding organic halide with one-carbon chain elongation:

1) R–Hal + Mg → R–MgHal

2) R–MgHal + CO_2 → R–COOMgHal

3) R–COOMgHal + H_3O^+ → R–COOH

This method is universally applicable and therefore it is quite appropriate to represent it in the terminology of a 'black box' as follows, without any reference to the details:

R–Hal → [Grignard Method] → R–COOH

This example may also serve as a good illustration of the importance of a reaction's generality as a major criterion for the evaluation of its merits as a synthetic method. *Generality* of the method implies that the reaction is feasible not only for a limited number of the compounds but can be efficiently applied to the vast majority of compounds containing a definite structural element, the functional group. In other words, generality suggests that the results of a reaction affecting a given functional group are minimally sensitive to the variations in the remainder of the molecule. It is the generality of a reaction that enables one to predict, with confidence, the results of the application of this method to a previously unexplored system, a situation frequently encountered in total syntheses. It is this very generality that allows one to describe a sequence schematically with the use of the symbol 'R', which tacitly implies that the final outcome does not depend on the nature of the group hidden behind this symbol.

The discovery of a new reaction may occur merely owing to a lucky chance or rigorous reasoning and typically it refers to a few isolated examples. However, it is always followed by a series of systematic investigations aimed at elucidation of the scope and limitations of the applicability of this reaction to a wide array of compounds. Without these studies it is hardly possible to make any assessments about the true value of the newly discovered reaction as a synthetic method.

Let us consider, for example, the origin and evolution of the well-known Diels–Alder reaction. The net outcome of this process is the cycloaddition of conjugated dienes with alkenes, leading to the formation of cyclohexene derivatives (Scheme 2.7). Isolated examples of this cycloaddition, for example

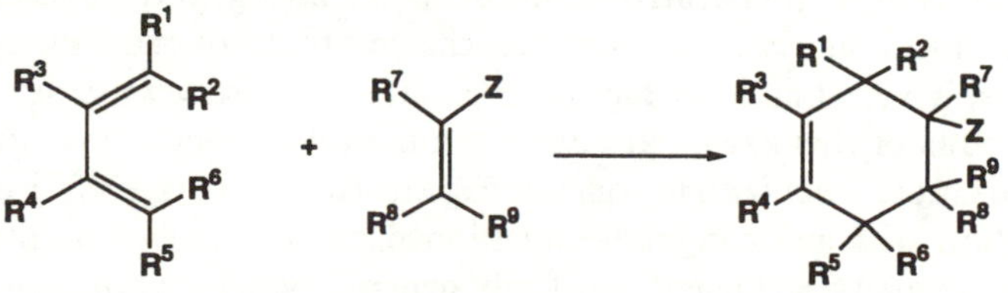

R¹ R² R³ R⁴ R⁵ R⁶ + R⁷ Z R⁸ R⁹ →

Z = an electron withdrawing group

Scheme 2.7

the dimerization of isoprene, were known for a long time but passed virtually unnoticed for lack of data on the course and scope of the reaction. A true discovery of this reaction was made in 1928, owing to the pioneering and insightful studies of Diels and Alder.[2a] From the very beginning of their investigations, Diels and Alder recognized the tremendous synthetic potential of the newly discovered cycloaddition and targeted their efforts at the elaboration of this reaction into a method applicable for the synthesis of polycyclic compounds. In a few years, an enormous amount of data was compiled which allowed them to elucidate the main peculiarities of the reaction course and delineate the limitations of its application. Owing to these efforts, the initial observations of an interesting, but limited value, transformation led, ultimately, to the elaboration of one of the most powerful methods in organic synthesis. This work was properly crowned in 1950 with a Nobel Prize and was perhaps even more fittingly recognized when it became generally known in the chemical community as the 'Diels–Alder reaction'.

Here is another example, different, but related to the same idea. The synthesis of carboxylic acids, mentioned earlier, is one of several powerful and general methods of synthesis based on the use of Grignard reagents (organomagnesium compounds). It was Grignard who discovered in 1900 that diethyl ether serves as both solvent and catalyst for the formation of these reagents from various organic halides and magnesium metal. At that time, the Grignard reaction represented one of the very few reliable methods for the creation of novel **C–C** bonds. The scope of its applicability, however, suffered from some limitations, most importantly the inability to prepare Grignard reagents from vinyl halides, **C=C–Hal**. This was especially unfortunate since vinylic Grignard reagents could have been of significant synthetic value. The solution to this problem, which turned out to be amazingly simple, was found by Normant[2b] in the 1950s. It was discovered that the preparation of vinylic Grignard reagents could be accomplished merely by using tetrahydrofuran as the solvent instead of diethyl ether. With this discovery it was possible to utilize vinylic Grignard compounds fully as valuable intermediates and significantly broaden the scope of the Grignard reaction into a nearly universal synthetic method.

While the generality of a reaction is primarily determined by its chemical mechanism (by the nature of the transformations involved), it may happen that the limitations of its preparative scope arise not owing to the chemistry but to purely 'technical' causes. For example, the solubility of reacting components may impose severe limits on the scope of an otherwise very general reaction. Complications of this nature are quite common when water-insoluble organic compounds must react with inorganic reagents (water, salts, *etc*.). For example, the oxidation with permanganate mentioned earlier, as well as other related oxidations, constitute efficient and fairly general methods. Potassium permanganate and other inorganic oxidants, however, are insoluble in the majority of organic solvents and their typical organic substrates are only marginally soluble in water. Because of this, the classical oxidation with permanganate required a heterogeneous system, conditions far from optimal. In the 1960s, however, a spectacular solution to this and related problems of incompatibility was discovered, the essence of which will now be considered.

Let us consider why potassium permanganate is insoluble in benzene but soluble in water. In an aqueous solution the ions derived from the dissolution of $KMnO_4$, the K^+ and MnO_4^- ions, exist not as free species but rather as complex aggregates formed by multiple tiers of the polar molecules of the solvent, water, surrounding the central ion. The energy gained in the formation of such a 'coat' more than compensates for the energy required to disrupt the crystal lattice of the solid salt with its transfer into the solution. A non-polar organic solvent such as benzene is unable to solvate charged species effectively and thus cannot dissolve inorganic salts. This dilemma was overcome with the aid of a third component, a compound soluble in benzene but at the same time serving to fulfill the role of a 'coat' for the ions.

In particular, the compounds capable of doing this job are the macrocyclic polyethers of type **17** (Scheme 2.8), the so-called crown ethers discovered in the 1960s by Pedersen.[2c] The interior of a crown ether contains a cavity into which an unsolvated ion of potassium can fit. Six atoms of oxygen coordinate strongly with the central potassium ion and thus effectively replace the hydration shell. As a result, complexes such as **18** are quite soluble in a number of organic solvents. Therefore, adding a small quantity of crown ether **17** to a two-phase system of crimson-colored aqueous $KMnO_4$ and colorless benzene immediately turns the benzene crimson due to the transfer of $KMnO_4$ into the organic phase. It is not surprising that the oxidation of an organic compound in this system proceeds incomparably more efficiently than in the absence of crown ether. Owing to this simple trick, oxidation reactions of organic compounds with inorganic salts, previously not very useful because of the phase barrier, were transformed into a set of valuable synthetic tools.

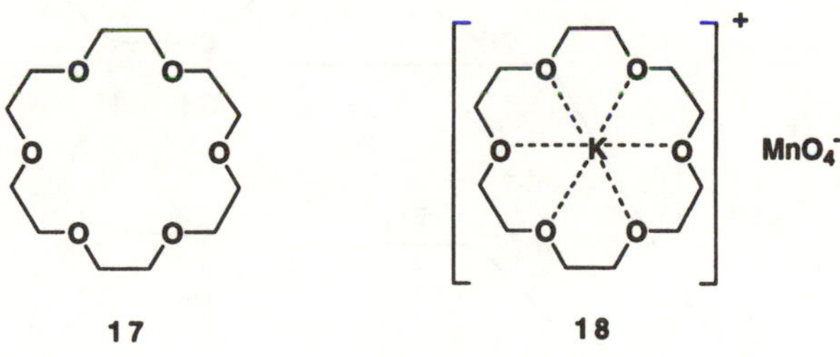

Scheme 2.8

A multitude of other organic reactions imply the utilization of inorganic reagents insoluble in regular organic solvents. These reactions became viable owing to the use of new and unusual catalysts, such as **17**, which are usually referred to as phase transfer catalysts.[2d] The ramifications of phase transfer phenomena are numerous and some of the most important will be discussed later in Section 4.2. For the moment we see that the solution of a purely technical problem resulted in a tremendous enrichment of the synthetic arsenal by broadening the applicability of old and well-known reactions.

The general applicability of a reaction, however, while being a necessary requirement, is far from being sufficient to attest to a given reaction as being a good synthetic method. Classical organic chemistry is abundant with numerous reactions that were very promising and rather general in their time, but upon closer scrutiny turned out to be unsuitable as synthetic methods.

In this respect, the Wurtz reaction seems to represent an especially instructive example. This reaction, coupling alkyl iodides, R^1–I and R^2–I, under the action of sodium metal, was discovered in the very dawn of organic chemistry (1855). It was probably the first example of an organic transformation leading to the formation of novel **C–C** bonds and its synthetic value seemed to be indisputable. In fact, textbooks in organic chemistry present the Wurtz reaction as a standard method for hydrocarbon preparation. In practice, though, this reaction was rarely used for that purpose (for the exceptions, see the synthesis of cyclic derivatives in Section 2.7.1). The basic reason for its 'unpopularity' lies in the fact that, in its classical version, the Wurtz reaction is efficient in the coupling of identical alkyl groups ($R^1 = R^2$), but if it were tried for non-identical groups, a mixture of all possible products would usually arise. Furthermore, the utilization of sodium metal as a coupling reagent precluded the use of this reaction with functionally substituted derivatives (owing to their reactivity with sodium metal). These complications were eventually resolved with the later modifications of the Wurtz reaction (see below), and as a result the Wurtz reaction was properly reinstituted as a practical synthetic method.

Thus the obligatory characteristic of a synthetically meaningful reaction is an unambiguity of its reaction course. It is not enough that a given transformation yields the desired product. It is mandatory that the desired product predominates in a mixture of products or, even better, is formed as an exclusive

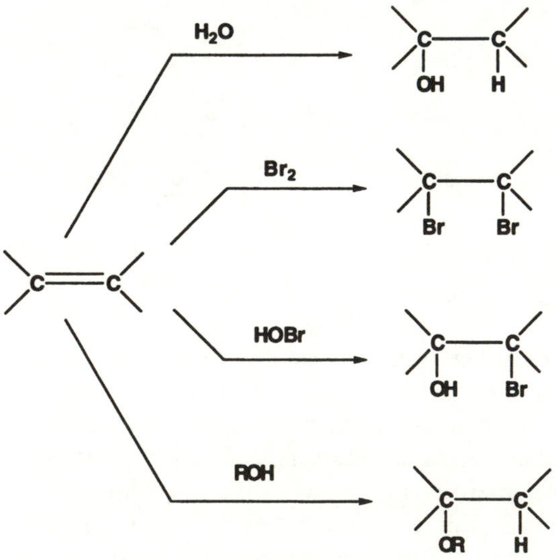

Scheme 2.9

product in a reaction. In fact, *cleanliness* is a rather severe criterion that must be met if a reaction is to be considered a useful synthetic method. This very important point deserves some additional examples from the history of organic chemistry.

Recognized more than century ago as one of the basic properties of alkenes is their ability to undergo a number of addition reactions to the double bond with such reagents as water, bromine, inorganic and organic acids, alcohols, *etc.* (Scheme 2.9, opposite). However, in spite of the fact that these reactions were thoroughly studied and are general in their scope, only few of them (like the addition of bromine) have been accepted into the synthetic arsenal of contemporary organic chemistry due to one simple reason: they lack cleanliness. Even a simple hydration reaction, for example the addition of water to ethylene in the presence of sulfuric acid, leads not only to the formation of the expected product, ethyl alcohol, but also diethyl ether, ethyl sulfate, and some other minor products (Scheme 2.10).

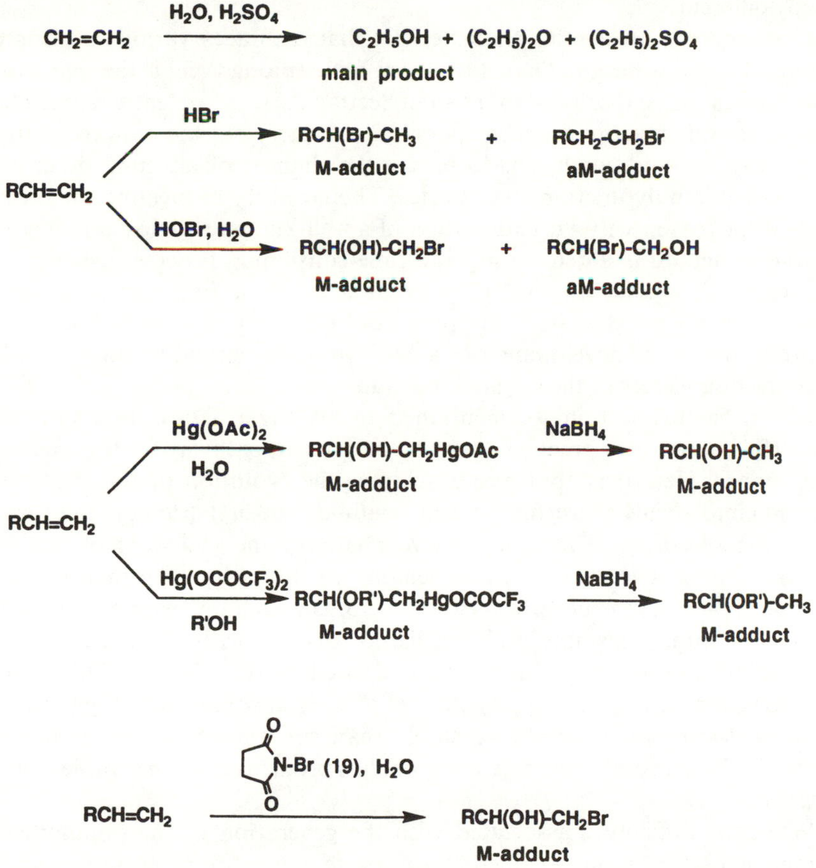

$$CH_2{=}CH_2 \xrightarrow{H_2O,\ H_2SO_4} \underset{\text{main product}}{C_2H_5OH} + (C_2H_5)_2O + (C_2H_5)_2SO_4$$

$$RCH{=}CH_2 \xrightarrow{HBr} \underset{\text{M-adduct}}{RCH(Br)\text{-}CH_3} + \underset{\text{aM-adduct}}{RCH_2\text{-}CH_2Br}$$

$$RCH{=}CH_2 \xrightarrow{HOBr,\ H_2O} \underset{\text{M-adduct}}{RCH(OH)\text{-}CH_2Br} + \underset{\text{aM-adduct}}{RCH(Br)\text{-}CH_2OH}$$

$$RCH{=}CH_2 \xrightarrow[H_2O]{Hg(OAc)_2} \underset{\text{M-adduct}}{RCH(OH)\text{-}CH_2HgOAc} \xrightarrow{NaBH_4} \underset{\text{M-adduct}}{RCH(OH)\text{-}CH_3}$$

$$RCH{=}CH_2 \xrightarrow[R'OH]{Hg(OCOCF_3)_2} \underset{\text{M-adduct}}{RCH(OR')\text{-}CH_2HgOCOCF_3} \xrightarrow{NaBH_4} \underset{\text{M-adduct}}{RCH(OR')\text{-}CH_3}$$

$$RCH{=}CH_2 \xrightarrow{\text{N-Br (19)},\ H_2O} \underset{\text{M-adduct}}{RCH(OH)\text{-}CH_2Br}$$

Scheme 2.10

This problem is even worse for alkenes more complex than ethylene. Equally simple reactions (in a formal sense!) such as the addition of hydrogen bromide or hypobromous acid are also far from being unambiguous and as a rule give rise to a mixture of isomers containing both the predominant Markovnikov (M) as well as anti-Markovnikov (aM) adducts.

To convert these reactions into useful synthetic methods it was necessary either to change dramatically the reaction conditions or to elaborate an entirely new protocol, based on different elementary steps and leading to the same overall result. The second approach turns out to be extremely fruitful. For example, it is possible to add water or alcohol across a double bond with the help of a slightly more complicated but reliable protocol *via* a sequence of two clean reactions: solvomercuration and reduction.

To achieve a clean addition of the elements of hypobromous acid (Br^+ and OH^-) it is advantageous to use reagents such as *N*-bromosuccinimide **19** as the source of Br^+ in an aqueous medium. So we see, the reactions given in the textbooks to illustrate the characteristic reactivity patterns of functional groups and the synthetic methods elaborated to realize their potential in practice can be vastly different.

Cleanliness is an overall characteristic that embraces various aspects of a chemical transformation. One of the most basic among them is the *selectivity* of the reaction (to be discussed in detail in Section 2.5). Other features include the absence of side reactions and the possibility to carry out the required transformation under mild conditions to reduce the chances of affecting other active centers in the polyfunctional substrates. The necessity of meeting these conditions is the reason why the elaboration of a well-known reaction into a reliable synthetic method is a demanding and time-consuming process. Devoid of the excitement of pioneering work, it is nonetheless a highly crucial step in developing a workable synthetic protocol. Therefore it is only fair that the contributors to the development of a new synthetic method be no less credited than the discoverers of the original reaction.

Earlier in this section we mentioned the Wurtz reaction as a potentially powerful but not very practical (in its original form) pathway for the creation of **C–C** bonds. Here it is appropriate to trace the evolution of this century-old reaction into a truly powerful synthetic method. The first attempts to overcome the drawbacks of the Wurtz coupling were based on the separation of the overall reaction into a two-step protocol, leading to the ultimate formation of the product R^1–R^2 (Scheme 2.11). However, organosodium compounds such as R^1Na are rather unstable and are liable to react with the starting halide, R^1–Hal, to produce immediately the undesired product R^1–R^1. This problem was solved, at least partially, by the use of Grignard reagents (organomagnesium analogs), which are more stable than sodium derivatives and, under carefully chosen conditions of generation, less prone to the undesirable coupling reactions.

While the difficulties associated with the generation of the organometallic component (step 1) were avoided by this route, the entire problem was still not solved. It was observed that, along with the desired coupling with R^2–Hal in

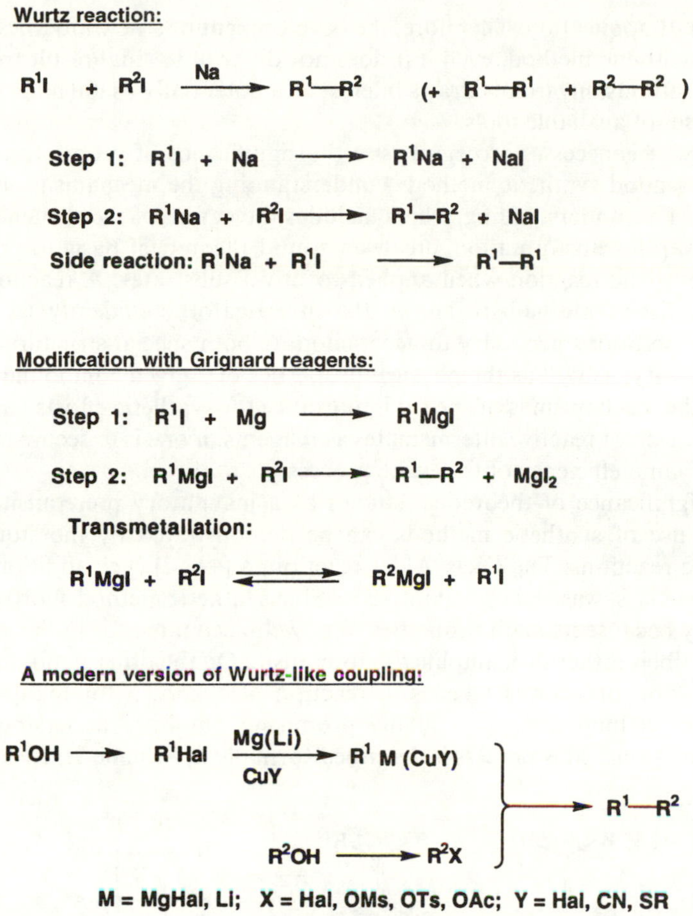

Wurtz reaction:

$$R^1I + R^2I \xrightarrow{Na} R^1{-}R^2 \quad (+\ R^1{-}R^1 + R^2{-}R^2\ )$$

Step 1: $R^1I + Na \longrightarrow R^1Na + NaI$

Step 2: $R^1Na + R^2I \longrightarrow R^1{-}R^2 + NaI$

Side reaction: $R^1Na + R^1I \longrightarrow R^1{-}R^1$

Modification with Grignard reagents:

Step 1: $R^1I + Mg \longrightarrow R^1MgI$

Step 2: $R^1MgI + R^2I \longrightarrow R^1{-}R^2 + MgI_2$

Transmetallation:

$$R^1MgI + R^2I \rightleftharpoons R^2MgI + R^1I$$

A modern version of Wurtz-like coupling:

$$R^1OH \longrightarrow R^1Hal \xrightarrow[CuY]{Mg(Li)} R^1M(CuY)$$

$$R^2OH \longrightarrow R^2X$$

$$R^1M(CuY) + R^2X \longrightarrow R^1{-}R^2$$

M = MgHal, Li; X = Hal, OMs, OTs, OAc; Y = Hal, CN, SR

Scheme 2.11

step 2, the Grignard reagents might undergo transmetallation, to some extent, and thus form a mixture of products. A truly workable synthetic method came only after the development of a new generation of organometallic reagents, namely cuprate derivatives. We will encounter these reagents several times further on and here we will note only that their tendency to undergo transmetallation is greatly reduced. For the first time it was possible to carry out a highly specific coupling as a general reaction, leading exclusively to the formation of the desired product $R^1{-}R^2$ (Scheme 2.11). Moreover, it was found that the cuprate reagents are able to alkylate not only alkyl halides, but also alkyl sulfonates and acetates. Of great value as well is the inertness of cuprate reagents toward many functional groups. Cuprate-mediated coupling reactions have become the method of choice for the creation of **C–C** bonds between variously functionalized moieties.

The target organic compound can contain a different combination of functional groups. Consequently, an organic chemist should have a variety of synthetic methods, each with a different and clearly defined pattern of selectivity and area of application. Therefore, the development of a new modification of a known synthetic method, even if it does not differ in its final result from other known methods, invariably draws interest as a potentially useful addition to the existing set of available tools.

Finally, it is necessary to emphasize the significance of yet another requirement of a good synthetic method – understanding the mechanism of its basic reaction. This understanding, which includes the nature of the elementary steps of the overall transformation, creates a sound theoretical basis to predict the outcome of the reaction when applied to novel substrates. A reaction with a well-established mechanism allows the investigator confidently to seek the optimal conditions necessary to accommodate both special structural features and reactivity, as well as the physical properties of a given compound. Knowledge of the mechanism is of special importance for synthetic efforts based upon the utilization of reactive intermediates as reagents, if one is to secure maximum selectivity and efficiency of the entire process.

The significance of theoretical studies as a mandatory prerequisite for the rational use of synthetic methods can be demonstrated by the story of the pericyclic reactions. The Diels–Alder reaction, a [4 + 2] cycloaddition belonging to this class, was elaborated into a reliable synthetic method shortly after its discovery because its main properties were well-accounted for in the terms of a unified (albeit rather oversimplified) mechanism. On the other hand, the [2 + 2] cycloaddition of various alkenes, a reaction also known for many decades, stayed for a long time as a highly promising but little understood set of transformations. This process is described formally in Scheme 2.12.

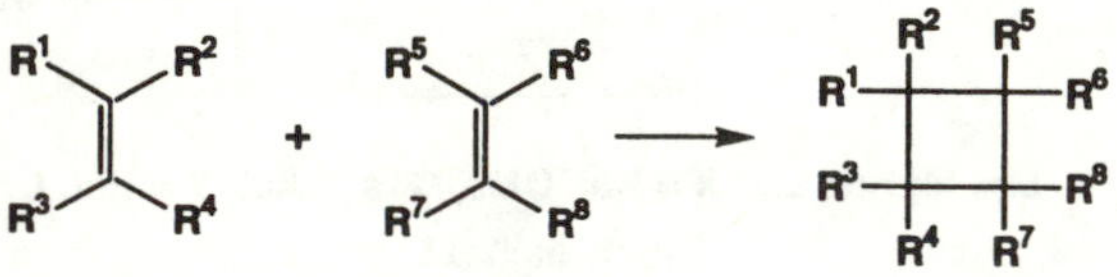

Scheme 2.12

A plethora of experimental data referring to these cycloadditions with various substrates under various conditions gave contradictory information and was difficult to interpret from the point of view of a single mechanism. For example, sometimes these cycloadditions were induced thermally, but sometimes only photochemical induction gave efficient results. Information regarding the stereochemistry of the reactions was insufficient and confusing. Needless to say, these uncertainties created an almost insurmountable obstacle in the way of utilizing [2 + 2] cycloadditions in a total synthesis. Problems pertaining to the interpretation of these and other pericyclic reactions were so puzzling that they were frequently referred to in textbooks published in the 1950s as 'no-mechanism' reactions, *i.e.* processes for which it was not possible to give a truly consistent mechanistic interpretation. The situation changed dramatically in the 1960s, however, when Woodward and Hoffmann found a generalized and rational interpretation of the course of various pericyclic reactions.[2a] The basics of the Woodward–Hoffmann theory can be found in any modern textbook on organic chemistry. For our purposes, it is essential to emphasize that according to their classification there are two distinct and clearly defined classes of pericyclic reactions: thermal reactions occurring between the reactant in the ground state, and photochemically induced reactions which proceed in the excited state. The Woodward–Hoffmann theory makes it possible to predict accurately both the conditions necessary for carrying out a particular pericyclic reaction and its stereochemical outcome. In particular, it became clear that the photochemical dimerizations (alkenes to cyclobutane derivatives, a [2 + 2] pericyclic reaction) and the thermally induced dimerizations are reactions of different classes, and proceed by entirely different mechanisms. Subsequent studies based on these mechanistic concepts elucidated the structural demands of the substrates, the optimal reaction conditions, and the selectivity pattern of these reactions. As a result of these pursuits, a previously rather obscure process was transformed into a popular and powerful synthetic method. In fact pericyclic reactions, which enable us to create ring systems in 'one stroke' with unusual ease and selectivity, have replaced, in many cases, the tedious and unreliable multiple step sequences. Their broad application over the past 20 years has contributed, to a significant extent, to the achievements of organic syntheses in the creation of complex polycyclic systems.

In concluding this section we wish to emphasize one more time that not a single one of the most elaborated modern synthetic methods can be considered to be universal. Every method has its area of application and its limits defined both by the individual peculiarities of the reactivity pattern of the substrate and its possible liability to undergo some undesirable transformations under the conventional conditions of the method. Therefore, while considering the applicability of this or that method, the entire 'entourage' of the system must be taken into account and hence it is still necessary to pursue the seemingly endless task of modifying even the most excellent synthetic method.

In the following parts of this chapter we will examine the basics of the main synthetic methods. We begin with the most important among them, those leading to the creation of the carbon skeleton of an organic molecule.

PART II THE FORMATION OF A C–C BOND: THE KEY TACTICAL PROBLEM OF ORGANIC SYNTHESIS

2.5 PRINCIPLES OF C–C BOND ASSEMBLAGE. HETEROLYTIC REACTIONS

The framework of a typical organic molecule is its carbon skeleton. It is composed of a system of directly linked carbon atoms. Not surprisingly, the fundamentals of organic synthesis are represented in methods for the creation of **C–C** bonds. To discuss the possible routes for construction of these bonds it is convenient to start with an analysis of the *reverse operation*, an imaginary disconnection of the molecule. Such an operation, designated by double arrow symbols, enables one to identify the immediate precursors required to form the desired bonds.

The carbon–carbon single bond can be broken down into either two ions (heterolytic cleavage) or into two radicals (homolytic cleavage). A bond in a molecule exists because its formation is energetically favorable. Consequently, the assemblage from either pair of precursors should be a thermodynamically allowed process. The charged species shown in Scheme 2.13 contain a large

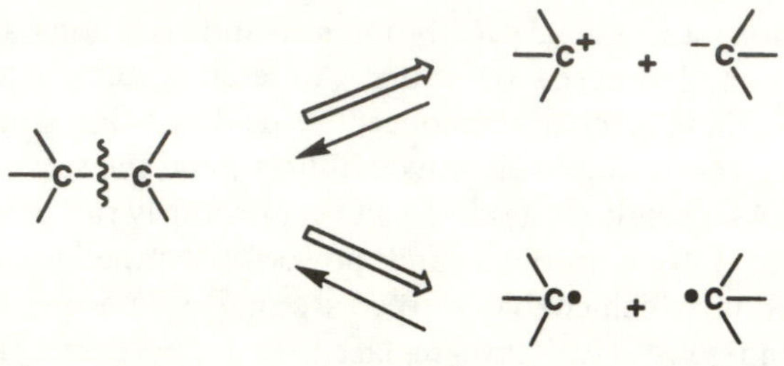

Scheme 2.13

excess of energy and in principle the process of their recombination should occur rapidly, almost without an activation barrier. For example, the reaction between two ionic species, the triphenylmethyl cation **20**, and the cyanide anion **21**, takes place readily to form triphenylacetonitrile **22** with a new **C–C** covalent bond. This is one of a very few organic reactions which proceeds with the ease of simple ionic reactions, reactions normally encountered in inorganic chemistry:

$$\underset{\mathbf{20}}{Ph_3C^+} + \underset{\mathbf{21}}{CN^-} \rightarrow \underset{\mathbf{22}}{Ph_3C\text{–}CN}$$

However, because ionic and radical organic species are extremely reactive, once generated they are liable to undergo numerous side reactions. Therefore it is necessary to moderate the reactivity of these intermediates to reduce their excessive and indiscriminate activity while at the same time preserving their ability to participate easily in the desired transformation.

Since there are more opportunities to exert control over the generation and the activity of organic cations and anions than radical species, heterolytic reactions are more widely used in organic synthesis. Therefore, while there exist many important synthetic methods based on radical reactions (especially in the synthesis of polycyclic compounds), we have chosen the area of heterolytic processes for the discussion of the basics of organic synthesis in the beginning sections of this chapter.

2.6 ORGANIC IONS AND FACTORS GOVERNING THEIR STABILITY. POLARIZATION AND ION-LIKE REACTIVITY

The high reactivity of carbocations and carbanions is associated primarily with the strong coulombic interactions between opposite charges. A point charge concentrated on one carbon atom creates an electrostatic field that greatly influences its surroundings. These species strongly attract ions of the opposite charge that are also present in the solution and ultimately, this attraction may result in the formation of a covalent bond. For example, it is nearly impossible to create conditions that permit the methyl cation and the chloride anion pair, **23**, to coexist in a solution since this pair would immediately collapse into the covalent molecule, methyl chloride **24**, regardless of the refined procedure used for the generation of **23**:

$$\underset{\mathbf{23}}{Me^+ + Cl^-} \rightarrow \underset{\mathbf{24}}{Me\text{–}Cl}$$

Furthermore, the interaction of a charged species with a non-ionic molecule increases the polarization in the bonds of the latter and thus creates conditions that ultimately may lead to a reaction between ions and polar compounds. These processes account for the high and diverse reactivity of organic ions including, of course, their liability to participate in undesirable reactions. This important facet imposes rather severe limits in one's choice of solvents as they must be inert toward the ions. Further limited is the set of functional groups permissible in the molecule used as a substrate in reactions with charged organic intermediates. For example, as we already mentioned, the acetyl cation can be easily generated. Reactions with this cation, however, cannot be performed in hydroxylic solvents as they will immediately undergo acetylation. If a substrate contains a hydroxyl substituent, as is the case with benzyl alcohol **25**, acetylation at the carbon atom of the substrate (as in a Friedel–Crafts reaction) is not possible owing to the ease of the competing *O*-acylation:

$$MeCO^+SbCl_6^- + \underset{\mathbf{25}}{C_6H_5CH_2OH} \rightarrow C_6H_5CH_2OCOMe + HSbCl_6$$

The polarization of bonds leading to a chemical reaction occurs not only in molecules surrounding an ion, but in the remaining organic moiety of an ion as well. Thus, a relatively stable ion, the *tert*-butyl cation **26**, readily loses a proton in the presence of even weak bases to form isobutene **27** (Scheme 2.14). In the course of this elimination reaction the electron pair of the **C–H** bond is completely shifted to the positive charge to form a carbon–carbon double bond. The shift is accompanied by a complete dissociation of the polarized **C–H** bond with the net loss of a proton.

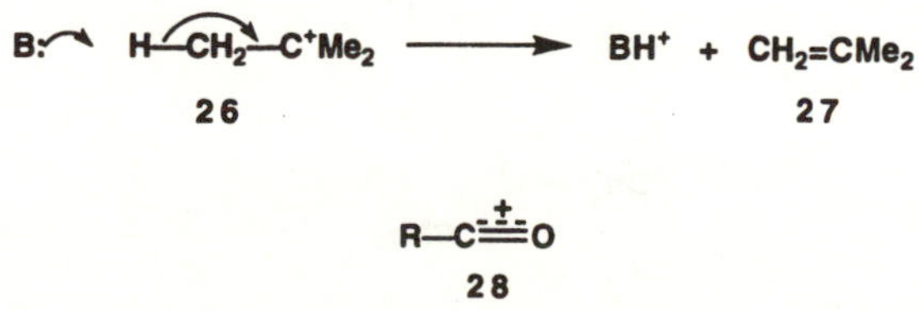

$$B{:}\ \ \underset{\mathbf{26}}{H\text{–}CH_2\text{–}C^+Me_2} \longrightarrow BH^+ + \underset{\mathbf{27}}{CH_2{=}CMe_2}$$

$$\underset{\mathbf{28}}{R\text{–}\overset{+}{C}{\equiv}O}$$

Scheme 2.14

To decrease the reactivity of an organic ion responsible for the undesirable side reactions it is necessary to weaken its electrostatic interaction with other molecules. There exist at least three pathways to achieve this goal. The first and most general approach is based upon the modification of the ion structure *via* introduction of polar or easily polarizable groups that are able to participate in the delocalization of the charge over several atoms. The net charge, of course, does not change but its effective radius increases and hence the coulombic interaction with this dispersed charge is much weaker (the electrostatic interactions are inversely proportional to the square of the distance). Quite different groups are required for the stabilization of carbocations and carbanions. In the first case, these groups must be capable of providing electrons (*electron donating groups*), whereas the stabilization of carbanions requires the presence of electron acceptors (*electron withdrawing groups*).

The charge in the *tert*-butyl cation **26**, due to the polarization along the **C–H** bonds, is spread over the central carbon atom and the nine hydrogen atoms. As a result of this significant delocalization, the cation, while still reactive, is sufficiently stable to exist in the appropriate solvent at low temperature. In acylium ions, the source of the stabilizing effect is different and so strong that in many cases the crystalline salts of acylium cations can be handled as stable species even at ambient temperature. Here the stabilization is achieved by a partial shift of the lone electron pair from oxygen toward the positively charged carbon. The shift produces a species like **28**, with an additional partial bond between carbon and oxygen and the charge likewise distributed between these atoms. In a similar fashion, carbocations can be stabilized by the electron release of other heteroatoms attached either directly to the cationic center or at even somewhat more remote centers.[3]

The classical example of a stabilized carbanion is the enolate anion, formed from carbonyl compounds upon treatment with a base. Such an ion can be represented as a resonance hybrid of two canonical structures, namely **29**, with the charge on carbon, and **30**, with the charge on oxygen (Scheme 2.15). The structure is perhaps more properly described by the hybrid structure **31**, where the dashed line and the charge sign indicate that the electron pair is spread over

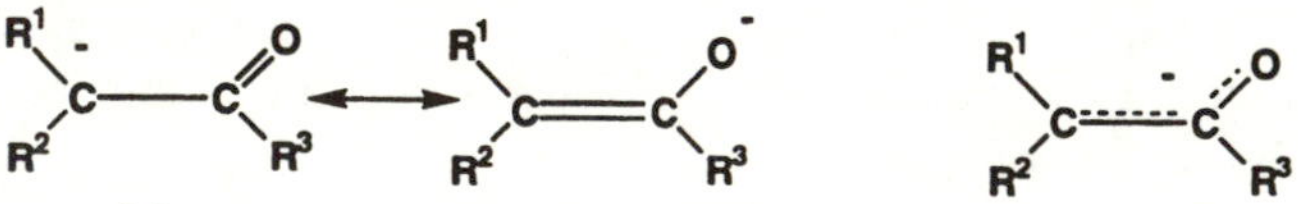

R¹ R² C⁻–C(=O)R³ ⟷ R¹ R² C=C(O⁻)R³ R¹ R² C⋯C⋯O R³

29 **30** **31**

Li^+ CH_2⋯C(⋯O)OCMe$_3$

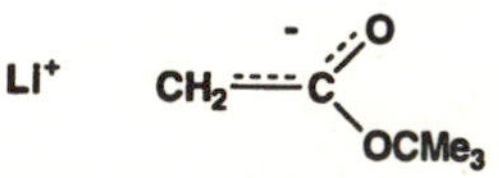

32

Scheme 2.15

the three atoms of this system, with the charge centered at the terminal atoms of this triad.

Quite a number of other electron withdrawing groups containing multiple heteroatomic bonds, such as the ester carbonyl, nitrile, carboxamide, *etc.*, can likewise stabilize carbanions.[4] The lithium salt of *tert*-butyl acetate **32** is an example of an enolate anion sufficiently stable as a salt to be used as a shelf reagent. Substituents containing easily polarizable atoms, such as sulfur or selenium, are also capable of stabilizing an adjacent anionic center.

A universally effective 'buffer' of charge of either sign is the aromatic ring, which behaves like some sort of molecular condenser. Its closed system of π-electrons is readily shifted toward a positive charge and away from a negative charge. Because of this, both the benzyl cation **33** and benzyl anion **34** are relatively stable species:

$PhCH_2^+$	$PhCH_2^-$	$Ph_3C^+BF_4^-$	$Ph_3C^-Na^+$
33	**34**	**35**	**36**

As might be expected, the contribution is cumulative for several polarizable groups that can participate in the charge delocalization. For example, exceptionally high stability is exhibited by triarylmethyl ions. Both triphenylmethyl tetrafluoroborate **35** and triphenylmethylsodium **36** exist as purely ionic salts, something not very common for an organic species with a charge residing at the carbon center.

Extremely efficient stabilization can be observed for dipolar ions if the atoms or groups of atoms bearing the opposite charges are in the close vicinity. Such is the case with ylides, dipolar ions in which the carbanionic center is stabilized by the positive charge located on a neighboring multiple-electron atom such as phosphorus, sulfur, arsenic, *etc.* (onium centers). Typical ylides are represented in the structures **37** and **38**:

$Ph_3P^+—CH_2^-$	$Me_2S^+—CH_2^-$
37	**38**

In modern synthesis it is becoming more and more common to encounter fairly 'exotic' ionic reagents, those never listed earlier among the useful reagents of traditional organic chemistry. These reagents may contain fragments bearing atoms with easily polarizable electronic shells (such as the transition metals cobalt, chromium, nickel, or iron) and endowed with an increased capacity for charge stabilization.[5]

The other two options for stabilizing organic ions are related to the external media. In this regard, the nature of the reagent's counterion might be of significant importance. Especially suitable are counterions with completely filled valence shells that are unable to participate in the formation of the covalent bonds. Good counterions for a carbocation are anions like tetrafluoroborate **39**, hexachloroantimonate **40**, hexafluorophosphate **41**, perchlorate **42**, and triflate **43**. The most common counterions for carbanions are very stable and unreactive cations of the alkali metals, especially those complexed with ligands (such as a crown ether). As a convenient alternative, one may also utilize stabilized and coordinately saturated organic cations such as tetra-butylammonium **44**:

BF_4^-	$SbCl_6^-$	PF_6^-	ClO_4^-	$CF_3SO_3^-$	Bu_4N^+
39	**40**	**41**	**42**	**43**	**44**

The nature of the solvent is the other external factor that plays an important role in the stabilization of the charged organic species. The influence of a solvent is multifaceted but for simplicity can be reduced to two essential aspects. A polar solvent, a liquid with a high dielectric constant, will decrease the coulombic attraction between charged ions. This effect can be very significant. For example, in going from the non-polar solvent hexane to the polar solvent acetonitrile the coulombic interactions may decrease 21-fold. On the other hand, a solvent can interact with ions of either sign owing to charge–dipole interactions, hydrogen bonding, the formation of complexes with variable stoichiometry, and, in short, everything generally implied by the term 'solvation'. This solvation can lead to significant shielding of the charged center with delocalization of charge into the solvent shell.

No matter what sort of trick is employed for ion stabilization, its effectiveness depends upon the extent of electron displacement towards the cationic center or away from the anionic one. Such shifts, however, must not exceed certain limits that would lead to the cleavage of an old and/or formation of a new covalent bond. The first case is exemplified in the ease of proton elimination from the *tert*-butyl cation **26** (Scheme 2.14). Under certain conditions the entire charge can be shifted to the proton, and with proton abstraction by an external base the organic cation is completely destroyed. Related events can occur at an anionic center. For example, the chlorine atom is very electronegative and would seem, therefore, to be good at stabilizing a carbanion. It is reasonable to expect, then, that the trichloromethyl anion **45** should be a stable anion. In fact, **45** can be used as a $-CCl_3$ transfer agent in a number of reactions. At the same time, however, this species may readily lose a chloride anion, resulting in the formation of dichlorocarbene **46**. This elimination occurs simply because chlorine is too effective as an electron acceptor and readily accepts an electron pair to become the stable chloride anion:

$$Cl_3CH + B{:} \rightarrow \underset{\mathbf{45}}{CCl_3C^-} \rightarrow \underset{\mathbf{46}}{Cl_2C{:}} + Cl^-$$

Typical carbocations and carbanions, even with significant charge delocalization, are still usually quite reactive. This property leads to two very important consequences from the point of view of organic synthesis.

(1) Carbocations are able to react effectively not only with anions but also with polarized covalent molecules by attacking the negative end of the dipole. Likewise, carbanions are capable of attacking the positive end of a polarized

molecule. Thus typical carbanions, for example **47** (Scheme 2.16), readily react with alkyl halides by attacking the halogen-bearing carbon, the positive end of the dipole, with formation of a **C–C** bond.

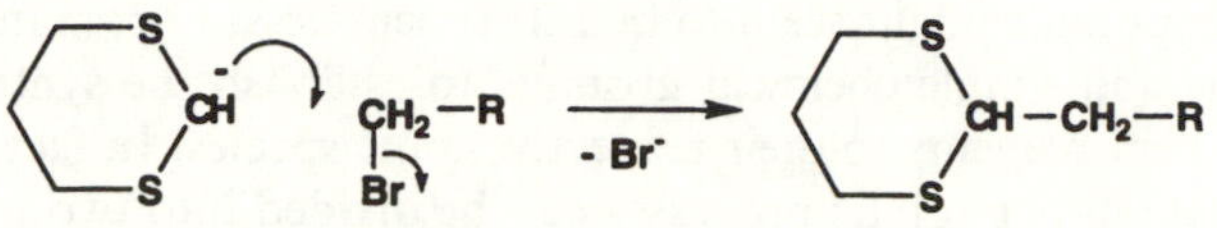

47

Scheme 2.16

(2) In certain organic compounds, covalent bonds can be so strongly polarized that their structure and reactivity may be approximated by the corresponding ionic formulas. For example, the covalent bonds in methyl triflate **48** and methyllithium **49** are so strongly polarized that they behave as if they were fully ionized compounds, sources of the methyl cation and methyl anion respectively:

$$\underset{\mathbf{48}}{CH_3^{\delta+}\text{–}OSO_2CF_3^{\delta-}} \qquad \underset{\mathbf{49}}{CH_3^{\delta-}\text{–}Li^{\delta+}}$$

In a practical sense, these features have important ramifications. First of all, a heterolytic reaction leading to the formation of a **C–C** bond and described by a formal interaction between a carbocation and a carbanion does not require both components to be represented as truly ionic species. It is sufficient for one to be present as a stabilized ionic species and the second to be merely polarizable. Furthermore, there are many reactions that can be performed with neither of the components being truly ionic, provided at least one of them contains sufficiently polarized bonds. In either case, the net result of the reaction corresponds exactly to the idealized scheme above, which envisages the retrosynthetic cleavage of a covalent bond, leading to two ionic precursors.

Such a 'dethroning' of carbocations and carbanions can be carried another step further. Certain purely covalent compounds possessing weakly polarized bonds can serve as ionic reagents if their electronic system is highly polarizable. In such molecules the approach of a charged particle, or even a dipole, induces a significant displacement of electrons. The electron displacement is to such an extent that the reaction intermediates may become almost fully ionized, as if an ionic reagent had been actually used. A typical example is represented by the previously mentioned reactions of electrophilic substitution in aromatic series, where a neutral molecule of an aromatic hydrocarbon, ArH, behaves as an efficient equivalent to a carbanion, Ar^-.

2.7 ELECTROPHILES AND NUCLEOPHILES IN C–C BOND-FORMING REACTIONS

Many organic reactions that are described as interactions between charged reactants to form ionic intermediates have initial reactants that are actually covalent and hence play the role of *synthetic equivalents* to carbocations or carbanions. Now, with all the verbal calisthenics to explain that even a covalent compound may serve as an ionic reagent, the attempt to suggest a rational classification of reactions and reagents involving ionic or synthetically equivalent ionic compounds might seem to be rather senseless. Fortunately, there are unambiguous and simple chemical grounds to establish the synthetic equivalency of covalent reagents to their respective ionic species. In fact, all reagents that participate in heterolytic processes can be divided into two broad classes: *electrophiles and nucleophiles*.

Electrophiles (**E**) are reagents that are able to accept a pair of electrons whereas nucleophiles (**Nu**) are reagents that can donate a pair of electrons. Both processes ultimately form a covalent bond. The simplest electrophile is the proton, $\mathbf{H}^+$, and the simplest nucleophile is the hydride anion, $\mathbf{H}^-$. Carbocations containing a vacant orbital, and carbanions with their unshared pair of electrons, are the most obvious examples of electrophiles and nucleophiles.

For a compound to behave as an electrophile it is not necessary, however, to have this orbital vacancy *per se*. It is sufficient to develop this vacancy in the course of the formation of the transition state for a given reaction. Similarly, nucleophilic reactivity does not require a reagent to have a pre-existing pair of unshared electrons. Again, it is sufficient for this electron pair to become available to an attacking electrophile in the course of the reaction.

The reactivity of electrophiles and nucleophiles can vary within broad limits. Their ability to mimic the actual behavior of carbocations and carbanions strongly depends upon the structure of the reagents, the nature of the reaction, and, of course, the reaction conditions. It is possible to arrange reagents roughly in order of their reactivity as electrophiles (*electrophilicity*) or as nucleophiles (*nucleophilicity*). However, it must be recognized that not every imaginable combination of electrophile and nucleophile would result in a viable reaction. Therefore, the heterolytic disconnection of a carbon–carbon bond in the course of the retrosynthetic analysis, while being extremely useful, may serve only to identify tentatively the structure of the prototypes required for the creation of this bond. Next it is necessary to consider the nature of suitable reagents and reactions to be used to achieve the synthetic goal within the limitations imposed by the structural context of a target molecule.

Typical examples that illustrate how one can plan an actual assemblage of carbon–carbon bonds, using most usual heterolytic reactions, follow below.

2.7.1 The Wurtz Reaction. Allylic and Related Couplings

Earlier we mentioned the Wurtz reaction as being one of the simplest approaches to the formation of **C–C** bonds. In this reaction, the alkyl halide serves as the electrophile (carbocation equivalent) and the organometallic derivative plays the role of the nucleophile (carbanion equivalent). We have also seen that this old reaction has recently become a feasible route for the creation of **C–C** bonds due to the elaboration of reagents which suppress the formation of unwanted by-products. In view of this, it is possible to plan a model synthesis corresponding to a certain set of alternative options for disconnection of the target molecule. For example, consider the case of *n*-butane:

$$n\text{-}C_4H_{10}\left\{\begin{array}{l}\Rightarrow CH_3^+ + C_3H_7^- \\ \Rightarrow C_2H_5^- + C_2H_5^- \\ \Rightarrow C_3H_7^+ + CH_3^-\end{array}\right.$$

All three of these possibilities are equivalent and viable since the required electrophiles, CH_3I, C_2H_5I, and *n*-C_3H_7I, and the corresponding nucleophiles, CH_3MgI, C_2H_5MgI, and *n*-C_3H_7MgI, are readily available. The synthesis of *n*-butane for its own sake is not an appealing task since it can be obtained in unlimited amounts from petroleum and natural gas sources. However, if the need arises for *n*-butane containing a labeled atom, for example ^{13}C, at a given position for research applications, then this reaction would become useful. In this case, the particular selection of reaction partners depends upon the availability of the respective labeled precursors.

More often than not, when planning a synthesis one encounters solutions that are not equivalent. Let us examine the synthesis of the hydrocarbon **50** (Scheme 2.17). This molecule has two structural elements, the benzyl and methylacetylene fragments, that are related to well-known and readily available compounds. It would seem reasonable to synthesize **50** from precursors containing these fragments. Hence, we have the option of disconnecting at two alternative **C–C** bonds. Disconnection at the benzyl–ethynyl bond (**a**) produces two pairs of ions, **51** and **52** or **53** and **54**. Path $\mathbf{a_1}$ leads to the benzyl anion **51**, for which it is easy to find synthetic equivalents such as benzyllithium or benzylmagnesium chloride. No obvious reagent, however, serves as an equivalent cationic counterpart for **52**. On the other hand, path $\mathbf{a_2}$ produces ions **53** and **54**, equivalents easily found in benzyl chloride and sodium acetylide. With these reagents the synthesis of **50** is readily accomplished, as shown in Scheme 2.17.

Cleavage of the phenyl–propargyl bond (**b**) can also be carried out by either of two ways, leading to pairs **55** and **56** or **57** and **58**. It is not difficult to find an equivalent to the propargyl anion **56**, but equivalents to the phenyl cation **55** are less obvious. Hence we put path $\mathbf{b_1}$ aside and examine $\mathbf{b_2}$. Here the Grignard reagent, phenylmagnesium bromide **59**, serves as a good equivalent of the phenyl anion **57**, with the corresponding halide **60** an equivalent to the carbocation **58**. With this approach we have another realistic sequence leading to the target hydrocarbon **50**, also shown in Scheme 2.17.

In the above scheme we utilized the Grignard reagents and acetylides as the equivalents of carbanions and the benzyl and propargyl halides as the carbocation equivalents. The latter reagents, as well as related allyl halides, share the feature of having a π-system adjacent to the potential carbocation center, and reagents of general formula **61–63** are among the most commonly used covalent electrophiles.

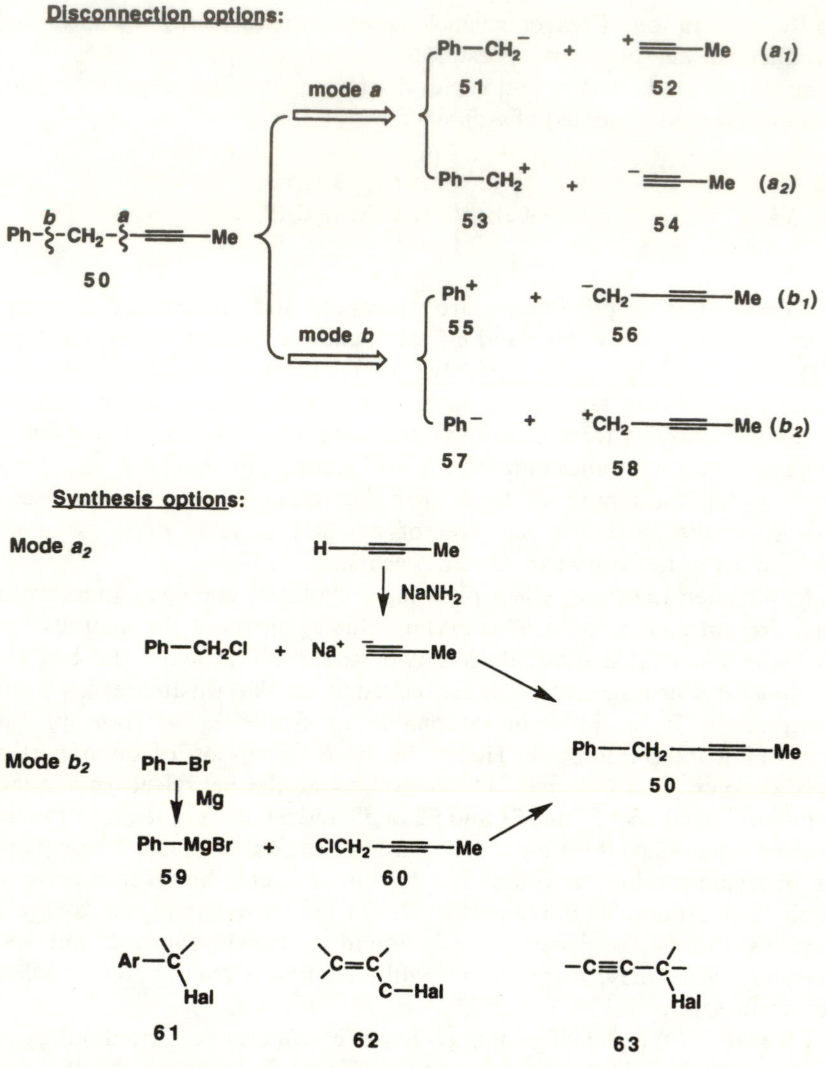

Disconnection options:

$Ph{-}CH_2^-$ + $^+{\equiv}{-}Me$ (a_1)
51, 52

$Ph{-}CH_2^+$ + $^-{\equiv}{-}Me$ (a_2)
53, 54

Ph-§(b)-CH_2-§(a)-≡-Me
50

mode *a*, mode *b*

Ph^+ + $^-CH_2{-}{\equiv}{-}Me$ (b_1)
55, 56

Ph^- + $^+CH_2{-}{\equiv}{-}Me$ (b_2)
57, 58

Synthesis options:

Mode a_2: H—≡—Me → ($NaNH_2$); $Ph{-}CH_2Cl$ + Na^+ $^-{\equiv}{-}Me$ → $Ph{-}CH_2{-}{\equiv}{-}Me$ **50**

Mode b_2: Ph—Br → (Mg); Ph—MgBr **59** + $ClCH_2{-}{\equiv}{-}Me$ **60** → **50**

Ar—C(Hal) **61**; C=C—C—Hal **62**; —C≡C—C—Hal **63**

Scheme 2.17

Therefore if a target molecule contains an aromatic ring or a double or triple bond, it is useful to initiate a retrosynthetic analysis of such a structure by cleavage of the bonds at the benzylic, allylic, or propargylic positions, respectively. The aim is to arrive at the pair consisting of the stabilized cation and the required anion and then to analyse the comparative availability of the corresponding reagents.[6]

A typical example of such an approach can be found in the synthesis of the sex attractant **64** (Scheme 2.18) of the apple worm *Laspeyresia pomonella*, a common pest of apple orchards.[7] It is not surprising to see that retrosynthetic disconnection was done across the allylic **C–C** bond to give the allylic electro-

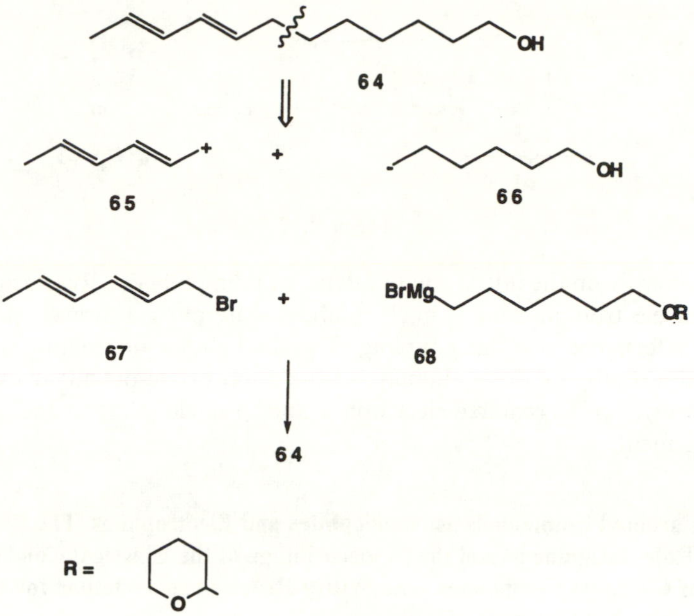

64

65 + 66

67 Br + BrMg 68 OR

64

R =

Scheme 2.18

phile **65** and the carbanion **66**. Obvious equivalents of these ions include the bromide **67** and Grignard reagent **68** (with temporary protection at the hydroxyl group). Both reagents are made from easily available precursors. Coupling of **67** and **68** (in THF in the presence of HMPA) leads to the formation of the necessary **C–C** bond. Subsequent removal of the protecting group produces the final product **64**.

The mobile π-electrons of unsaturated systems, responsible for the stabilization of the carbocations, provide equally efficient stabilization for carbanions. Consequently, a retrosynthetic cleavage of a benzylic, allylic, or propargylic **C–C** bond has additional merits since the resulting fragments can be visualized as either an electrophile or a nucleophile. The dual synthetic value of the allylic moiety has been extensively utilized in the synthesis of a large number of natural acyclic isoprenoids.* The structures of many of these compounds look like they were purposely tailored for this type of retrosynthetic analysis. In fact, the 1,5-diene system, usually present in their structure (Scheme 2.19), immediately suggests the cleavage of its central **C-3–C-4** bond, which leads to two allylic precursors.

Either of these fragments can be regarded as a potential cation or anion. The formation of **C–C** bonds *via* the coupling of two allylic fragments is considered to be dependable and there are numerous and well-elaborated procedures to achieve this goal.[6] This is the reason why so many isoprenoid syntheses are

*A common structural feature of these natural compounds is the presence of several isoprene units joined in the order determined by the routes of their biosynthesis. This structure can be further modified by cyclization and/or rearrangement processes.

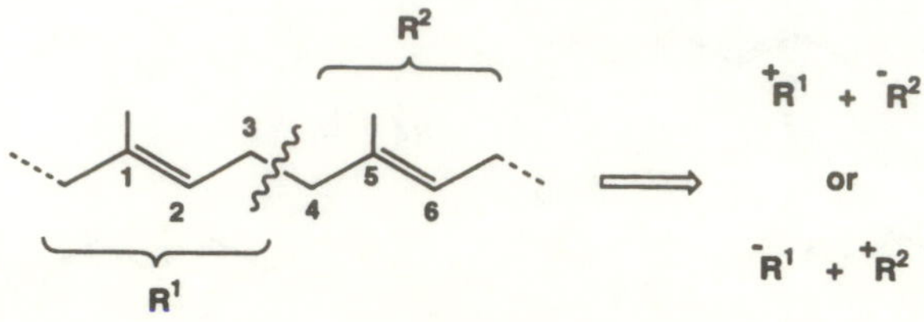

Scheme 2.19

based primarily on the utilization of allylic 'building blocks'. Rearrangement in the allylic electrophilic and/or nucleophilic counterparts, however, can jeopardize the effectiveness of this coupling. The final choice of reacting partners is governed not only by the availability of the appropriate precursor but also by the propensity of the required electrophile and/or nucleophile to undergo allylic rearrangement.

2.7.2 Carbonyl Compounds as Nucleophiles and Electrophiles. The Problem of 'Role Assignment' and the Modern Image of the Classical Condensations of Carbonyl Compounds. The Wittig Reaction as a Method for the Controlled Synthesis of Alkenes

In some respects the carbonyl-containing moiety **C–C=O** can be considered as as an analog to an allylic carbon system, **C–C=C**. In contrast to the latter system, however, only a negative charge can be effectively stabilized on the carbon adjacent to the carbonyl group. The inherent partial positive charge at the carbonyl carbon precludes any opportunity for the stabilization of an additional positive charge in this triad. On the other hand, owing to this basic carbonyl polarization, the respective anionic species, enolates, are especially stable.

As we have seen, the charge in the enolate anion is distributed between the oxygen and the α-carbon. Therefore, its reactions with electrophiles can occur as either **C**- or **O**-attack (Scheme 2.20). Fortunately, the direction of an electrophilic attack can usually be controlled by varying the nature of the reagents and the reaction conditions. As a result, there are a number of methods for constructing **C–C** bonds which take advantage of the simplicity of the *C*-alkylation of enolates with various *C*-electrophiles. Yet, until recently, the preparative scope of these methods was limited owing to certain complications which are quite common in the practice of organic synthesis.

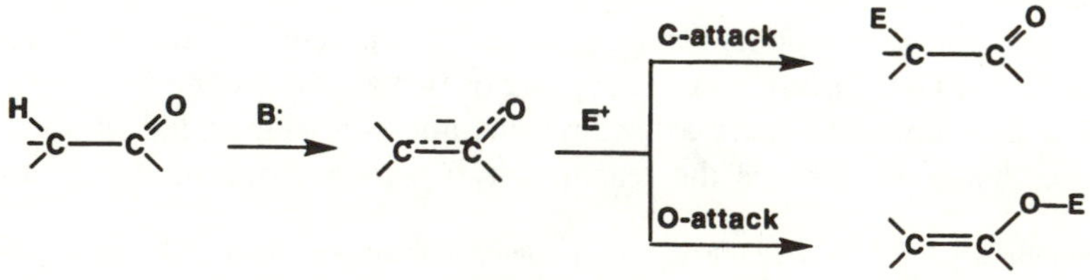

Scheme 2.20

Typical carbonyl compounds behave as weak **C–H** acids. If, for example, one generates the enolate anion of acetone using even a relatively strong conventional base, such as sodium ethoxide, the resultant equilibrium will be shifted far to the left (Scheme 2.21). Carbonyl compounds themselves, as we soon will see, are active electrophiles due to the presence of the partially positive carbonyl carbon. Hence, the nucleophilic enolate generated in the above system can react with non-ionized acetone molecules abundantly present in this equilibrium. This reaction is the well-known aldol condensation (Scheme 2.21). Although useful in its own right, its ease of occurrence creates serious obstacles for the use of an *in situ* generated enolate anion as a nucleophile in reactions with other electrophiles.

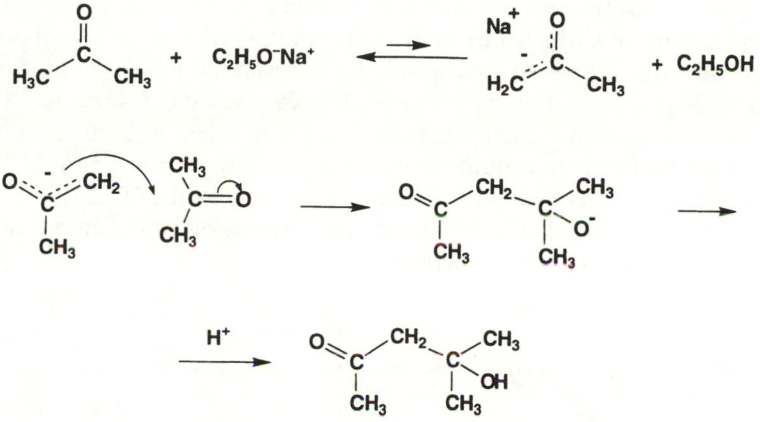

Scheme 2.21

Classic synthetic methods based upon the alkylation of enolates were therefore limited to cases where especially stable enolates could be generated. Usually β-dicarbonyl compounds such as acetoacetic ester or malonic ester were used as precursors in these reactions. For example, alkylation of the stable enolate derived from malonic ester served as a routine and totally reliable method to achieve $\mathbf{C_2}$ chain elongation, as shown in the standard sequence below:

$$CH_2(COOEt)_2 \xrightarrow{EtONa} {}^-CH(COOEt)_2 \xrightarrow{R\text{–}Hal} R\text{–}CH(COOEt)_2 \xrightarrow{H_2O,\ ^-OH}$$
$$R\text{–}CH(COOH)_2 \xrightarrow[-CO_2]{} R\text{–}CH_2COOH$$

In this sequence, malonic ester was used as a synthetic equivalent of the enolate anion derived from acetic acid. The presence of an additional carboxyl substituent served as an auxiliary tool to stabilize the enolate species. This approach was extended to the alkylation of enolates of more complicated structure, but here it was mandatory to create first the required β-dicarbonyl system by supplementing the initial structure with an additional carbonyl substituent. This auxiliary operation, while being generally viable, noticeably increased the complexity of the synthetic scheme and thus limited the application of the whole approach.

Thus the polarization in the carbonyl group, $C^{\delta+}=O^{\delta-}$, secures the opportunity to generate enolates from carbonyl compounds and to employ them as nucleophiles in organic syntheses.[8] This very polarization also accounts for the property of carbonyl compounds $\mathbf{R^1R^2C=O}$ to serve as electrophilic reagents, synthetic equivalents to the cation $\mathbf{R^1R^2C^+–OH}$.

One of the most important reactions taking advantage of this polarization is the Grignard reaction, the addition of organomagnesium reagents across the carbonyl group. (Organolithium compounds are equally efficient reagents for this purpose.) The net result of this process is the creation of a new **C–C** bond accompanied by the transformation of carbonyl into hydroxyl.[9a] Until recently, the application of this reaction was limited owing to a rather high basicity of the **Mg** and **Li** reagents, which are capable of promoting the enolization of carbonyl compounds susceptible to ionization. As a result of this complication, the classical Grignard reaction with ketones like **69** gives the desired adducts in a very low yield (Scheme 2.22). Fortunately, this problem was finally solved. It was found that organocerium derivatives (formed *in situ* from **Li** or **Mg** derivatives upon treatment with a stoichiometric amount of $CeCl_3$) are devoid of basic properties and their reactions with even enolizable ketones (*e.g.* **69**) proceed in an uncompromised manner.[9b]

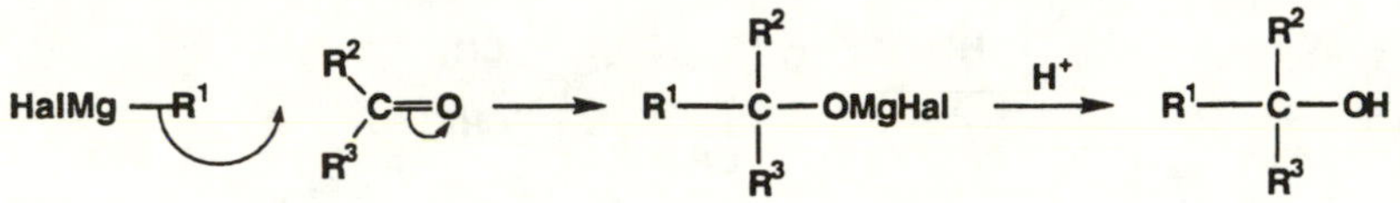

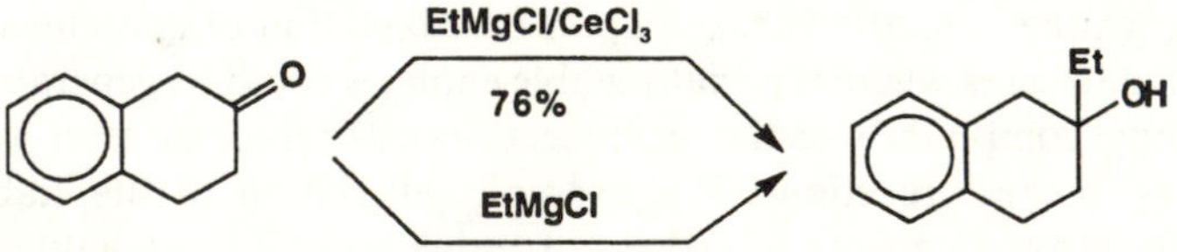

HalMg—R^1 + R^2R^3C=O → R^1—C(R^2)(R^3)—OMgHal —H^+→ R^1—C(R^2)(R^3)—OH

69 —EtMgCl/$CeCl_3$, 76%; EtMgCl, 8%→ Et, OH

Scheme 2.22

The reliability of the Grignard reaction and its broad scope of application warrants utilization of the retrosynthetic disconnection of **C–C** bonds adjacent to the hydroxyl-bearing carbon as a routine procedure in the analysis of practical synthetic pathways.

The electrophilic character of the carbonyl group makes it an appropriate substrate for interaction with a variety of nucleophiles. It is by no means limited to simple alkyl- or aryl-magnesium or -lithium reagents of the types considered above. Of special importance are the numerous condensations where carbonyl-containing compounds undergo interaction with enolates. The spectrum of these transformations embraces such classical reactions as the aldol condensation (*e.g.* Scheme 2.21), Claisen ester condensation, Darzens, Stobbe, Perkin, and Knoevenagel reactions, *etc.*[10] While the nature of precursors, the reaction conditions, and the products may vary substantially, the key step in all of these condensations, the formation of a **C–C** bond, can be described by a common scheme (Scheme 2.23). As is shown in this scheme, both nucleophilic and electrophilic partners are carbonyl compounds. If the same compound serves as the common precursor for both reactants, the reaction proceeds in an unambiguous way, as is seen in the synthesis of acetoacetic ester by the Claisen ester condensation (Scheme 2.24).

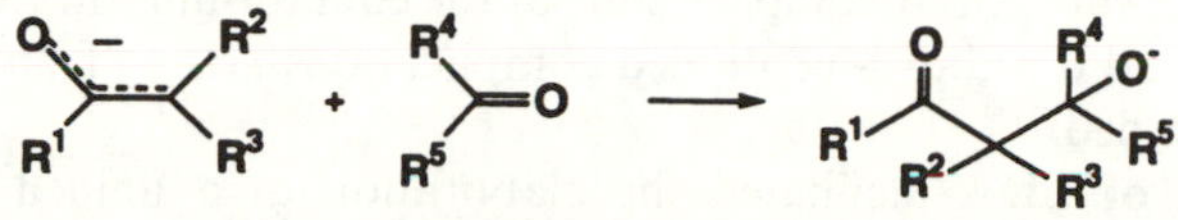

Scheme 2.23

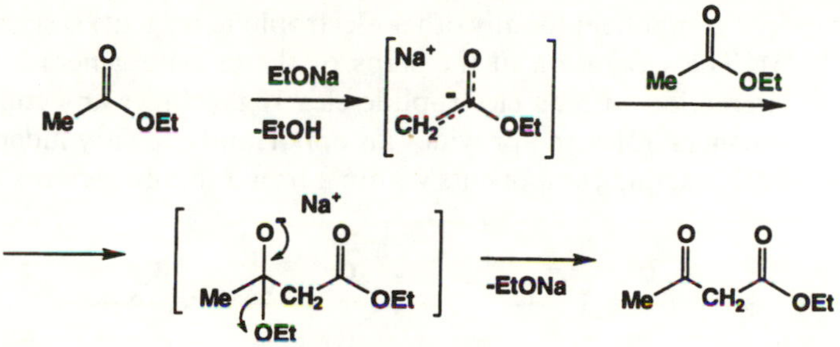

Scheme 2.24

In the case of two non-identical substrates, four different products of Claisen ester condensation may be formed. In fact, that is exactly what happens if ethyl acetate and ethyl propionate are employed as components in the Claisen ester condensation under classical conditions (treatment of the mixture of the esters with sodium ethoxide). The synthetic potential of such condensations depends entirely upon how efficiently the 'casting' between the potential nucleophilic and electrophilic partners is achieved.

Classically, this separation of roles was accomplished by utilizing substrates that differed drastically in their ability to form enolate anions. In fact, the various name reactions mentioned above are distinguished not by basic differences in mechanism, but by the nature of the components preferentially utilized as substrates. For example, in the Perkin reaction, the condensation of an aromatic aldehyde with an aliphatic acid anhydride is based on the fact that the electrophilic component, *e.g.* benzaldehyde, lacks α-hydrogens and therefore is incapable of forming an enolate. At the same time the utilized electrophile, acetic anhydride, contains a carbonyl group with reduced propensity to interact with nucleophiles. Hence, it is incapable of undergoing self-condensation. As a result, the reaction can be directed to form a single product of the cinnamic acid type with no cross-over possible:

$$\mathrm{ArCHO + (CH_3CO)_2O \xrightarrow{B:} ArCH(OH){-}CH_2COOAc \xrightarrow{H_2O} ArCH{=}CHCOOH}$$

Fortunately, a more general solution is now available to achieve the 'role assignments' in these reactions. A new generation of strong bases, such as lithium diisopropylamide (LDA), are capable of generating an enolate anion under very mild conditions from practically any carbonyl derivative containing at least one α-hydrogen. A complete shift of the equilibrium towards the ionized form ensues and the above-mentioned complications arising from self-condensation are avoided.

Utilization of LDA facilitated the elaboration of a unified protocol for controlled condensations of carbonyl compounds. The protocol generally involves two separate steps. Initially, one of the carbonyl components is transformed entirely into the enolate under the action of LDA. Then, the second carbonyl component (or any other electrophilic reagent) is simply added (Scheme 2.25). The separation in the steps of the enolate generation and its subsequent interaction with an electrophile clearly precludes any confusion of the reacting partners' roles and provides an opportunity to vary independently the nature of the reacting components within a broad set of reactants.[4,10]

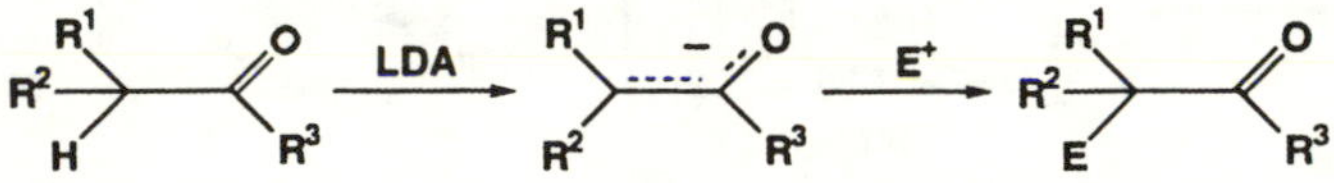

Scheme 2.25

It is exactly this protocol that resolved the lack of selectivity control in the crossed Claisen ester condensation of dissimilar esters, as is shown below for coupling of the esters of acetic and propionic acids:

$$\mathrm{CH_3COOEt \xrightarrow{LDA} \underset{\mathbf{Nu}}{{}^{-}CH_2COOEt} \xrightarrow[\mathbf{E}]{CH_3CH_2COOEt} CH_3CH_2CO{-}CH_2COOEt}$$

$$\mathrm{CH_3CH_2COOEt \xrightarrow{LDA} \underset{\mathbf{Nu}}{{}^{-}CH(CH_3)COOEt} \xrightarrow[\mathbf{E}]{CH_3COOEt} CH_3CO{-}CH(CH_3)COOEt}$$

The differences among the previously distinct classical condensation reactions of carbonyl compounds have actually been eliminated by new and excellent methods developed to generate selectively a variety of stable enolates. In fact, the use of authors' names to identify these reactions (except to serve as an acknowledgement of our deep respect to the outstanding achievements of the classics of organic chemistry) may only obscure the logical relationships among various carbonyl condensations.

One cannot help but notice that the above reactions lead to related products characterized by the presence of two oxygens in a 1,3-relationship as either a β-hydroxycarbonyl (if both components are aldehyde or ketones) or a β-dicarbonyl system (in the case of esters). Both of these functionalities are useful in subsequent conversions and we see that the synthetic utility of the reactions used to prepare these adducts is broadened further. Typical transformations are shown in Scheme 2.26 for the product **70** of an acetone aldol condensation. Oxidation of **70** leads to the formation of the corresponding β-dicarbonyl compounds **71**, while the 1,3-diol **72** is formed as a result of reduction of **70** and the α,β-unsaturated carbonyl compound **73**, formed *via* dehydration of **70**.

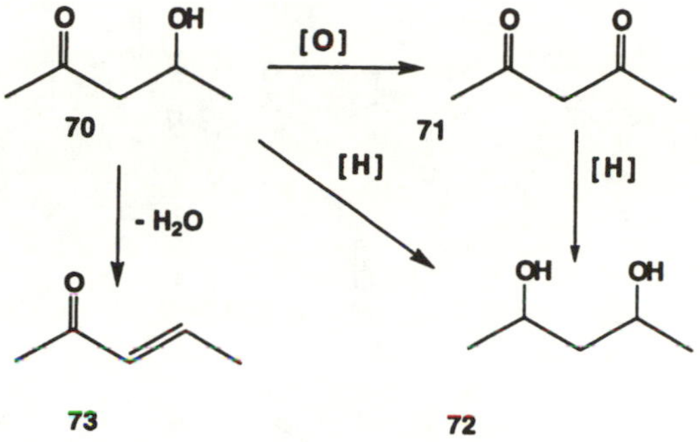

Scheme 2.26

Thanks to the reliability of these conversions, compounds like **70**–**73** can all be regarded as products of a condensation between carbonyl components described in terms of an interaction between an electrophile and a nucleophile. Hence, an important recommendation in retrosynthetic analysis is to identify the presence of fragments identical to **70**–**73** (or easily derivable from them). Retrosynthetic cleavage of the respective **C–C** bond will then reveal the structures of possible carbonyl precursors. The retrosynthetic analysis of **74**, a basic fragment of the complex macrolide antibiotic 6-deoxyerythronolide B, provides a good example of how workable this principle might be (Scheme 2.27).[11]

In this fragment one encounters a set of 1,3-positioned oxygen-containing substituents that almost automatically dictate a series of consecutive disconnections. As indicated in the scheme, these disconnections (one of the last is based upon the Michael addition which will be discussed later) eventually lead to rather simple starting compounds. The real synthesis of **74**, however, also had to address a formidable stereochemical problem. Every single **C–C** bond-forming step should be carried out with absolute control over the stereochemistry of the arising centers. Once this problem was solved for a pair of simple carbonyl compounds at the initial step, the whole synthesis became viable, as all the constructive reactions along the sequence belong to the same class of reliable aldol-type reactions.

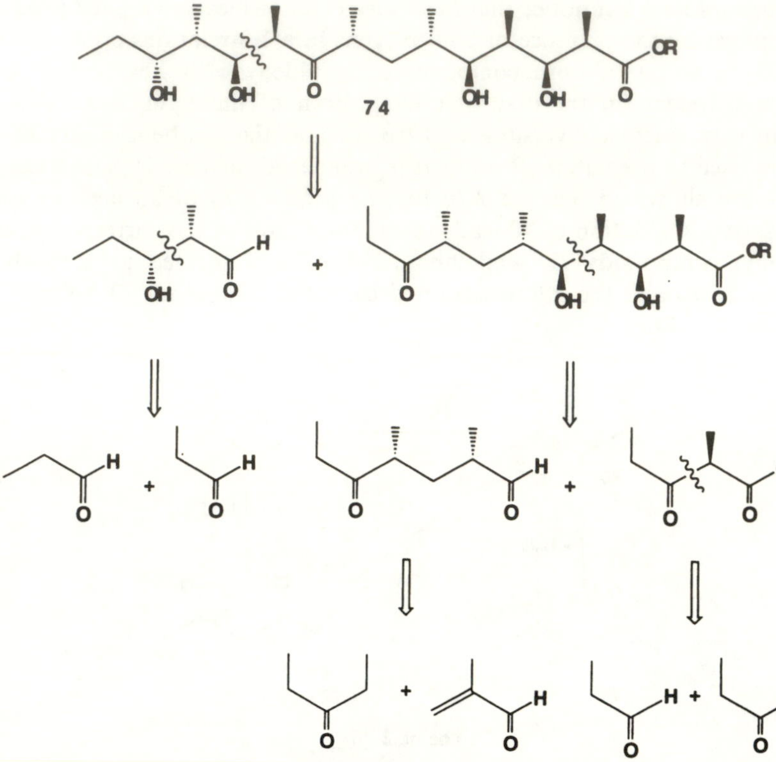

Scheme 2.27

Another series of closely related reactions also proceeds with nucleophilic addition to the carbonyl group, but involves the interaction of *both* the carbon and oxygen centers. In these cases the carbonyl group is employed as an equivalent to the doubly charged ion $R^1R^2C^{2+}$. One of the most important transformations of this type is the Wittig reaction[12a,b] between phosphorus ylides **75** and the carbonyl group of an aldehyde or ketone. The initial step of this reaction is the nucleophilic addition of the carbanion to the carbonyl group, which gives a dipolar ion, betaine **76** (Scheme 2.28). The structure of this dipole is ideally suited for interaction between opposite charges of oxygen and phosphorus, which eventually results in the elimination of triphenylphosphine oxide with simultaneous formation of a double bond between the participating carbon atoms. Since the required ylide is easily obtained from the alkyl halide **77** *via* the phosphonium salt **78**, the net synthetic outcome of the Wittig method is a creation of a double bond between the carbons of an alkyl halide and aldehyde or ketone carbonyls. This result can also be achieved by the Grignard addition/dehydration sequence, but the latter generally proceeds with the formation of isomeric mixtures and thus cannot compete with the Wittig reaction. For example, methylenecyclohexane **79** is formed as a nearly inseparable mixture with its isomer *via* the Grignard addition–

elimination route but as a single product in the Wittig reaction with methylenephosphorane **80**.

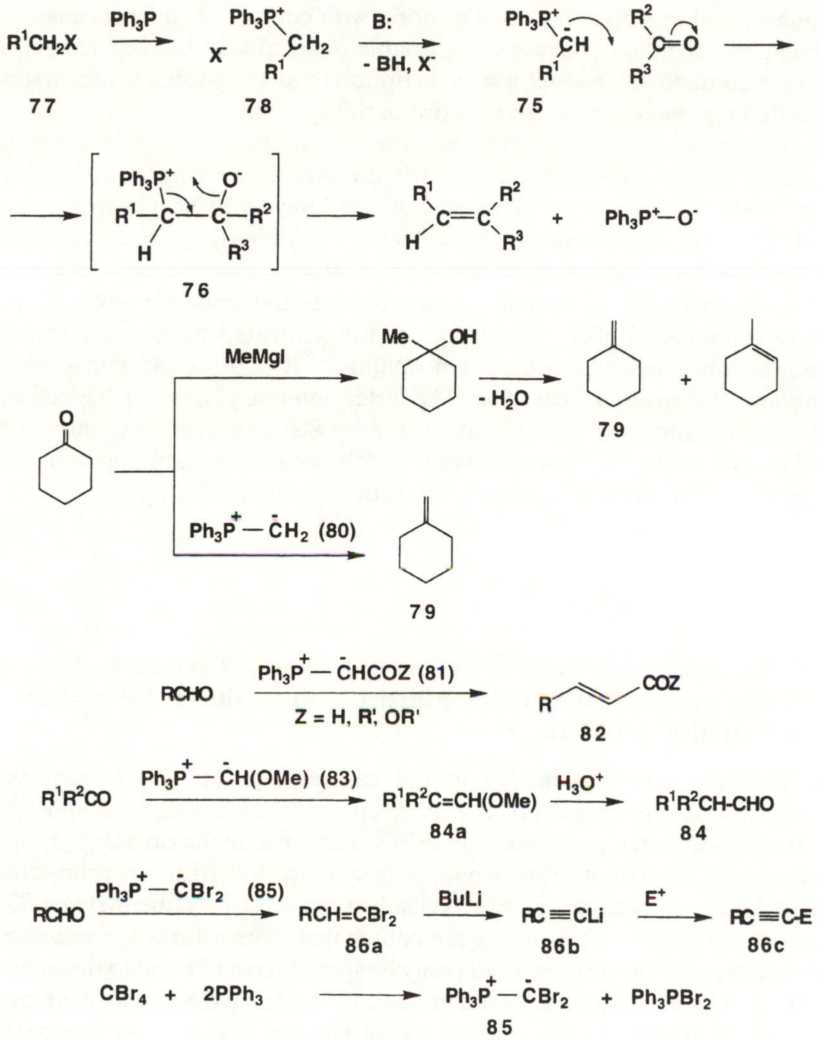

R^1CH_2X (**77**) → (Ph_3P) → X^- $Ph_3P^+CH_2R^1$ (**78**) → (B:, - BH, X^-) → $Ph_3P^+\bar{C}HR^1$ (**75**) + $R^2R^3C{=}O$ →

→ [Ph_3P^+–CHR^1–$C(O^-)R^2R^3$] (**76**) → $R^1HC{=}CR^2R^3$ + $Ph_3P^+–O^-$

Cyclohexanone → (MeMgI) → 1-methylcyclohexanol → (- H_2O) → methylenecyclohexane (**79**) + 1-methylcyclohexene

Cyclohexanone → ($Ph_3\overset{+}{P}$—$\bar{C}H_2$ (**80**)) → methylenecyclohexane (**79**)

RCHO → ($Ph_3\overset{+}{P}$—$\bar{C}HCOZ$ (**81**); Z = H, R', OR') → RCH=CHCOZ (**82**)

R^1R^2CO → ($Ph_3\overset{+}{P}$—$\bar{C}H(OMe)$ (**83**)) → $R^1R^2C{=}CH(OMe)$ (**84a**) → (H_3O^+) → R^1R^2CH-CHO (**84**)

RCHO → ($Ph_3\overset{+}{P}$—$\bar{C}Br_2$ (**85**)) → $RCH{=}CBr_2$ (**86a**) → (BuLi) → RC≡CLi (**86b**) → (E^+) → RC≡C-E (**86c**)

CBr_4 + $2PPh_3$ → $Ph_3\overset{+}{P}$—$\bar{C}Br_2$ (**85**) + Ph_3PBr_2

Scheme 2.28

Alkylidenephosphoranes like **80** are rather unstable and require *in situ* generation. An electron accepting group introduced into an alkyl moiety greatly enhances the stability of these species and derivatives like **81** are available as commercial reagents. These derivatives are extensively used to synthesize a variety of α,β-unsaturated carbonyl compounds, such as **82**, shown in Scheme 2.28. This method is especially advantageous owing to the opportunity to achieve rigorous control over the stereochemistry of the newly formed double bond.

The discovery of the Wittig reaction in 1953[12b] (which merited a Nobel Prize) greatly enriched the arsenal of organic synthesis. In fact, the Wittig reaction and its later modification[12c] turned out to be the first general method for establishing a double bond in a predetermined position with controlled stereochemistry. As a result, retrosynthetic cleavage of a double bond to lead to alkyl halides and carbonyl components became a reliable option to identify suitable precursors in the search for the optimum constructive pathway.

It is also worthwhile to add that the versatility of the Wittig reaction suggested entirely new opportunities for the synthetic utilization of carbonyl compounds. Thus, the employment of methoxymethylenephosphorane **83** (easily prepared from chloromethyl ether) as the ylide component represents a standard protocol for the transformation of aldehydes or ketones into homologous aldehydes **84**, *via* the intermediate formation of enol ether **84a**.[12a]

A rather unusual ylide, **85**, can be readily generated by the interaction of carbon tetrabromide with triphenylphosphine.[12d] Reaction of **85** with aldehydes furnishes 1,1-dibromoalkenes **86a**. The latter compounds, under the action of BuLi, are transformed into lithium acetylides **86b** and ultimately into substituted acetylenes **86c**.[12e] This sequence is widely used as a reliable method for the preparation of various acetylenes from readily available aldehydes.

2.7.3 Conjugate Addition to α,β-Unsaturated Carbonyl Compounds. The Robinson Annulation and the Michael Addition with the Independent Variation of Addends

The synthetic potential hidden in the carbonyl group is far from being exhausted in its pronounced role as an electrophile in the Grignard and Wittig reactions. Additional possibilities arise in systems where the carbonyl group, or another electron-withdrawing group, is in conjugation with a carbon–carbon double bond. In such structures, typically represented by the systems **87–89** (Scheme 2.29), the π-electrons of the conjugated system form a single moiety over which polarization effects can easily be spread from one end to the other. In reactions with nucleophiles, such compounds can behave either as the familiar carbonyl electrophiles or as a new type of electrophile with the electrophilic center located at the β-carbon, as shown in bipolar structures **87a–89a**.

Owing to this dichotomy, α,β-unsaturated aldehydes, ketones, or esters can undergo a nucleophilic attack at either the carbonyl carbon or the β-carbon atom (Scheme 2.29). The first of these reactions is a familiar addition to the carbonyl group (1,2-addition) which leads, in this case, to the valuable allylic alcohols. Even more intriguing synthetic options, however, are offered by the alternative pathway, the 1,4-addition generally known as the Michael reaction.[13] The classic version of this reaction employed stable carbanions such as those generated *in situ* from malonic ester or nitromethane under the action of bases and in the presence of Michael acceptors, *e.g.* methyl vinyl ketone **90**:

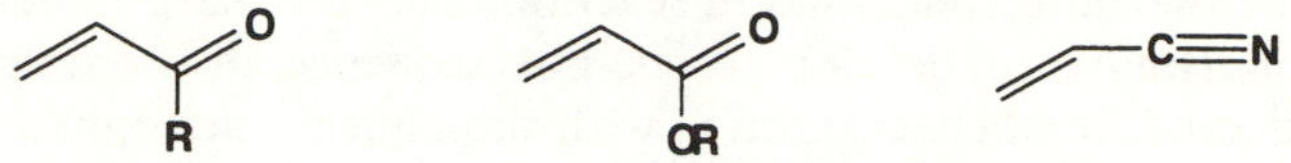

O R 87   O OR 88   C≡N 89

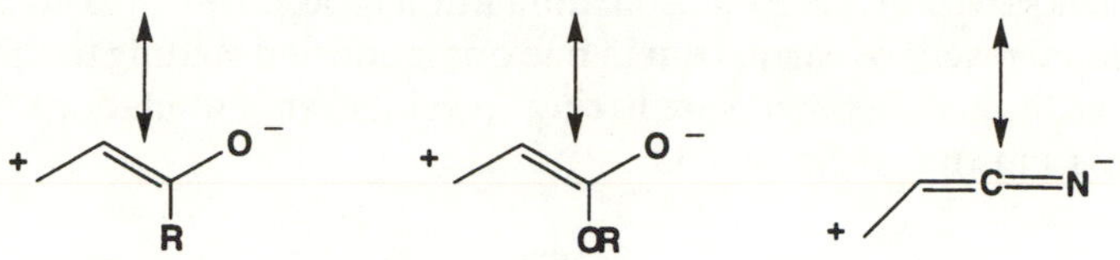

$^{+}$ O^{-} R 87a $^{+}$ O^{-} OR 88a $^{+}$ =C=N^{-} 89a

1,2-Addition:

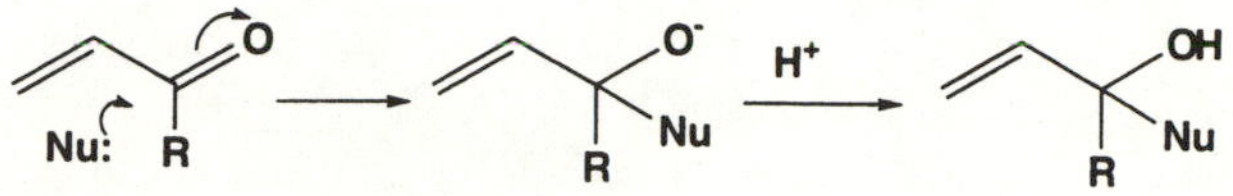

Nu: R O → O^{-} Nu R —H^{+}→ OH Nu R

1,4-Addition:

Nu: O R → Nu O^{-} R —H^{+}→ Nu O R

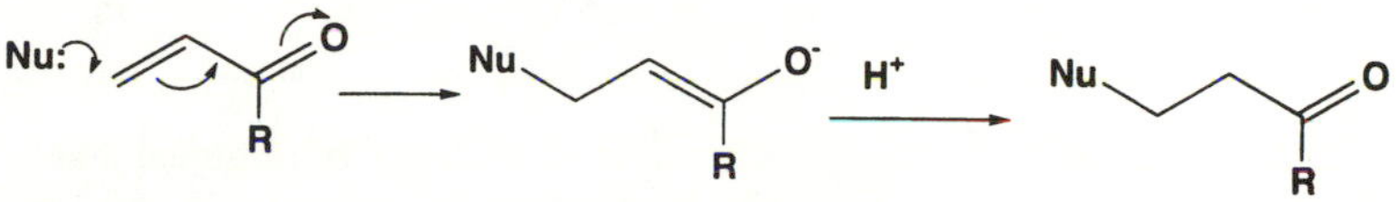

Scheme 2.29

$$CH_2(COOEt)_2 \xrightarrow{EtONa} {}^{-}CH(COOEt)_2 \xrightarrow[90]{CH_3COCH=CH_2}$$

$$CH_3CO\bar{C}H\text{–}CH_2CH(COOEt)_2 \xrightarrow{H^+} CH_3COCH_2CH_2CH(COOEt)_2$$

$$CH_3NO_2 \xrightarrow{R_4NOH} {}^{-}CH_2NO_2 \xrightarrow[90]{CH_3COCH=CH_2} CH_3CO\bar{C}H\text{–}CH_2CH_2NO_2$$

$$\xrightarrow{H^+} CH_3COCH_2CH_2CH_2NO_2$$

The Michael addition represents an extremely efficient synthetic method for achieving chain elongation by adding a three (or more) carbon fragment electrophile to a nucleophilic moiety. Notice that the typical Michael electrophiles (*e.g.* **90**) are products of condensation of carbonyl compounds and can be easily formed *via* the aldol-like condensation, the Wittig reaction (with ylides like **81**), the Perkin reaction, or the Mannich reaction (see below).

It must also be emphasized that the typical nucleophilic component of the Michael reaction is an enolate, derived from the respective carbonyl compound. Thus, the conditions required for the preparation of Michael acceptors might be identical to or at least very close to those optimal for effecting either the Michael addition or the aldol condensation reaction. This peculiarity offered an interesting opportunity to develop a one-pot sequence of steps involving the combination of the Michael addition with subsequent nucleophilic additions at the carbonyl carbon. A prominent example showing the fruitfulness of such an approach is served by the Robinson annulation procedure,[14a,b] a standard series of sequential reactions carried out in one operation and leading to the formation of an additional six-membered ring (cyclohexanoannulation). A typical example is shown in Scheme 2.30.

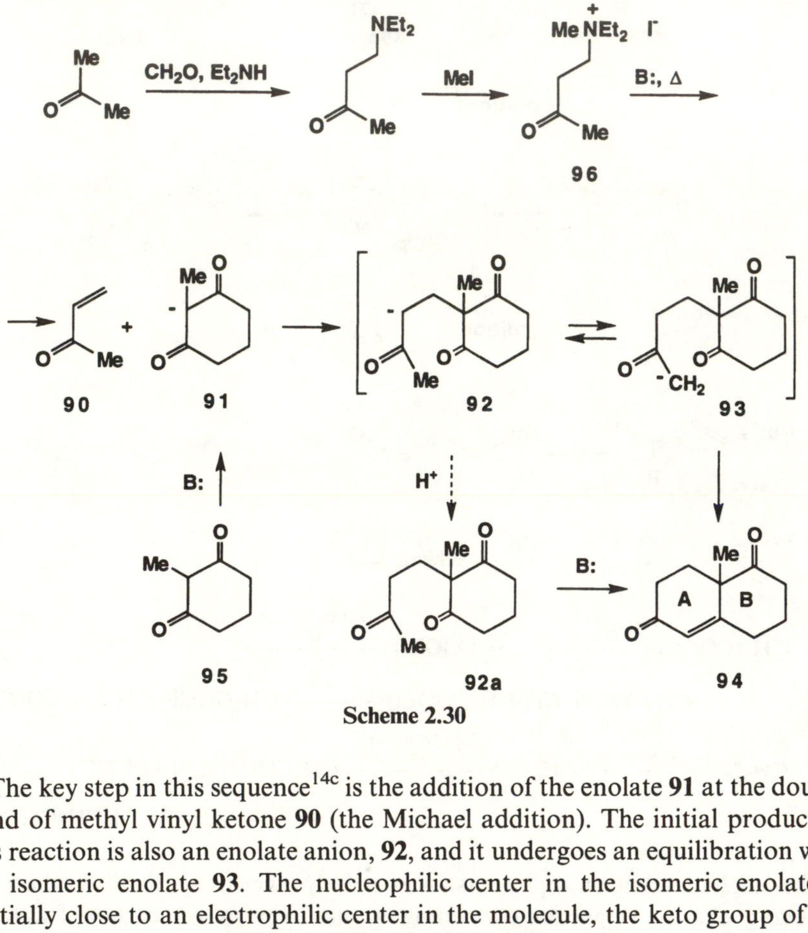

Scheme 2.30

The key step in this sequence[14c] is the addition of the enolate **91** at the double bond of methyl vinyl ketone **90** (the Michael addition). The initial product of this reaction is also an enolate anion, **92**, and it undergoes an equilibration with the isomeric enolate **93**. The nucleophilic center in the isomeric enolate is spatially close to an electrophilic center in the molecule, the keto group of the cyclohexane fragment. Owing to this proximity, a facile intramolelecular aldol condensation followed by dehydration (*i.e.* crotonic condensation) occurs to form the bicyclic diketone **94**. The bicyclic system present in **94** represents one of the basic structural features common to a plethora of triterpenoids and steroids (A and B rings of their framework). Thus diketone **94** may serve as an advanced intermediate in the synthesis of these compounds. Especially valuable is the fact that it already contains the angular methyl group and conveniently placed functional groups.

Preparative values of the Robinson procedure are obvious. The sequence **90** + **91** → **94** proceeds as a single synthetic operation. Moreover, the formation of enolate **91** from ketone **95** and the preparation of the Michael acceptor **90** might also be carried out in the same vessel. The latter opportunity is especially welcome owing to the instability of **90** upon prolonged storage. Generation of this Michael acceptor occurs by treating its stable equivalent, salt **96**, with a strong base.[14a] Thus when the mixture of **95** and **96** is heated in the presence of a strong base, a sequence of reactions is triggered leading ultimately to the bicyclic product **94**. It must be also added that salt **96** can be easily prepared from trivial starting compounds (acetone, formaldehyde, and diethylamine) *via* the Mannich reaction,[14d] a reaction essentially similar to the above-described aldol condensation.

In essence, the annulation shown in Scheme 2.30 merely represents a sequence of two trivial reactions, the Michael addition and the aldol condensation. Why, then, is this protocol specifically labeled as the Robinson annulation? Most of all, because Sir Robert Robinson was able to recognize the benefits of synthetical solutions gained upon exhaustive utilization of the potential hidden in the chemical reactivity of simple functional groups. He designed a novel strategy to construct six-membered rings based upon a tandem sequence of reliable and simple reactions. In fact, the spectacular efficiency of the Robinson annulation rests upon the fact that the carbanionic intermediate **92**, formed at the first step of Michael addition, is not intercepted by a proton (as is usually the case with classical Michael reactions, see above). It serves as a nucleophile (*via* equilibration to **93**) in the reaction with an electrophilic carbon of the initial Michael acceptor. Therefore, free energy accumulated in the course of generation of the 'hot' intermediate **92** is not wasted on the formation of a regular covalent Michael adduct, **92a**, but is used directly to facilitate the next step leading to the formation of new ring. An absence of proton donors in the media is an obvious prerequisite for this reaction course. The difference in the reaction pattern of the Robinson annulation and the classical Michael addition which eventually leads to the same product **94**, *via* formation of covalent product **92a**, is shown schematically in Fig. 5.

Owing to numerous studies, a fairly diversified set of compounds have been recognized as potential components of the Robinson annulation and this protocol has been adjusted and utilized to assemble complex frameworks of terpenes, steroids, and alkaloids.[14b,d]

It seems appropriate to inquire whether or not it is possible to carry out other Michael reactions and, generally, other nucleophilic additions to unsaturated compounds as a sequence of kinetically independent steps using one's choice of nucleophiles and electrophiles? The answer is definitely 'yes'. A rationale similar to that used to describe the Robinson annulation provides us with the key to how this goal may be attained. First of all, the initial step of the reaction, addition of the nucleophilic component across a double (or triple) bond, needs to be carried out in the absence of the external electrophiles (preferably in aprotic solvents). Secondly, a carbanionic intermediate, incipiently formed at this step, requires sufficient stabilization to survive as a chemical entity under

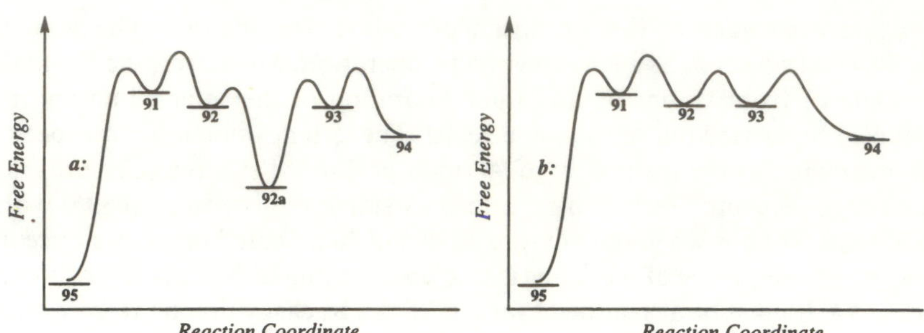

Figure 5 *Schematic free-energy profile for two modes of formation of* **94** *from the reaction of* **90** *with* **91**: *(a) conventional sequence of independent Michael addition–crotonic condensation via formation of covalent* **92a**; *(b) a one-pot Robinson annulation without proton quenching of carbanionic intermediate* **92**

the ambient conditions of the reaction until the required electrophile is introduced into a reaction mixture. In addition, this intermediate must also be unreactive toward the starting unsaturated substrate.

This reasoning, actually nothing more than a mere ramification of the classic mechanistic description of Ad_N reactions, turned out to be truly productive in the elaboration of an efficient and general approach to the solution of many synthetic problems. One of the first examples of the successful utilization of a sequence of independent additions of **Nu** and **E** to Michael acceptors was described by Stork,[15a] who took advantage of this stepwise plan to achieve a short-route preparation of **97** (Scheme 2.31), the advanced intermediate in the synthesis of the polycyclic alkaloid lycopodine.

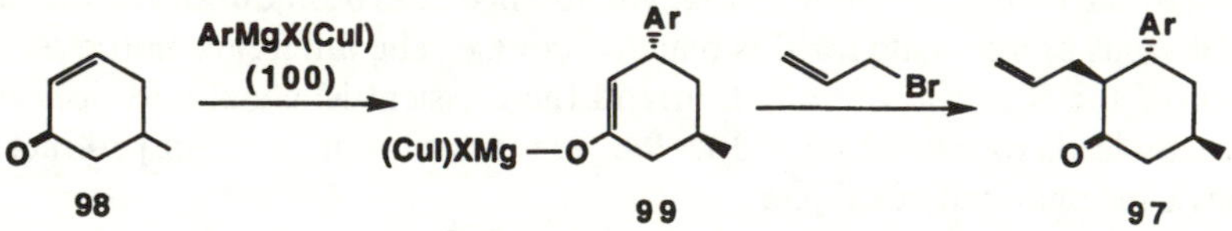

ArMgX(CuI) (100)
(CuI)XMg—O
Ar
Br
98 → 99 → 97

Scheme 2.31

In essence, the ideology of this synthesis is similar to that employed in the Robinson annulation. In fact, here again the carbanionic intermediate **99** (formed upon the initial addition of arylmagnesium cuprate reagent **100** at the double bond of Michael acceptor **98**) is treated with a carbon electrophile (allyl bromide) to give the final adduct **97** with two new **C–C** bonds. The only essential difference lies in the fact that the quenching of the enolate intermediate **99** with the electrophile occurs as an intermolecular reaction (in contrast to the Robinson annulation where this step proceeds intramolecularly).

The synthetic applications of this reaction pattern are tremendous since they provide the opportunity to vary, independently, the nature of both nucleophiles and electrophiles and thus possibly develop a fairly general protocol for the preparation of many diverse compounds. It should also be emphasized that the utilization of cuprate reagents, like **100**, is crucial for the viability of this approach. The corresponding Mg or Li organic compounds most usually exhibit low regioselectivity in reactions with α,β-unsaturated ketones. In addition, the lowered basicity and reduced nucleophilicity of the resulting Cu enolates (*i.e.* **99**) minimizes typical complications such as self-condensation and/or indiscriminate reactions with electrophiles. The generality and usefulness of this one-pot, three-component coupling is well documented in a plethora of published data related to various areas of total synthesis[15b–d] and we shall often refer to this method later in this text.[15e]

2.7.4 Alkyne Carbometallation as a Versatile Method for the Stereoselective Synthesis of Alkenes

The fruitfulness of the idea of a stepwise addition with an independent variation of the addends was brilliantly illustrated by Normant's studies, which resulted in the elaboration of a general method of alkene synthesis based on the reaction of alkyne carbometallation.[16a] Basically this reaction represents a case of the well-known nucleophilic addition to a carbon–carbon triple bond. In the Normant reaction, however, the initial addition of a nucleophile (an organometallic reagent) across the triple bond results in the formation of a stabilized carbanion-like intermediate equivalent to a vinyl carbanion. This intermediate can similarly be further reacted with an external electrophile. Most typically, copper-modified Mg or Li reagents, which are unable to react with acidic acetylenic hydrogens, are used in this sequence.

As a result of this process, the sequential addition of a carbon nucleophile and a carbon electrophile across a triple bond is achieved. For example, addition of cuprate **101** (Scheme 2.32) across the triple bond of acetylene produces the vinylcuprate intermediate **102**. Quenching of the latter with electrophile **103** gave acetoxydiene **104**,[16b] the active constituent of the pheromone *Cossus cossus*. The sequence exemplified in Scheme 2.32 enables independent variations in the structure of all participants involved, namely the alkyne, the organometallic nucleophile, and the electrophile. Therefore this approach can serve as a unified protocol for the one-pot assemblage of various alkenes from simple precursors.[15c,16a]

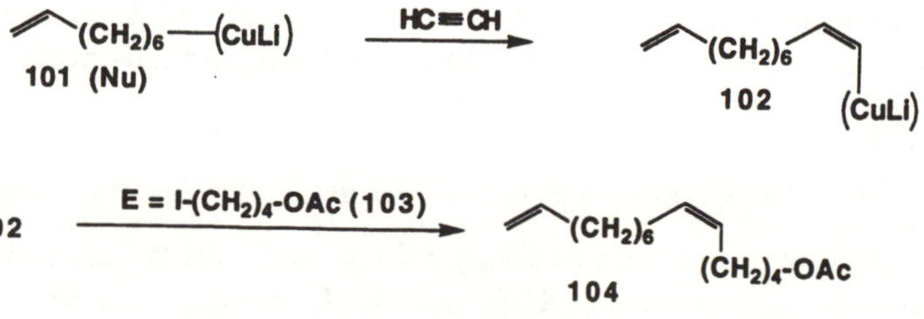

$(CH_2)_6$—(CuLi) **101** (Nu) $\xrightarrow{HC\equiv CH}$ $(CH_2)_6$ (CuLi) **102**

102 $\xrightarrow{E\,=\,I\text{-}(CH_2)_4\text{-}OAc\ (103)}$ $(CH_2)_6$ $(CH_2)_4$-OAc **104**

Scheme 2.32

The carbometallation of alkynes is also unique in its stereochemical pattern, since it is executed as an exclusive *cis* addition. For example, the stereoisomeric purity of pheromone **104**, prepared as shown, is higher than 99.9%! This feature is of special importance in the synthesis of natural pheromones, as the biological activity of these compounds is dramatically affected by the presence of an even negligible amount (less than 0.5%!) of the undesirable stereoisomer. Not surprisingly, this reaction has found numerous areas of application, especially in the stereospecific synthesis of the tri- and tetrasubstituted alkenes,[16c] a goal difficult to achieve by other methods. To illustrate the effectiveness of this approach, the synthesis of faranal **105**, the trail-marking pheromone of the ant species *Monomorium pharaonis*,[16d] is shown in Scheme 2.33.

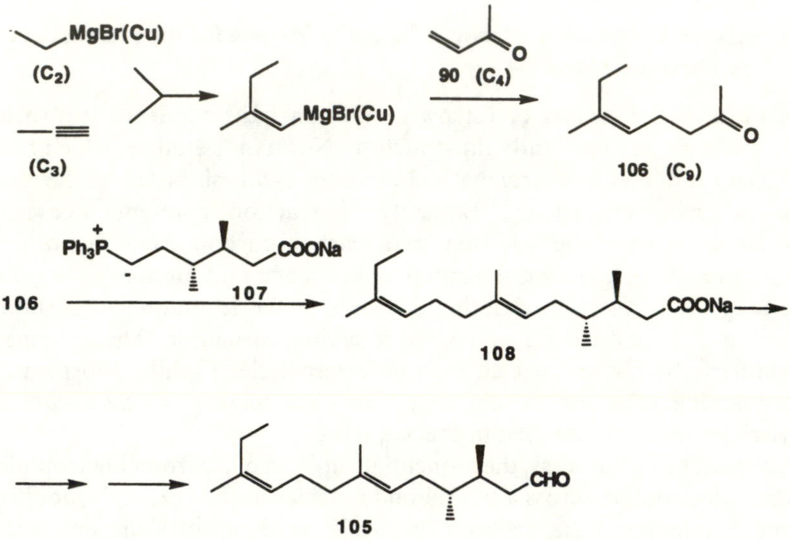

Scheme 2.33

The key step in this scheme is the stereospecific assemblage of the C_9 ketone **106** from three smaller fragments: methylacetylene (C_3), ethylmagnesium bromide (C_2), and methyl vinyl ketone **90** (C_4). A subsequent Wittig reaction between **106** and **107**, followed by the transformation of the carboxylic group in the product **108**, completes the preparation of faranal **105**. It is also worth noting that the immediate result of the nucleophilic addition to methylacetylene is the formation of an organocuprate reagent, which, as we already know, is especially suitable for conjugated addition reactions with α,β-enones such as **90**.

2.7.5 Retrosynthetic Analysis of Acyclic Target Molecules. Key Leads

In this section we have considered only a few of the most typical and frequently used methods for assembling C–C and C=C bonds. This selection, while limited, is nevertheless useful in formulating a set of guidelines for rational disconnection of a relatively simple molecule in the course of a retrosynthetic analysis. These guidelines are given below with reference to the structural peculiarities of the target compounds.

2.7.5.1 A single C–C bond with no nearby functional groups present

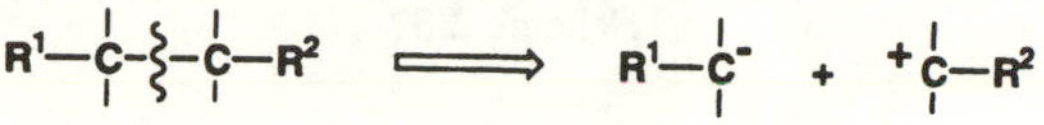

$$R^1{-}\overset{|}{\underset{|}{C}}{-}\!\wr\!{-}\overset{|}{\underset{|}{C}}{-}R^2 \Longrightarrow R^1{-}\overset{|}{\underset{|}{C}}{}^{-} + {}^{+}\overset{|}{\underset{|}{C}}{-}R^2$$

Scheme 2.34

Here, mixed cuprates prepared from organolithium or organomagnesium compounds serve as a carbanion, while derivatives like alkyl halides, alkyl sulfonates, or alkyl acetates are suitable equivalents for the carbocationic counterpart (Scheme 2.34).

2.7.5.2 A single C–C bond where one of the carbons bears an oxygen substituent

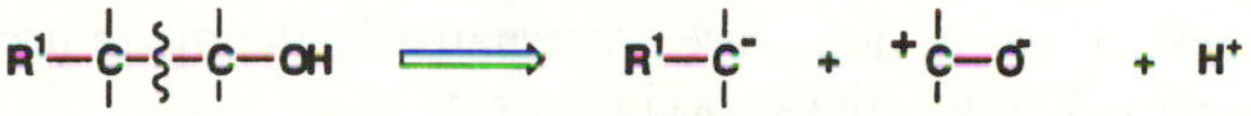

$$R^1{-}\overset{|}{\underset{|}{C}}{-}\!\wr\!{-}\overset{|}{\underset{|}{C}}{-}OH \Longrightarrow R^1{-}\overset{|}{\underset{|}{C}}{}^{-} + {}^{+}\overset{|}{\underset{|}{C}}{-}O^{-} + H^{+}$$

Scheme 2.35

This disconnection represents the retro-Grignard reaction. It leads almost automatically to organomagnesium or lithium compounds as equivalents of carbanions and aldehydes or ketones as equivalents to the carbocation in the synthesis of alcohols (Scheme 2.35), or carbon dioxide in the synthesis of carboxylic acids.

2.7.5.3 A single C–C bond at an allylic position

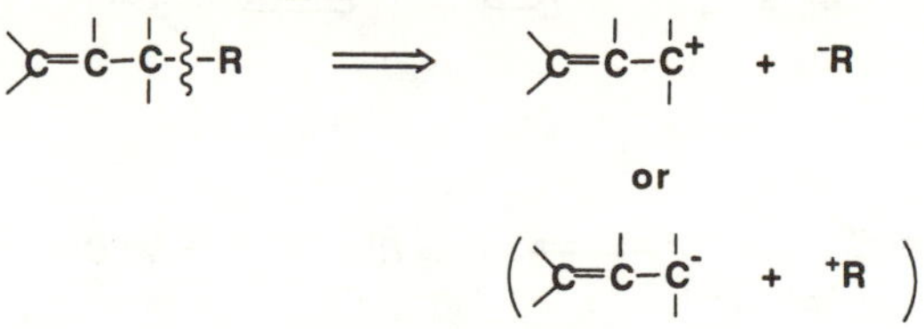

$$>C{=}\overset{|}{C}{-}\overset{|}{\underset{|}{C}}{-}\!\wr\!{-}R \Longrightarrow >C{=}\overset{|}{C}{-}\overset{|}{\underset{|}{C}}{}^{+} + {}^{-}R$$

or

$$\left(>C{=}\overset{|}{C}{-}\overset{|}{\underset{|}{C}}{}^{-} + {}^{+}R\right)$$

Scheme 2.36

This scheme refers to a very general approach as the respective allylic electrophiles (*e.g.* halides) are readily available, easy to handle, and capable of reacting with various carbanionic species (Scheme 2.36). Below we will consider also a less common, but nevertheless very useful, disconnection of this moiety which leads to reversed polarity of the components.

2.7.5.4 A single C–C bond in the fragment containing two oxygen substituents in a 1,3 relationship

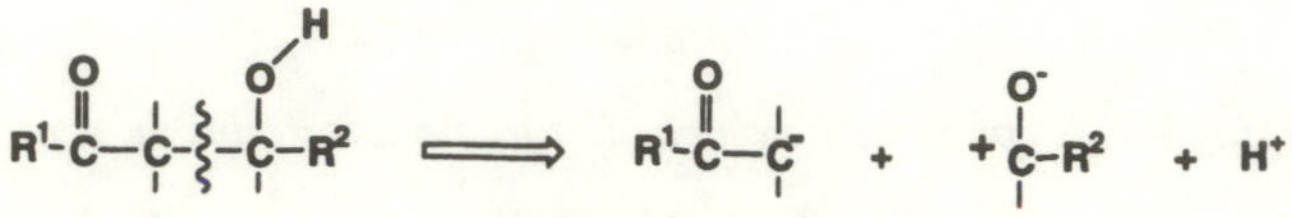

Scheme 2.37

This is a typical retro-aldol disconnection using the enolate as an equivalent of the carbanion and a carbonyl electrophile (Scheme 2.37).

2.7.5.5 Carbon–carbon double bond in α,β-unsaturated carbonyl compounds

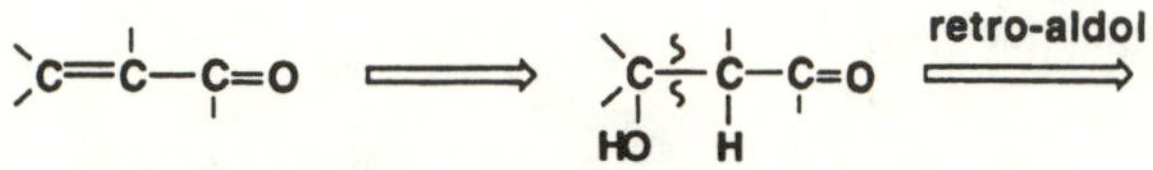

retro-aldol

Scheme 2.38

The retrosynthetic hydration of the double bond as indicated (Scheme 2.38) brings us back to the previous case. Alternatively, the disconnection can be carried *via* a retro-Wittig pathway, as in the next case.

2.7.5.6 Carbon–carbon double bond in any context

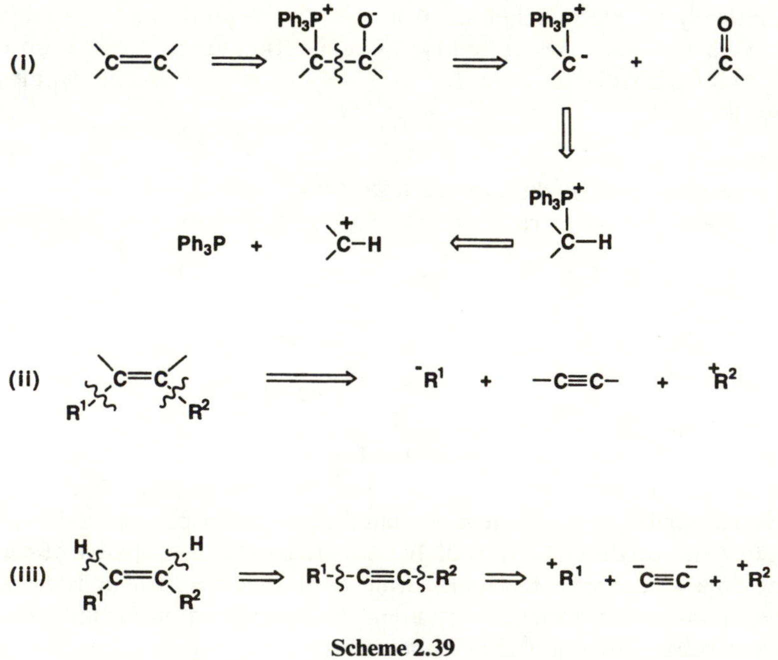

(i)

(ii)

(iii)

Scheme 2.39

The retro-Wittig reaction (**i**) (Scheme 2.39), with disconnection across the double bond, seems to be an obvious candidate for models containing disubstituted double bonds. In these cases, the steric outcome of the reaction can be easily controlled. An alternative route of disconnection at the vinyl bonds (**ii**) corresponds to a retro-carbometallation reaction. This disconnection is generally applicable to double bonds with any substitution pattern and is especially useful owing to the high stereoselectivity of the carbometallation step. Route (**iii**) involves a retrosynthetic dehydrogenation leading to the immediate acetylenic precursor, which can be conventionally disassembled into a parent acetylide and a pair of electrophiles as shown.

2.7.5.7 An ordinary C–C bond in a system containing two oxygen substituents in a 1,5 relationship

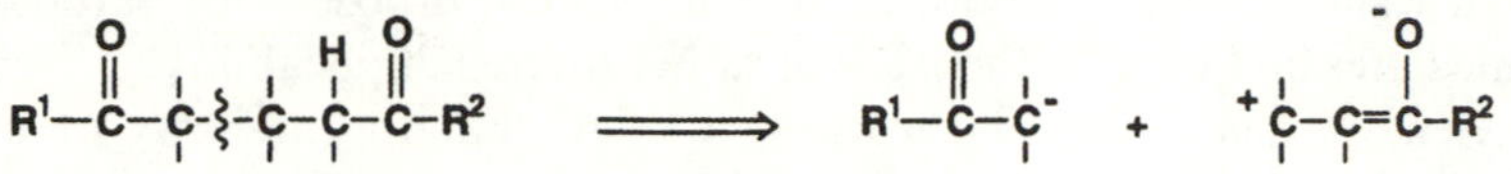

$$R^1-\overset{O}{\overset{\|}{C}}-C-C-\overset{H}{C}-\overset{O}{\overset{\|}{C}}-R^2 \Longrightarrow R^1-\overset{O}{\overset{\|}{C}}-C^- + {}^+C-C=\overset{^-O}{\overset{|}{C}}-R^2$$

Scheme 2.40

The retro-Michael addition is shown in Scheme 2.40. The standard components are generated, with the enolate acting as an equivalent to a carbanion and an α,β-unsaturated carbonyl compound as an equivalent to the cationic counterpart.

Of course, plenty of other ways can be devised to accomplish a retrosynthetic analysis of these systems. Nevertheless, the above guidelines are useful as leads to initiate an analysis along the potentially most promising pathways. It is imperative to consider the *entire* set of requirements when evaluating the merits of an alternative synthetic route in a retrosynthetic analysis. These requirements may refer to the relative stability and reactivity pattern of the ions (or their equivalents) that are generated, the generality of the scope and effectiveness of the corresponding synthetic methods, the availability of starting compounds, the possibility of complications due to the presence of interfering functionalities, *etc.*

2.7.6 Carbocationic *vs*. Carbanionic Reagents. Some Novel Options for C–C Bond-forming Reactions

In the beginning of this section we did not make any special comments regarding the relative synthetic importance of carbocations and carbanions, and treated them as species equally applicable for the creation of C–C bonds. This is not exactly true, as might be seen in the above discussion of important synthetic methods. Carbanionic counterparts of the heterolytic disconnections were usually represented by near-ionic species (enolates, acetylides, or ylides) or highly polarized (organolithium and organomagnesium compounds) reagents. At the same time, electrophiles conventionally designated in retrosynthetic schemes by the corresponding carbocationic species were represented by fully covalent reagents with an electrophilic pattern of reactivity. In reality we did not even examine the reverse situation, in which the carbocation-like intermediate serves as an active electrophile while a purely covalent and comparatively inactive nucleophile is employed as an equivalent to the carbanion. This was a deliberate choice. It reflects the fact that carbanions are generally more stable than carbocations and are easier to generate and to handle.[4,8] Furthermore, a carbanion, with its complete octet of electrons, is much less susceptible to rearrangements or other side reactions common to carbocationic species (because of the presence of only a sextet of electrons in their valence shell).[3] Additionally, there are many options available to stabilize carbanionic species of diverse structure. Such options include the utilization of appropriate solvents, counterions, additional ligands, and/or the introduction of special structural fragments to achieve this goal. In comparison, relatively few approaches have been elaborated for the same purpose with carbocationic intermediates and most of them are of rather limited applicability.[3,17a]

In view of this, it is not surprising to find that the majority of modern synthetic methods based on the heterolytic reactions of **C–C** bond formation rely mainly on the combination of the formally ionic nucleophile and covalent electrophile, with the reverse options being much less common. A notable exception is in the case of electrophilic substitution in the aromatic series (Friedel–Crafts type reactions), which serve as extremely powerful synthetic

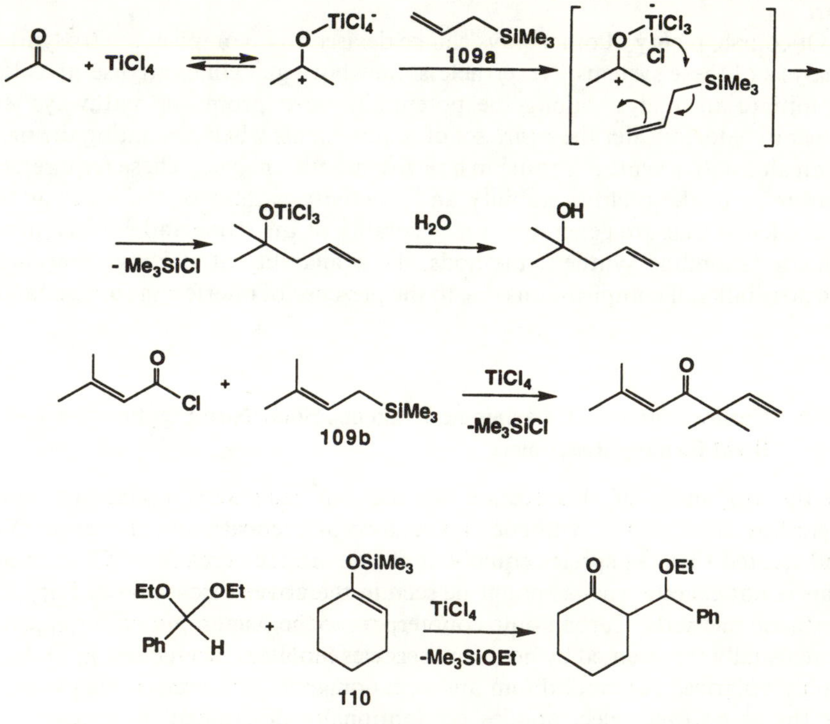

Scheme 2.41

methods. It should be noted, however, that the classical options in Friedel–Crafts chemistry are of limited usefulness outside the area of aromatic compounds since the reaction conditions are too severe and generally unapplicable for acid-sensitive compounds. We believe, however, that with the development of effective and general methods for the stabilization of carbocations, a new cadre of complementary synthetic methods utilizing the tremendous synthetic potential of these species will emerge eventually. While this area remains largely to be explored, we shall consider below several examples attesting to the promise of this approach.

As we mentioned before, a classical Grignard reaction is formally described by the coupling of a covalent (albeit polarized) electrophile with an anionic nucleophile. Reactions shown in Scheme 2.41 (opposite) exemplify the alternative approach involving an interaction between cationic intermediates generated from carbonyl compounds (or their derivatives) under the action of Lewis acids and a purely covalent nucleophile, an allylsilane such as **109a** or **109b**.[17b,c] Similar electrophiles used in reactions with covalent silyl enolates such as **110** result in the formation of the aldol-like products (the Mukaiyama reaction[17d]).

Both silyl enolates and allylsilanes are excellent nucleophiles for alkylation by other stabilized carbocations such as the tertiary alkyl cations **111** or **112** (Scheme 2.42).[17c,e] Similarly, Michael-like additions, for example, the coupling of **113** with silyl ketene acetal **114**, can be also achieved.[17b,f,g] Owing to the high electrophilicty of the enone system, this reaction proceeds smoothly in polar solvents, even in the absence of Lewis acids.[17f]

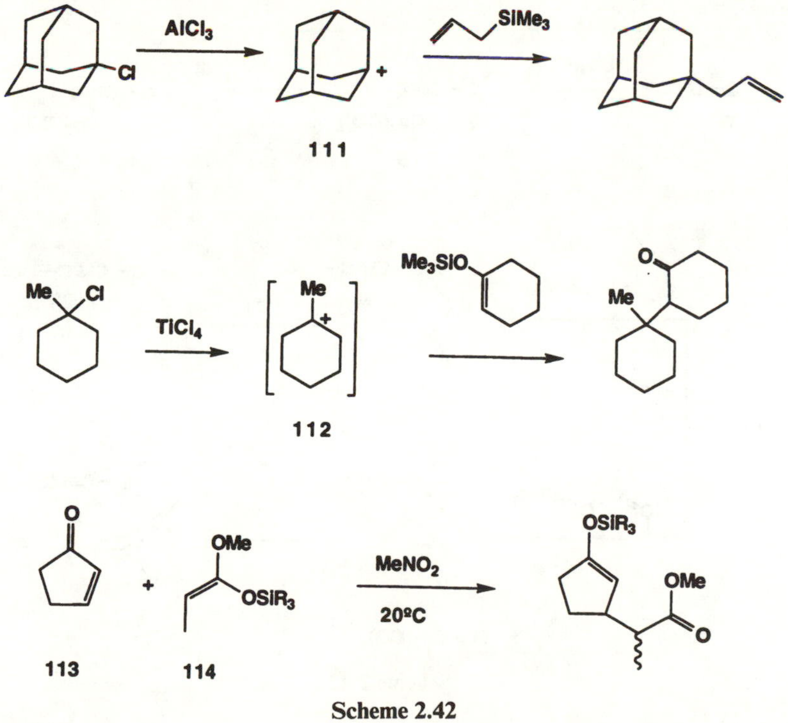

$AlCl_3$ → **111** + $\diagup\!\!\diagdown$SiMe$_3$ →

111

Me Cl → $TiCl_4$ → [Me +] → Me_3SiO → O Me

112

O + OMe OSiR$_3$ → $MeNO_2$, 20ºC → $OSiR_3$ OMe O

113 **114**

Scheme 2.42

The opportunity to achieve results similar to those described earlier in this section, but under entirely different conditions, greatly enhances the area of preparative utilization of formally similar transformations. For example, the alkylation of ionic enolates is a fairly general reaction if various primary alkyl halides are used. It inevitably fails, however, for tertiary alkyl halides owing to a propensity of these compounds to undergo elimination in the presence of basic reagents. On the other hand, the reaction of silyl enolates with tertiary alkyl halides, like that shown in Scheme 2.42, proceeds smoothly and in high yields.[17e] Also of significant synthetic importance are the ramifications related to the absolutely novel options to control the steric outcome of reactions such as the aldol condensation owing to the elaboration of methods based on the use of covalent enolates.[17g]

During the last decade, a new area of carbocationic chemistry has emerged because of the discovery of a strikingly efficient use of transition metal complexes to stabilize the positive charge of an adjacent carbon atom. For example, Nicholas discovered[18a] that the formation of the dicobalt hexacarbonyl (DCHC) complex at a triple bond, *e.g.* **115** (Scheme 2.43), ensures an effective stabilization for propargylic cations **116**, which are otherwise absolutely unattainable as even transient intermediates. The stabilization provided by the DCHC facilitated the development of a general method for the propargylation of various neutral nucleophiles, as is shown in Scheme 2.43. Ease in preparation of the required DCHC complexes, from almost any compound containing an alkyne fragment, and in their oxidative removal from the final products makes this protocol truly convenient. It should be emphasized

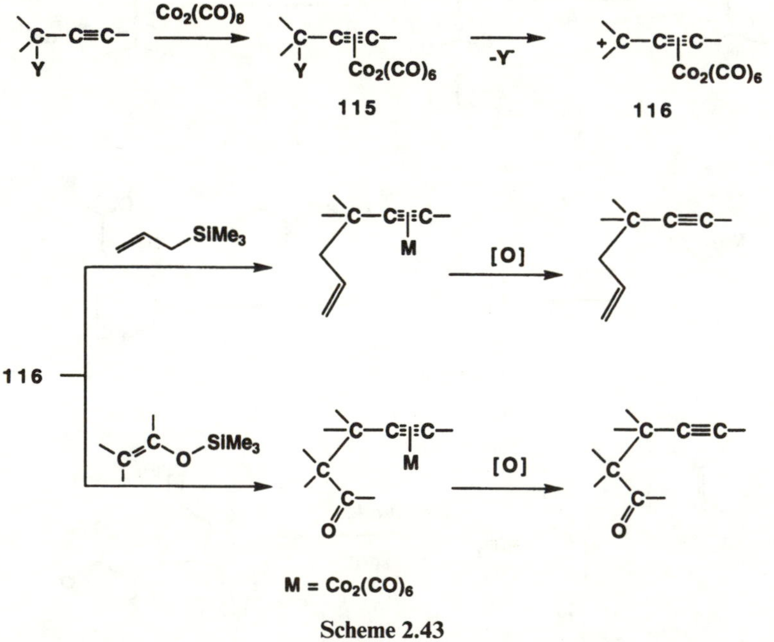

$Co_2(CO)_8$ 115 $-Y^-$ 116 $Co_2(CO)_6$

116 $SiMe_3$ [O] M = $Co_2(CO)_6$

Scheme 2.43

that salts of primary, secondary, and tertiary propargylic cations of the type **116** can be readily made from available precursors and handled as stable, shelf reagents. In this respect, the reactions involving these species differ from those shown in Schemes 2.41 and 2.42 (in the latter cases the reacting cationoid electrophiles can be usually generated only as transient intermediates).

Previously known propargylation methods based upon the utilization of Grignard reagents were applicable only for the introduction of primary propargylic residues. Their scope was additionally limited owing to the ease of the propargyl–allenyl rearrangement. By contrast, the reactions shown in Scheme 2.43 proceed without rearrangement even for tertiary propargyl cations and their respective adducts are usually formed in excellent yields.

The opportunity to achieve efficient stabilization of carbocationic intermediates like those exemplified by the propargylic species **116** suggested an interesting opportunity for the development of a truly stepwise mode of electrophilic addition to the double bond of conjugated enynes (*e.g.* **117**, Scheme 2.44). The viability of this sequence of independent additions of **E** and **Nu** was proven by the results of studies by the Smit–Caple group, and is represented in Scheme 2.44.[18b,c]

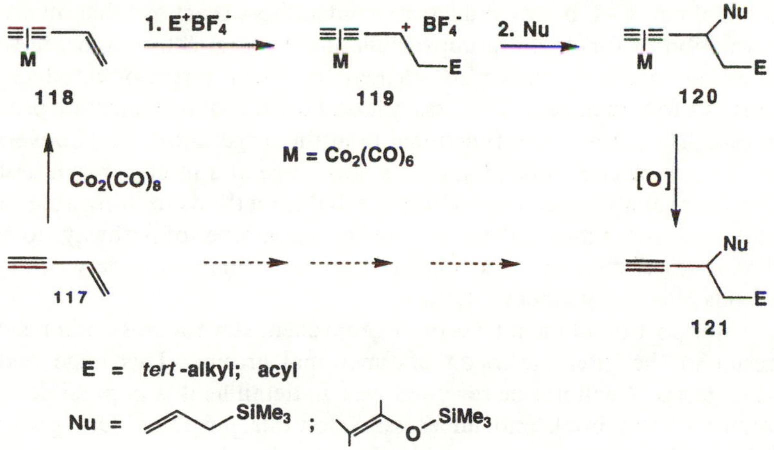

Scheme 2.44

The key element of this protocol is the initial addition of cationic electrophiles such as *tert*-alkyl or acyl cations to the double bond of a DCHC complex of the conjugated enyne **118**, which results in the formation of the substituted propargylic cation intermediate **119**. Subsequent reaction with pre-selected external nucleophiles, for example allylsilanes or silyl enol ethers, leads to the formation of the final adducts **120**. The reaction is carried out as a one-pot, three-component coupling and can be used for the creation of two novel **C–C** bonds. It is a process somewhat complementary to the stepwise Michael addition described earlier (Scheme 2.31), with a reverse order of **E** and **Nu** addition. Oxidative decomplexation of **120** yields the product **121**. The overall conversion, **117** →→ **121**, corresponds to the controlled regiospecific addition of **E** and **Nu** at the double bond, with independent variations in the nature of the addends.

Thus the utilization of the approach based upon the coupling of neutral nucleophiles with ionic electrophiles, while still not as popular as the coupling of ionic nucleophiles with covalent electrophiles, has already produced a set of results clearly demonstrating its specific merits and promises.

PART III FUNCTIONAL GROUP INTERCONVERSIONS. THEIR ROLE IN ACHIEVEING SYNTHETIC GOALS

2.8 THE OXIDATION STATE OF THE CARBON CENTER IN FUNCTIONAL GROUPS. TRANSFORMATIONS WITHIN AND BETWEEN THE OXIDATION LEVELS. SYNTHETIC EQUIVALENCY OF FUNCTIONAL GROUPS

Until now, we have examined only constructive reactions that result in the formation of new **C–C** bonds and have excluded those reactions that involve the transformation of functional groups. Functional group conversions, however, constitute an extremely important element of every target-oriented organic synthesis as a tool to make the necessary modifications of intermediate products and to establish the required functionality in the target molecule. Conversions of this type can be encountered at almost any stage of a multistep synthesis. In fact, it is the availability of a collection of reliable methods to bring about these transformations that makes the relatively limited number of pathways to create **C–C** bonds so effective in the synthesis of a nearly limitless variety of organic compounds of entirely different classes.

The major portion of most texts on organic chemistry focuses upon reactions that result in the interconversion of functional groups. This huge body of factual material[19a] will not be reviewed here in detail as it is impossible within the volume of this book and unnecessary for our purposes. Our goal is to highlight the importance of these interconversions in a total synthesis. The immense diversity of transformations can be actually reduced to a few types that we hope are sufficient to provide the reader with an understanding of the principles necessary to select the conversions for a chosen synthetic plan.

We will start with some general comments about the term 'functional group'. The carbon skeleton is the basic element of structure for any organic compound. It is for this reason that organic chemistry texts usually begin with a discussion of the saturated hydrocarbons, *i.e.* the alkanes and cycloalkanes. The replacement of the hydrogens in these hydrocarbons and/or the subsequent introduction of unsaturation gives rise to 'functionalized' derivatives like alkenes, alkynes, ketones, alcohols, esters, *etc.* In fact, even the hydrogen atoms in a hydrocarbon can be considered as a functional group since they can be substituted (*via* chlorination, nitration, oxidation, *etc.*) to yield a hydrocarbon derivative. In a sense, the notion of 'functional group' is a mere convention. However, it is generally understood that this notion refers to some specific moiety present in the structure. The nature of this moiety and its location determines the propensity of the molecule to interact with reagents and the selectivity pattern of the reactions with various agents.

Various approaches may be used to classify functional groups and pathways for their interconversion. For our purposes it seems most appropriate to use an approach based upon the oxidation state of the carbon present in the functional group.[19b] We will begin with an analysis of the oxidation states of carbon in various functional groups.

2.8.1 The Oxidation Level of the Carbon Center and the Classification of Functional Groups and their Interconversions

By definition, oxidation reactions are associated with a loss of electrons from an atom or molecule. Changes in the oxidation states of reacting partners are easy to identify for purely ionic reactions. However, conversions of covalent organic compounds rarely can be described in the terms of 'oxidation' or 'reduction', unambiguously, without some additional provisos. Certainly, in the case of the conversion of a primary alcohol into a carboxylic acid (or the reverse process) it is clear that there is a net oxidation (reduction). There is no ambiguity in defining the hydrogenation of an alkene as a reduction or the epoxidation of an alkene as an oxidation. However, the application of these terms to other alkene additions, such as hydration or bromination, or to the respective eliminations yielding alkenes, is far from obvious. Nevertheless, these reactions can be reliably (albeit formally!) classified in the terms of oxidation or reduction if one applies a certain set of formal criteria and ascribes a zero-oxidation level to the carbon atom of alkanes.

Consider the **C–H** bond in alkanes. Carbon is a more electronegative element than hydrogen. Consequently, the electron pair that forms this bond is shifted towards the carbon atom. In the extreme, an ionic representation of this bond can be given as pictured in **122** (Scheme 2.45). Within these conventions the carbon atom in an alkane can be approximated as a carbanion (oxidation level 0 by definition). Using this definition it becomes possible to apply oxidation–reduction terminology to the processes as if they occurred to ion pair **122**. Thus, oxidation of **122** with the loss of one electron leads to the radical **123**. With the loss of two electrons, the oxidation leads to carbocation **124**. Similarly, the conversion of an alkane to an alcohol and the alcohol into an aldehyde and the aldehyde eventually to a carboxylic acid can unambiguously be classified as an oxidation sequence with the loss of two, four, and six electrons. The oxidation levels 1, 2, and 3 are ascribed respectively to these functional derivatives. The conversion of an alkane to an alkene or alkyne can be interpreted in an analogous fashion.

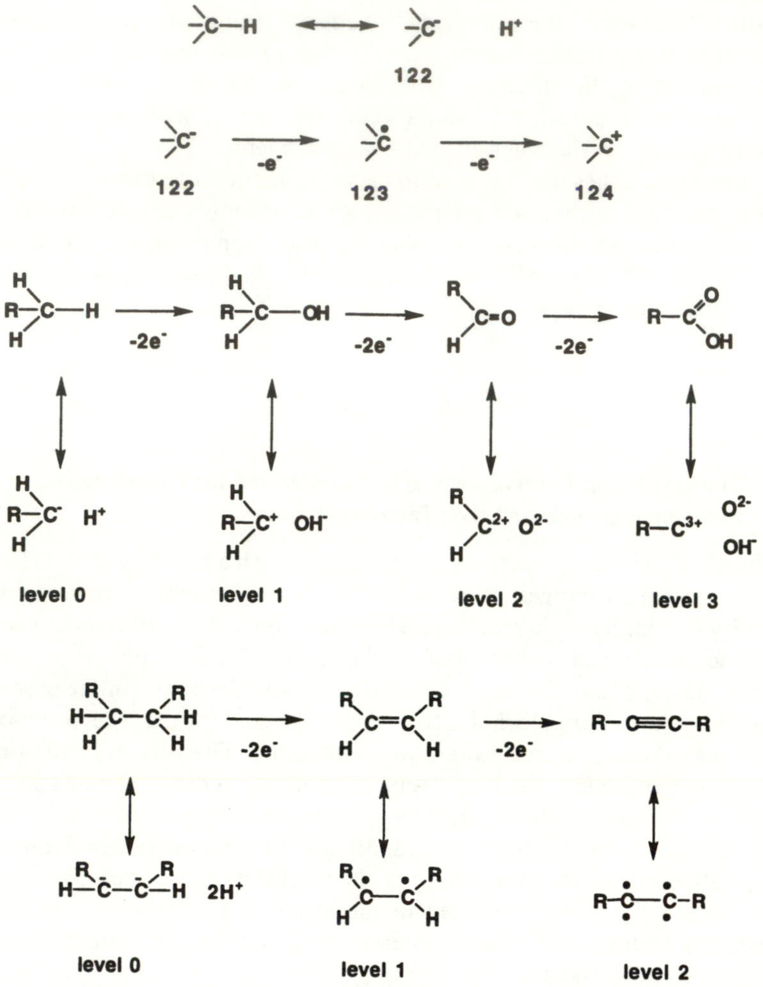

Scheme 2.45

This approach provides a basis for classifying important functional groups formally derived from alkanes, as shown in Scheme 2.46. One could extend this classification to even more complex polyfunctional compounds, but the principle should be already clear.

This classification of functional groups provided the opportunity to identify clearly two major types of their transformations:

A. *Isohypsic reactions*. Conversions occurring without a change in the oxidation level of the carbon atoms.

B. *Non-isohypsic reactions*. Conversions occurring with a change in the oxidation level of the carbon atoms. These conversions may proceed as oxidations, leading to an increase in the oxidation level, or as reductions, resulting in the decrease of the oxidation level.

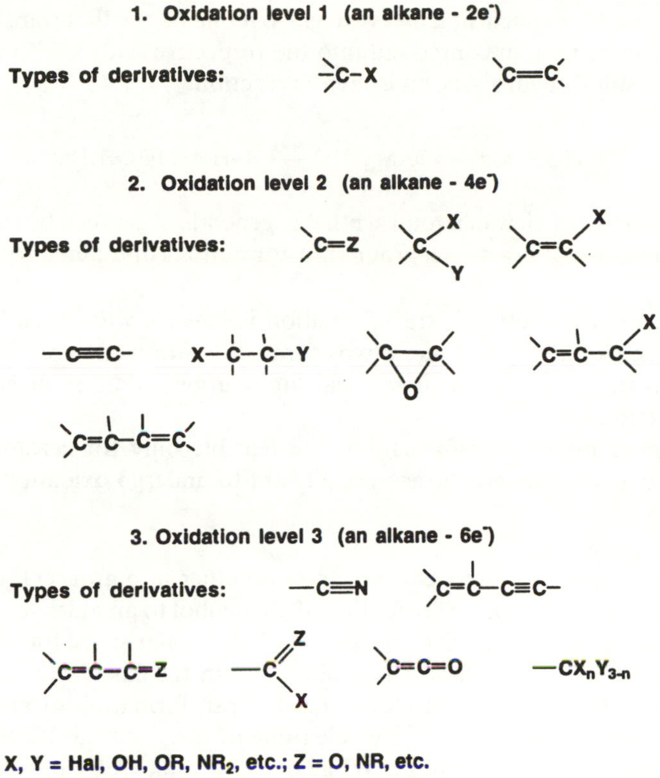

1. Oxidation level 1 (an alkane - 2e⁻)

Types of derivatives: C–X   C=C

2. Oxidation level 2 (an alkane - 4e⁻)

Types of derivatives: C=Z   C(X)(Y)   C=C–X

–C≡C–   X–C–C–Y   epoxide   C=C–C(X)

C=C–C=C

3. Oxidation level 3 (an alkane - 6e⁻)

Types of derivatives: –C≡N   C=C–C≡C–

C=C–C=Z   –C(=Z)X   C=C=O $-CX_nY_{3-n}$

X, Y = Hal, OH, OR, NR_2, etc.; Z = O, NR, etc.

Scheme 2.46

The changes in the oxidation level for a given reaction can be easily assessed by simply following the change in the oxidation status of the inorganic reagent used. Thus, for example, the formation of alcohols by alkene hydration, and the reverse elimination, are clearly isohypsic in that they involve water with no changes in its oxidation level. In contrast, all hydoxylation reactions of alkenes leading to the formation of glycols correspond to the formal addition of the elements of hydrogen peroxide (H_2O_2) and unquestionably should be treated as a non-isohypsic transformation (oxidation). Likewise, reactions such as the addition of hydrogen (reduction) or bromine (oxidation) to double and triple bonds (and the reverse processes of dehydrogenation or debromination) are non-isohypsic as well.

Following this logic, we should classify the formation of organolithium compounds or Grignard reagents *via* the interaction of metals (reducing agents) with alkyl halides as non-isohypsic reactions, which transfer substrates from oxidation level 1 into oxidation level 0 status. Thus we arrive at the apparently paradoxical conclusion that highly reactive functional compounds such as organometallics have the same zero-oxidation status as the parent hydrocarbons. This conclusion is easy to grasp if one recalls that transformation of an organomagnesium compound into the respective hydrocarbon proceeds easily as a result of hydrolysis, an isohypsic reaction:

$$\text{R–Hal} + \text{Mg} \longrightarrow \text{R–Mg–Hal} \xrightarrow{H_2O} \text{R–H} + \text{Mg(OH)Hal}$$

From the point of view of a total synthesis, generalizations can be made about the characteristics of functional group transformations of different types:

1. Almost any isohypsic transformation is feasible within the limits of a given oxidation level, as isohypsic transformations do not affect oxidation status and imply most usual substitution, addition or elimination reactions.
2. Non-isohypsic transformations are feasible only for certain types of derivatives, namely those especially apt to undergo oxidation or reduction.

Thus, for example, the direct conversion of an ether into an acetal or ketal is difficult to achieve whereas the oxidation of an alcohol to an aldehyde or ketone (or the reverse process) is a trivial transformation. Similarly, the transition from an oxidation level of 2 to level 1 is problematic in the case when one tries to convert dihalides into monohalides while the transformation of alkynes into alkenes may be safely considered a viable route to carry out this transition.

Consider the following analogy. While one can walk freely into and out of various rooms located on the same floor ('oxidation level'), there is no way to get directly from any room on one floor to a room on another floor without the help of special passage (stairway or elevator) that specifically serves as the tool for communication between the floors. Such an analogy, all its schematics notwithstanding, represents an accurate description of the possibilities and limitations for the transformations of functional groups. It enables us to focus our attention in the following sections on only a very limited number of the most important transformations which, as we believe, can best serve to illustrate the scope and limitations of functional group interconversions in a synthesis.

2.8.2 Isohypsic Transformations. Synthetic Equivalency of Functional Groups of the Same Oxidation Level

As we have seen above, alcohols and alkenes are produced routinely in the numerous reactions utilized in the formation of carbon–carbon bonds. These two functionalities are extremely useful for both the same-level oxidation interconversions and reduction–oxidation transitions (as 'stairways' connecting different 'floors'). Therefore, it is not surprising to see an enormous number of methods developed to capitalize on the diverse options of alkene and alcohol transformations. These methods occupy a key place among the reactions known for isohypsic conversions of level 1 functionalities.

Among the numerous isohypsic transformations of alcohols is a set of reactions leading to the formation of esters, alkyl halides, or sulfonates, especially valuable for synthetical purposes. These derivatives are widely used as electrophilic reagents, the synthetic equivalents of the carbocation $\mathbf{R}^+$ in **C–C** bond-forming reactions with carbon nucleophiles. Reactions used to convert alcohols into ethers, alkyl halides, *etc*., are trivial, elementary transformations. Therefore it may seem surprising to see an unceasing flow of publications describing 'new and efficient methods' to achieve these conversions. One can grasp the intensity of pursuits in this field from the fact that more than 40 methods are already listed in Larock's monograph for the specific conversion of an alcohol to an alkyl chloride.[19c] The list of reagents includes conventional inorganic compounds like HCl or $SOCl_2$ as well as various Cl^- transfer agents (*e.g.* Ph_3P/CCl_4, Me_3SiCl, $(PhO)_3P/PhCH_2Cl$, *etc.*). Equally diversified and numerous procedures are elaborated for other isohypsic transformations of alcohols. The necessity to have all these seemingly similar methods, however, is justified by the needs of synthesis. In fact, thanks to this arsenal of available tools, practically any of the hydroxyl group conversions of the type $ROH \rightarrow ROAc$, RHal, or $ROSO_2R'$ can be efficiently accomplished regardless of the presence of various complicating factors in a given structural context.

Especially important are the transformations of alcohols into alkyl halides or sulfonates. The enhanced leaving ability of the latter derivatives secures a nearly universal applicability of sulfonates as electrophiles for the ultimate transformation of alcohols into oxidation level 1 functionalized derivatives in accordance with a general reaction:

$$ROSO_2R^1 \xrightarrow{Nu} R\text{–}Nu$$

$$R^1 = p\text{-}MeC_6H_4,\ Me,\ CF_3;\ Nu = Hal,\ OR^2,\ OCOR^2,\ SR^2,\ N(R^2)_2,\ N_3,\ NO_2,\ etc.$$

Alkyl halides, as well as sulfonates, besides being useful for the preparation of derivatives as shown above, are widely utilized for the preparation of phosphorus derivatives such as $R\text{–}PPh_3^+Hal^-$, precursors for the generation of Wittig reagents. Alkyl halides are also extremely important as intermediates in the preparation of the most valuable derivatives of oxidation level 0, the organometallic compounds. The required reduction step is usually carried out *via* direct reaction of the alkyl halides with an active metal such as lithium or magnesium.

Isohypsic reactions of alkenes, like electrophilic additions of H_2O or HX, represent a conventional pathway for the preparation of alcohols and alkyl halides from alkenes. The scope of their application was originally limited as unsymmetrical alkenes (*e.g.* **125**) gave product mixtures composed of both Markovnikov (**M**) adducts and anti-Markovnikov (**aM**) adducts. As was already mentioned above (see Scheme 2.10), an efficient and general method for the conversion of alkenes into alcohols or ethers **126** (Scheme 2.47), with a nearly complete **M** selectivity, was elaborated using mercury salts as electrophiles in conjunction with the reduction of the formed adducts. It is also possible to convert the same alkenes into anti-Markovnikov alcohols **127**, using a different set of reactions, namely hydroboration of the double bond followed by oxidation of the intermediate alkylborane with hydrogen peroxide. A convenient method for the selective transformation of **125** into **aM** adduct **128** then implies the utilization of homolytic addition of HBr or, alternatively, a hydroboration–bromination sequence.

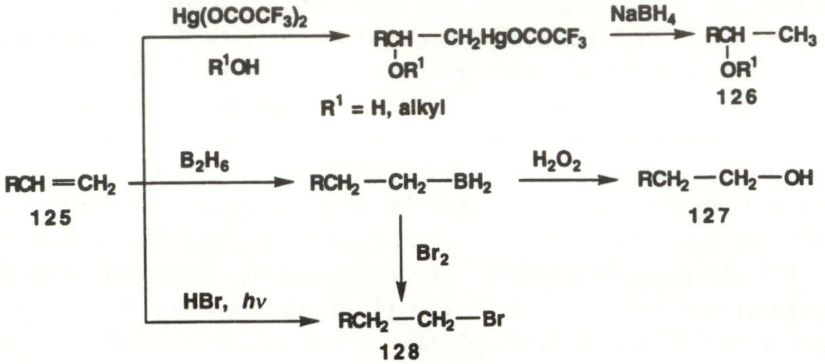

$RCH{=}CH_2$ (**125**) $\xrightarrow[R^1OH]{Hg(OCOCF_3)_2}$ $RCH(OR^1){-}CH_2HgOCOCF_3$ $\xrightarrow{NaBH_4}$ $RCH(OR^1){-}CH_3$ (**126**); R^1 = H, alkyl

$RCH{=}CH_2$ (**125**) $\xrightarrow{B_2H_6}$ $RCH_2{-}CH_2{-}BH_2$ $\xrightarrow{H_2O_2}$ $RCH_2{-}CH_2{-}OH$ (**127**)

$RCH_2{-}CH_2{-}BH_2$ $\xrightarrow{Br_2}$ $RCH_2{-}CH_2{-}Br$ (**128**)

$RCH{=}CH_2$ (**125**) $\xrightarrow{HBr,\ h\nu}$ $RCH_2{-}CH_2{-}Br$ (**128**)

Scheme 2.47

As was amply demonstrated in the preceding sections of this chapter, numerous **C–C** bond-forming reactions are applicable for the preparation of products with a terminal double bond. Thus the sequence (i) introduction of the terminal alkene moiety, (ii) isohypsic double bond transformation leading to the derivatives like **127** or **128**, and (iii) the **C–C** bond-forming step, may be considered as a reliable operation for carbon chain elongation.

The reverse reactions, such as the elimination of HX or H_2O leading to the formation of alkenes, are also feasible and a set of methods is available to carry out these transformations. Here again the main limitations are due to the non-selectivity of the reaction in unsymmetrical systems. In many cases, however, this problem can be alleviated by the utilization of appropriately tuned conditions.

In view of myriad of options available for the interconversions of functional groups at oxidation level 1, one can safely consider all of these functions to be synthetically equivalent. In essence, this means that if it is necessary to introduce a certain functional group into a given structure, the task can be considered achievable if a constructive reaction chosen to create the **C–C** bond leads to the formation of a double bond or hydroxyl group at the desired location.

At oxidation level 2 we will first examine the carbonyl compounds and alkynes, which are readily formed either as a result of reactions utilized for creating **C–C** bonds or with the help of functional groups transformations.

Perhaps the most notable type of isohypsic transformation of carbonyl compounds and alkynes is their conversion into the synthetic equivalents of carbanions according to Scheme 2.48. We have already seen the pronounced role played by these carbanions and their covalent equivalents in constructing

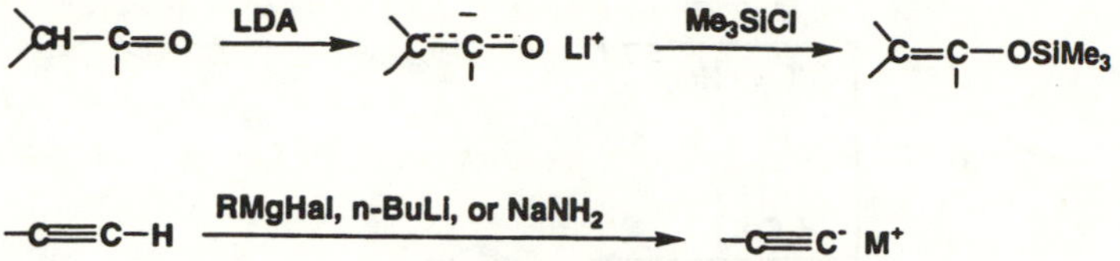

$$\gt CH-\overset{|}{C}=O \xrightarrow{\text{LDA}} \gt C\overset{-}{\cdots}\overset{|}{C}\cdots O\ Li^+ \xrightarrow{\text{Me}_3\text{SiCl}} \gt C=\overset{|}{C}-OSiMe_3$$

$$-C\equiv C-H \xrightarrow{\text{RMgHal, n-BuLi, or NaNH}_2} -C\equiv C^-\ M^+$$

Scheme 2.48

C–C bonds. It needs only to be emphasized that, in addition to the silyl enol ethers shown in the scheme, modern organic synthesis extensively employs enolates of boron, tin, titanium, zirconium, and other elements, as their utilization offers additional opportunities to achieve higher efficiency and exert control over the selectivity of the corresponding reactions.

Important in laboratory and industrial syntheses is the addition of alcohols, carboxylic acids, and hydrogen halides to alkynes, leading to the corresponding vinyl derivatives (Scheme 2.49). These compounds may also be regarded as the derivatives of enols and in many cases they can be more conveniently obtained from carbonyl compounds (Scheme 2.50).

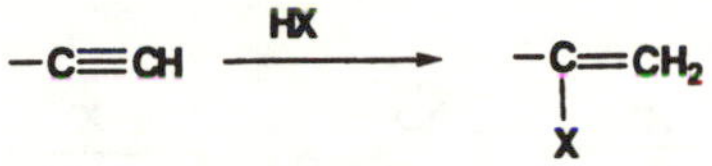

$$-C\equiv CH \xrightarrow{\text{HX}} -\underset{X}{\underset{|}{C}}=CH_2$$

X = OR, OCOR, Hal

Scheme 2.49

$$\gt\underset{H}{C}-\overset{O}{\overset{\|}{C}} \longrightarrow \gt C=C\lt^{OR}, \quad \gt C=C\lt^{OCOR}, \quad \gt C=C\lt^{Hal}$$

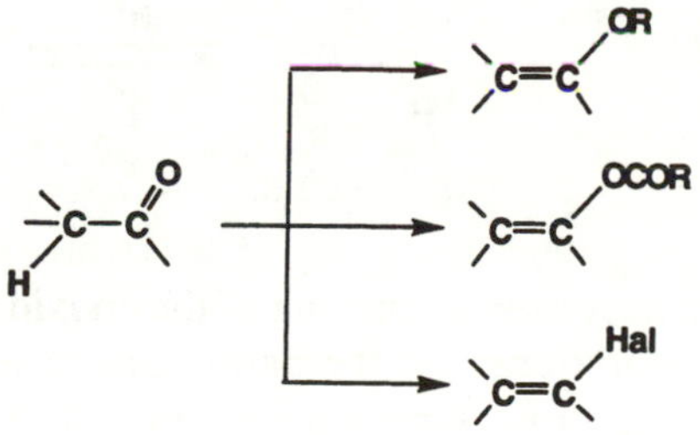

Scheme 2.50

The initial interest in vinyl derivatives was due to their importance in the production of polymers. At present, however, vinyl halides are also widely used in organic syntheses as both precursors for the generation of nucleophilic reagents like vinyllithium or -magnesium derivatives, and as formal equivalents of the vinyl cations in couplings with various organocuprates (Scheme 2.51).

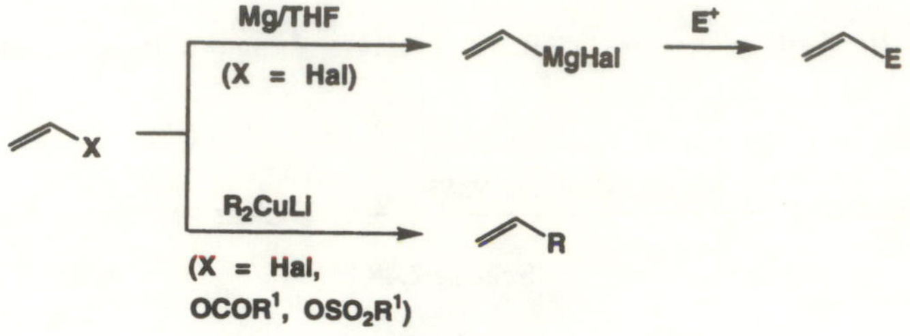

Scheme 2.51

Derivatives containing two functional groups of oxidation level 1 can also be assigned to oxidation level 2. When such functional groups are separated in the molecule by a sufficient distance, then each one can be considered separately as a monofunctional derivative of level 1. A different situation arises, though, when two such functionalities are on adjacent carbons and in essence form a single functional group. Epoxides (oxiranes), 1,2-disubstituted (vicinal) alkane derivatives, and allylic compounds are typical in this regard. Epoxides and 1,2-difunctional alkane derivatives are closely related compounds whose interconversions are achieved easily with the help of standard isohypsic substitution reactions, as shown in Scheme 2.52.

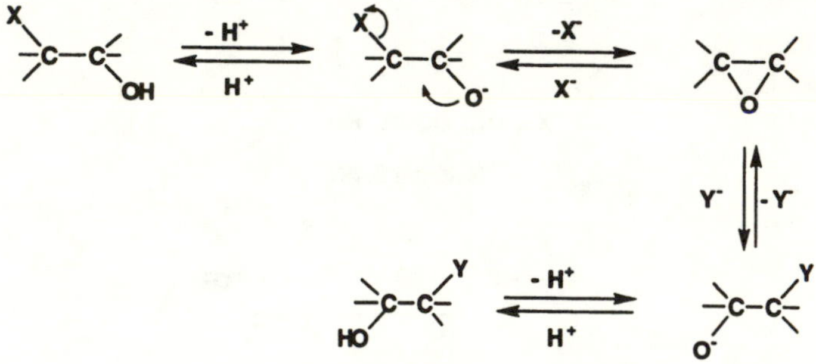

Scheme 2.52

The direction of the nucleophilic opening of the epoxide can be controlled by varying the reaction conditions and the nature of the nucleophile. Therefore many permutations of 1,2-substituents can be achieved with the intermediacy of epoxides, as shown in Scheme 2.52. The opening of the epoxide ring usually involves a predominant or exclusive attack of the reagent from the side opposite to the oxygen bridge, and thus the steric outcome of the process is efficiently controlled.

The basic route to the synthesis of epoxides and, in general, vicinal bifunctional derivatives is a non-isohypsic (oxidative) transformation of alkenes (to be discussed below). Epoxides can also be formed directly as a result of certain **C–C** bond-forming reactions such as the Darzens reaction (modification of a classical aldol-like reaction with α-chloro esters **130** as a methylene component) or addition of a sulfur ylide, dimethylsulfonium methylide **131**, to carbonyl compounds (Scheme 2.53). Both these methods are based on the ease of intramolecular nucleophilic substitution in the intermediates **130a** or **131a**, which are formed as a result of the initial nucleophilic addition to the carbonyl group.

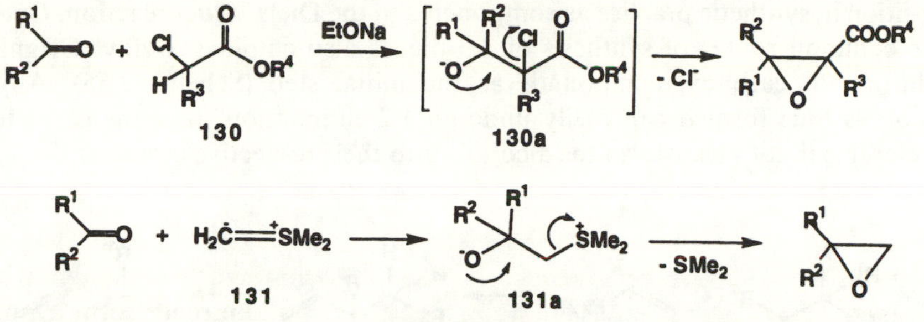

Scheme 2.53

Among isohypsic transformations of oxidation level 2 functional groups, the most important are those which are carried out with allylic compounds **132a** and **132b** (Scheme 2.54). Nucleophilic substitution of the group **X** in these systems can occur either directly at **C**-1 or at **C**-3, with a migration of the double bond (allylic rearrangement). In many cases, either of these options can be accomplished selectively with the proper choice of reaction conditions. Ease of substitution in these system allows one to treat isomers **132a** and **132b** as synthetically equivalent reagents.

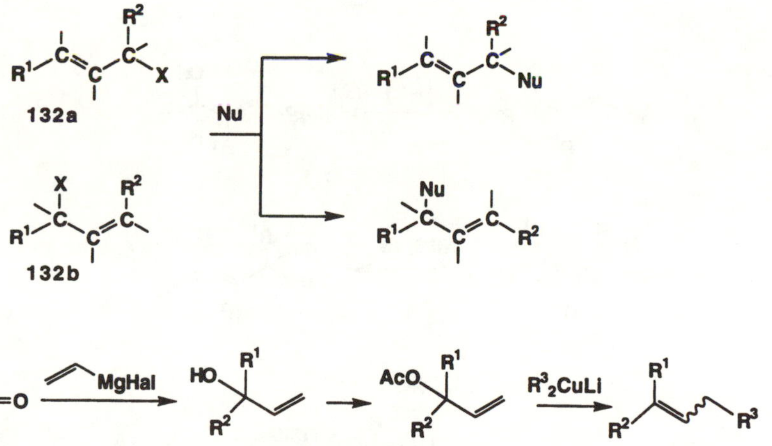

Scheme 2.54

The ability of tertiary allylic compounds to undergo a nucleophilic substitution with an allylic rearrangement allows their utilization as electrophiles in a general chain-lengthening protocol involving (i) a conversion of a carbonyl compound into an allylic alcohol (by reaction with vinylmagnesium bromide) and (ii) a coupling of the respective esters with lithium alkylcuprates (see Scheme 2.54).

Another isohypsic transformation of special significance involves elimination of **H–X** elements from allylic derivatives to form 1,3-dienes. Besides being extremely important compounds as monomers, 1,3-dienes occupy a unique position in synthetic practice as components in the Diels–Alder reaction. One of the common routes of synthesis of 1,3-dienes also employs a vinyl Grignard addition to carbonyl compounds as the initial step (Scheme 2.55). Allylic alcohols thus formed can easily undergo 1,2-elimination (in some cases it is preferable first to transform the alcohols into their respective acetates).

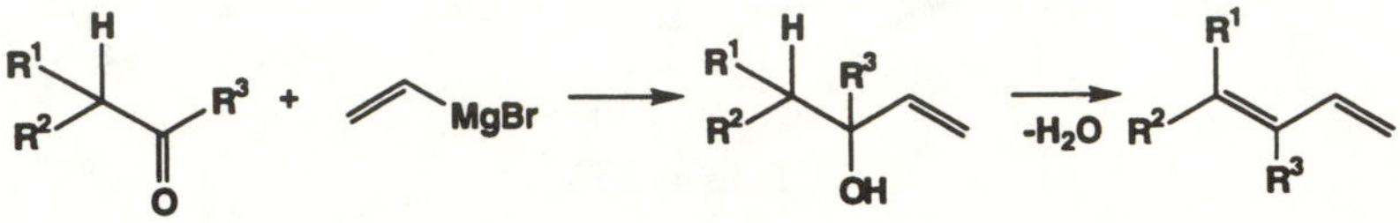

MgBr -H_2O

Scheme 2.55

Beside the Grignard and other **C–C** bond-forming reactions, a number of functional group transformations may also serve as an entry into allylic systems. Some of them, namely the reduction of α,β-unsaturated carbonyl compounds (products of crotonic condensation), halogenation of alkenes at the allylic position with *N*-bromosuccinimide (NBS) and epoxide isomerization,[19d] are shown in Scheme 2.56.

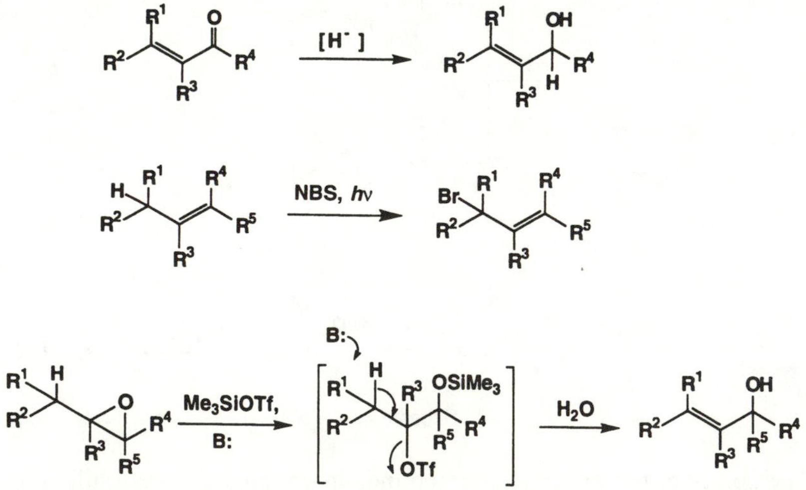

[H^-]

NBS, $h\nu$

Me_3SiOTf, B:

H_2O

Scheme 2.56

At oxidation level 3, acid chlorides occupy a key position, since they may serve as a nearly universal substrate for an isohypsic transformation into any kind of carboxylic acid derivative. Acid halides are electrophiles that are synthetically equivalent to acyl cations (**RCO$^+$**). In this capacity they are used for the synthesis of such important compounds as esters, amides (and hence, nitriles), thioesters, *etc.* (see Scheme 2.57), and for the formation of **C–C** bonds in the Friedel–Crafts reaction (see above). Acid chlorides may readily lose HCl upon treatment with triethylamine. This isohypsic conversion leads to ketenes, important reagents widely employed in [2 + 2] cycloadditions, as we will see later.

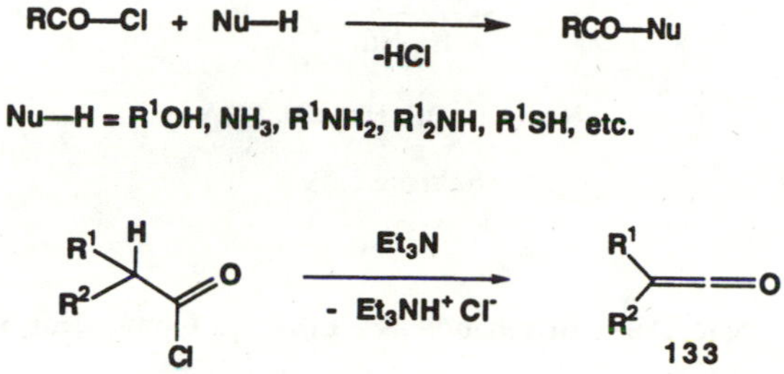

$$RCO{-}Cl + Nu{-}H \xrightarrow[-HCl]{} RCO{-}Nu$$

$Nu{-}H = R^1OH, NH_3, R^1NH_2, R^1_2NH, R^1SH$, etc.

$$R^1R^2CH{-}COCl \xrightarrow[-\ Et_3NH^+\ Cl^-]{Et_3N} R^1R^2C{=}C{=}O\ \ \mathbf{133}$$

Scheme 2.57

Ethynyl carbinols (propargylic alcohols) such as **134** (Scheme 2.58) represent another important group of oxidation level 3 compounds. Their preparation involves nucleophilic addition of acetylides to the carbonyl group, a reaction that is nearly universal in its scope. Elimination of water from **134** followed by hydration of the triple bond is used as a convenient protocol for the preparation of various conjugated enones **135**. Easily prepared *O*-acylated derivatives are extremely useful electrophiles in reactions with organocuprates, which proceed with propargyl–allenyl rearrangements to furnish allene derivatives **136**.

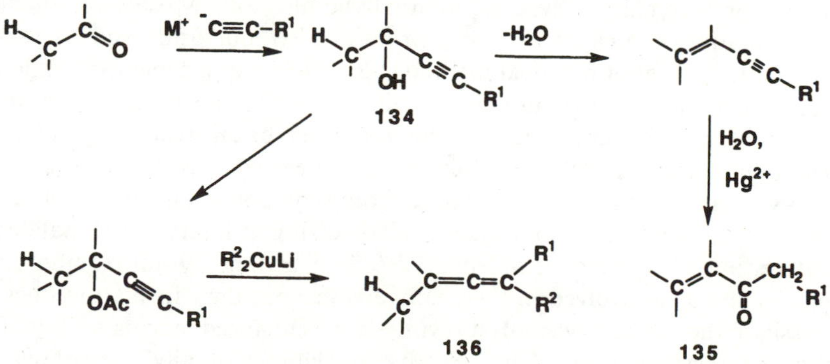

$M^+\ {}^-C{\equiv}C{-}R^1$; $-H_2O$; H_2O, Hg^{2+}; R^2_2CuLi

134 **135** **136**

Scheme 2.58

α,β-Unsaturated aldehydes and ketones are among the most notable representatives of polyfunctional derivatives of oxidation level 3. Their synthetic significance as carbon electrophiles or as dienophiles in the Diels–Alder reaction has already been referred to in Section 2.3. Isohypsic functional group transformations of these derivatives are based upon their ability to add various nucleophiles, **NuH**, at the double bond (Michael addition). This reaction is the most efficient way to prepare a variety of β-substituted functional derivatives of carbonyl compounds. For the majority of cases it is also possible to accomplish the reverse transformation *via* the elimination of **NuH** (Scheme 2.59).

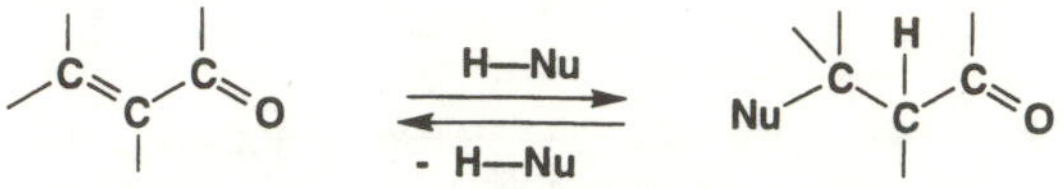

H—Nu

- H—Nu

(Nu = OH, OR, Hal, SR, NR_2)

Scheme 2.59

2.8.3 Non-isohypsic Transformations as Pathways Connecting Different Oxidation Levels

The oxidation of alcohols to carbonyl compounds or carboxylic acids (and the corresponding reverse reductive transformations) are among the most significant routes for transitions between oxidation levels. A tremendous amount of effort has been spent to develop infallible methods that accomplish these conversions. These efforts have not been in vain and it is now possible to carry out virtually any of these conversions selectively even when complicating factors such as the lability of substrates or products, the presence of other reactive groups, and stereochemical or other issuess are involved (Scheme 2.60).

The oxidation of alcohols seems to be an especially popular exercise for all those who are interested in the development of new methods. More than 140 procedures for the oxidation of alcohols are mentioned in Larock's monograph,[19e] including several dozen marked specifically for the oxidation of primary or secondary, allylic or homoallylic alcohols. Most typically these transformations are carried out by chromium(VI) anhydride and its various complexes.[19f] A set of practical methods also widely applicable for a mild and efficient oxidation of alcohols employs dimethyl sulfoxide as an oxidant in the presence of various Lewis acids.[19g] The initial step of this reaction involves the formation of a dimethylalkoxysulfonium ion intermediate, which subsequently reacts with a base to give the required carbonyl compound and dimethyl sulfide. Dimethyl sulfoxide is also capable of oxidizing primary alkyl halides or tosylates into the respective aldehydes.[19g] Similar dimethylsulfoxonium intermediates are also involved in these conversions, but their formation does not necessitate the use of any catalyst owing to the enhanced leaving ability of the halide or tosylate group. The selective oxidations of allylic alcohols into aldehydes are often carried out with MnO_2.[19f] A variation of this method,

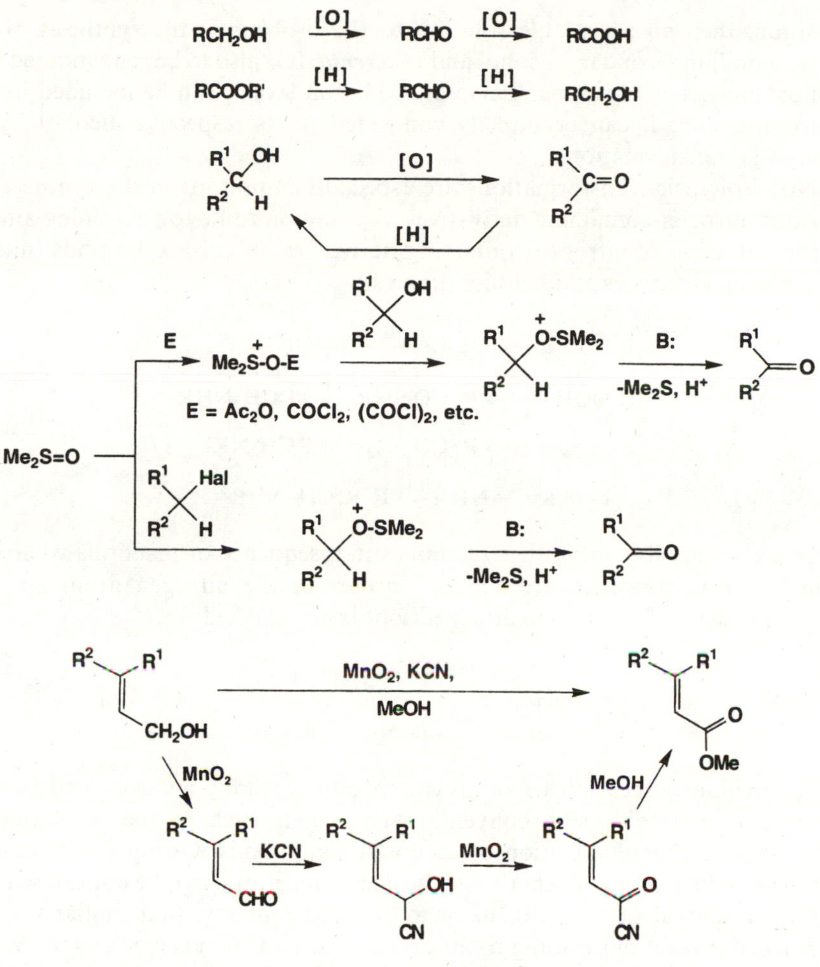

$RCH_2OH \xrightarrow{[O]} RCHO \xrightarrow{[O]} RCOOH$

$RCOOR' \xrightarrow{[H]} RCHO \xrightarrow{[H]} RCH_2OH$

$R^1R^2CH(OH) \underset{[H]}{\overset{[O]}{\rightleftharpoons}} R^1R^2C{=}O$

$Me_2S{=}O \xrightarrow{E} Me_2\overset{+}{S}\text{-}O\text{-}E \xrightarrow{R^1R^2CH(OH)} R^1R^2CH\text{-}\overset{+}{O}\text{-}SMe_2 \xrightarrow[-Me_2S,\ H^+]{B:} R^1R^2C{=}O$

$E = Ac_2O, COCl_2, (COCl)_2$, etc.

$Me_2S{=}O \xrightarrow{R^1R^2CH(Hal)} R^1R^2CH\text{-}\overset{+}{O}\text{-}SMe_2 \xrightarrow[-Me_2S,\ H^+]{B:} R^1R^2C{=}O$

$R^2R^1C{=}CH\text{-}CH_2OH \xrightarrow{MnO_2,\ KCN,\ MeOH} R^2R^1C{=}CH\text{-}C(O)OMe$

$R^2R^1C{=}CH\text{-}CH_2OH \xrightarrow{MnO_2} R^2R^1C{=}CH\text{-}CHO \xrightarrow{KCN} R^2R^1C{=}CH\text{-}CH(OH)CN \xrightarrow{MnO_2} R^2R^1C{=}CH\text{-}C(O)CN \xrightarrow{MeOH} R^2R^1C{=}CH\text{-}C(O)OMe$

Scheme 2.60

suggested by Corey,[19h] was specifically adjusted for a one-pot conversion of allylic alcohols *via* intermediately formed cyanohydrins into methyl esters of the respective acids (see Scheme 2.60). Aldehydes themselves can be easily oxidized to carboxylic acids by a variety of inorganic reagents, such as O_2, Ag_2O, $KMnO_4$, $NaClO_2$, *etc.*

The reverse conversions of carbonyl compounds into alcohols are typically achieved with the utilization of complex hydrides such as $LiAlH_4$, $NaBH_4$, *etc.* Activity and selectivity of these reagents can be attenuated within wide limits by varying the nature of the hydride donor. Thus, one can finely tune the selectivity to a particular structural pattern (see discussion in the next section).

Thanks to reliable methods for these conversions, all oxygen-containing functionalities of different oxidation levels can be regarded as synthetically equivalent. In other words, if the target molecule bears a keto group at a certain position, then an acceptable solution to the problem is the synthesis of the corresponding secondary alcohol and *vice versa*. It is also to be remembered that almost any other functional group of oxidation level 1 can be included in this reasoning since it can be directly connected to its respective alcohol by an isohypsic transformation.

Non-isohypsic transformations are especially important in the syntheses of various nitrogen-containing derivatives. A common route for obtaining amines is the reduction of nitrogen-containing derivatives of carboxylic acids (nitriles or amides), aldehydes and ketones (imines):

$$\mathrm{RCOOH} \rightarrow \mathrm{R{-}CN} \xrightarrow{[H]} \mathrm{RCH_2NH_2}$$

$$\mathrm{RCOOH} \rightarrow \mathrm{R{-}CONHR'} \xrightarrow{[H]} \mathrm{RCH_2NHR'}$$

$$\mathrm{RCOOH} \rightarrow \mathrm{R{-}CONR'_2} \xrightarrow{[H]} \mathrm{RCH_2NR'_2}$$

$$\mathrm{R^1R^2C{=}NR^3} \xrightarrow{[H]} \mathrm{R^1R^2CH{-}NHR^3}$$

It is also possible to synthesize amines in a sequence of reactions where the non-isohypsic conversion (reduction) occurs at the nitrogen atom and the oxidation state of the carbon attached to it is not affected:

$$\mathrm{R{-}Hal} \xrightarrow{NO_2^-} \mathrm{R{-}NO_2} \xrightarrow{[H]} \mathrm{RNH_2}$$

$$\mathrm{R{-}Hal} \xrightarrow{N_3^-} \mathrm{R{-}N_3} \xrightarrow{[H]} \mathrm{RNH_2}$$

As previously mentioned, an alkyne moiety is readily incorporated into an assembled molecule using conventional methods such as the alkylation of acetylides or the ethynylation of carbonyl compounds. Owing to the ease of stepwise reductive transformations, the acetylenic group can be considered as a synthetic equivalent to an alkene or to an alkane moiety. In a similar way, all constructive reactions leading to the creation of **C=C** bonds (such as the Wittig reaction) are applicable for the preparation of saturated systems as a plethora of reliable methods exists that reduce the double bond.

Among the oxidative transformations of alkenes, the conversion to epoxides is especially significant and again is evidenced by an impressive number (more than 80) of recommended methods[19i]. On an industrial scale, this oxidation is carried out by oxygen in the presence of metal catalysts. Laboratory procedures typically involve the utilization of peroxy acids, such as *m*-chloroperbenzoic acid. Epoxides can be easily cleaved into their respective 1,2-diols. Since the latter reaction proceeds with an inversion of configuration at the epoxide ring carbon atom, the net outcome of alkene → 1,2-diol conversion corresponds to *anti* addition. *Syn* addition can also be efficiently accomplished but it requires the use of inorganic oxidants such as $KMnO_4$ or OsO_4. This difference in the steric outcome of the formally similar process is nicely accounted for in terms of the mechanisms shown in Scheme 2.61.

Other oxidizing systems have been designed to achieve highly specific oxidation of the double bond within a particular structural context. Basic

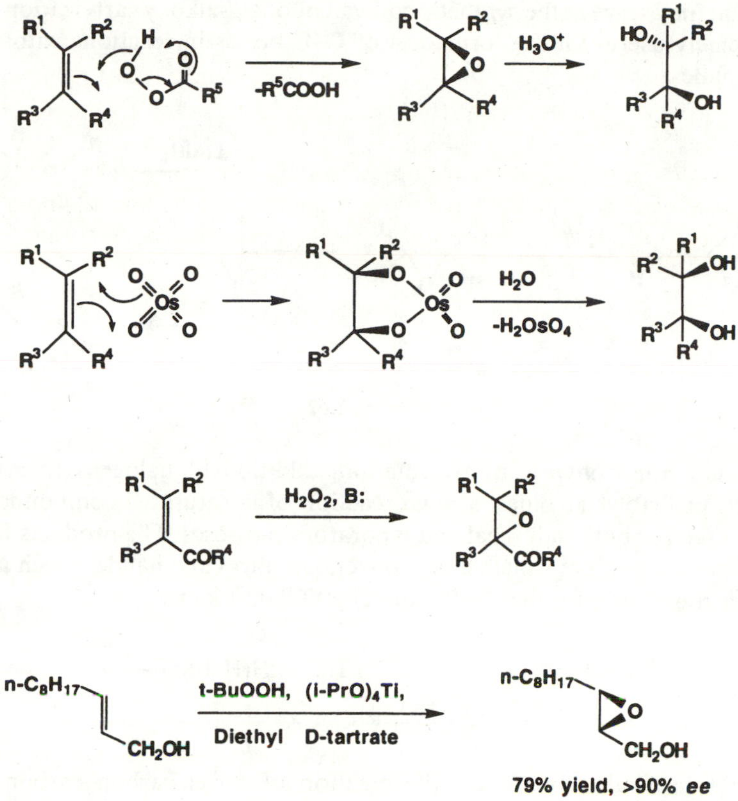

Scheme 2.61

hydrogen peroxide is an especially convenient and selective oxidant for unsaturated compounds having an electron withdrawing group conjugated with a double bond. Immense efforts were spent to elaborate methods for enantioselective epoxidation of double bonds. One of the most significant achievements in this field is due to the studies of Sharpless's group. It was found that the system *tert*-butyl hydroperoxide, titanium tetraisopropoxide, and D- or L-diethyl tartrate is capable of oxidizing allylic alcohols into the respective epoxides with exceptionally high stereoselectivity and can furnish optically active products in 90% or better enantiomeric excess.[19j]

The conversion of alkenes into epoxides is important not only because it is one of the most reliable routes leading from oxidation level 1 to level 2, but also because reactions of non-symmetrical epoxides with nucleophiles invariably proceed as an attack at the less substituted carbon with inversion of configuration. Thus, hydride reduction of epoxides represents an additional option for the preparation of alcohols (Scheme 2.62), especially valuable for the synthesis of optically pure isomers from epoxides obtained by the Sharpless oxidation. It is also of merit that as a result of alkene–epoxide conversion, a nucleophilic moiety (double bond) is transformed into an electrophilic epoxy ring. The latter fragment may serve as the synthetic equivalent of a β-alkoxy carbocation, which is extremely useful for the formation of **C–C** bonds in reactions with carbon nucleophiles.

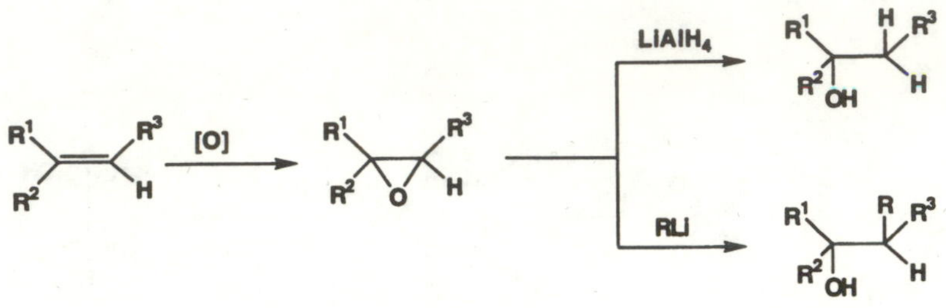

Scheme 2.62

Another non-isohypsic transformation, addition of halogens to a double bond, is probably the oldest known reaction of unsaturated compounds. It is widely used for both industrial and laboratory purposes. The products formed, 1,2-dihaloalkanes, are valuable for conversion into vinyl halides (such as vinyl chloride monomers for the production of PVC) or alkynes:

$$R^1CH{=}CHR^2 \xrightarrow{Hal_2} R^1CH(Hal){-}CH(Hal)R^2 \xrightarrow{-HHal}$$
$$R^1CH{=}C(Hal)R^2 \xrightarrow{-HHal} R^1C{\equiv}CR^2$$

As we have already shown, the creation of novel carbon–carbon bonds usually results in the formation of functionalized derivatives (the Wurtz coupling and Friedel–Crafts alkylation are probably the only exceptions). That is why a set of special reductive methods was devised to remove residual functionality that is unwanted in the final structure. The well-known hydrogenation of alkenes and alkynes belongs to this group of non-isohypsic transformations. Several other pathways available for the reductive removal of various functions will briefly be considered below.

Direct reductive elimination of the hydroxyl group can be achieved readily only for tertiary and benzylic alcohols, which are especially liable to form carbocationic intermediates in acidic media. These intermediates are able to abstract hydrogen from such hydride donors as triethylsilane. This procedure, ionic hydrogenation, was elaborated into a preparatively useful protocol:[19k]

$$tert\text{-}R{-}OH \xrightarrow{H^+} tert\text{-}R^+ \xrightarrow{Et_3Si{-}H} tert\text{-}R{-}H$$

Primary and secondary alcohols should first be converted into halides or sulfonates, followed with treatment by complex hydrides such as $LiAlH_4$ in order to achieve hydrogenolysis:[19l]

$$R{-}OH \longrightarrow R{-}Hal(OTos) \xrightarrow{[H^-]} R{-}H$$

The method of choice for the hydrogenolysis of primary allylic alcohols involves treatment with a pyridine–SO_3 complex (to form a sulfoester) followed by reduction with $LiAlH_4$.[19m] This method eliminates complications encountered for the reactions in allylic systems, such as double bond migration or stereoisomerization:

$$>C{=}C\text{–}CH_2\text{–}OH \xrightarrow{C_5N_5N\cdot SO_3} >C{=}C\text{–}CH_2\text{–}OSO_3H \xrightarrow{LiAlH_4} >C{=}C\text{–}CH_2\text{–}H$$

Sulfur-containing functions are especially prone to undergo reductive removal. Hydrogenolysis of the thio group proceeds easily in the presence of Raney nickel and this procedure is ideally suited for the reduction of functional groups like thiols, sulfides, or thioacetals. The sulfonyl group can be reductively removed under the action of agents like Na in liquid ammonia:

$$R\text{–}SH \xrightarrow{\text{(Raney Ni)}} R\text{–}H$$

$$R\text{–}S\text{–}R' \xrightarrow{\text{(Raney Ni)}} R\text{–}H + R'\text{–}H$$

$$R^1R^2C(SR)_2 \xrightarrow{\text{(Raney Ni)}} R^1R^2CH_2$$

$$R^1SO_2R^2 \xrightarrow{Na/NH_3} R^1H + R^2H$$

These transformations are of special importance since various sulfur derivatives are widely utilized as synthetically useful precursors, owing to the ability of sulfur substituents to stabilize both carbanionic and carbocationic adjacent centers.

Direct transition from level 2 to level 0 can be achieved by way of the Wolff–Kishner reaction (treatment of the respective hydrazones with alkali), a classical pathway for the reduction of carbonyl compounds. At the same time, a direct conversion of aldehydes and ketones into alkenes is also feasible *via* reductive cleavage of their tosylhydrazones under the action of MeLi, the Shapiro reaction (Scheme 2.63).[19n]

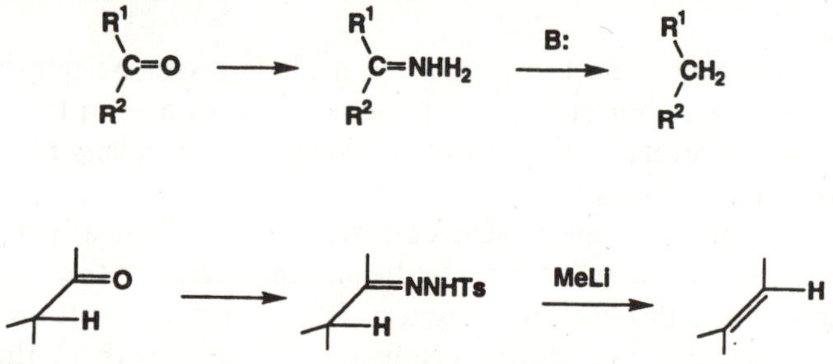

$$R^1R^2C{=}O \longrightarrow R^1R^2C{=}NNH_2 \xrightarrow{B:} R^1R^2CH_2$$

$$\text{(H–C–C=O)} \longrightarrow \text{(H–C–C=NNHTs)} \xrightarrow{MeLi} \text{(C=C–H)}$$

Scheme 2.63

In spite of the abundance of existing methods for functional group transformations, investigations in this area are by no means withering. Efforts are still devoted to the refinement of already known methods to achieve a broader scope, higher efficiency, and greater selectivity of transformations. At the same time, special emphasis is given to the search for entirely novel methods, which may present the opportunity to perform unusual transitions, both within the same oxidation level and between the levels. The benefits gained from pursuits in this area can be illustrated by a set of randomly chosen representative examples:

$$\textbf{(a)} \quad RCOCl \xrightarrow{(PNCl_2)_n} RCN + POCl_3$$

$$\textbf{(b)} \quad RNH_2 \xrightarrow{[O]} RNO_2$$

$$\textbf{(c)} \quad R^1R^2CHNO_2 \xrightarrow[2.\ H_2SO_4/MeOH]{1.\ MeONa/MeOH} R^1R^2C(OMe)_2$$

$$\textbf{(d)} \quad R^1R^2C{=}O \xrightarrow{HNR_2,\ NaBH(OAc)_3} R^1R^2NR_2$$

$$\textbf{(e)} \quad R{-}OAc \xrightarrow{[H],\ Ni\ boride} R{-}H$$

$$\textbf{(f)} \quad R^1{-}O{-}R^2 \xrightarrow{Me_3SiI} R^1{-}I + R^2{-}OSiMe_3$$

The synthetic importance of reaction **(a)**[20a] comes from the fact that it reduces to one step the pathway for conversion of an acid chloride into a nitrile (instead of the classical and rather inconvenient two-step route *via* an acid amide). Reaction **(b)**[20b] is an example of a new transformation for aliphatic amines. Previously, there were no methods available for the direct transformation of an amino into a nitro group and the stepwise procedures were too cumbersome to be of practical use. Transformation of a nitro group into a carbonyl is a well-known reaction. Its modification, shown in reaction **(c)**,[20c] represents a welcome opportunity to obtain a protected carbonyl group as the immediate result of such a transformation. The viability of the sequential reactions **(b)** plus **(c)** enables the employment of a $>CHNH_2$ moiety as a synthetic equivalent to a protected carbonyl group. A one-pot sequence of imine formation and its reduction with sodium triacetoxyborohydride **(d)**[20d] represents a convenient option for the preparation of various amines from readily available aldehydes and ketones.

Catalytic reduction of acetates (tertiary or allylic) with Ni boride (reaction **(e)**)[20e] may serve as a shortcut option to achieve the transition from oxidation level 1 to oxidation level 0. This pathway eliminates the necessity first to prepare sulfur-containing derivatives.

The ether group has never been considered a synthetically manipulable moiety owing to the rather severe conditions required for the release of the hydroxyl group. The discovery of reaction **(f)**[20f] changed this situation dramatically and, at present, the alkoxy group can be dealt with in the synthetic operations as an equivalent of the respective alcohol (the silyl group can be easily hydrolysed) or alkyl iodide.

The tendency to establish equivalency relationships between functional groups, previously treated as unrelated, is one of the major motivations in the development of transformation methods. These pursuits are targeted at the creation of a bank of standard procedures that allow one to perform the mutual interconversion of any functional groups in one or two operations. It should be emphasized that, while the present situation is far from this ideal state, there nevertheless exists an adequate synthetic arsenal for the interconversions of the majority of functional groups. At least one can confidently rely upon the following simple rules in planning a synthesis.

1. *Practically all functional groups, and especially those of the same oxidation level, can be considered synthetically equivalent from the point of view of a retrosynthetic analysis.* If, for example, the target molecule bears a hydroxyl group at a given position, then in the course of a retrosynthetic analysis it is permissible to transform it, say, into a halide or carbonyl moiety or double bond or epoxide, *etc.* All structures that appear as a result of these retrosynthetical transformations can be considered as likely subgoals and the synthesis of any structure from this set would represent a solution to the problem. Obviously, the greater the allowable diversity of the functions present in these intermediates, the greater the choice of the preparative options suitable for the synthesis of the required carbon framework.

2. *Since any functionality can be removed to produce a hydrocarbon fragment, functional group(s) can be inserted at almost any position in the carbon framework in the course of a retrosynthetic analysis in order to identify the appropriate synthetical transformation to create the* ***C–C*** *bond in this fragment.*

A thoughtful reader would have noticed that, while plenty of methods are available for the reductive transformation of functionalized moieties into the parent saturated fragments, we have not referred to the reverse synthetic transformations, namely oxidative transformations of the **C–H** bond in hydrocarbons. This is not a fortuitous omission. The point is that the introduction of functional substituents in an alkane fragment (in a real sequence, not in the course of retrosynthetic analysis) is a problem of formidable complexity. The nature of the difficulty is not the lack of appropriate reactions – they do exist, like the classical homolytic processes, chlorination, nitration, or oxidation. However, as is typical for organic molecules, there are many **C–H** bonds capable of participating in these reactions in an indiscriminate fashion and the result is a problem of selective functionalization at a chosen site of the saturated hydrocarbon. At the same time, it is comparatively easy to introduce, selectively, an additional functionality at the saturated center, provided some function is already present in the molecule. Examples of this type of non-isohypsic (oxidative) transformation are given by the allylic oxidation of alkenes by SeO_2 into respective α,β-unsaturated aldehydes, or α-bromination of ketones or carboxylic acids, as well as allylic bromination of alkenes with NBS (Scheme 2.64).

It is also appropriate to mention that considerable progress has been achieved during the last decades in solving the problems of selective functionalization of **C–H** bonds in the absence of nearby activating groups. Several reactions can be utilized for this purpose but, in general, selectivity is ensured by using an intramolecular process (rather than intermolecular) and a specific design of

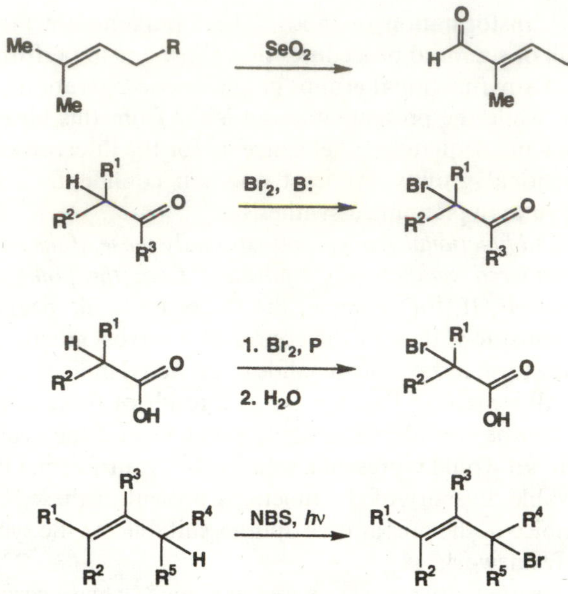

Scheme 2.64

auxiliary elements to secure the proximity of the reacting functionality with a given **C–H** bond at the saturated carbon center. These methods will be discussed in the next chapter.

Until now, we have portrayed functional group transformations as secondary tools, somehow associated with the more important task of creating a carbon skeleton. There does exist, however, a wide class of tasks in which the interconversions of functional groups constitute the very essence of a synthetic problem.

2.9 FUNCTIONAL GROUP INTERCONVERSIONS AS STRATEGIC TOOLS IN A TOTAL SYNTHESIS

In certain cases it is possible to plan the synthesis of a compound from already available precursors which contain the required carbon skeleton and only a change in the nature and location of the functional groups is required to arrive at the target structure. A great deal of classical organic synthesis developed along this line. A good example of this is the first synthesis of cyclooctatetraene **137** by Wilstatter in 1911.[21a] As a starting compound for this synthesis, the natural alkaloid pseudopellterin **138** (isolated from the roots of pomegranate trees) was chosen for its eight-membered carbocyclic framework (Scheme 2.65). The challenge to Wilstatter was to use the available functionality in the ring to introduce four double bonds. He accomplished this by a sequence of conversions: a reduction of the carbonyl group; dehydration; and an iterative series of simple reactions, such as exhaustive methylation, Hofmann elimination, bromine addition, *etc*., which eventually led to the target molecule, **137**. All ten steps of this remarkable synthesis were actually functional group transformations.

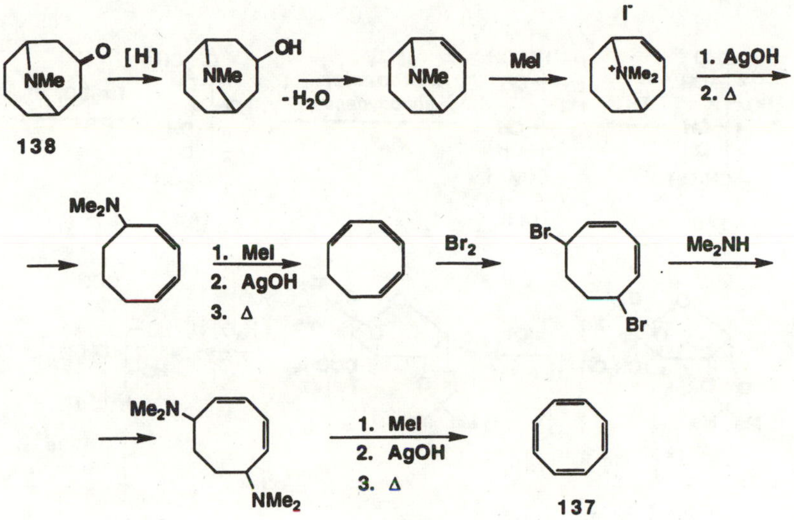

Scheme 2.65

Another area largely based on functional group transformations is the synthetic chemistry of carbohydrates. There are usually two synthetic goals in this field. The first is the synthesis of natural monosaccharides and their analogs. The second is the assemblage of oligosaccharides and polysaccharides from the monosaccharides. Natural monosaccharides vary considerably in their structure but the main differences between them lie in the location and nature of the functional groups and the configuration of the chiral carbons. The majority of monosaccharides have similar, if not identical, carbon skeletons consisting of C_5 or C_6 non-branched carbon chains. Many natural monosaccharides such as D-glucose or L-arabinose are readily available. For their conversion into other monosaccharides it is usually sufficient to change the character of just a few functional groups. It might be necessary, for example, to transform a hydroxyl group into an amino group or a primary alcohol into a carboxyl group, or change the configuration of one or more chiral carbons. There is no need to create a carbon skeleton from scratch or to repeat steps Mother Nature has already done for us in the course of biosyntheses.

For illustration purposes, let us examine the industrial synthesis of ascorbic acid **139** from D-glucose **140** (Scheme 2.66). The catalytic hydrogenation of **140** produces the hexa-atomic alcohol D-sorbitol **141**. The latter is subjected to microbial oxidation which selectively introduces a keto group at position 2 (formerly the C-5 position in **140**). The resulting isomer of glucose, L-sorbose **142**, is converted into the protected derivative **143**, which contains only one unprotected alcohol function at C-1 (corresponding to C-6 in the staring glucose structure). This group is readily oxidized to give a carboxyl function. The removal of the protecting groups from the resulting acid **144** leads to the open-chain form of ascorbic acid **139a**, which is converted spontaneously into the enol form of lactone **139**.

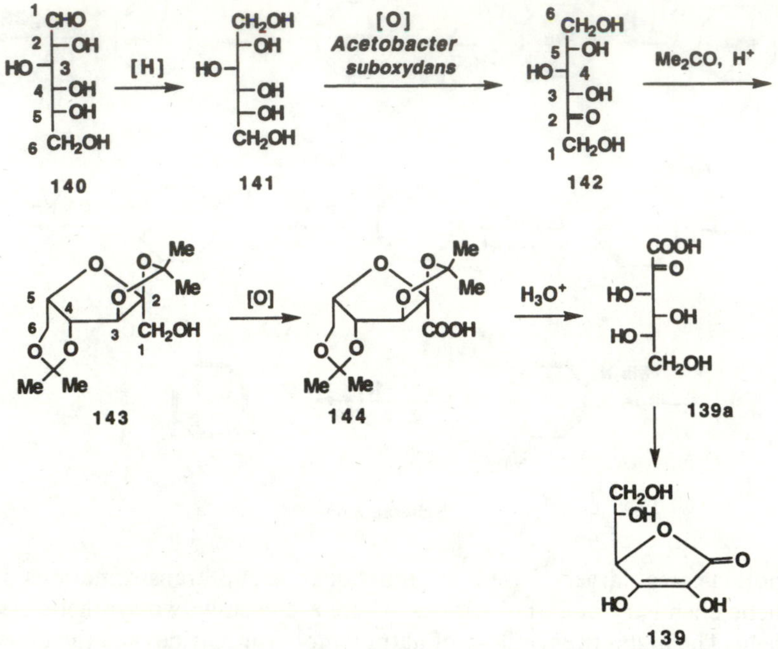

Scheme 2.66

As illustrated, the major steps in the conversion of **140** to **139** correspond to non-isohypsic transformations of functional groups: the reduction of an aldehyde to a primary alcohol, the oxidation of a secondary alcohol to a ketone, and the oxidation of a primary alcohol to a carboxylic acid. The introduction and removal of the isopropylidene protecting groups and the use of the bacterium *Acetobacter suboxydans* (a non-typical oxidizing agent) ensures selectivity in the reactions of the polyfunctional intermediate compounds.

Oligo- and polysaccharides are constructed from monosaccharide units connected *via* glycosidic linkages. The key step in the synthesis of these systems is the creation of the glycosidic bond between the individual monosaccharide units. The formal scheme for the creation of such a bond is shown for the disaccharide lactose **145** (known as milk sugar) from the monosaccharide precursors D-galactose **146** and D-glucose **140** (Scheme 2.67). Here again there is no need to worry about the creation of a new **C–C** bond. Our only concern is with the formation of the **O**-glycosidic bond between the two sugar moieties. From the point of view of general organic chemistry, this is an elementary functional group transformation. It is very far from being trivial, however, when a *glycosidic* bond is considered. The stereoselective formation for this type of bond is difficult and remains a central concern in carbohydrate chemistry. Hundreds of publications, including several monographs, deal exclusively with this subject.[21b]

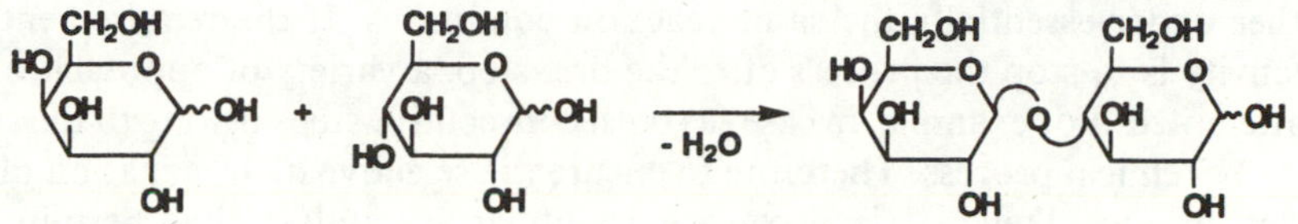

146 140 145 ($-H_2O$)

Scheme 2.67

The synthesis of two other important biopolymers, proteins and nucleic acids, also involves a sequence of functional group transformations. Simple (amidic or phosphodiester) bonds are formed between readily available monomeric units (amino acids or nucleotides). Almost all synthetic efforts in this area are centered around the elaboration of an optimal method to achieve an efficient formation of this bond. Given the complexity of the final structure, this task is never too simple.

It should now be clear that functional group transformations play more than an auxiliary role in synthesis. The importance of these reactions, especially with respect to the chemistry of natural compounds, makes it imperative to have a multitude of diverse and, at times, rather sophisticated methods to effect these often apparently trivial transformations.

PART IV HOW TO CONTROL THE SELECTIVITY OF ORGANIC REACTIONS

2.10 FORMAL CLASSIFICATION OF SELECTIVITY PROBLEMS

The question of the selectivity of a reaction is so critical to organic synthesis that a detailed discussion is most certainly warranted. We will start by examining some general aspects of this problem.

Reliability, for a given synthetic method, implies that it can be employed to achieve the given synthetic transformation efficiently and cleanly with no undesirable conversions occurring under the chosen conditions. Even if this condition is met, however, the problems associated with selectivity are far from being fully solved. Frequently a substrate may contain not just one but several functional groups that are capable of interacting with the same reagent(s). The synthetic task at hand often demands the involvement of only one of them. Furthermore, even the reaction of a single functional group carried out with the help of an otherwise 'clean' reaction may result in the formation of a mixture of products.

The problems related to selectivity are diverse. Therefore, we will only examine a few typical cases to illustrate some of the main facets of this problem. General reasons that cause the non-selectivity of an organic reaction course can be classified in terms of the formal kinetics of the overall process.

Type 1. *Consecutive reactions.* The common feature of these examples (Scheme 2.68) is that the product formed in the first step is capable of reacting further under essentially the same reaction conditions. If the requirement for selectivity is to stop the process after the first step, a variety of approaches can be attempted. For example, in case **(a)** both consecutive steps belong to the same type of chemical process. Therefore to ensure the selective hydrogenation of the alkyne to the alkene, it is necessary to utilize a catalyst that permits the reduction of the triple bond but not the double bond. This requirement is met in Lindlar's catalyst, a palladium metal catalyst adsorbed on a carbonate that is partially deactivated with lead (Pd–$CaCO_3$–PbO).

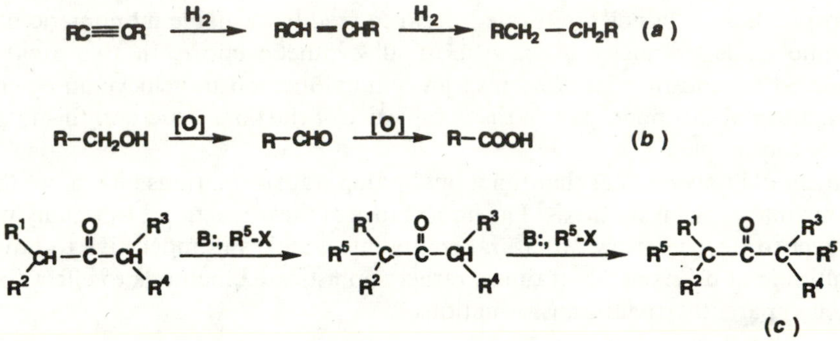

$$RC{\equiv}CR \xrightarrow{H_2} RCH{=}CHR \xrightarrow{H_2} RCH_2{-}CH_2R \quad (a)$$

$$R{-}CH_2OH \xrightarrow{[O]} R{-}CHO \xrightarrow{[O]} R{-}COOH \quad (b)$$

$$R^1R^2CH{-}C(=O){-}CHR^3R^4 \xrightarrow{B:,\ R^5\text{-}X} R^5R^1R^2C{-}C(=O){-}CHR^3R^4 \xrightarrow{B:,\ R^5\text{-}X} R^5R^1R^2C{-}C(=O){-}CR^3R^4R^5 \quad (c)$$

Scheme 2.68

In contrast, the chemistry of the oxidation of a primary alcohol to an aldehyde differs sharply from the oxidation of an aldehyde to a carboxylic acid (case **(b)**). Advantage, in this case, must be taken of the difference in the mechanisms of these steps. Among the reagents which can effectively oxidize alcohols and remain rather inert toward aldehydes are pyridinium chlorochromate (a chromium trioxide–hydrogen chloride complex of pyridine) or dimethyl sulfoxide–Lewis acid.

Ensuring the selective monoalkylation of ketones [case **(c)**] is of special importance in synthetic practice and numerous approaches are elaborated for this purpose. This problem deserves special comment and will be considered later (Section 2.13).

Type 2. *Parallel reactions.* In these examples (Scheme 2.69), a mixture of closely related products may arise owing to the availability of several competing pathways for a given reaction. The challenge here is to direct the reaction exclusively (or at least predominantly) along one specific pathway. In case **(d)** the initial step, attack of $\mathbf{Br^+}$ leading to the formation of a cationoid intermediate, may occur both at C-1 and C-2. The preference of this attack determines the ratio of positional isomers (**147** + **148**):(**149** + **150**). The second step of the reaction is the interaction of the intermediate with the nucleophile

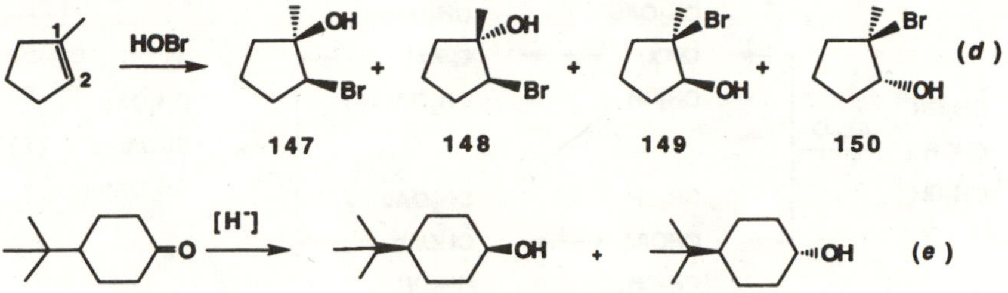

HOBr (*d*)

147 148 149 150

[H⁻] (*e*)

Scheme 2.69

HO$^-$. The orientation of the approach of the nucleophile determines the ratio of *cis* and *trans* isomers in the resulting mixture, (**147** + **149**):(**148** + **150**). In actuality, electrophilic addition of **Br**$^+$ is directed almost exclusively at C-2 and hence products **147** and **148** are formed preferentially. While the ratio of these isomers is very sensitive to the reaction conditions, it is rather difficult to achieve a high stereoselectivity in the reaction and hence this method cannot be recommended for the preparation of the pure stereoisomers **147** or **148**.

In a somewhat related example, case **(e)**, the stereochemistry of the product is determined by the direction of approach of the hydride reagent to the carbonyl group, which can occur either 'from above' or 'from below' the plane of the ring. The steric course of hydride reductions can easily be controlled by a careful choice of the reagent. Related examples will be considered in Section 2.12.

Type 3. *Consecutive-parallel reactions*. In these examples (Scheme 2.70) we have to deal with the problems of the first two types combined. The starting compounds are polyfunctional. An initial reaction can occur at any of the available functions and therefore the reaction is likely to produce isomeric products. Then, as is the case with consecutive reactions, the intact functional groups still present in these products can be subject to additional transformations.

It is obvious that the task of ensuring selectivity in these situations is by far more troublesome than in the preceding cases. The challenge here is to carry out a reaction selectively at one of the available functional groups and at the same time employ an efficient 'block' to prevent the second reaction from occurring [*e.g.*, the selective mono- or biacetylation of glycerol can be achieved in exactly this manner, reaction **(f)**]. This rather formal approach may be flawed, as it suggests that the reactivity of the functional group retained in the first product remains unchanged. In general this is not very likely the case. For example, in the course of the alkylation of toluene by the Friedel–Crafts process, reaction **(g)**, the addition of the first alkyl group *increases* the nucleophilicity of the aromatic nucleus and as a result the second alkylation occurs more rapidly than the first one. Likewise, the third alkylation may occur even faster. The mutual interaction of functional groups is a very common phenomenon and its effect can be quite significant if the interacting functionalities are in close proximity in the molecule or separated by a system of conjugated double bonds. Such an influence can either accelerate or decelerate a given reaction. Taking advantage

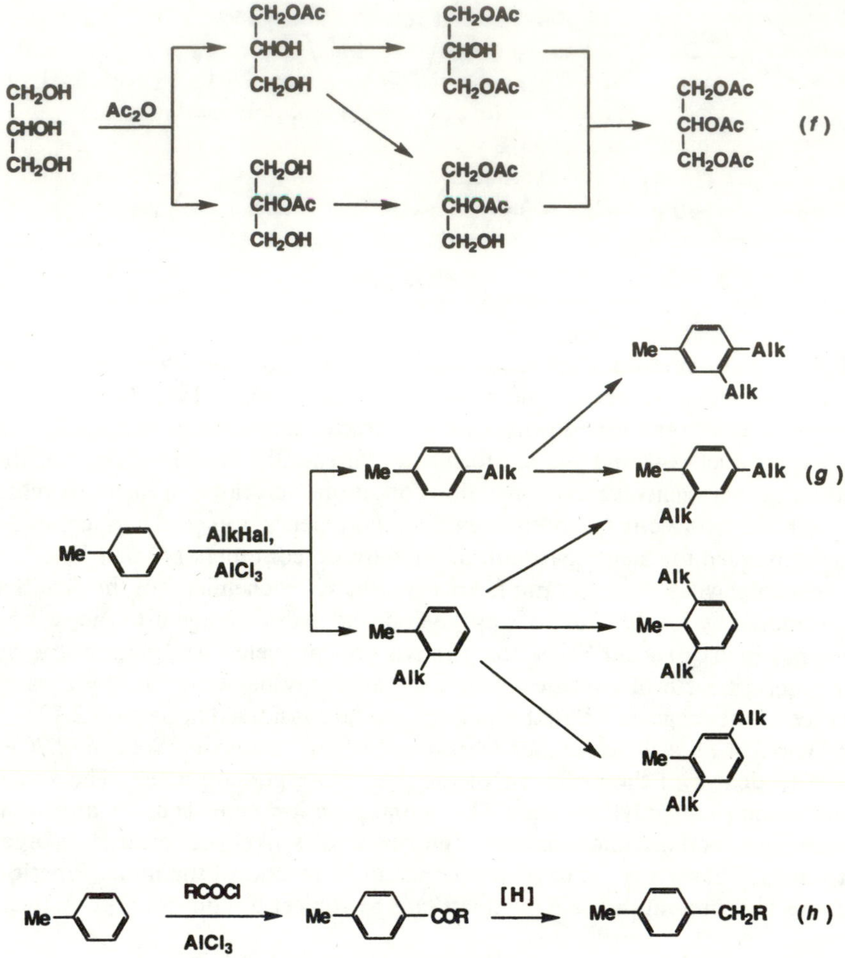

(*f*)

(*g*)

(*h*)

Scheme 2.70

of these mutual interactions can secure the overall selectivity of the process. For example, if toluene undergoes a Friedel–Crafts acylation instead of an alkylation, the exclusive formation of a monoacylated product (predominantly the *para* isomer) would be observed as the presence of one acyl group deactivates the aromatic molecule toward further electrophilic substitution. The carbonyl group in the acylation product can be easily reduced and, thus, the overall monoalkylation of the starting toluene can be achieved in two highly selective steps [reaction (**h**), Scheme 2.70].

The examples shown above clearly illustrate the multifaceted problems related to the control of selectivity. It must also be added that, in principle, every organic compound is polyfunctional. Even methane, the simplest of organic molecules, can produce four different products upon chlorination, from CH_3Cl to CCl_4. Not surprisingly, the question of selectivity receives top priority in the planning of an organic synthesis. Even our cursory examination demonstrates how diverse the obstacles may be on the route to complete selectivity. Equally varied are the approaches to overcome these obstacles.

Type 1 selectivity problems are addressed in the above discussions relating to the fundamental requirements for a reaction to be considered useful as a synthetic method. We will concentrate our attention in the following sections mainly upon approaches utilized to solve type 2 and, to a lesser extent, type 3 selectivity problems. The discussion will cover some of the most common pathways based upon varying the nature of reagents and/or substrates, as well as on the changing the reaction mechanism. These are the major, but by no means the exclusive, options available for solving the problem of selectivity. It must not be overlooked that in certain cases a dramatic enhancement in selectivity can be also achieved by the thoughtful use of purely physical approaches, such as removing the main product from an equilibrating mixture, or careful control over the kinetics of the process by the proper choice of reaction parameters.[22a]

Before going further, a few words about the terminology are necessary. *Chemoselectivity* refers to the selective reaction at one center in a substrate containing several non-identical functional groups. The term *regio* (or *site*) *selectivity* is applicable to reactions which may lead to the formation of positional isomers. The term *stereoselectivity* is applied to reactions that form stereoisomers. In cases where total selectivity is achievable, the term *specificity* is used to describe chemo-, regio-, and stereospecificity (as opposed to selectivity).

The final aspect of selectivity is related to forming optically active mirror image isomers (enantiomers). The problem of *enantioselectivity* is extremely important in organic synthesis, but will not be addressed in this text since this is an independent topic in its own right[22b] (see, however, discussion of some aspects of this problem in Chapter 4).

2.11 THE CHOICE OF REACTION FOR THE REQUIRED SELECTIVITY PATTERN

In the bromination of toluene (see Section 2.1) we examined a rather simple example of the approach to securing the selectivity of a reaction by choosing the appropriate reaction conditions and reacting partners. Toluene, in fact, has two functional groups capable of reacting with bromine, the methyl group and the aromatic nucleus. Nevertheless, as we observed, one can chemoselectively direct the bromination to either the methyl group (by radical bromination) or to the aromatic ring (*via* an ionic pathway). Another example of this approach to securing the selectivity of a reaction is the chemoselective reduction of toluene. Catalytic hydrogenation can reduce all three double bonds to produce a saturated system. In contrast, the Birch reduction leads to the selective reduction of only one of them.

In the same manner, the regioselectivity of many other transformations can be controlled by the correct selection of a reaction type. Let us begin with a consideration of model structure **151** (Scheme 2.71) that contains two isolated double bonds, differing only by the substitution pattern. How can one selectively reduce bond **a** (or bond **b**)? Catalytic hydrogenation on metal catalysts such as palladium is very sensitive to steric hindrance. This method can be used to reduce selectively the less hindered disubstituted double bond **a** to yield alkene **152**. The selective reduction of bond **b**, leading to the isomeric alkene **153**, can be achieved by ionic hydrogenation,[19k] reduction in a system of trifluoroacetic acid–triethylsilane. The mechanism of this reaction is fundamentally different from that of catalytic hydrogenation. The key step of the process is the formation of a carbocationic intermediate trapped by a hydride transfer from the silane. In this case, site selectivity is determined by the ease of formation of alternative carbocation intermediates. The protonation of diene **151** will occur almost exclusively at the bond **b**, as it produces the most stable tertiary carbocation **154**. Therefore, the overall selectivity of the reduction of this bond is safely ensured under these conditions.

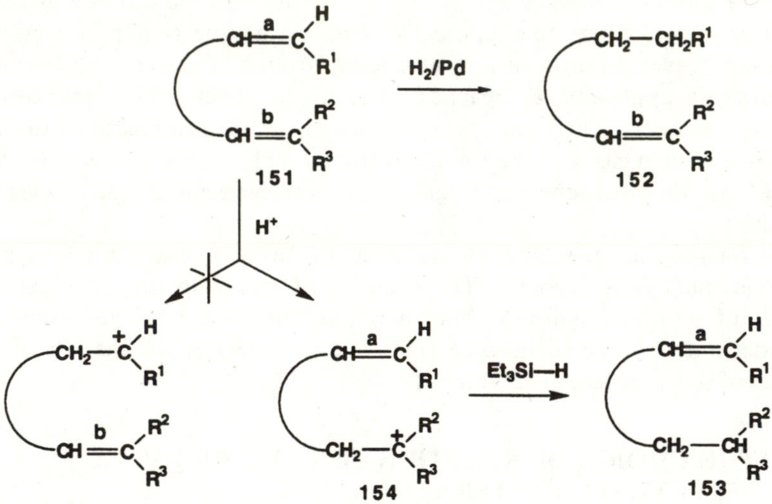

Scheme 2.71

In the above example the substrate contained two non-related functional groups with the task of performing a reaction at one of them. A situation of a different nature is also commonly encountered when more than one option is available for the interaction of a given reagent with a monofunctional substrate. A typical example is shown in Scheme 2.72. Conversion of alkenes into alkanes generally does not represent any problem and can be efficiently achieved by either catalytic or ionic hydrogenation. This task, however, becomes non-trivial if we need to prepare an alkane selectively labeled by deuterium. Both catalytic and ionic hydrogenation are unsuitable for this purpose. An efficient and practical solution to this problem requires the utilization of an entirely different protocol, a sequence involving a hydroboration reaction leading to alkylborane intermediates **155a** and **155b** followed by protolysis. The net outcome of this process is also a reduction of the double bond, but the first added hydrogen originates from borane, while the second is provided by the acid used for the protolysis. Thus the use of either deuterated borane or deuterated acid leads to the selective formation of monodeuterated alkanes.

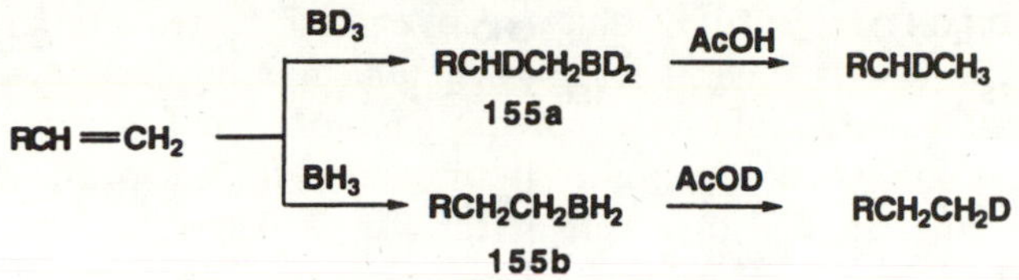

$$RCH{=}CH_2 \xrightarrow{BD_3} \underset{\textbf{155a}}{RCHDCH_2BD_2} \xrightarrow{AcOH} RCHDCH_3$$

$$RCH{=}CH_2 \xrightarrow{BH_3} \underset{\textbf{155b}}{RCH_2CH_2BH_2} \xrightarrow{AcOD} RCH_2CH_2D$$

Scheme 2.72

As was already mentioned, the standard procedure for acid catalyzed alkene hydration exhibits a rather low selectivity. On the other hand, the use of a hydroxymercuration–reduction sequence leads to the exclusive formation of Markovnikov's alcohols. A nearly exclusive anti-Markovnikov's hydration is achieved *via* a hydroboration–oxidation reaction (see Section 2.4). The result in both these cases is the net addition of H_2O, but the basic differences in the reaction mechanisms unambiguously determine a reversed regioselectivity pattern.

An expedient solution to selectivity problems is not always found, however, in varying the nature of the reactions. In many cases a lot can be achieved within the limits of the same reaction by merely varying the nature of the reagents.

2.12 VARYING A REAGENT'S NATURE AS A TOOL TO CONTROL SELECTIVITY

It is well known that the selectivity pattern of any reaction can be noticeably or even significantly altered depending upon the nature of the reagents and, to a lesser extent, the reaction conditions. The rational selection of reagents or a special design of new reagent may turn out to be the most efficient way to achieve the required selectivity. Below are some typical examples.

Towards the end of the 1940s an extremely effective new reagent was introduced to the practice of organic chemistry. This reagent was lithium aluminum hydride, a powerful reducing agent for many functional groups. Without going into the details of the reduction mechanism, it can be said that the essence of the reaction consists of attack of the $\mathbf{H^-}$ nucleophile on the substrate. Not surprisingly, any substrate that behaves as an electrophile should be potentially subject to reduction by lithium aluminum hydride. For example, a model system arbitrarily drawn as **156** (Scheme 2.73), bearing three typical electrophilic groups, can in principle be reduced with lithium aluminum hydride at all three centers. These three groups, however, are easily arranged in an order of decreasing electrophilicity: **CHO > COOMe > $\mathbf{CH_2Cl}$**. The reaction of

$LiAlH_4$ occurs rapidly with the first two groups and rather sluggishly with the last. Therefore, preparation of chlorodiol **157** would not cause any problem.

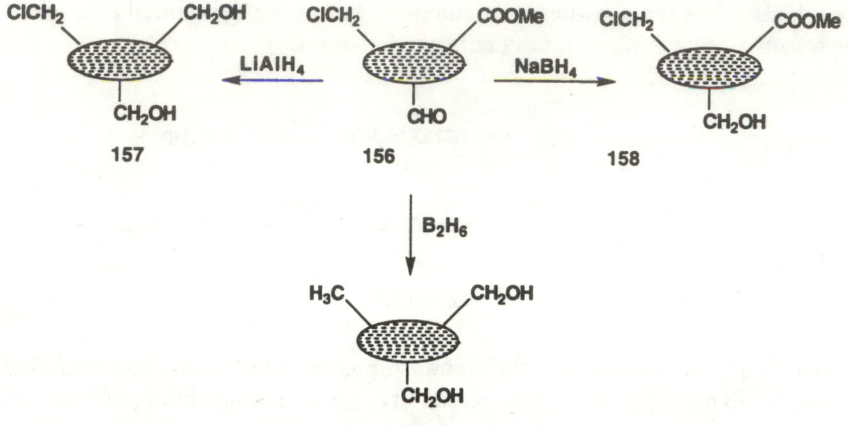

Scheme 2.73

At the same time, the selective reduction of only the aldehyde group cannot be achieved that easily. Both the aldehyde and ester groups are prone to react with this reagent and the difference in the reaction rates is not sufficient to achieve an acceptable level of chemoselectivity. If lithium aluminum hydride was the only source of the hydride anion, then it would be difficult or even impossible to reduce selectively the aldehyde group in a system like **156**. Fortunately, a set of hydride reagents emerged that differed both in activity as hydride donors as well as in general reactivity pattern.[23a] One of these analogs was sodium borohydride. This hydride source is a much weaker nucleophile than lithium aluminum hydride. Thanks to this peculiarity, the difference in the rate of reduction of the aldehyde *versus* the ester by borohydride reagents is preparatively significant and the selective transformation of **156** into **158** is a viable procedure.

Finally, if it is necessary to carry out an exhaustive reduction of **156**, one would have to use yet another reagent, in this case a very powerful donor of the hydride anion. This reagent is diborane, B_2H_6, which was already mentioned as the reagent for hydroboration of alkenes. Diborane is able to reduce (under more drastic conditions) even the relatively unreactive chloromethyl group, permitting substrates of the type **156** to be converted to the diol **159**. Even more efficiently, this goal can be achieved with the help of Super Hydride, $LiEt_3BH$.[191] Rather surprisingly, the selective reduction of the alkyl halide fragment, R–Hal (Hal = Br, I), in the presence of COOR or CHO groups can also be carried out with another complex hydride, $NaBH_3CN$.[23b]

The presently available wide selection of complex hydrides greatly simplifies the problem of control over the chemoselectivity of reduction in polyfunctional compounds. For example, a 1,2-reduction of the carbonyl group in α,β-

conjugated aldehydes and ketones is often accompanied by a 1,4-reduction of the whole system. This complication can be avoided if a standard reducing agent, $NaBH_4$, is modified by the addition of $CeCl_3$[23c] or substituted by a more selective reagent, $Zn(BH_4)_2$ (Scheme 2.74).[23d] The latter reagent also revealed a unique property to reduce selectively (under carefully controlled conditions) the isolated keto group in the presence of the enone moiety.[23e] A recently introduced new class of powerful (and moisture stable!) reducing reagents, lithium aminborohydrides ($LiABH_3$; A = cyclic amines) exhibited a rather peculiar selectivity pattern in reducing selectively the carbonyl group without affecting the double bond for both conjugated enals and enones.[23f] A smooth 1,4-reduction requires the utilization of a different set of reagents such as $LiAlH_4$ in the presence of copper(I) complexes[23g] or more conventional reagents such as Li in $EtNH_2$.[23h] A total reduction of the conjugated system to yield saturated alcohols requires powerful hydride donors like KBH(*sec*-Bu)$_3$,[23i] or in certain cases can be carried out under conditions of ionic hydrogenation with the Et_3SiH–CF_3COOH system.[23j]

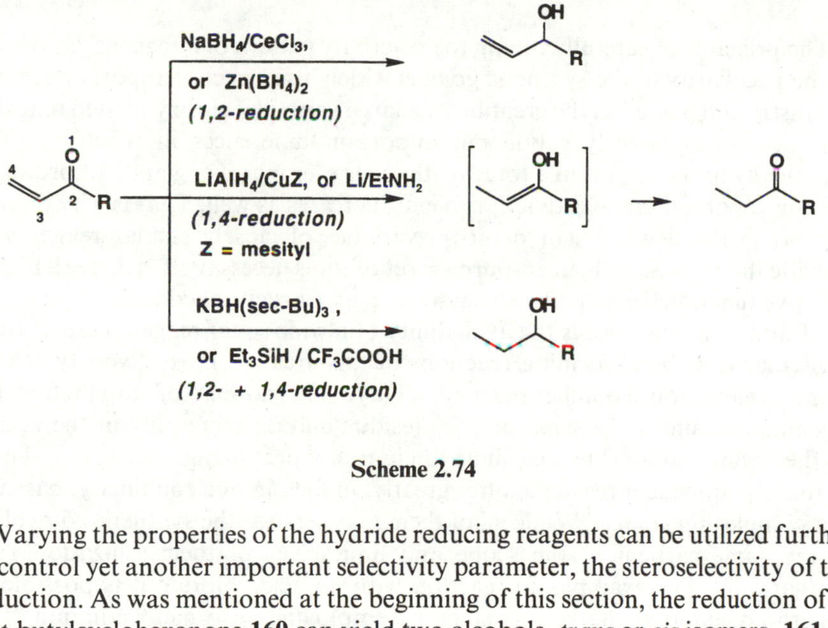

$NaBH_4/CeCl_3$, or $Zn(BH_4)_2$ *(1,2-reduction)*

$LiAlH_4$/CuZ, or Li/$EtNH_2$ *(1,4-reduction)* Z = mesityl

KBH(sec-Bu)$_3$, or Et_3SiH / CF_3COOH *(1,2- + 1,4-reduction)*

Scheme 2.74

Varying the properties of the hydride reducing reagents can be utilized further to control yet another important selectivity parameter, the steroselectivity of the reduction. As was mentioned at the beginning of this section, the reduction of 4-*tert*-butylcyclohexanone **160** can yield two alcohols, *trans* or *cis* isomers, **161** or **162** respectively (Scheme 2.75). If $NaBH_4$ is used as the hydride donor, the major product is the more stable *trans* isomer **161**. If, however, the reduction is carried out with a hydride complex containing a bulky alkyl group such as in LiBH(*sec*-Bu)$_3$,[23i] it is possible to reverse this selectivity and to obtain preferentially the *cis* isomer **162**. This dichotomy can be explained reasonably well in terms of the steric control of the approach of the reagent. Two axial hydrogens, present at C-3 and C-5, effectively block the 'from the above' approach of the sterically demanding LiBH(*sec*-Bu)$_3$ reagent but do not interfere significantly with the addition of the smaller molecule of $NaBH_4$. Paradoxically, the above mentioned lithium aminoborohydrides, $LiABH_3$, behave as unhindered reagents regardless of the size of the amine moiety and in reaction with **160** gave 99% of the *trans* isomer **161**.[23f]

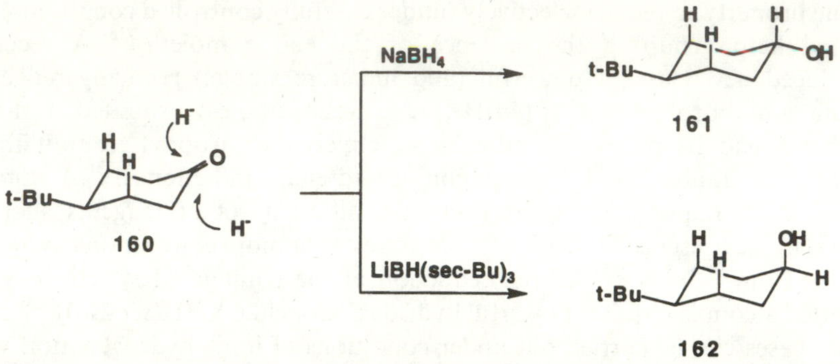

Scheme 2.75

The principle of carefully tuning the reactivity pattern of a reagent, according to the peculiarity of the synthetic goal, is widely used in contemporary organic chemistry and has led to the creation of a large variety of closely related reagents that possess apparently subtle but important differences in reactivity. The complexity of the targets in a total synthesis is ever increasing and, accordingly, the complexity of the selectivity problems increases as well. That is why there is no end to the development of novel varieties of closely similar reagents to provide the chemist with an all-purpose set of tools necessary to achieve a highly selective functional group transformation in any structural context.

Of no less importance is the availability of a wide set of reagents useful for a particular **C–C** bond-forming reaction. In this area the most diversity can be found among nucleophilic reagents. There are dozens of organometallic reagents containing the same organic residue and differing only in the nature of the metal (lithium, magnesium, calcium, copper, manganese, *etc.*). These seemingly equivalent reagents differ greatly, in fact, in nucleophilicity, basicity, and complexing ability.[4] While all of them may serve as the synthetic equivalent of the same carbanion, the scope and limitations of their utilization vary dramatically. For example, in reaction with an acid chloride it is possible to halt the reaction at the ketone stage by employing an organocadmium[24a] or organomanganese[24b] reagent, or to go one step further to prepare the symmetrical tertiary alcohol by using a classical organomagnesium reagent. It is also remarkable that the sequential utilization of organomanganese and then organomagnesium compounds may be employed as a one-pot protocol for the conversion of acyl chlorides into non-symmetrical tertiary alcohols:[24c]

$$\mathrm{RCOCl} \xrightarrow{\mathrm{R^1_2Cd\ or\ R^1MnHal}} \mathrm{RCOR^1}$$

$$\mathrm{RCOCl} \xrightarrow{\mathrm{R^1MgHal}} [\mathrm{RCOR^1}] \xrightarrow{\mathrm{R^1MgHal}} \mathrm{RC(OH)R^1_2}$$

$$RCOCl \xrightarrow{R^1MnI} RCOR^1 \xrightarrow{R^2MgHal} RC(OH)R^1R^2$$

Additional opportunities to control the chemoselectivity of these reactions come from the elaboration of methods to affect the reactivity pattern of a given organometalic reagent by using specific ligands or, in a broader sense, some modifying additives.

Among the variety of modified carbanionic equivalents, the organocopper reagents are probably the most popular. These reagents may be employed in organic synthesis in the simple form of RCu or, more often, as mixed cuprates with the composition varying from R_2CuLi to R_3Cu_2Li.[24d] They can be additionally modified by complexation with ligands of the type Me_2S, Ph_3P, CuCN, *etc.*[15c,d] In a similar way, classical Grignard compounds can be used in the presence of copper salts and ligands, for example as $RMgBr/CuBr/Me_2S$.

The historical background leading to the recent surge of these somewhat exotic reagents is rather instructive and will be briefly reviewed here. The first organocopper compound, dimethylcopper (Me_2Cu), was prepared by Gilman in 1936.[24e] His project was aimed merely at the expansion of organometallic species to study further the peculiarities of their properties, with no relevance to the general problems of organic synthesis. Somewhat later, and completely independent of this work, it was accidentally discovered that the addition of certain salts, including those of copper, to classical Grignard reagents, RMgBr, could significantly affect the pattern of their reactivity. In the 1960s there emerged a pressing need to develop general preparative methods to secure rigorous control over the selectivity of reactions of carbanionic reagents with various polyfunctional electrophiles. The key to the solution was found in the above mentioned and apparently rather particular observations. In a matter of a few months, several publications appeared almost simultaneously that described preparative procedures using copper-modified organometallic compounds as the reagents of choice for selective reactions with a number of electrophiles.[24f] Subsequent studies revealed also that the very troublesome problem of the specific addition of organometallic reagents to the double bond of a conjugated enone system was easily solved by using lithium alkylcuprates.[15c] Owing to these findings, the classic preparative method of the Grignard addition to the carbonyl group in various compounds was complemented by an equally reliable method, general in scope, for conjugated addition. The selectivity of both processes is easily switched 'on/off' by simply adding cerium or copper salts (for 1,2- and 1,4-additions, respectively; Scheme 2.76) or by utilizing specific solvents.[4,15b–d]

As already mentioned, the coupling of alkyllithium (or magnesium) reagents with electrophiles of the R–X type is likely to proceed non-selectively owing to a rather high basicity and propensity to undergo transmetallation. Here again the problem of selectivity was successfully solved when cuprate reagents were introduced, as opposed to the classical organometallic species (see discussion on the Wurtz reaction in Section 2.2). It was also discovered shortly after that while cuprates are almost unreactive toward carbonyl electrophiles, they reveal

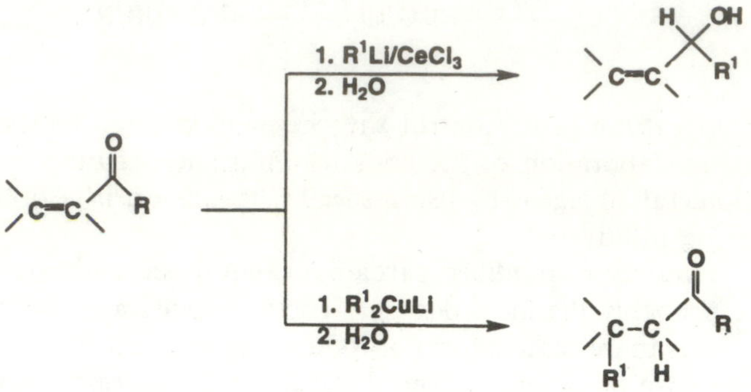

Scheme 2.76

a unique property to react with aryl or vinyl halides.[24g] Aryl or vinyl halides had never before been considered a reactive electrophilic species. As a result, it is now possible to effect a number of selective transformations, such as those shown in Scheme 2.77.[15b,c]

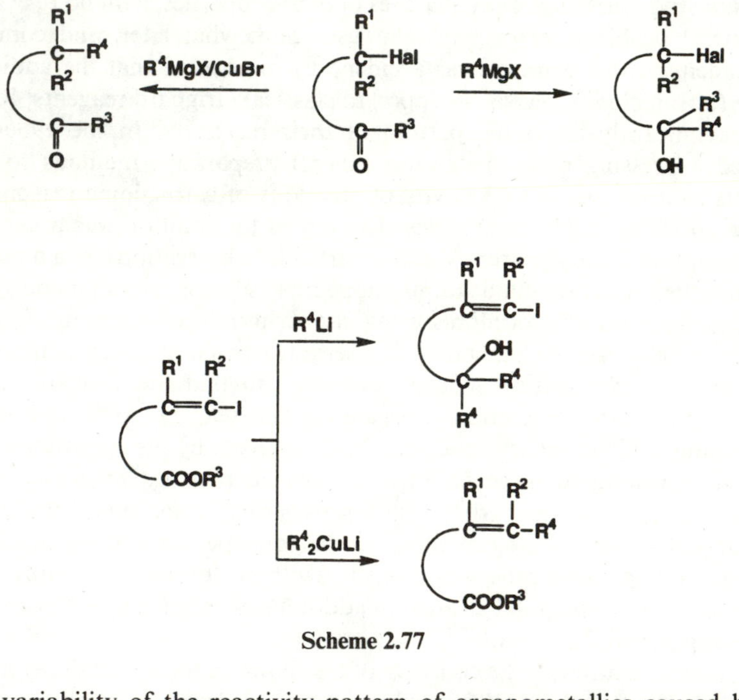

Scheme 2.77

The variability of the reactivity pattern of organometallics caused by the seemingly insignificant modifications in their nature was further illustrated in a spectacular way by the studies of Knochel's group.[24h] Taking advantage of the higher reactivity of the **C–Cu** bond toward electrophiles (compared to the **C–Zn** bond), they designed an efficient protocol for the one-pot sequential coupling of 1,*n*-heterobimetallic reagents, such as **163**, with a pair of different electrophiles (Scheme 2.78).

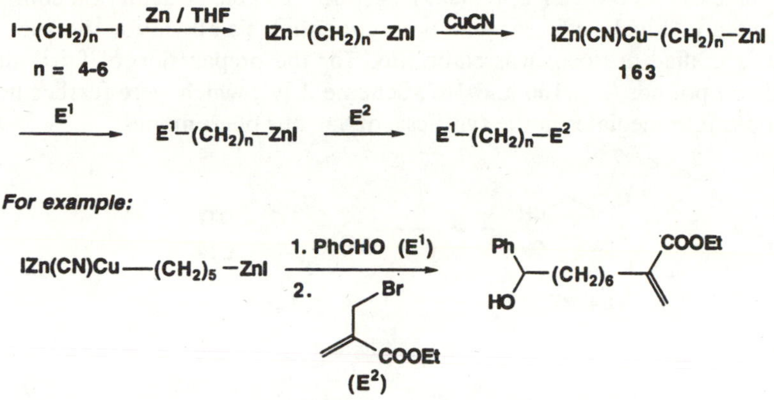

$$I{-}(CH_2)_n{-}I \xrightarrow{Zn\ /\ THF} IZn{-}(CH_2)_n{-}ZnI \xrightarrow{CuCN} IZn(CN)Cu{-}(CH_2)_n{-}ZnI$$

n = 4-6 ; **163**

$$\xrightarrow{E^1} E^1{-}(CH_2)_n{-}ZnI \xrightarrow{E^2} E^1{-}(CH_2)_n{-}E^2$$

For example:

IZn(CN)Cu—$(CH_2)_5$—ZnI → (1. PhCHO (E^1); 2. $CH_2{=}C(COOEt)CH_2Br$ (E^2)) → PhCH(OH)—$(CH_2)_6$—C(=CH_2)COOEt

Scheme 2.78

It is also important to note that the utilization of cuprate reagents, rather than conventional organometallics such as Li or Mg derivatives, opened an entry into the elaboration of novel and extremely efficient protocols for the creation of **C–C** bonds based upon the addition of these carbanion equivalents across a triple bond (Section 2.3.3).

Additional flexibility in the control over the selectivity of heterolytic reactions is provided in the diversity of electrophilic reagents that formally correspond to the same electrophile. For example, reagents such as $RCO^+BF_4^-$, $RCOOSO_2CF_3$, RCOCl, and $(RCO)_2O$ are employed in synthesis as equivalents of the acyl cation RCO^+. However, a tremendous difference in the reactivity of these acylating species enables one to choose a reagent specifically adjusted to the peculiarity of the nucleophilic counterpart. In a similar way, such unlike compounds as trialkyloxonium salts, $R_3O^+BF_4^-$, alkyl halides, tosylates, or acetates can serve as transfer agents of the same alkyl cation, R^+, but they differ drastically in their activity and pattern of selectivity toward various nucleophiles.

The judicious choice of proper reacting partners among seemingly almost identical reagents can be illustrated by the spectacular results of Kotsuki's studies[24i] in the synthesis of chiral pheromones, shown in Scheme 2.79. Reasoning that the readily available compound **164** could be utilized as a chiral precursor for the preparation of a set of useful intermediates, Kotsuki's group anticipated that strict control could be exerted over the selectivity of the **C–C** bond-forming reactions at two competing electrophilic centers. To achieve this goal, diol **164** was initially transformed into the mixed tosyl–triflate derivative **165** with the aim of creating two electrophilic centers differing in their activity toward nucleophiles. The next step was to find nucleophilic reagents capable of reacting selectively with either of these centers. Standard organocuprate reagents were shown to attack indiscriminately at both centers. However, the enhanced leaving ability of triflate made it prone to react with the less nucleophilic alkylmagnesium reagents in the presence of a copper catalyst, whereas the tosylate group stayed untouched under these conditions. By taking advantage of this difference, Kotsuki's group was able to achieve a completely selective and highly efficient one-pot sequential bis-alkylation of **165**. As a result, a unified protocol was elaborated for the preparation of fairly diverse chiral compounds (*e.g.* **166** and **167**, Scheme 2.79), which were further used as advanced intermediates in the synthesis of various pheromones.

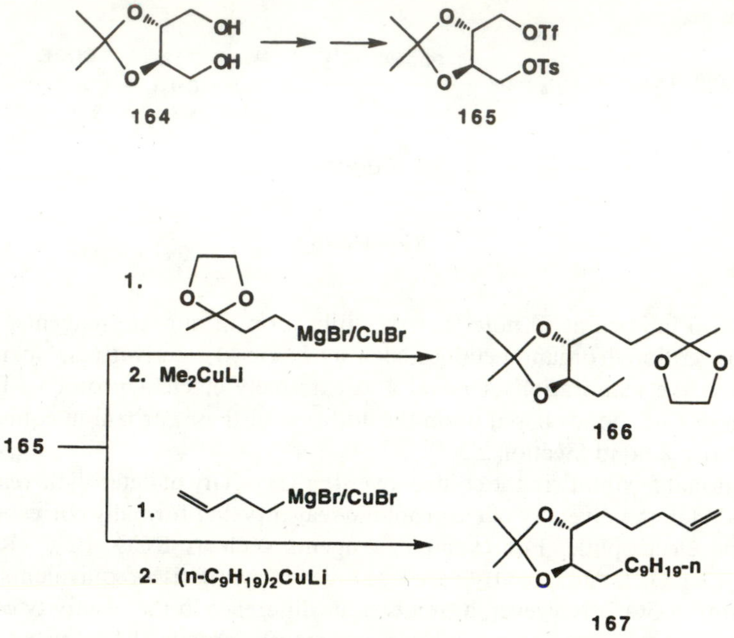

Scheme 2.79

A number of synthetic problems are greatly simplified owing to the fact that, for almost any electrophilic and/or nucleophilic fragment that may emerge in the course of retrosynthetic analysis, there is a rich choice of reagents that enable the chemist to select candidates finely adjusted for the selectivity requirements of the given synthetic task.

2.13 THE SELECTIVE ACTIVATION OF ALTERNATIVE REACTION SITES IN SUBSTRATES

The classical example of the selective activation of a reaction site in a substrate with more than one reactive center is acetoacetic ester **168** (Scheme 2.80). Its reactive form is the enolate **169**, which reacts with a variety of electrophiles selectively at the central carbon atom. A subsequent hydrolysis and decarboxylation of the product **170** leads to the formation of ketone **171**. The structure of **171** corresponds to the coupling of the electrophile with the carbanion **172**, or, in other words, with deprotonated acetone. Thus acetoacetic ester is actually employed in this sequence as a synthetic equivalent to **172**.

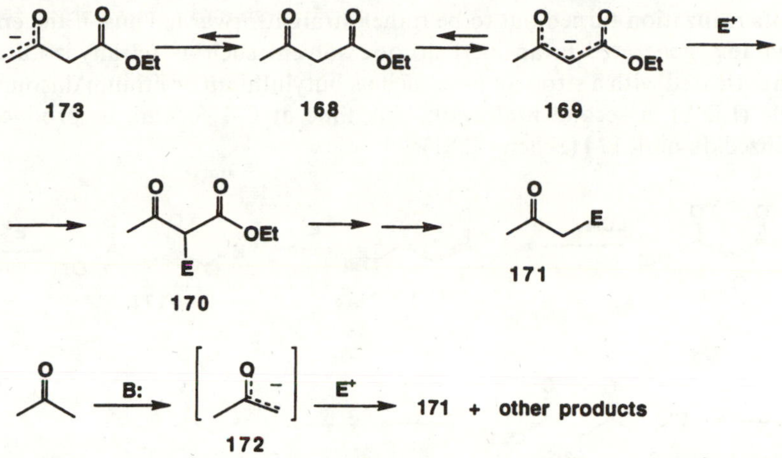

173 168 169 E⁺

170 171

B: 172 E⁺ 171 + other products

Scheme 2.80

Why, really, is such a sophisticated replacement for the simple carbanion **172** needed? It would appear that a much more straightforward synthesis of **171** is possible by the reaction of **172** (formed directly from acetone) with an electrophile. There are a couple of reasons, however, that make this route a poor choice. First of all, acetone itself is an active electrophile and would react with its own enolate, leading to a self condensation (see Section 2.3). A second and perhaps more significant reason is apparent in the light of our discussion of selectivity. Product **171** contains the $-COCH_3$ fragment that differs little from the same fragment in the starting acetone and, therefore, would be expected to compete with acetone for enolate formation. Consequently, this route is non-selective and would lead to a mixture of products.

This is not the case with the acetoacetic ester pathway. Compound **168** also has two groups potentially capable of enolate formation, the CH_3 and CH_2 groups. However, owing to the combined effect of two carbonyl substituents, the acidity of the protons on the CH_2 group of **168** is several orders of magnitude higher than the those on the CH_3. For the same reason, the resulting enolate **169** is much more stable than **173**. Therefore it is not only possible to generate the enolate from **168** under milder basic conditions than those required for the enolization of acetone, but this reaction also proceeds almost exclusively with the formation of enolate **169** to secure the selective formation of the desired alkylation product, **170**. Thus owing to the presence of the auxiliary ethoxy-carbonyl group, the selective activation of one of the two alternative α-positions was achieved. This is a classic utilization of acetoacetic ester as a versatile synthetic intermediate. Though used for several decades, it is still extensively utilized.

More thoughtful analysis revealed additional opportunities hidden in the structure of acetoacetic ester. In fact, the non-equivalence of the two α-positions to the carbonyl group suggests an alternative mode of selectivity based on preferential attack at the C-1 carbon. This idea seems to be paradoxical, but in fact its realization turned out to be rather straightforward. Thus, if the enolate anion **169**, generated in an inert aprotic solvent such as tetrahydrofuran, is further treated with a stronger base such as butyllithium or lithium diisopropylamide (LDA), a second ionization, this time at C-1, occurs to produce the stabilized dianion **174** (Scheme 2.81).

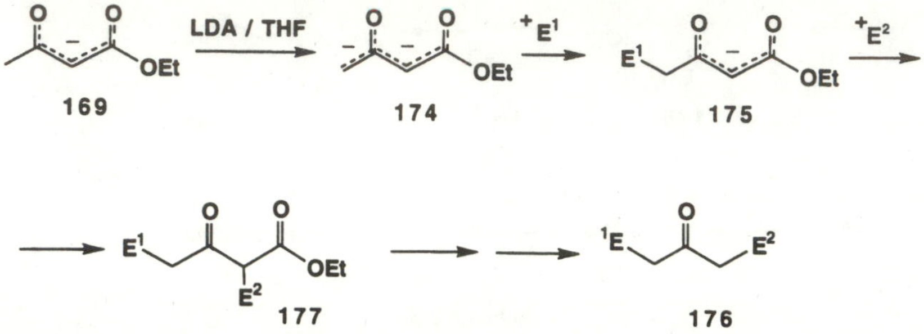

Scheme 2.81

The presence of two negative charges in close proximity makes this new reagent **174** extremely reactive. Its carbanionic sites, at **C-1** and **C-3**, however, differ sharply in their nucleophilicity and reactivity. The different surroundings of the carbanionic centers in this system makes the carbanion at **C-3** better stabilized than at **C-1**. Therefore, electrophilic attack should be directed primarily at **C-1**. In fact, the addition of one equivalent of an electrophile to a solution of **174** leads to a highly selective attack at the terminal carbon atom. The product of this reaction, **175**, still retains a carbanionic center and with the addition of another electrophile the formation of a second bond occurs selectively at **C-2**. In this manner, the dianion **174** is an excellent three-carbon building block for the synthesis of ketones of type **176** or four-carbon building block for the synthesis of esters of the type **177**.

Doubly (and even triply) charged anions appeared on the horizon of organic chemistry relatively recently. They were immediately recognized as extremely useful intermediates due to the synthetic opportunities offered by their unique pattern of reactivity. Besides the dianion **174**, the dianions of carboxylic acids **178** (Scheme 2.82), propargylic dianion **179**, and the dianion **180** derived from propargyl alcohol are now widely used as nucleophiles. Selectivity in electrophilic attack for these species is governed in much the same way as was described for **174**: the least-stabilized anionic center is the preferential site of attack by the electrophile (marked with an asterisk in Scheme 2.82).

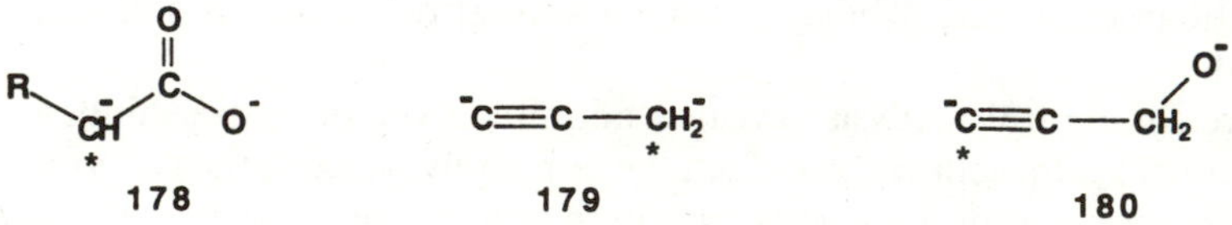

Scheme 2.82

Many synthetic problems requiring a reverse in the traditional mode of selectivity have been solved by the application of these and other polyanions. It is worthwhile to note that this novel aspect of carbanion utilization was brought to light owing to the elaboration of the conditions (strong bases, polar aprotic solvents) which secured completeness in carbanionic species generation.

The selective monoalkylation of ketones at the α-position had no general solution for a long time. A classical approach to this problem, based upon the selective activation of this site *via* the introduction of additional electron withdrawing substituents (*e.g.* alkoxycarbonyl group), is applicable only for symmetrical ketones. Non-symmetrical compounds react non-selectively in this auxiliary step. The synthetic importance of this problem triggered a thorough study of enolate chemistry in the 1960s and, as a result, at present the selective substitution at any of the α-positions of carbonyl compounds can be achieved *via* a number of routes.

The initial procedure, while reliable, was not very convenient. A mixture of enolates generated from the interaction of a ketone, such as **181** (Scheme 2.83), with a base in an aprotic medium was treated with chlorotrimethylsilane, an electrophile that attacks exclusively at the enolate oxygen. The resulting mixture of regioisomeric silyl enol ethers (*e.g.* **183a** and **183b**) was then separated by distillation. Treatment of the individual silyl enol ethers with CH_3Li led to the scission of the **O–Si** bond and *in situ* generation of the individual Li enolates. Further treatment with the electrophile produced the required products, **182a** or **182b**, in a highly selective way.

In the course of these investigations it was noticed that the isomeric composition of the resultant mixture of silyl enol ethers depended upon the 'pre-history' of the reaction. If the enolate reaction mixture was quenched with Me_3SiCl immediately after the carbonyl precursor was treated with a strong base at low temperature, isomer **183a** was formed preferentially. If the initially generated enolate was maintained for several hours before quenching, then the predominate formation of isomer **183b** was observed.[25a] The implications of these observations were obvious. The regioselectivity of the initial ionization is determined by the relative ease of proton abstraction (kinetic control). The presence of a methyl substituent at the α-position makes this site less accessible to base attack. Therefore, the kinetically controlled deprotonation is more likely to happen at the less substituted site, producing enolate **181a** trapped as **183a**. In the absence of the quencher, an equilibrium of isomeric enolates is established and the most substituted enolate, **181b** trapped as **183b**, dominates this mixture owing to its higher stability (thermodynamic control). The formation of *O*-silyl derivatives proceeds much faster than the interconversion of enolates and hence Me_3SiCl acts like a 'fixing' reagent.

These observations resulted in the elaboration of a reliable and fairly general procedure to prepare both kinetically and thermodynamically controlled silyl enol ethers without the painstaking procedure of isomer separation (see Scheme 2.83). In this method, the kinetically controlled product, such as **183a**, is obtained by applying strong and sterically demanding bases, such as lithium bis(trimethylsilyl)amide (LBSA) and immediately quenching the resultant

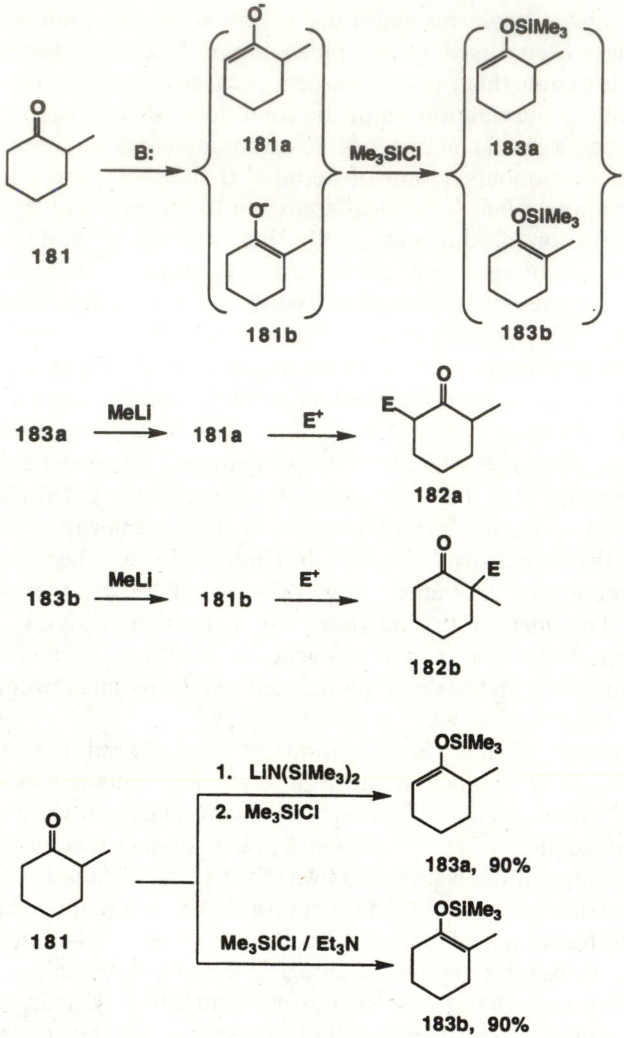

Scheme 2.83

enolates with chlorotrimethylsilane. In contrast, the conditions for generating the thermodynamically controlled product (for example, **183b**) involve the use of conventional bases, such as triethylamine and a longer time, as required for complete equilibration.[25b]

In modern organic chemistry, silyl enol ethers, as well as the corresponding titanium, tin, boron, or zirconium derivatives, are widely employed as nucleophilic components in enolate alkylation reactions. Their usefulness prompted the elaboration of numerous methods for the selective production of isomerically pure enol ethers from almost any type of carbonyl compounds.

There is another feature of silyl enolates worthy of special comment. Silyl enolates are less reactive toward electrophiles than their respective lithium enolates. As a result, Lewis acids are generally required to generate carbonium ion-like electrophiles capable of interacting with silyl enolates (*cf.* data in Schemes 2.41–2.44). An alternative way to promote this reaction under essentially neutral conditions, however, is to use the unsolvated fluoride anion to activate a silyl ether.[17c] The **Si–F** bond is an extremely strong covalent bond and its formation is an exothermic process. The addition of the fluoride anion promotes a ready scission of the **O–Si** bond accompanied by interaction with covalent electrophiles, as shown in Scheme 2.84.

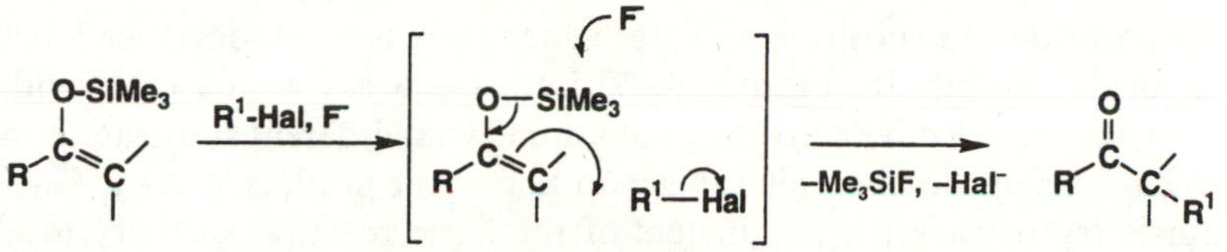

Scheme 2.84

Among other important options available for the regioselective monoalkylation of ketones at the α-position are methods employing ketone derivatives such as imines, hydrazones, or oximes.[4] In these cases, additional factors to control the regioselectivity in the carbanion generation are in place due to (i) the nitrogen atom's ability to serve as a ligand for a lithium cation and (ii) the steric requirements of substituents on a nitrogen. Nearly complete selectivity in substitution at the less substituted position suggests interesting ramifications for synthetic practice.[25c] The one-pot assemblage of the complicated polyfunctional compound **184** (Scheme 2.85) from three simple precursors, **185**, **186**, and **187**,[25d] is given as a representative example.

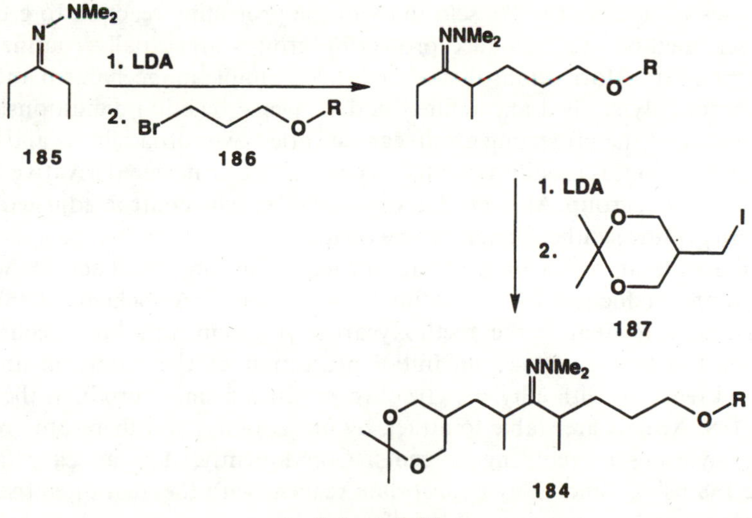

Scheme 2.85

2.14 PROTECTION OF FUNCTIONAL GROUPS AS AN ULTIMATE TOOL IN SELECTIVITY CONTROL

In the previous sections we examined options to achieve selectivity based upon variations in reaction conditions and/or reagents or even in the mechanism of the process. While this approach is an efficient way to achieve a goal, it does not come without a price tag. An 'adjustment' of the basic and general method is required to meet the needs of a particular task. In practice, it frequently turns out to be more efficient to utilize a totally different approach to selectivity problems. We will explain this in terms of a generic example.

Let us consider the substrate **A–X**, for which there is a well-developed method for its conversion into the product **A–Y**. Let us now say we are faced with the task of making a selective conversion of a somewhat different substrate **Z–A–X**, where **Z** is a group closely similar to group **X**, into the product **Z–A–Y**. One can, of course, try to vary the parameters of the basic reaction so that only **X** is affected. However, it might be rather difficult if not impossible to achieve, especially if it requires time-consuming investigations in order to modify the course of an already finely tuned reaction. Furthermore, to acquire a slightly modified target, **Z′–A–X**, additional changes to the process would have to be investigated.

An alternative possibility to resolve such problems is temporarily to 'sideline' **Z** from the game while applying standard procedures to convert group **X** to **Y**. Once complete, group **Z** can be 'sent back' into the game without any ill effects. The practical way to achieve this sequence uses a simple and reversible transformation to convert temporarily group **Z** into a functionality that is inert to the conditions needed to convert **X** to **Y**. Such a protection or 'masking' of a functional group is widely used in the practice of organic chemistry. While this generic example suggests how the question of selectivity for the main reaction can be solved by protecting the interfering group **Z**, the same dilemma now arises with respect to the selectivity of the protecting reaction to group **Z**. However, methods to introduce protecting groups are usually routine functional group transformations that are relatively simple in mechanism and have been thoroughly studied and refined under diverse reaction conditions. Also, the structure of the protecting group can be varied over broad limits and thus a number of options is usually available to prepare the protected derivative for the given functional group. As a result, a versatile selectivity control, adjusted to the peculiarity of the synthetic goal, can be assured.

To illustrate an application of the 'protecting group' approach we will re-examine the reduction of the trifunctional model **156** (Scheme 2.86). The simplest way to reduce the methoxycarbonyl group without affecting the aldehyde function envisages an initial protection of the latter *via* an acid-catalysed reaction with ethylene glycol to produce a single product, the cyclic acetal **188**. Acetals are stable to attack by nucleophiles and therefore are inert toward hydrides as reducing reagents. Consequently, it is an easy task to reduce **188** using almost any appropriate reagent with the guarantee that only the methoxycarbonyl group will be affected. Subsequent removal of the acetal protecting group from product **189** using mild acid hydrolysis gives the desired hydroxy aldehyde **190**. This sequence enables us to achieve a reverse mode of selectivity in the reduction of **156** compared with that achieved by the approach of varying the reagents (choosing the proper reducing agent) mentioned earlier.

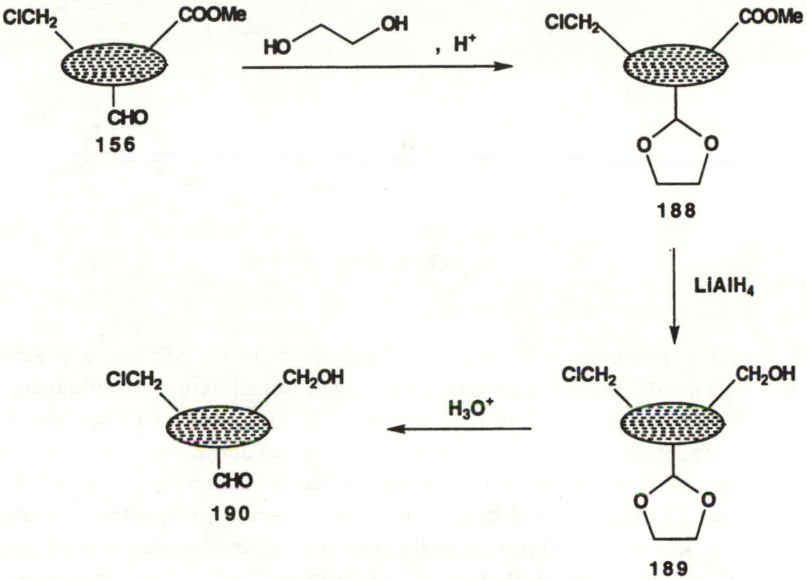

ClCH$_2$ COOMe CHO **156** HO OH, H$^+$ → ClCH$_2$ COOMe O O **188** LiAlH$_4$ → ClCH$_2$ CH$_2$OH O O **189** H$_3$O$^+$ → ClCH$_2$ CH$_2$OH CHO **190**

Scheme 2.86

Starting with carbonyl compounds, we will now examine more specific examples of selectivity problems solved by selecting an appropriate protecting group. The above mentioned acetal protection of an aldehyde group can, in principle, be applied to any carbonyl compound and a number of different alcohols can be used for this purpose. The rate of these reactions can vary by several orders of magnitude, depending upon the structure of the substrates. This allows, in particular, an easy discrimination between aldehyde and ketone functionalities, since the aldehyde is generally more reactive and can be selectively protected in the presence of the ketone. The example below may serve as a good illustration of how this difference can be exploited in a total synthesis.

Keto aldehyde **191** (Scheme 2.87) was prepared as a common precursor in the synthesis of a series of isoprenoid pheromones.[26a] One of the synthetic options required the selective reduction of a ketone carbonyl in this compound. Under mild conditions of acetalization (weak acid, methanol), only the aldehydic function of **191** was affected to form a mono-protected derivative, **192**. Reduction of the keto group in the derivative with sodium borohydride and subsequent removal of the acetal protecting group gave the desired hydroxy aldehyde **193**.[26a]

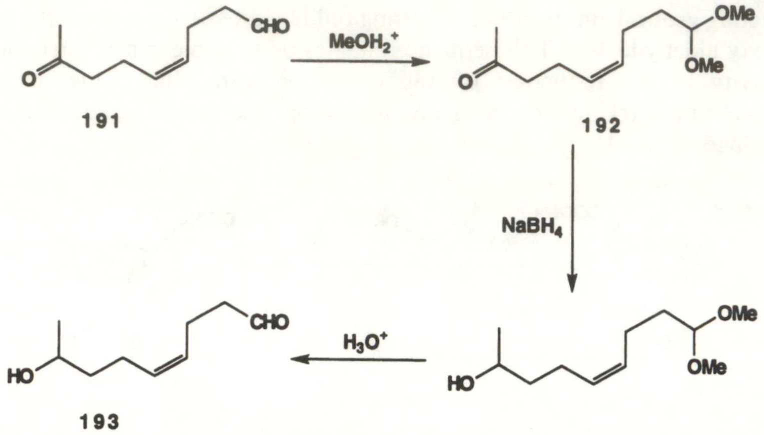

Scheme 2.87

This example can also be used to illustrate how to achieve a reverse in selectivity. The aldehyde group is first converted, selectively, into thioacetal **194** (the sulfur analog of **192**) by interaction with ethanedithiol (Scheme 2.88). Because thioacetals are stable and rather resistant to acidic reagents, protection of the keto group remaining in **194** can be easily achieved. As a result, the doubly protected derivative **195** is formed. An unusual property of thioacetal moieties, not shared by ordinary acetals or ketals, is the ease with which they are hydrolysed in the presence of mercury or cadmium salts. Thus it is relatively easy to obtain the derivative **196** containing a protected keto group and a free aldehyde function, which is further amenable to react with nucleophiles such as hydride reducing agents, Wittig, or Grignard reagents.

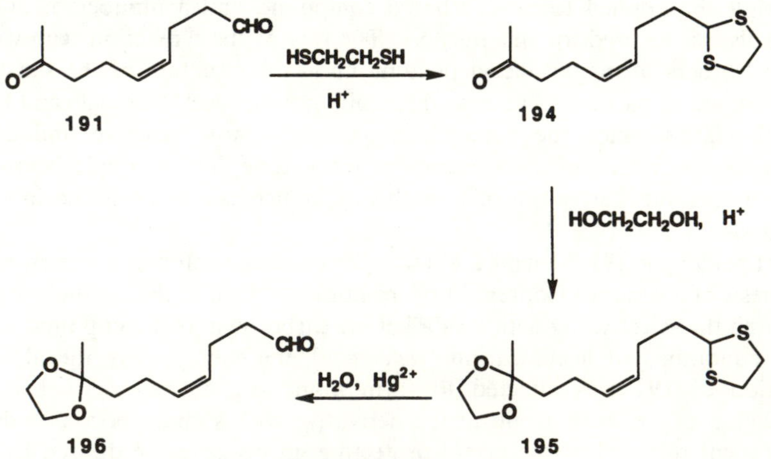

Scheme 2.88

The presence of an α,β-conjugated double bond noticeably reduces the electrophilic activity of a carbonyl carbon. This effect allows one to protect, selectively, an isolated keto group as a ketal in the presence of a conjugated enone moiety.[26b] Thus, the selective transformations of conjugated enones, a situation frequently encountered in steroid chemistry, are achieved in this manner.

The selective protection of hydroxyl groups is a problem of immense importance in the chemistry of polyhydroxylated compounds such as carbohydrates. Let us suppose that we need to carry out a selective reaction at the primary hydroxyl on **C-6** of α-methyl-D-glucopyranoside **197** (Scheme 2.89). Obviously, the protection of the other three hydroxyl groups present in this substrate should somehow be secured. The direct conversion of **197** into the partially protected derivative, triacetate **198**, is not feasible as acetylation would occur faster at the primary hydroxyl group than at the secondary ones. Therefore, one must first protect the primary hydroxyl group using the reagent capable of reacting selectively with this function. This goal is usually achieved with the help of a bulky reagent, like triphenylmethyl chloride (trityl chloride). Treatment of **197** with trityl chloride in a pyridine solution selectively yields **199**. With the primary hydroxyl group protected, the conversion of **199** into triacetate **200** becomes a routine operation, usually carried out in the same reaction vessel. There are two types of *O*-protecting groups in **200** and the difference in their properties is very noticeable. A mild acid hydrolysis is usually employed for the selective scission of the **O–Tr** bond and the desired triacetate **198** can be thus prepared without complications.

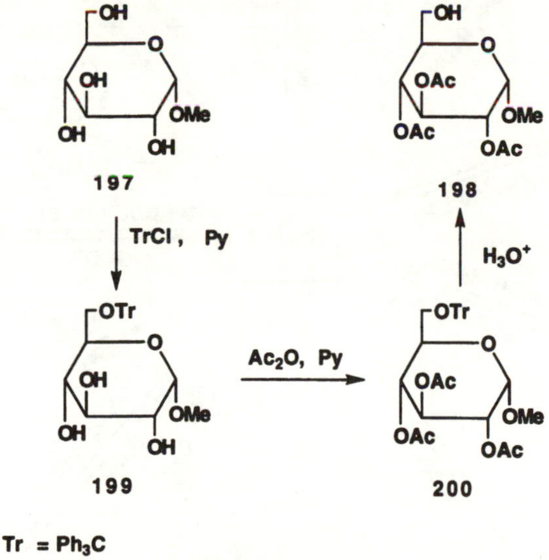

OH O OH OMe OH OH
197

OH O OAc OMe OAc OAc
198

TrCl, Py

H_3O^+

OTr O OH OMe OH OH
199

Ac_2O, Py

OTr O OAc OMe OAc OAc
200

Tr = Ph_3C

Scheme 2.89

A closer look at the above example might be instructive as to the general methodology in the use of protecting groups. The overall selectivity in this sequence was primarily achieved by the proper choice of the first protecting group, installed with a reagent capable of discriminating between the available reacting sites. At the same time, the difference in the nature of the utilized protecting groups permits the selective removal of the first one. The selectivity with respect to the introduction and the selectivity associated with removal of the protecting group are controlled by entirely different factors. In essence, this provides two powerful and independent methods of controlling the selectivity in the entire sequence.

The necessity to protect hydroxyl groups selectively is frequently encountered in a total synthesis. Therefore an 'all cases scenario' set of reactions has been created for this important functionality. Some of the more commonly encountered methods are summarized in Scheme 2.90. The list of protected compounds includes such derivatives of alcohols as esters (**201–203**), acetals (**204**, **205**), ethers (**206–209**), and organosilicon ethers (**210**, **211**).[26c,d] The methods used to introduce these protecting groups vary tremendously. All of them actually represent different versions of the same reaction type, electrophilic substitution of the hydrogen of the hydroxyl group. The main differences are in the reaction conditions, *e.g.* acidic, neutral, or basic, required to form these derivatives. The

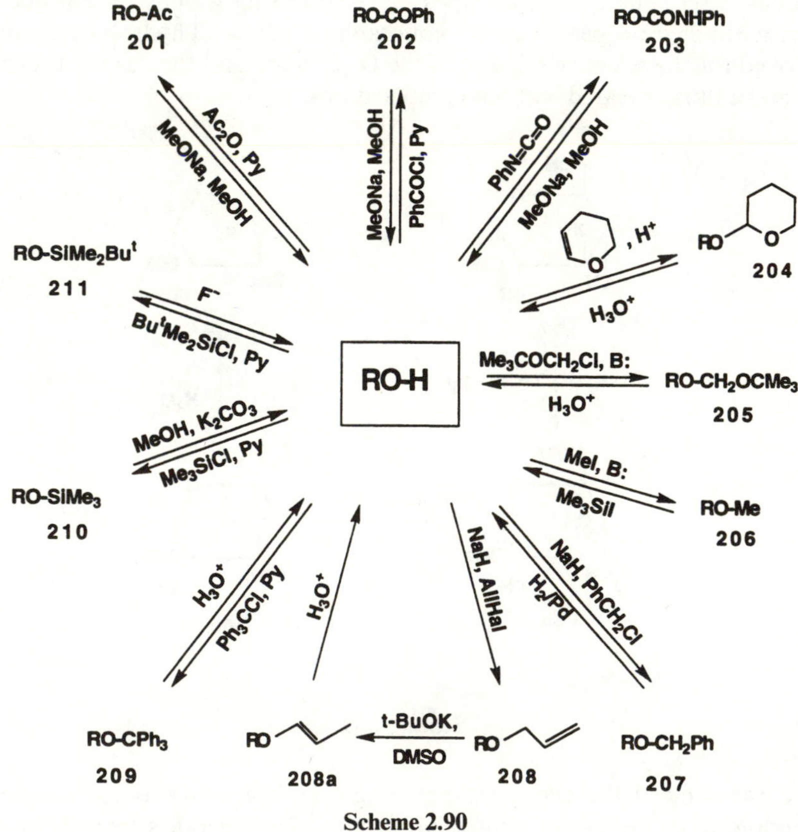

Scheme 2.90

ease of the reaction is sensitive to the specific structure of the alcohol. For example, the reactivity of alcohols decreases in the series: *n*-AlkOH > *sec*-AlkOH > *tert*-AlkOH; (equatorial)-ROH > (axial)-ROH. As we have just seen (Scheme 2.89), advantage can be taken of these differences. The diversity in alcohol protecting groups is so immense that one may safely expect to find an appropriate protective group to tolerate almost any type of reaction condition (perhaps with the exception of superacid media) necessary for the transformation of other reacting centers. Generally, ethers, acetals, and ketals are stable to basic and nucleophilic reagents as well as oxidation–reduction conditions. Esters can tolerate electrophiles, oxidants and acids, while silyl ethers are totally inert toward oxidants and reducing agents and relatively stable in the presence of electrophiles.

Release of the free hydroxyl from a protected derivative can also be carried out under a wide range of conditions, which include acid or base hydrolysis, catalytic hydrogenolysis, and the use of specific agents such as the fluoride anion in the solvolysis of silyl ethers, or trimethylsilyl iodide for the cleavage of otherwise very stable methyl ethers. Within a given class of protecting group the susceptibility to cleavage varies substantially. For example, the resistance to basic solvolysis for esters can be readily arranged in the following series: $Cl_3CCOO–R$ < $ClCH_2COO–R$ < $CH_3COO–R$ < $C_6H_5COO–R$ < $C_6H_5NHCOO–R$. Likewise, the stability of silyl ethers under solvolysis conditions can be arranged in the series: $Me_3Si–O–R$ < $Me_3CSiMe_2–O–R$ < $Me_3CSiPh_2–O–R$. Silyl protection offers an additional advantage as it can be specifically removed by unsolvated F^- under conditions that would leave other protection groups unaffected. The ethers also sharply differ in the conditions needed for the removal of the protecting group as this group changes from alkyl to allyl to benzyl to trityl. Thus, cleavage of allyl ethers is carried out in a stepwise manner, first by base-catalysed isomerization into the respective vinyl ethers, followed by hydrolysis by weak acids. The benzyl group can be removed under neutral conditions by hydrogenolysis over Pd catalysts or by one-electron reduction (Na/NH_3). Trityl and *p*-methoxytrityl derivatives, while very similar in most respects, differ so much in the rate of acid-catalysed solvolysis that it is possible to remove selectively the *p*-methoxytrityl in the presence of the trityl group.

The abundance and diversity of alcohol-protecting groups, matched to nearly all possible requirements, has given organic chemists the upper hand in dealing with syntheses that involve alcohols as intermediates and/or target molecules. Moreover, a carefully selected set of protecting groups can be installed in the initial polyfunctional compound in order to tune the whole system to a synthetic protocol involving a series of sequential reactions at more than one center.

To show how this is done in practice, we will examine Nicolaou's synthesis of zoopatenol **212**, a biologically active natural diterpenoid (Scheme 2.91).[26e] A retrosynthetic analysis of this structure envisions the disconnection of bonds **a**, **b**, and **c**, to lead to rather simple precursors, the bromo ketone **213** and the triol **214**. A formal pathway for the assemblage of **212** from these precursors involves several steps, as outlined in Scheme 2.91. This plan appears strategically

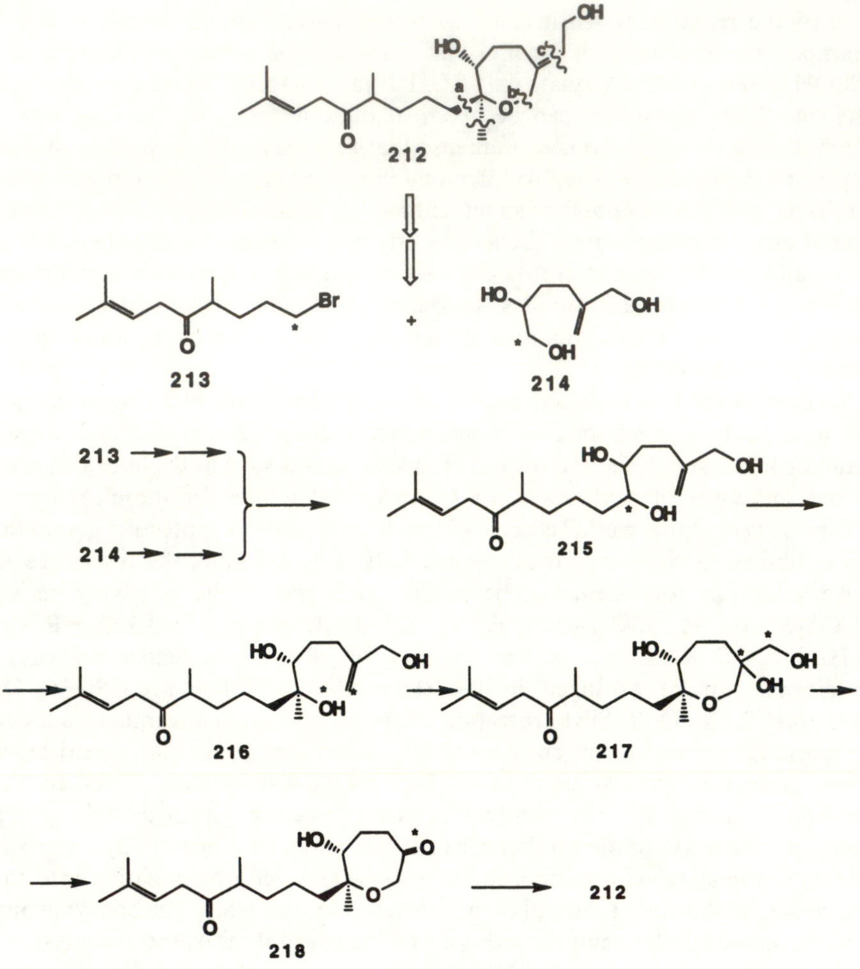

Scheme 2.91

appealing and concise. However, its viability critically depends upon the solution of problems arising from the specific functionality pattern of the target molecule. The main roadblocks one must by-pass to achieve the planned scheme include the following. (1) The coupling of **213** and **214** to give **215** should involve the formation of a Grignard reagent from the bromo ketone **213** and the preparation of the aldehyde from triol **214** by selective oxidation of one of the two primary hydroxyl groups. (Throughout this scheme an asterisk is placed to identify the center(s) affected at the particular step.) (2) Only one of the two secondary hydroxyl groups in the keto triol **215** should be oxidized. The resulting diketo diol should undergo a selective Grignard reaction with MeMgI to produce the keto triol **216**. (3) Product **216** must be oxidized selectively at the terminal double bond to give an epoxide. The intramolecular nucleophilic ring opening in the latter should involve a selective interaction with a tertiary hydroxyl group to lead to the formation of the required seven-membered ring in **217**. (4) The *vic*-glycol moiety in **217** must be cleaved with periodate and a Wittig reaction must be carried out at the keto group located in the ring fragment of **218**.

While the methods needed to achieve any of these conversions are well developed and reliable, it would be sheer madness even to think about their direct application to polyfunctional compounds such as **213–218**. The sequence of selective reactions indicated in the scheme is, of course, completely fictitious. Nevertheless, the real synthesis was carried out in full accord with this scheme, but only after the needed polyfunctional substrates were appropriately modified by a very judicial use of protecting groups (see Scheme 2.92).

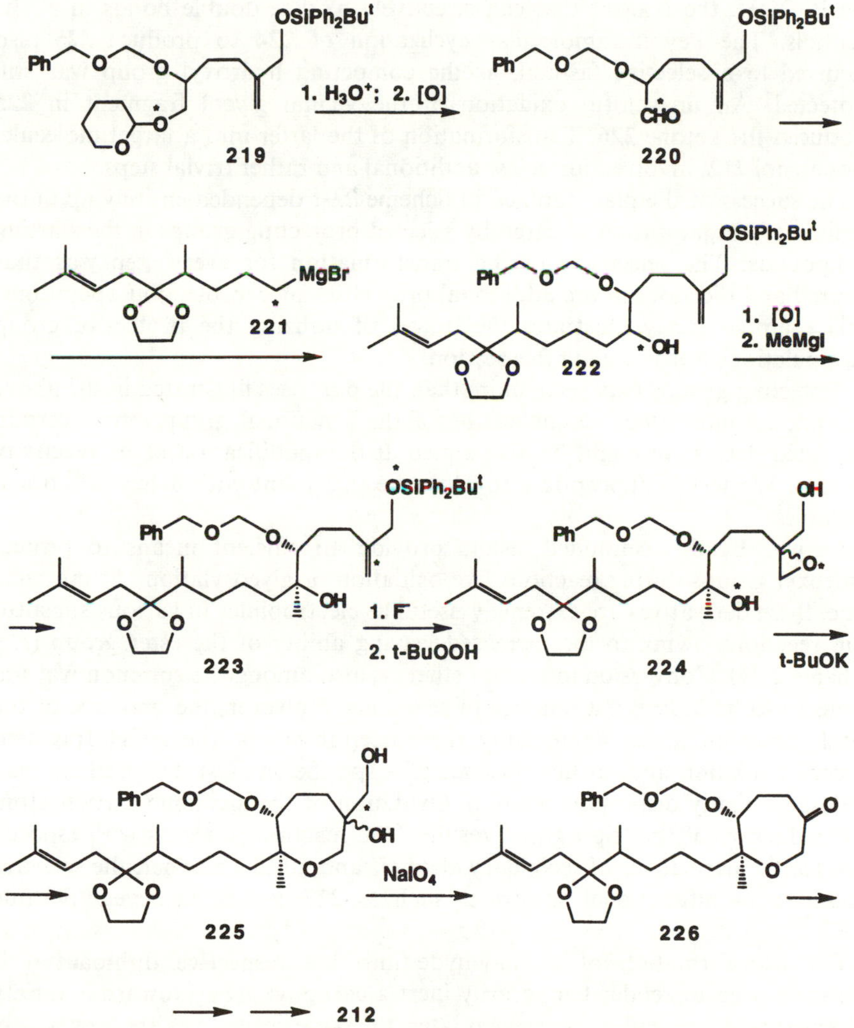

219 → 1. H_3O^+; 2. [O] → **220** → **221** → **222** → 1. [O] 2. MeMgI → **223** → 1. F^- 2. t-BuOOH → **224** → t-BuOK → **225** → $NaIO_4$ → **226** → → → **212**

Scheme 2.92

The triply protected derivative **219** was prepared as a synthetic equivalent to the triol **214**. Selective removal of the tetrahydropyranyl protecting group from **219** liberated the primary hydroxyl, to be easily oxidized to the required aldehyde **220**. Keto bromide **213** obviously cannot be used directly to form Grignard reagents owing to the presence of the carbonyl group. To remove this obstacle, **213** was converted to the respective ketal and used successfully to generate the Grignard reagent **221**. The coupling of **220** with **221**, oxidation of the resultant alcohol **222** into the respective ketone, and subsequent reaction of the ketone with MeMgI to produce **223** did not present any problems. Adduct **223** contains two double bonds but only one of them must be converted into an epoxide. To achieve this selectivity, the silyl protection was removed (by fluoride anion catalysed cleavage of the O–Si bond) and the product treated further with *tert*-BuOOH, the reagent that can selectively oxidize double bonds in allylic alcohols. The key intramolecular cyclization of **224** to produce **225** also occurred in a selective fashion, as the competing hydroxyl group was still protected. An uneventful oxidation of the vicinal glycol fragment in **225** produced the ketone **226**. Transformation of the latter into a target molecule, zoopatenol **212**, involved just a few additional and rather trivial steps.

The success of the plan outlined in Scheme 2.91 depended entirely upon the preliminary organization of carefully selected protecting groups in the starting compounds. The selectivity of the transformation for every step was thus secured and did not require additional protection or deprotection operations. This example nicely illustrates the power of utilizing the protective group manipulations introduced in this section.

Protecting groups may serve more than the purposes illustrated in the above examples. Quite often the conversion of the functional group into a certain protected derivative might be also aimed at the modification of its reactivity pattern. We feel it appropriate to illustrate this point with a few additional examples.

As we already mentioned, esters provide an efficient means to protect hydroxyl groups during reactions like oxidation or glycosylation. At the same time, these derivatives are extremely useful as electrophiles in various substitution reactions owing to the increased leaving ability of the ester group (*e.g.* Scheme 2.79). Conversion into trityl ethers is also among the common ways to protect alcohols during a number of reactions. However, the presence of the trityl group enhances significantly the susceptibility of the α-CH fragment toward oxidation and, in the presence of a specific catalyst, trityl ethers may undergo a ready disproportionation. Oxidation of the alcoholic carbon atom and reduction of the trityl group results. The reaction proceeds with especial ease for the derivatives of secondary alcohols and, in this manner, the selective oxidation of bifunctional substrates such as **227** can be achieved (Scheme 2.93).[26f]

The transformation of an aldehyde into the respective dithioacetal is routinely used to render temporarily inert a carbonyl group toward a variety of reagents. As we will see somewhat later, the same derivatives are now widely employed in synthetic practice as reactive intermediates owing to the enhanced

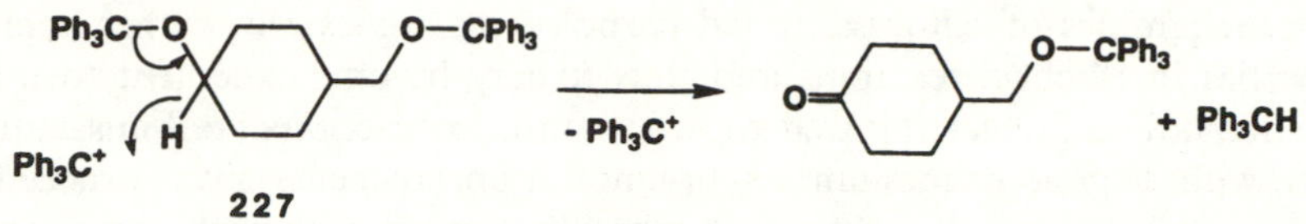

Scheme 2.93

ability of the dithioacetal moiety to stabilize efficiently an adjacent carbanionic center.

In certain cases, the selective protection of the closely similar sites offers an opportunity to achieve a reversal in selectivity from a common precursor. Thus, the monoketal derivative **228** (Scheme 2.94) can easily be prepared from the respective diketone owing to the steric shielding of the **C-17** carbonyl. The opportunity, now, to reduce the non-protected carbonyl in **228** to form alcohol **229** should not be a surprise. However, it is truly remarkable that a selective reduction can also be achieved at the **C-3** protected carbonyl. This paradoxical result is due to the utilization of the reagent H_2SiI_2, which selectively attacks the ketal moiety and induces its removal coupled with a reduction to form the iodo derivative **230**. A successful and nearly quantitative reductive conversion at either **C-17** or at **C-3** is achieved in this manner.[26g] In this example, the protected carbonyl functionality served as a non-conventional functional group with a pattern of reactivity sharply differing from that of the unprotected group.

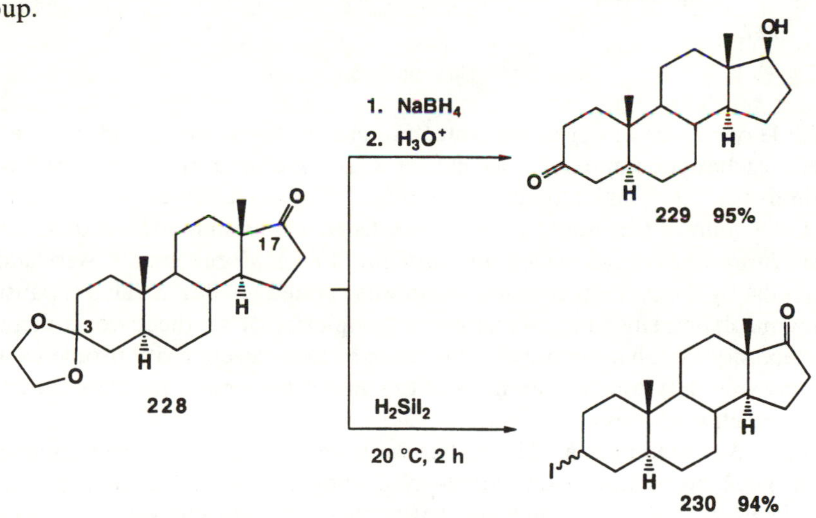

Scheme 2.94

Carboxylic acid derivatives, like esters or acid chlorides, behave as reactive electrophiles, acylating species. This activity is greatly reduced for acid amides. Thus the latter derivatives can be utilized for the protection of carboxyl groups in a number of reactions.[26c] In general, it is not considered to be a convenient protecting group because its removal may require rather drastic conditions.

However, amides of α,β-unsaturated carboxylic acids exhibit rather peculiar properties in Michael reactions and thus it may become expedient to utilize these derivatives. In fact, it is well known that the interaction of α,β-unsaturated esters with organomagnesium or organolithium reagents may proceed as competitive 1,2- and 1,4-additions. In many (but not in all!) cases, this problem can be solved by the use of cuprate reagents. The situation is changed dramatically if dimethylamides of the same substrates, for example **231**, were used (Scheme 2.95). In this case, the presence of the substituted amide group completely blocks the approach of the nucleophile to the carbonyl group and an exclusive Michael addition occurs for the set of the most diversified carbon nucleophiles.[26h] Moreover, a stepwise addition of **E** and **Nu** addends is also feasible with these Michael acceptors without the necessity of using cuprate reagents. Similar results are described for related derivatives, such as carboxylic acid trimethylhydrazides **232**.[26i]

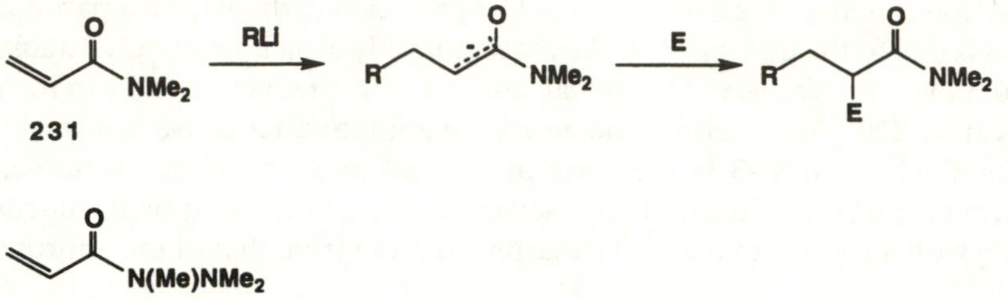

Scheme 2.95

While our focus has been on protecting groups for alcohols and, to a lesser extent, carbonyl compounds, one can present a plethora of similar systems of methods for other major functional groups.[26c,d] About 500 protecting groups for five common functional groups were listed in the first edition of Green's book, *Protective Groups in Organic Synthesis* (1981). More than 200 were added to this list by the time the second edition was printed in 1991.[26c] This expansion is not surprising since the ever-growing complexity of synthetic goals dictates the necessity of solving more and more sophisticated selectivity problems and thus provides a permanent stimulus to the searches aimed at the elaboration of new protective groups.

The background presented in the last sections enables us to resume discussion of the problems of the carbon framework assemblage, but at a somewhat higher level. Now we should comprehend that a real synthesis must solve a matrix of three major tasks: assembling the carbon skeleton, introducing the needed functionality and, finally, providing an efficient selectivity control of the reactions involved.

PART V REAGENTS. EQUIVALENTS. SYNTHONS

2.15 AN IDEAL ORGANIC SYNTHESIS. A FANTASY OR AN ACHIEVABLE GOAL?

Let us consider how we would like to see an ideal organic synthesis. By definition, a synthesis implies the construction of a molecule. Therefore it is logical to search for an analog in the area of construction of objects that are larger and more familiar to us than molecules, like a mechanical gadget or some piece of electronic equipment. In the epoch preceding the so-called technological era, inventors were supposed to be master craftsmen in order to materialize their idea in the simplest available materials, in very much the same way as Cyrus Harding had to do in Jules Verne's 'Mysterious Island'. Galileo had to grind the lenses for his telescope and Benjamin Franklin had to make his own kite. As we would say now, the innovations of our predecessors were 'patently pure', as there were no prototypes for their construction or even the basic elements needed for their purpose. In contrast, a person working these days in science or engineering meets an entirely different situation. A large selection of prefabricated details and ready solutions for particular problems, standard and highly developed technical procedures, and a wealth of all types of materials are at the constructor's disposal. Thus a scientist or inventor need not be concerned with the rudimentary details of their project and may concentrate all mental efforts on the creative part of the endeavor.

An ideal organic synthesis can be thought of in much the same light. In fact, an organic synthesis tends to evolve in this direction. Ideally, a synthetic chemist should have a stockpile of readily available reagents corresponding to standard molecular fragments and a list of standard, diversified, and reliable synthetic procedures which, taken together, enable the chemist to 'construct' any required molecule. Well, when this ultimate goal is achieved, the future chemist will not have to worry about the details of the tactics and will be able to concentrate exclusively on more challenging problems, such as the strategy of the synthesis and the design of the target structures. As a matter of fact we are still very far from that state of art and it can be even argued whether this goal is ever achievable in principle, because of the ever-increasing complexity of synthetic targets. Nevertheless, there are good reasons to believe that the gap between the complexity of the problems and the inadequacy of available tactical methods will tend to become smaller. Some basic ideas highlighting the possibilites of progress in this direction have already been formulated and they will be discussed below.

As mentioned in the beginning sections, a synthetic method could be compared to a 'black box' or an operator with the help of which one could carry out certain transformation(s) of structures. Our attention up until now has been focused on the reactions that lead to these transformations, with very little said about the 'building blocks' that could be 'installed' into the assembled molecule. We will now discuss this very important topic.

2.16 SYNTHONS AS UNIVERSAL (BUT ABSTRACT!) BUILDING BLOCKS IN ASSEMBLING A MOLECULAR FRAMEWORK AND THEIR REAL SYNTHETIC EQUIVALENTS

2.16.1 Reagents and 'Installable' Synthetic Blocks

Of course, the analogy between reagents in a synthesis and the details used in a mechanical construction should not be taken too seriously, because a reagent usually does not become incorporated into the molecule as a whole entity. For example, the highly effective and popular reagent methylmagnesium iodide is used to introduce a methyl group into a molecule and not magnesium or iodine. Likewise, there is a whole series of reagents, $CH_3CO^+SbCl_6^-$, CH_3COCl, $(CH_3CO)_2O$, and CH_3COOH, that are used to insert the same acetyl fragment into a molecule. Clearly, methylmagnesium iodide serves as a synthetic equivalent of the carbanion CH_3^- and the derivatives of acetic acid serve as synthetic equivalents of the carbocation CH_3CO^+. The rest of the reagent molecule is merely an auxiliary part, with their peculiar structure not relevant to the structure of the assembling molecule.

We have already employed in our text the term 'synthetic equivalents'. 'Synthetic equivalency' implies that the result of the actual reaction with a given reagent is equivalent to that of a virtual reaction employing the reactive intermediate of the corresponding structure. The concept of synthetic equivalency can be applied to reactions of any type and gives one the opportunity to describe the net structural outcome in general terms without the necessity to specify exact reagents.

An even higher level of generalization can be achieved by considering the possibility of subsequent transformation of the initial fragments. Thus CH_3COCl, besides being equivalent to CH_3CO^+, may also be correctly viewed as a synthetic equivalent of the fragment CH_3CH^+OH insomuch as compounds of the type CH_3COR are readily reduced to $CH_3CH(OH)R$. Furthermore, if one takes into account an equally viable opportunity to reduce the keto group into the methylene fragment (by the Wolff–Kishner method), then CH_3COCl, as well as other acetylating species, is also synthetically equivalent to $CH_3CH_2^+$. On the basis of the same reasoning, acetaldehyde, CH_3CHO, can correctly be considered to also be a CH_3CO^+ equivalent as the products of its reactions with carbon nucleophiles, $CH_3CH(OH)R$, are easily oxidized to CH_3COR.

2.16.2 The Notion of Synthons. Trivial and Not-very-trivial C_1–C_4 Synthons and Reagents

The generalized interpretation of equivalency is of great pragmatic value in planning an organic synthesis since it expands substantially the list of reagents available in achieving the required transformations. In recognition of the importance of equivalency, a special term, 'synthon', was coined by Corey.[27a] This term was suggested to designate 'structural units within a molecule which are related to possible synthetic operations'. There is no truly rigorous definition of this term. Even in the excellent monograph by Corey and Cheng, *The Logic of*

Chemical Synthesis[27b] the glossary of terms does not contain any reference to this word. The best description we were able to find was given by March in his textbook:[27c] 'synthon is defined as a structural unit within a molecule that can be formed and/or assembled by known or conceivable synthetic operations'. The meaning of the term 'synthon' may seem much too abstract since it frequently relates to somewhat fictitious, non-existent species such as $^{+}COOH$ or ^{2-}CO. Therefore every synthon must be associated with a real reagent that ensures the possibility of inserting a given unit into the molecule. We believe, however, that there is no special necessity to strive for an exact definition in this case. Far more important is the comprehension that this new term appeared not as a fashionable catch-word. Its introduction reflects a novel mode of synthetic thinking, a mode that is dynamic in its essence and is oriented to the generalized description of the results of synthetic operations. The essence and the meaning of the term 'synthon' is probably better understood with real examples such as those that follow.

2.16.2.1 The synthon $^{-}COOH$

With the help of this synthon one can effect coupling of a carboxyl group with an electrophile, a reaction of the type:

$$E^{+} + {}^{-}COOH \rightarrow E\text{–}COOH$$

Obviously the $^{-}COOH$ species cannot exist, but nevertheless a conversion described by this formal equation has been known to organic chemistry for more than a century. A very simple and familiar reagent, the cyanide anion, serves as an equivalent of the described synthon in the reaction with electrophiles to give nitriles, E–CN. The latter can be hydrolysed to the corresponding carboxylic acids, E–COOH, as illustrated below:

$$PhCH_2Cl + NaCN \longrightarrow PhCH_2CN \xrightarrow{H_2O} PhCH_2COOH$$

2.16.2.2 The synthon $^{+}COOH$

This is also a non-existent entity, but here again a very trivial reagent, CO_2, can be utilized to do the job of the electrophilic carboxyl synthon to couple with organometallic nucleophiles to produce salts of carboxylic acids (see Section 2.4).

2.16.2.3 C_2 synthons based on acetylene

Acetylene, as is well known, is acidic enough to form acetylide salts with strong bases. A monoacetylide, $HC{\equiv}C^{-}M^{+}$ (M = metal), in reactions with electrophiles ($^{+}E^{1}$) yields a monosubstituted derivative of acetylene, $HC{\equiv}CE^{1}$. The latter can undergo the same sequence of reactions to produce compounds of the type $E^{2}C{\equiv}CE^{1}$. Consequently, acetylene can be regarded as a reagent equivalent to the synthons $HC{\equiv}C^{-}$ or $^{-}C{\equiv}C^{-}$. If one further considers the possibilities for the transformation of the acetylenic fragment (for example, hydrogenation to the alkene or alkane, hydration to yield ketones, and other addition reactions), it becomes obvious that acetylene is equivalent to a family of synthons C_2^- and C_2^{2-} with almost unlimited patterns of functionalization within the C_2 fragment.

A new aspect of synthon application in the use of acetylene (and, in general, alkynes) emerged with the development of methods for the carbometallation of triple bonds in accordance with the Scheme 2.96. As can be seen from the scheme, the net contribution of acetylene in this sequence of conversions corresponds to the insertion of a bipolar C_2 unit (synthon **233** with a *Z*-configuration of the double bond) between the nucleophilic and electrophilic fragments.

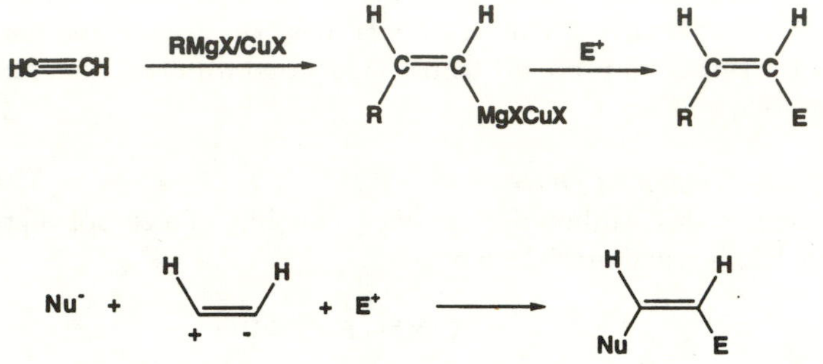

HC≡CH —RMgX/CuX→ (H)(R)C=C(H)(MgXCuX) —E^+→ (H)(R)C=C(H)(E)

Nu^- + H(+)C=C(−)H (**233**) + E^+ ——→ (H)(Nu)C=C(H)(E)

Scheme 2.96

The universal applicability of acetylene as this type of bipolar synthon and its efficiency in the synthesis of natural compounds has already been demonstrated (Section 2.3.3.).

2.16.2.4 C_4 synthons based on methyl vinyl ketone

Methyl vinyl ketone has traditionally been used in the Michael reaction as a reagent equivalent to the $^+CH_2CH_2COCH_3$ synthon, which can be combined with a variety of nucleophiles according to the scheme:

$$CH_2{=}CHCOCH_3 \xrightarrow{Nu^-;\,H^+} NuCH_2{-}CH_2COCH_3$$

Methods developed to carry out this reaction under aprotic conditions with an 'interception' of the intermediate carbanion by an external electrophile, however, allowed this same reagent to be used as an equivalent to the bipolar $^+CH_2{-}^-CHCOCH_3$ synthon, as is represented in the following general scheme:

$$CH_2{=}CHCOCH_3 \xrightarrow[2.\ E^+]{1.\ Nu^-} NuCH_2{-}CH(E)COCH_3$$

The well-known Robinson annulation reaction (Section 2.3.3) exploits a similar synthetic utilization of methyl vinyl ketone as an isomeric bipolar

$^{+}CH_2CH_2COCH_2^-$ synthon. While the term synthon was not known at the time this remarkable annulation protocol was elaborated (1937), this reaction is undoubtedly one of the earliest (and probably the first!) examples illustrating the value of a synthon designed and specifically tailored to achieve the assemblage of a complex structure.

2.16.2.5 C_2, C_3, and C_4 synthons based on the use of acetoacetic ester

Acetoacetic ester has been commonly used in the practice of organic chemistry almost for a century, again long before the term 'synthon' was conceived. However, the absence of this term did not prevent chemists from using acetoacetic ester in various situations in the role of a reagent equivalent to the nucleophilic C_2 ($^-CH_2COOH$) or C_3 ($^-CH_2COCH_3$) synthons. The development of a route to generate the dianion of this compound (Section 2.13) furthered the classical use of acetoacetic ester. In this case the reagent serves as an equivalent to the 1-C_4^- synthon or to the doubly charged synthons 1,3-C_3^{2-} or 1,3-C_4^{2-}. A list of the set of synthons corresponding to the common reagent acetoacetic ester is given in Scheme 2.97.*

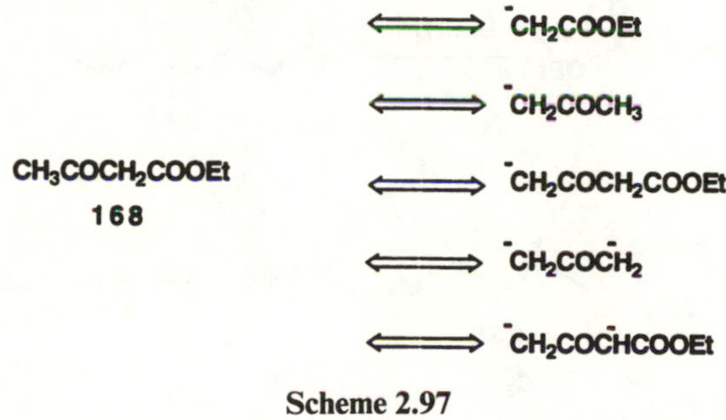

Scheme 2.97

Even with these developments, the synthetic potential of acetoacetic ester was still not completely exhausted. Notice in the transformations that not all four of the carbon atoms of this reagent are used. In the concluding step of the synthesis, the COOEt or CH_3CO group is usually removed as if it were simply an extraneous pendant. The strive to find a '100% utilization' of the acetoacetic ester carbon skeleton was realized with the development of a method for substitution at vinylic positions with the use of cuprate reagents (Section 2.12). It turns out that a similar reaction can be carried out with the enol esters of 1,3-dicarbonyl compounds.

This finding made it easy to carry out the following reaction sequence: (i) alkylation at the γ-carbon atom of a bis-anion of acetoacetic ester; (ii) *O*-acetylation of the resulting product; and (iii) reaction with a cuprate reagent to lead to the substitution of the **OAc** group by another alkyl group. This route now constitutes one of the most reliable methods to assemble stereoselectively

* We suggest the use of the symbol ⇔ to designate the equivalency relationships between reagent and synthon. In mathematical logic, this symbol means the equivalency of assertions.

molecules containing a trisubstituted double bond. This method is illustrated in the synthesis of the natural isoprenoid geraniol **234** (Scheme 2.98) by the sequence formally described as a $C_4 + C_5 + C_1$ coupling.[27d] In this scheme, 'good old' acetoacetic ester successfully plays the role of the equivalent of the totally exotic bipolar C_4 synthon **235**.

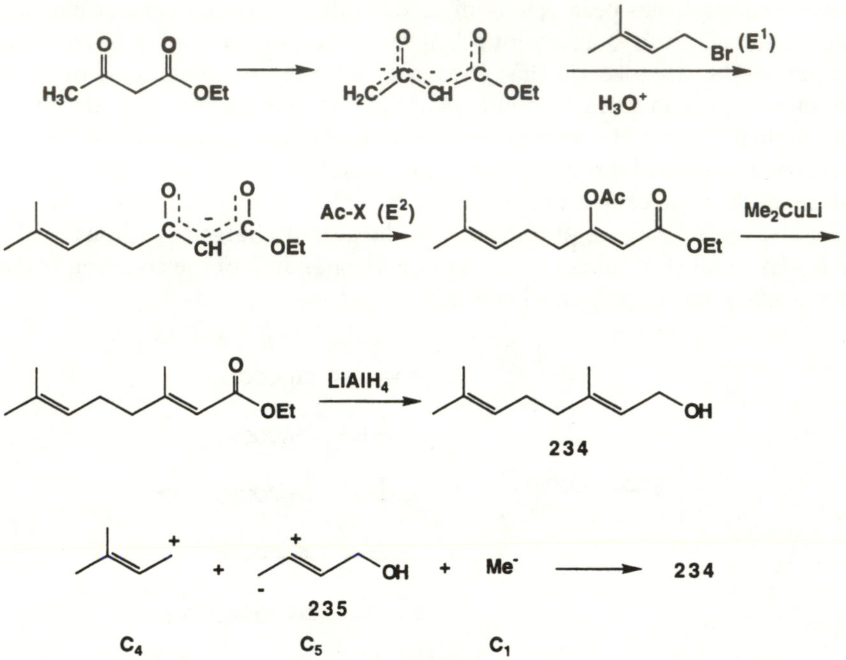

H_3C OEt → H_2C CH OEt; Br (E^1), H_3O^+ →

CH OEt; Ac-X (E^2) → OAc O OEt; Me_2CuLi →

O OEt; $LiAlH_4$ → OH **234**

+ + OH + Me^- → **234**

235

C_4 C_5 C_1

Scheme 2.98

Thus the term 'synthon' is a very capacious notion as it implies not only the *nature* of the reagents and respective synthetic methods, but also the options available for *further transformations of the groups* introduced into the molecule with the help of a given synthon. The term 'synthon' warrants some degree of appreciation. For a species to be 'worthy' of such a label it must satisfy certain criteria of synthetic significance. These criteria can be described, but again in rather general form. First of all, a truly good synthon should be able to serve for incorporation of a frequently encountered fragment into a molecule. Furthermore, the synthon must bear, in either a visible or latent form, a functional group to permit subsequent transformations and/or participation in subsequent steps of the synthetic scheme. Thus, for example, the Robinson annulation reaction with methyl vinyl ketone is a general route to create a new six-membered ring, 1,2-fused to the already existent cyclic system. This is a very important and widely encountered structural element of polycyclic natural compounds. The synthon used in this reaction has additional significance in that it contains a carbonyl group useful for further manipulations with the initially formed adducts of annulation.

Reactions applicable for a given synthon should meet the requirements of a well elaborated synthetic method. More often than not, one has to deal with a wide selection of fairly different reagents, all corresponding to the same synthon. For example, one can introduce the synthon $C_6H_5^-$ with the help of reagents such as benzene itself (*via* Friedel–Crafts reaction), phenylmagnesium bromide, phenyllithium, phenyltrimethylsilane, *etc*. This diversity implies that a corresponding set of different electrophilic reagents can be employed, and thus a number of pathways are available to insert the benzene nuclei into the desired position. It is worth noting that the utilization of this synthon is especially fruitful owing to the rich potential of the aromatic ring (which is easily converted into variety of substituted derivatives, hydrogenated, or reduced by the Birch protocol).

The diversity in nucleophilic and their complementary electrophilic reagents, in conjunction with the variability of methods for their coupling, simplifies the problem of finding an appropriate reagent for a particular structural context. In fact, the fruitfulness of retrosynthetic analysis in planning real synthetic pathways depends to a substantial extent upon the abundance of alternative synthons, corresponding reagents, and the reliability of the respective synthetic procedures.

2.16.3 The Synthon Approach as a Pragmatic Tool in Elaborating Viable Synthetic Pathways

In analysing a target structure the contemporary synthetic chemist first tries to recognize structural fragments that correspond to known synthons. Therefore, from the very beginning, retrosynthetic analysis can be directed at the most promising and economical pathways of structure assemblage, with the elimination of low probability variants. This type of synthetic planning is often referred to as the *synthon approach*. Further on we will see with real examples how effective this application can be in synthetic practice, but in the meantime we are going to examine an additional and extremely important aspect of the synthon approach: formulation of requirements for the structure of novel synthon provides a powerful impact toward the development of new reagents capable of serving as the synthetic equivalent of this synthon.

Frequently, a retrosynthetic disconnection of a structure leads to two fragments, one recognized as a well-known synthon with its corresponding set of reagents, while the other looks like a bizarre species with no obvious equivalent reagents. In such cases it makes sense to analyse carefully the latter fragment with the hope of identifying a real synthetic equivalent. Very often such a search may produce a simple, although not immediately obvious, solution. Here are some model examples.

The structure of *cis*-4-*tert*-butylcyclohexanecarboxylic acid **236** (Scheme 2.99) seems to be rather simple for retrosynthetic analysis. The most obvious options in this analysis involve the disconnection of bonds **a** or **b** as shown.

Synthons such as **237** or **238** might appear to be non-productive as the availability of reagents corresponding to these synthons may be questionable. Nevertheless, if one takes into account the options to synthesize *p*-substituted aromatic compounds and their ease of conversion into *cis*-1,4-disubstituted cyclohexanes (*via* catalytic hydrogenation), the proposed disconnections become immediately feasible. This reasoning leads to routes **A** and **B** as realistic pathways for the preparation of **236**.

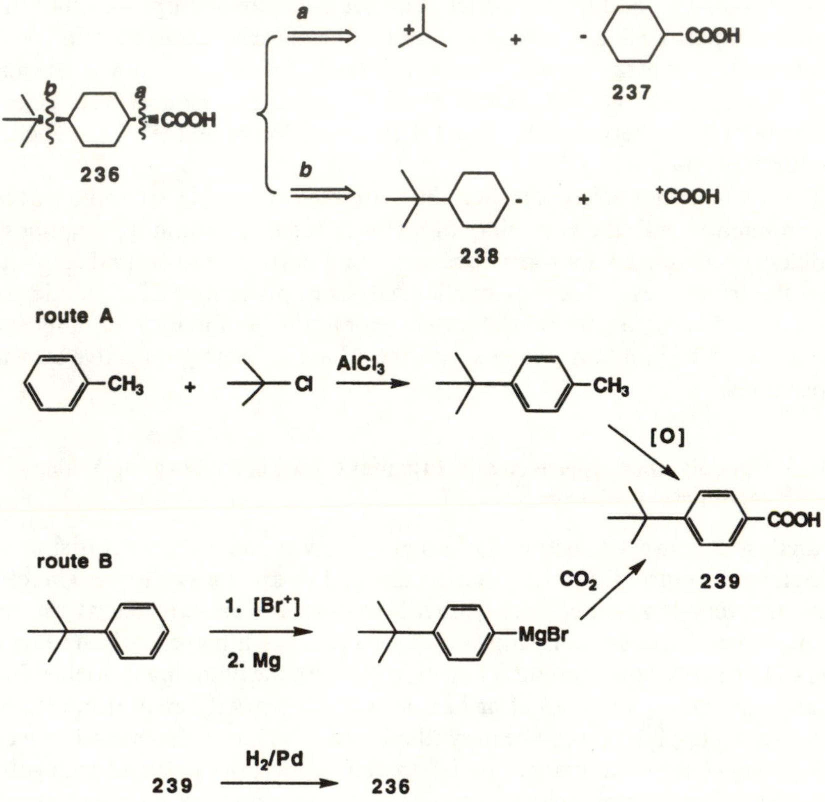

a b 236 237 238 COOH $^{+}$COOH

route A

CH_3 + Cl $AlCl_3$ CH_3 [O] COOH 239

route B

1. [Br^+] 2. Mg MgBr CO_2

239 H_2/Pd 236

Scheme 2.99

The electrophilic alkylation of toluene with *tert*-butyl chloride followed by oxidation yields carboxylic acid **239** (route **A**). Alternatively, the carboxylic acid can be prepared from *tert*-butylbenzene *via* electrophilic bromination, conversion into a Grignard reagent, and coupling with CO_2 (route **B**). Catalytic *cis*-hydrogenation of **239** proceeds without complications and leads to the target molecule **236**. In sequence **A**, toluene, C_7H_8, is actually employed as a synthetic equivalent to a paradoxical nucleophilic synthon bearing a carboxyl group, $^{-}C_6H_{10}COOH$ **237**, whereas in route **B** *tert*-butylbenzene serves as the equivalent of $C_{10}H_{19}^{-}$ synthon **238**.

Keeping in mind the availability of a Birch reaction for the partial reduction of an aromatic system, the retrosynthetic analysis of 4-*tert*-butyl-3-cyclohexenone **240** (Scheme 2.100) into the synthons $C_4H_9^+$ and **241** should not cause too much surprise. As is shown in this scheme, the readily available aromatic compound anisole **242** is a virtual equivalent to the fictitious species, the synthon **241**. All steps leading to the preparation of **240** from this precursor are absolutely reliable transformations.

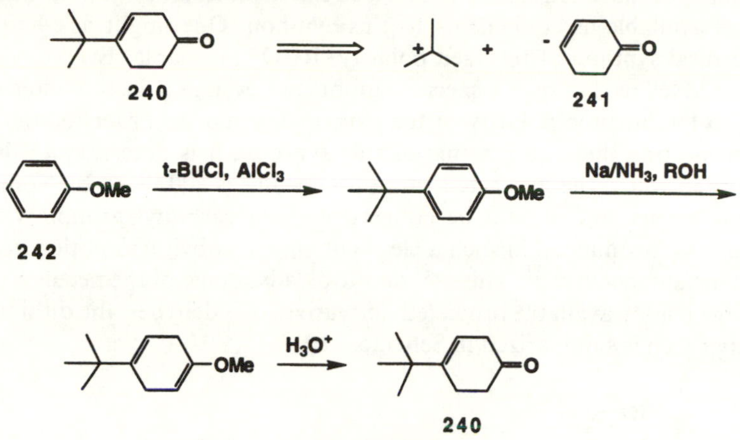

240 ⟹ + 241

242 —t-BuCl, AlCl₃→ —Na/NH₃, ROH→

—H₃O⁺→ 240

Scheme 2.100

2.16.4 Reversed Polarity Isostructural Synthons. New Horizons in the Synthetic Application of Carbonyl Compounds

The synthon approach offers an opportunity to plan a synthesis with the help of heterolytic reactions as a sort of assemblage of the target molecule from prefabricated standard polar units. The order of coupling is determined by their charges. The availability of the reagents equivalent to the structurally identical synthons, but bearing charges of the opposite signs, additionally expands the flexibility of this approach. This possibility was already illustrated in the pair of synthons $^-$COOH and $^+$COOH (see above).

Another illustration of *isostructural* synthons with reversed polarity comes from the allyl halide and allylsilane pair. Allyl halide, in reactions with standard nucleophiles, is an *electrophile*, equivalent to the synthon $CH_2{=}CH{-}CH_2^+$. Allylsilane is a *nucleophile*, equivalent to the synthon $CH_2{=}CH{-}CH_2^-$ (in Lewis acid mediated reactions with electrophilic reagents, examples in Scheme 2.41). The availability of both types of C_3 synthon permits the introduction of an allyl group to either a nucleophilic or electrophilic site and to treat the respective synthetic options as viable alternatives.

In much the same way it is possible to introduce the aryl group (Ar) either as the Ar^- or Ar^+ synthon. As we have already mentioned, hydrocarbons ArH, or organometallic reagents like ArMgBr and ArLi, can serve as the equivalents of Ar^- in reactions with appropriate electrophiles. For the less obvious electrophilic Ar^+ synthon, the aryl halides can serve as suitable equivalents under appropriate conditions. Since aryl halides are usually not prone to react with regular nucleophiles, it is necessary to use cuprate reagents, R_2CuLi, as nucleophiles or transition metal complexes as catalysts.

It is still premature to claim that the pairs of reversed polarity synthons can be envisaged for all important types of fragments. However, the trend is toward a dedicated search in this direction. Below we will illustrate the effectiveness of several synthetic solutions based upon the utilization of this approach.

The importance of the RCO^+ synthon is unquestionable for both the importance of the functionality introduced into a molecule and to the numerous reagents available as equivalents to this synthon. One might ask here if the isostructural synthon of reversed polarity, RCO^-, is viable. Everyday experience would tell us that such a species cannot exist even as a fleeting intermediate owing to the intrinsic polarity of the carbonyl group. In order to construct a reagent meeting the requirements of this synthon, it is necessary to design a structure in which the generated carbanion center is stabilized by 'something' and this 'something' should be convertible to a carbonyl group. When the problem was formulated in such a clear-cut way, an obvious solution appeared almost instantaneously.[27e] This solution took advantage of the peculiar properties of the readily available protected derivatives of aldehydes, the dithioacetals. This approach is summarized in Scheme 2.101.

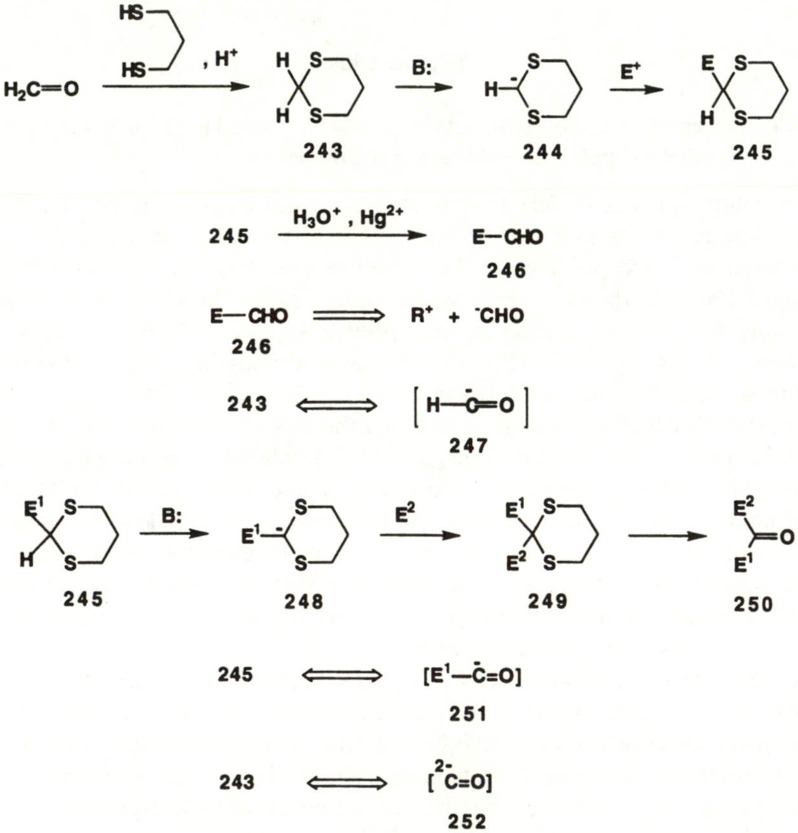

$H_2C{=}O$ —(HS–(CH₂)₃–SH, H^+)→ 243 —(B:)→ 244 —(E^+)→ 245

245 —(H_3O^+, Hg^{2+})→ E–CHO 246

E–CHO 246 ⟹ R^+ + ^-CHO

243 ⟺ [H–$\bar{C}$=O] 247

245 —(B:)→ 248 —(E^2)→ 249 → $E^1E^2C{=}O$ 250

245 ⟺ [E^1–$\bar{C}$=O] 251

243 ⟺ [^{2-}C=O] 252

Scheme 2.101

A common precursor to synthons of this type is 1,3-dithiane **243**, easily available from formaldehyde. Treatment of **243** with a strong base leads to the formation of the carbanion **244**, stabilized by the presence of two adjacent sulfur atoms. The transformation of formaldehyde into **243** is equivalent to converting a typical electrophile into a nucleophile with a carbanionic center located on the same carbon that used to be the electrophilic center in the parent compound. As might have been expected, **244** reacts readily with a variety of electrophiles to form derivatives of type **245**. Upon hydrolysis, these derivatives produce the corresponding aldehydes **246**. The described procedure exactly matches the seemingly paradoxical scheme to assemble higher aldehydes through the combination of an electrophilic E^+ group with the $H{-}C^-{=}O$ synthon **247** (formyl anion).

The intermediate dithioacetal **245** contains an ionizable hydrogen atom and the sequence: generation of carbanion **248**, interaction of **248** with an electrophile to give dithiane **249**, and hydrolysis will yield, finally, ketones, **250**. In the latter conversion the dithioacetal **245** is equivalent to the acyl anion synthon **251**. The overall conversion of formaldehyde into **250** can be described as a sequential coupling of two electrophiles, $^+E^1$ and $^+E^2$, with the doubly charged carbonyl anion **252**. The successful design of **243** as an equivalent of synthons **247**, **251**, and **252** triggered an avalanche of studies aimed at the elaboration of other acyl anion equivalents using various 1,1-bis-heterosubstituted derivatives, such as $MeSCH_2S(O)Me$, $PhSeCH_2SiMe_3$, $R^1OCH(R)CN$, *etc.* The compilation of various synthons listed in Hase's monograph[27f] contains more than 30 different reagents for the ^-CHO synthon alone! Various reagents equivalent to the same synthon may differ dramatically in their reactivity toward electrophiles, handling, and ease of preparation. The diversity of these tools makes the operation: $^+E^1 + {}^{2-}CO + {}^+E^2 \rightarrow E^1{-}CO{-}E^2$ a plausible and reliable method of general strategic importance (*umpolung strategy*).[27g]

The potential of the reverse polarity approach has been spectacularly demonstrated in a plethora of synthetic studies.[27g,h] A representative example can be found in Seebach's preparation of the antibiotic vermiculin.[27i] The key step of this synthesis involved the preparation of a polyfunctional intermediate **253** *via* the sequence shown in Scheme 2.102. The first stage of this sequence couples the formyl anion equivalent **244** with bromoepoxide **254**. The primary bromide is more active as an electrophile than epoxide and therefore, under carefully controlled conditions, the product **255** is formed selectively. Under somewhat more stringent conditions the epoxide ring present in the latter adduct reacts as an electrophile with the second acyl anion equivalent **256** to yield adduct **257**. In this sequence, **254** was used as an equivalent to the 1,4-doubly charged synthon $^+CH_2CH_2CH(OH)CH_2^+$. In the final step of this scheme, carbanion **258** was generated and reacted with dimethylformamide to produce the required product **253**. It is remarkable that all of these sequential operations are carried out in one reaction vessel without the isolation of any intermediate products. The overall yield of **253** is rather high (approximately 52%).

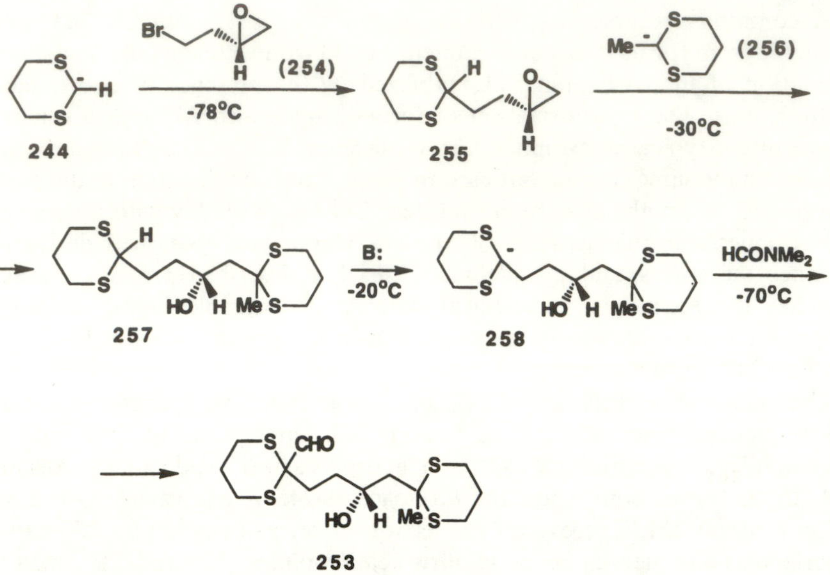

Scheme 2.102

It is instructive to consider some other types of synthons related to the chemistry of the carbonyl group. A classical procedure to introduce a substituent at the α-carbon atom of a carbonyl-containing molecule consists of converting the molecule into its respective enolate followed by alkylation with an electrophile. Modern versions of this transformation utilize various covalent enolates. In both versions, an enolate serves as an equivalent to the α-carbonyl carbanion synthon **259** (Scheme 2.103). An alternative protocol, the coupling of the α-carbonyl carbocation synthon **260** with a nucleophile, may be equally applicable to reach this goal. Here again, though, the intrinsic polarization of the carbonyl group is unfavorable for the stabilization of an adjacent positive charge. While the direct use of the parent carbonyl compounds for this purpose is prohibited, reagents equivalent to the synthon **260** have been successfully designed, for example dithioketene acetal **261**.[27j] These derivatives readily react with carbanions to give stabilized intermediates such as **262**. The reaction of the latter with electrophiles followed by removal of the dithioacetal protecting group leads to the formation of either a substituted aldehyde **263** or ketone **264** and thus demonstrates the preparative usefulness of the synthon **260**. In the preparation of aldehydes such as **263**, the reagent **261** is employed as an equivalent to the $R^1\text{–}\overset{+}{C}H\text{–}CHO$ synthon, while in the preparation of ketones such as **264**, the same reagent plays the role of the rather odd bipolar $R^1\text{–}\overset{+}{C}H\text{–}\overset{-}{C}{=}O$ synthon.

In terms of synthons, the Grignard addition to the carbonyl group corresponds to the coupling of $R^1R^2C^+(OH)$ and $^-R^3$ synthons. The reverse polarity approach can also be achieved with organometallic reagents providing they bear a protected hydroxyl group. The majority of the conventional

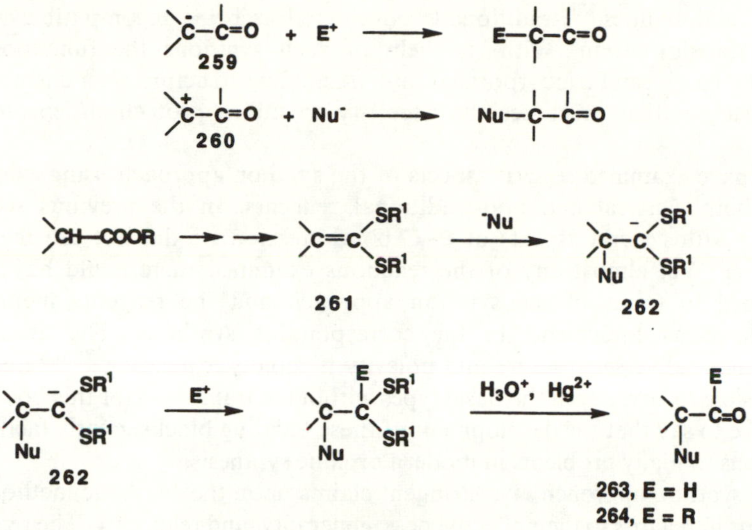

Scheme 2.103

reagents like lithium or magnesium organometallics of that type, however, are rather unstable. In addition, the high reactivity of these reagents toward electrophiles precludes the utilization of functionalized precursors for their generation. Therefore, of special interest is Knochel's finding that these limitations can be averted if one uses Zn–Cu reagents of the general formula, **265** (Scheme 2.104).[27k] Zn–Cu reagents are easily prepared from the respective aldehyde using the sequence of reactions shown.

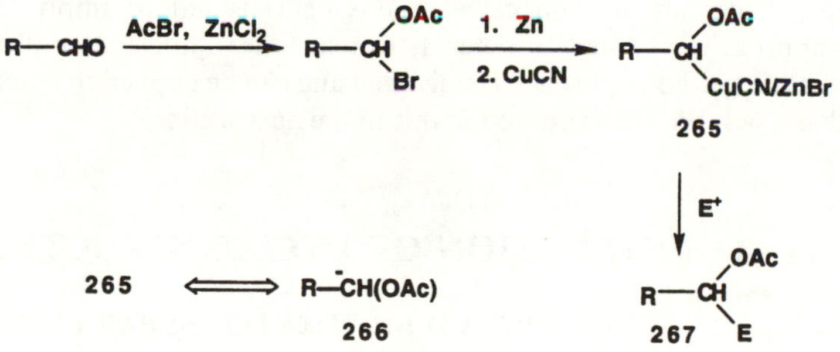

Scheme 2.104

These reagents serve as excellent transfer agents for R–$\bar{C}$H–(OAc) synthons **266** in reactions with electrophiles like alkyl triflates to give product **267**. At the same time, these mixed Zn–Cu reagents are totally inert toward electrophiles such CN or COOR^1 groups and therefore the reagent **265**, as well as its electrophilic partner, may also contain these functionalities. These examples illustrate one of the main trends in modern development of the synthon approach, namely the design of synthons and corresponding reagents which bear functionalities,[27l] traditionally considered as being incompatible with a given transformation. With the help of such synthons the functionalized building block can be incorporated into assembling structure *via* a chemoselective reaction that precludes the necessity to utilize protection/deprotection steps.

We have examined several aspects of the synthon approach using examples from both classical and non-traditional reagents. In the previous sections dealing with general aspects of **C–C** bond creation we did not use the term 'synthon'. Yet almost any of the reactions examined there could have been described in terms of the synthon approach and the reagents mentioned considered as equivalents to the corresponding synthons. The library of synthons of diverse structure and polarity is rapidly expanding[27f,m,n] and it is impossible to cover even the basic types within the framework of this book. Let it suffice to say that the development of these building blocks greatly facilitated solutions to many problems in modern organic synthesis.

The synthon approach lays stringent claims upon the synthetic method and reagents in regards to their effectiveness, generality, and reliability. The rigorous criteria associated with synthons often leads to the revision of existing synthetic methods and provides a strong motivation for their advancement or for the creation of totally different options. The present tendency to expand the list of synthons is not due to the desire to fill in 'gaps' in the chart of conceivable electrophilic and nucleophilic reagents. In its creative aspects this endeavor is closely related to modern searches toward the creation of short and flexible pathways to assemble target molecules from large standard building blocks, which requires the availability of a broad set of synthons with the most diversified structures.

Finally, it should be noted that in our discussion of synthons we used charged intermediates, carbanions and carbocations. This is not to imply that the synthon approach is limited to heterolytic reactions. Quite the contrary, the ideology of the synthon approach is universal and can be applied to reactions of all possible types. We are about to see this in the next sections.

PART VI CONSTRUCTION OF CYCLIC STRUCTURES

2.17 WHY THIS TOPIC SHOULD BE TREATED SEPARATELY

Generally speaking, the construction of a cyclic molecule composed of carbon atoms is nothing more than a particular case of carbon–carbon bond formation. Why is it necessary, then, to examine the ring-forming reactions in a special section? The answer to this question becomes apparent if we consider briefly the specific features of the transformations that should occur in order to form a cyclic structure.

Let us take a simple model situation (Scheme 2.105). Here the ring is formed from an acyclic precursor of the type **268**, where $\mathbf{C^a}$ and $\mathbf{C^b}$ bear functional groups, **X** and **Y**, capable of interacting with the formation of a new **C–C** bond.

It is easy to see that in a molecule of this type the interaction between the terminal groups may proceed in an *intra*molecular fashion, leading to the desired closed-ring product **269**, or as an *inter*molecular process to form oligomers **270**. Relative rates of reaction for these two processes may vary within broad limits, depending mainly upon the structure of the substrate. This difference can be easily understood if one takes into account some obvious ramifications of the intramolecular mode of interaction required for the ring closure.

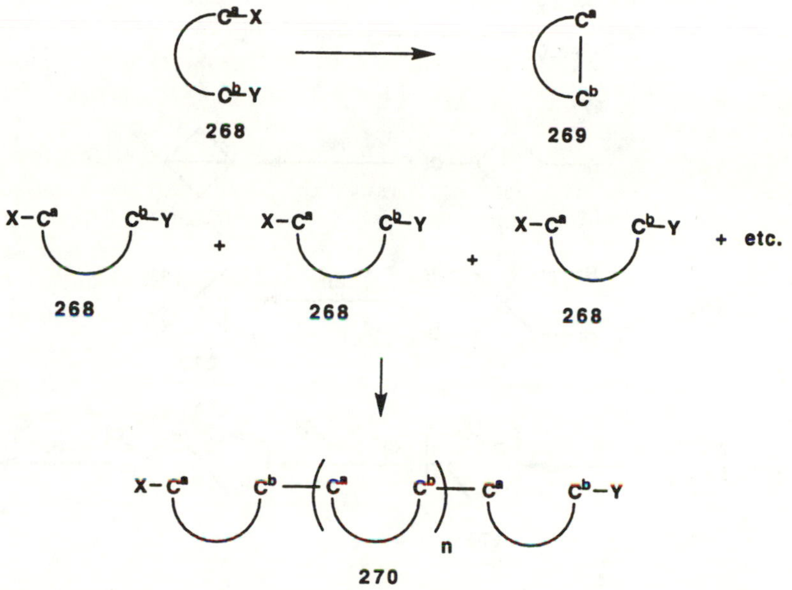

Scheme 2.105

In order for an *intra*molecular reaction to take place, groups $\mathbf{C^a{-}X}$ and $\mathbf{C^b{-}Y}$ of the same molecule must be brought together to within a proper reacting distance. The formation of the required transition state restricts the conformational mobility of the acyclic precursor and involves a substantial loss of entropy. This loss tends to increase with the length of the functional group tether. On the other hand, formation of the cyclic transition state may also involve an increase in angle strain and additional steric constraints. This enthalpy factor is also subject to changes caused primarily by the variations in the ring-size of the cyclic product.

Owing to these peculiarities, the relative ease of *intra- versus inter-*molecular processes may differ dramatically, depending upon the ring size of the target system. Below we will examine some specific problems and solutions which are typically encountered in the synthesis of various cyclic structures.

2.18 CONVENTIONAL METHODS OF ACYCLIC CHEMISTRY IN THE PREPARATION OF CYCLIC COMPOUNDS

2.18.1 Small Rings: Derivatives of Cyclopropane and Cyclobutane

The **C–C–C** bond angle for cyclopropanes is 60°, severely distorted from the preferred tetrahedral value of 109° 28′. It is not surprising, then, that the enthalpy of activation to convert a 1,3-bifunctional acyclic precursor such as **271** (Scheme 2.106) into a cyclopropane system would be rather high. This

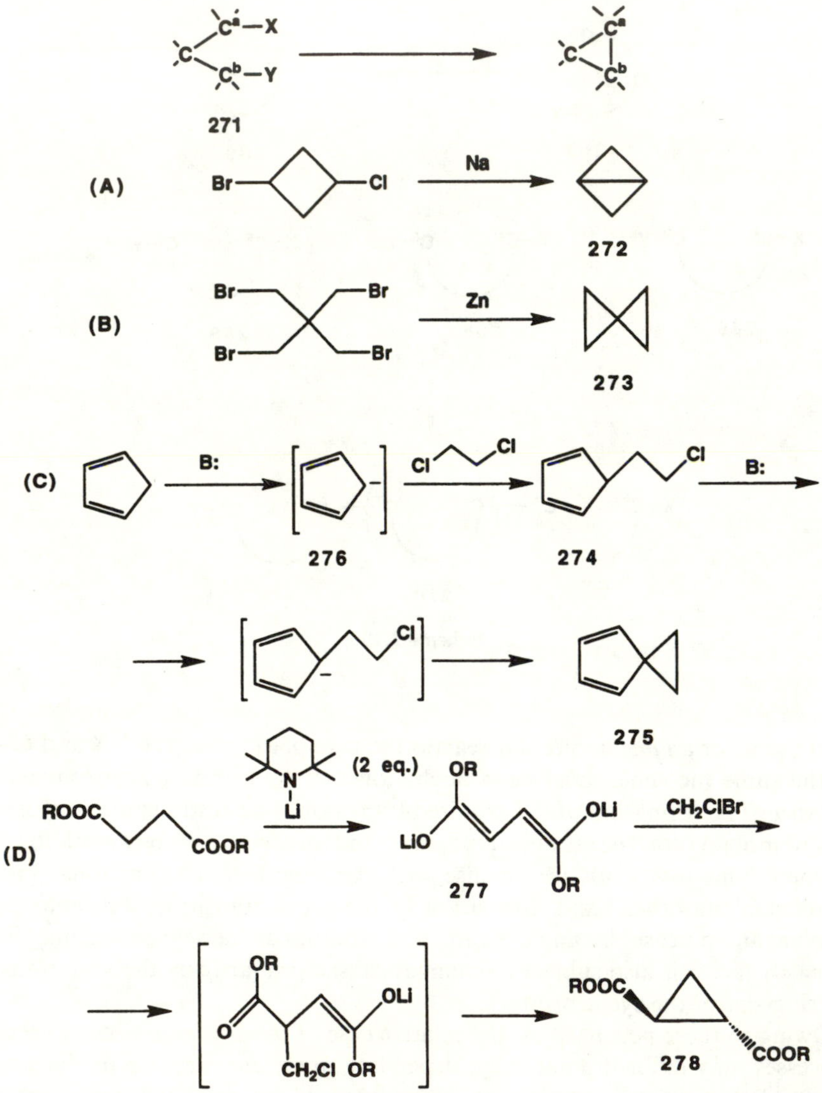

271

(A) Na 272

(B) Zn 273

(C) B: 276 274 B: 275

(D) (2 eq.) 277 CH_2ClBr 278

Scheme 2.106

cyclization may seem to be less favorable than the intermolecular coupling. However, the reaction centers in this type of precursor are in close proximity and the entropy loss to form a cyclic transition state is rather small. The latter factor is sufficient to overcome the unfavorable enthalpy changes. As a result, a number of conventional methods based upon the transformations of 1,3-disubstituted compounds can be applied to the synthesis of cyclopropane derivatives (see Scheme 2.106).

Reactions **A**[28a] and **B**[28b] represent typical examples of the Wurtz coupling. In these cases, the coupling proceeds with amazing efficiency even though the products, **272** and **273**, are rather strained compounds. The formation of the 1,3-bifunctional precursor **274** in reaction **C**[28c] is accomplished by the *inter*-molecular reaction of the carbanion **276** with one of the reaction centers of 1,2-dichloroethane. The subsequent formation of the three-membered ring is carried out using the same type of chemistry, but in its *intra*molecular version. The net outcome of this reaction sequence corresponds to the coupling of the 1,1-bis-carbanion with a 1,2-bis-carbocation to give **275**. A complementary sequence based upon the combination of 1,2-bis-carbanion with 1,1-bis-carbocation can also be used in cyclopropane synthesis. Thus a sequential alkylation of the 1,2-bis-enolate **277** (easily derived from succinic esters) with CH_2ClBr (reaction **D**) was elaborated for the stereospecific synthesis of the *trans*-1,2-cyclopropanedicarboxylic esters **278**.[28d] It should also be noted that three-membered rings of simple heterocyclic compounds, epoxides, are easily formed by an *intra*molecular version of an S_N2 reaction of the appropriate β-substituted alkanol derivatives.

The situation for cyclobutanes is similar except that the formation of a four-membered ring is energetically less demanding because of lesser angle strain in this cyclic system. The reacting groups at C^a and C^b, however, are also further apart in the acyclic precursor **279** (Scheme 2.107), and the probability of their encounter, required for an intramolecular interaction, correspondingly reduced. Therefore, while it is possible to make cyclobutanes by the same pathways as cyclopropanes, cyclobutane forming reactions tend to be less selective and likely to give more intermolecular products.[28e] The Wurtz reaction is especially applicable here for the formation of compounds containing a cyclobutane ring fused to an already present cyclic system, such as those in **280**[28f] and **281**.[28g] Synthesis of the latter compound is an example of the Wurtz coupling based upon *in situ* formation of labile silver organic intermediates.

The utilization of conventional **C–C** bond-forming reactions to construct cyclopropanes and cyclobutanes can be illustrated in a number of additional examples. However, there are many of totally different routes specifically targeted at the synthesis of these structures that are more efficient and versatile. We shall consider these reactions in the later sections.

2.18.2 Five- and Six-membered Rings

Distortions in valence angles and the strain resulting from the interaction of non-bonded groups in the cyclic transition states leading to the formation of

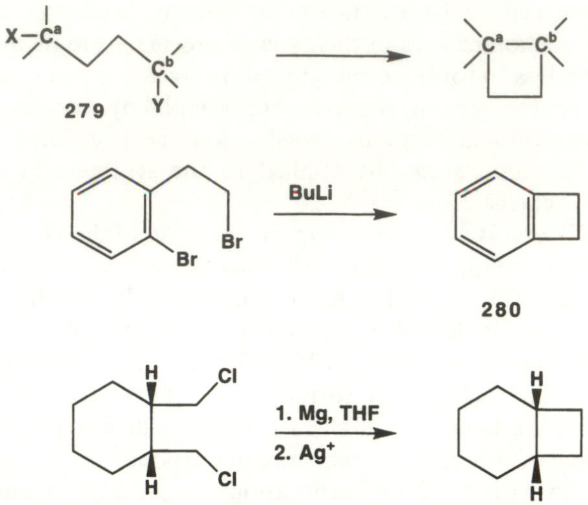

X–Cᵃ ... Cᵇ–Y 279 → Cᵃ–Cᵇ cyclobutane

BuLi → 280

1. Mg, THF 2. Ag⁺ → 281

Scheme 2.107

five- and six-membered rings are minimal. Therefore the corresponding enthalpy of activation is about equal to that of the corresponding intermolecular process. This is one of the main reasons explaining a well-known tendency of various organic molecules to undergo conversion into five- or six-membered rings almost spontaneously if given the slightest opportunity. In these cases, competitive intermolecular reactions are almost completely excluded, and nearly all of the usual methods for the creation of **C–C** bonds are applicable as cyclization methods. Below we shall consider only few of the practically useful examples.

The intramolecular interaction between an enolate and a carbonyl electrophile to form a six-membered ring is a well-known and general method (*e.g*. the Robinson annulation, see Section 2.3.3). This and related cyclization reactions involving interactions between 1,5-dicarbonyl moieties proceeds with high selectivity (**A**), and the alternative option, the formation of a four-membered ring (**B**), is much less favorable and rarely observed (Scheme 2.108). The usefulness of this method for the preparation of compounds containing the cyclohexenone moiety is abundantly documented in the literature.[14b]

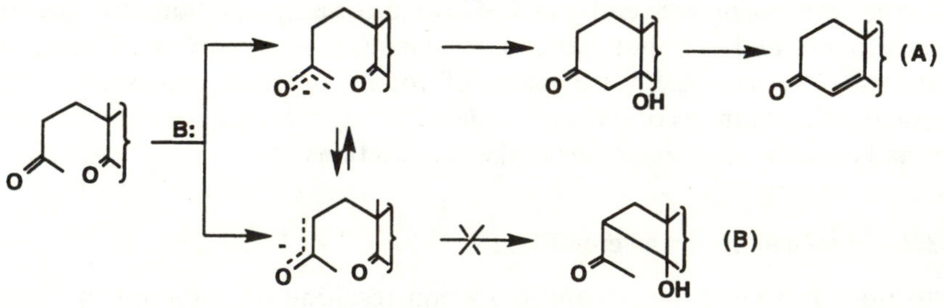

B: (A) (B)

Scheme 2.108

An equally high ring-size selectivity is observed in the cyclization of 1,4-diketones such as **282** (Scheme 2.109). Here again an intramolecular crotonic condensation may yield two products, containing either a three- or a five-membered ring. The reaction, however, invariably follows the latter pathway and serves as a reliable method to synthesize cyclopentenone derivatives.[29a] A similar reaction for 1,6-diketones, for example **283**, ultimately yields acylcyclopentenones. The competing pathway leading to the cycloheptenone derivative is virtually blocked.

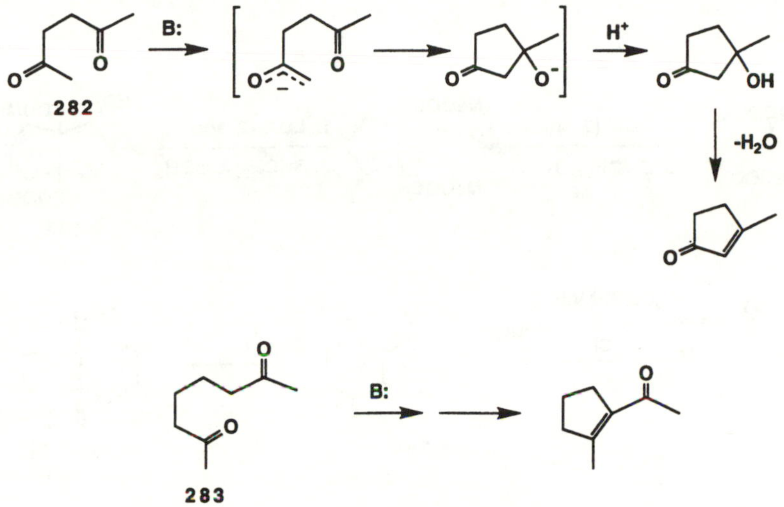

Scheme 2.109

Intramolecular versions of reactions other than aldols can also be considered as useful options to prepare five- or six-membered rings from their corresponding bifunctional precursors. Several examples to illustrate the diverse approaches to construct five-membered rings are given in Scheme 2.110.[29b] A high-yield method to prepare cyclopentenone **284**[29c] is given in the sequence: (i) alkylation of formyl-anion equivalent **285** to give **286**; (ii) Michael addition of the latter to methyl vinyl ketone; (iii) removal of the carbonyl protection; and (iv) intramolecular cyclization of the 1,4-diketone, **287**.

A remarkably short synthesis of the fused bicyclic system **288** was achieved through an ingenious exploitation of a bis-enolate alkylation.[29d] In general, this sequence corresponds to the coupling of two different 1,3-C_3^{2+} synthons with the tetradentate 2,3-C_4^{4-} synthon.

A Michael addition using the bifunctional reagent **289** was employed for the preparation of the bicyclic compound **290**. The initially formed Michael adduct **291** undergoes intramolecular enolate alkylation to give the target compound.[29e] The employment of **289** as a substituted 2,4-C_4^{-+} synthon in this sequence enables the formation of an *exo*-methylene derivative, otherwise difficult to prepare.

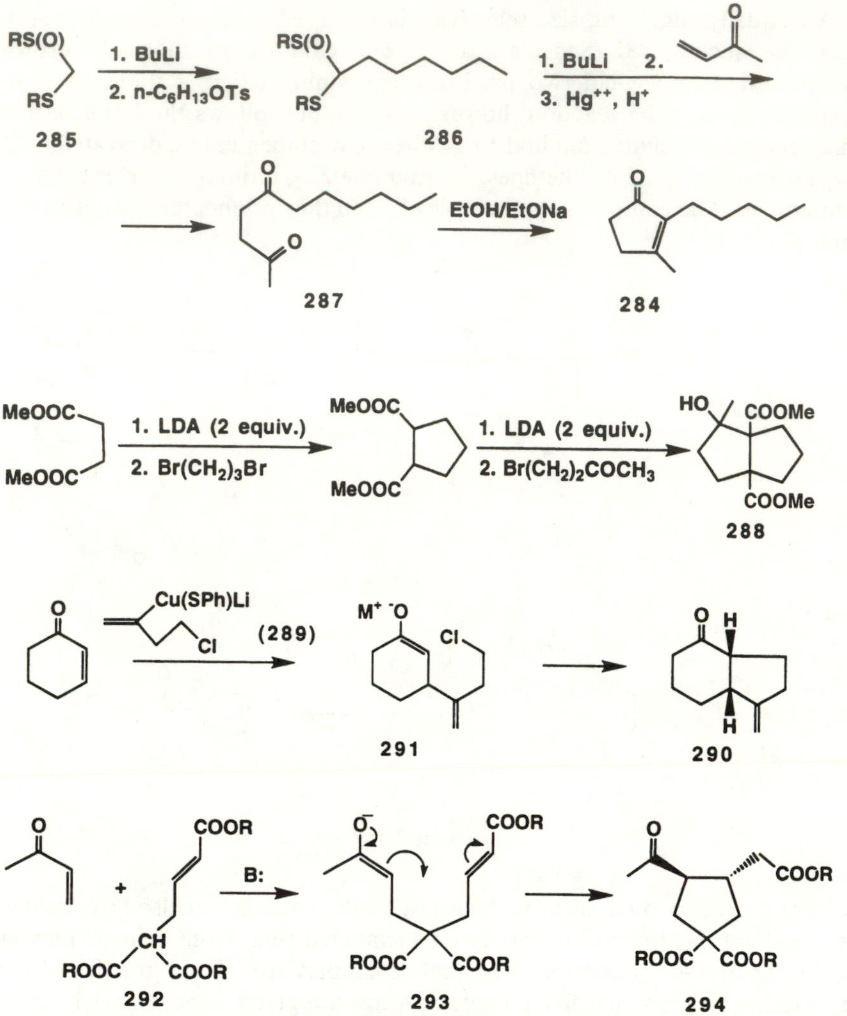

Scheme 2.110

The substrate **292** was specifically designed to prepare five-membered systems *via* a sequence of inter- and intramolecular Michael additions.[29f] This option is viable when active Michael acceptors such as methyl vinyl ketone are used, otherwise competing Michael reactions between two molecules of **292** are difficult to avoid. The reaction proceeds through intermediate **293**, which contains both a Michael donor and acceptor site and undergoes spontaneous conversion into the cyclic product **294**.

The transformation of 1,4-dien-3-ones such as **295** (Scheme 2.111) into cyclopentenone **296** represents an example of the preparative utilization of carbocationic chemistry for cyclopentenones. This reaction, known as Nazarov's cyclization, was discovered almost half a century ago.[29g] While the initial conversion was carried out with strong acids under rather drastic conditions, the synthetic promise of this method led to the elaboration of a number of milder methods to carry out this transformation.[29h] Among the methods available to prepare the required precursors, one of the most flexible involves a 1,2-vinyl Grignard addition to the respective enals followed by the oxidation of the resulting alcohols (Scheme 2.111).

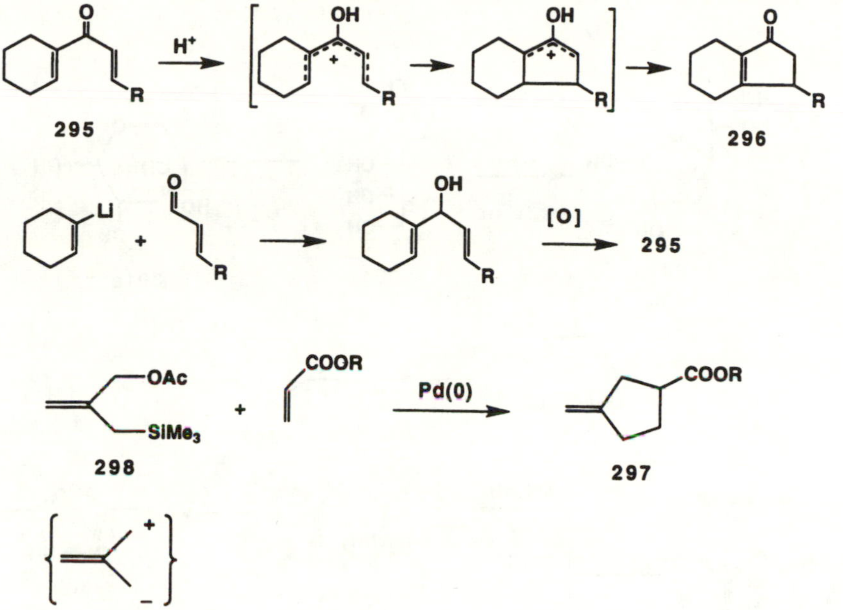

Scheme 2.111

Transition metal chemistry provides an especially rich source of methods for the construction of cyclic compounds. The idea to synthesize **297** from precursor **298** comes from the recognition of **298** as an equivalent to the 1,3-bipolar synthon **298a**, trimethylenemethane. It was reasoned that intermediates equivalent to **298a** might be formed as transient species and stabilized by complexation with transition metals, *e.g.* Pd(0). With reagents such as Michael acceptors they can be used for an efficient cyclopentanoannulation.[29i]

The enhanced tendency toward the formation of five- and six-membered rings is also widely exploited in methods developed to synthesize heterocyclic systems. In these cases, ring formation involves the creation of a carbon–heteroatom bond. A set of representative examples is shown in Scheme 2.112.

Formation of maleic anhydride **299** or valerolactone **300** proceeds easily upon moderate heating of the acyclic precursors in the presence of acid catalysts or even spontaneously. It is especially easy to form five- or six-membered hemiacetals from the respective hydroxycarbonyl compounds. Thus, in aqueous

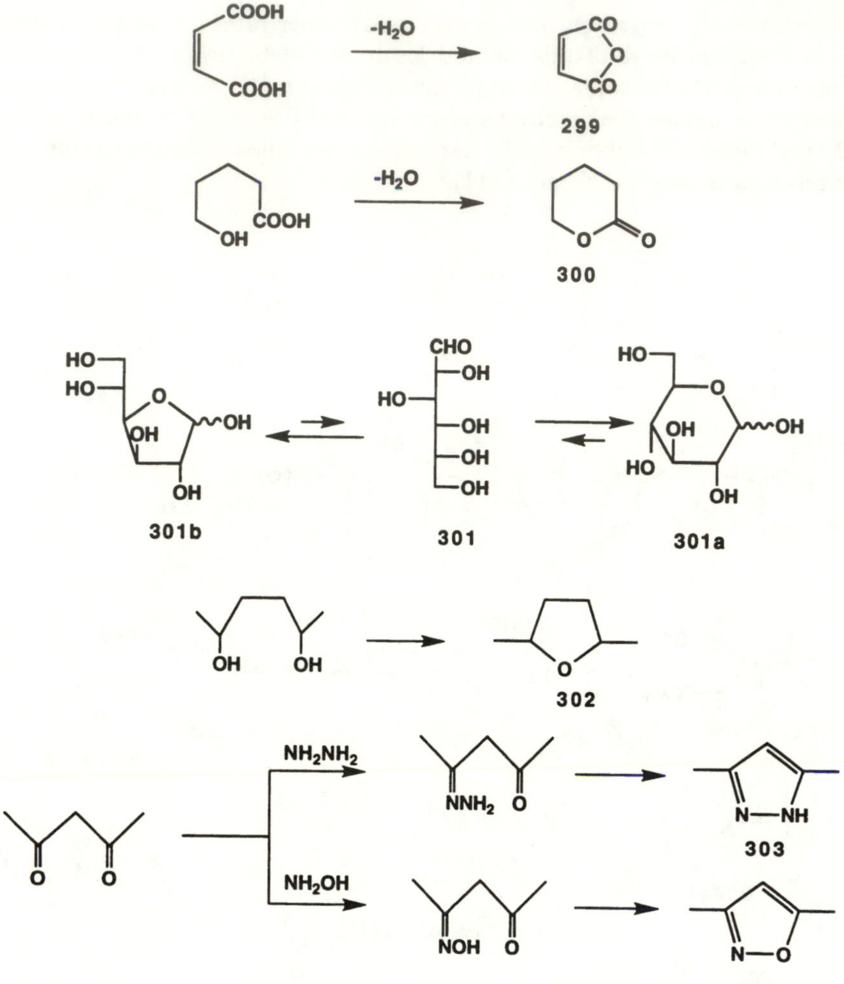

Scheme 2.112

solution, the acyclic form of D-glucose **301** is present only in insignificant amounts, while the equilibrium is shifted nearly completely toward the cyclic forms, **301a** and **301b**, with **301a** the predominant component. In fact, the acyclic forms of monosaccharides can only exist if either the aldehyde group or the hydroxyl groups at **C-4** and **C-5** are protected.

Conversions of 1,4-diols into the tetrahydrofuran derivatives, like **302**, take place easily and selectively under acid catalysis. An analogous intermolecular version, leading to oligomers, requires much harsher conditions.

The standard approach for the construction of a variety of five-membered rings containing **O**, **N**, or **S** heteroatoms is based upon the cyclization of the corresponding derivative of a 1,3-dicarbonyl compound. This is illustrated in the cyclization of the monohydrazone and monooxime of 1,3-pentanedione to form pyrazole **303** and isoxazole **304** derivatives, respectively.

It is also appropriate here to remind the reader that carbonyl compounds or diols can be protected *via* formation of cyclic acetals or thioacetals. These reactions, as we have already seen, occur under mild conditions owing to the ease of formation of five- or six-membered heterocyclic rings.

2.18.3 Rings of Larger Size. Principles of Macrocyclization. Effects of Multisite Coordination to a Binding Center

Cyclizations to form larger rings require the formation of a cyclic transition state from long-chain acyclic precursors which can adopt numerous conformations. This requirement implies a significant loss of entropy due to the coiling of the extended acyclic precursor. The increased entropy demands[30a] affect the rate of the intramolecular reaction substantially, as can be seen from the comparison of rate data for the formation of homologous lactones in the reaction of **305** → **306**[30b] (Scheme 2.113):

Ring size (n)	*Relative rate (at 50 °C)*
5	1.5×10^6
6	1.7×10^4
7	97.3
8	1.00
9	1.12
10	3.35
11	8.51
12	10.6
13	32.2
14	41.9
15	45.1
16	52.0
18	51.2
23	60.4

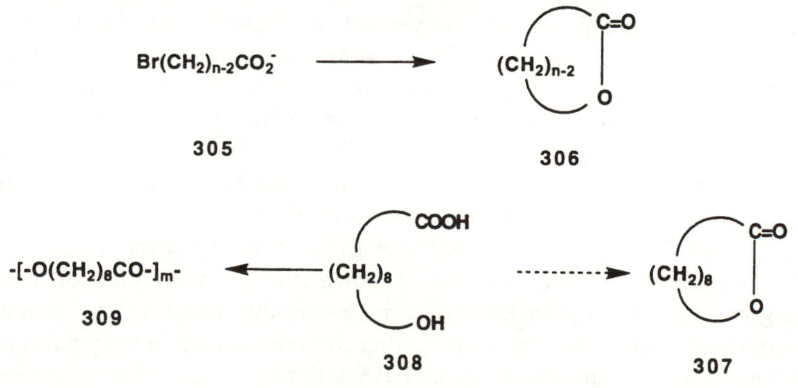

$Br(CH_2)_{n-2}CO_2^-$ → $(CH_2)_{n-2}$ lactone (C=O, O)

305 **306**

$-[-O(CH_2)_8CO-]_m-$ ← $(CH_2)_8$ (COOH, OH) ----→ $(CH_2)_8$ lactone (C=O, O)

309 **308** **307**

Scheme 2.113

As the rate of the intramolecular reaction decreases, the formation of oligomeric materials may become predominant. Thus, the attempt to form the ten-membered lactone **307** directly from the hydroxy acid **308**, under conditions optimal for the formation of a five- or six-membered lactone, would invariably lead to the formation of the oligomeric material **309** as the major product.

There are two obvious options to change the course of events and direct the reaction along the desired pathway: suppress the intermolecular reaction or facilitate the intramolecular one. Conventional approaches to control the selectivity of competing reactions, which were discussed in the preceding sections of this chapter, are not applicable here as both reactions are mechanistically very similar. Nevertheless, the problem is solvable.

The classical method to effect a macrocyclization was developed in Ziegler's studies in the 1930s.[30c] This method utilizes an approach based upon the *high dilution* technique. Under these conditions the probability of intermolecular collisions between the reacting partners is dramatically reduced and the formation of oligomers suppressed. At the same time, an intramolecular reaction does not necessitate the interactions between molecules and therefore high dilution does not affect the efficiency of cyclization. This method is rather universal and from the 1930s to the 1950s numerous compounds with medium to large rings were synthesized by this approach.[30c] The technical inconvenience of this method (small amounts of product and large volumes of solvent), plus its still low efficiency for the preparation of medium-ring sized systems, made it mandatory to develop alternative routes based upon the selective facilitation of the intramolecular reaction.

The problem of macrolactonization became especially crucial in the 1960s as intensive studies were directed at the synthesis of biologically active antibiotics containing the macrocyclic lactone (macrolide) moiety. As a result of these studies,[30d] the problem of macrolactonization, regardless of the size of lactone ring or the complexity of the acyclic precursor, was solved.

One of the most successful solutions takes advantage of double activation of both ends of the cyclizable substrate and is illustrated for a general case of the transformation of the ω-hydroxy acid **310** into lactone **311** (Scheme 2.114). In the initial step, **310** is converted into the respective 2-pyridinethioester **312**. The addition of **312** to refluxing xylene yielded the required lactones in generally good yields, even for n = 10–14. It was suggested that the observed tendency towards intramolecular esterification *versus* the intermolecular option is due to the ease of formation of betainic intermediates, **312a** and **312b**, which drive this electrostatically engineered cyclization.[30e] The preparation of the comparatively simple natural 12-membered macrolide recifeiolide **313** in 52% yield from the hydroxy acid **314** is given as an example of the efficiency of this protocol.[30f]

A fairly different approach to enhance the probability of the intramolecular pathway emerged with the discovery of crown ethers as a new class of complexing agents. Pedersen's pioneering experiments in this area revealed an unusual phenomenon. The 18-membered polyether **315** was formed in high yields by the condensation of two moles of catechol **316** with two moles of bis(2-chloroethyl) ether **317** (Scheme 2.115).[2c] The result was truly remarkable as, in this case, the

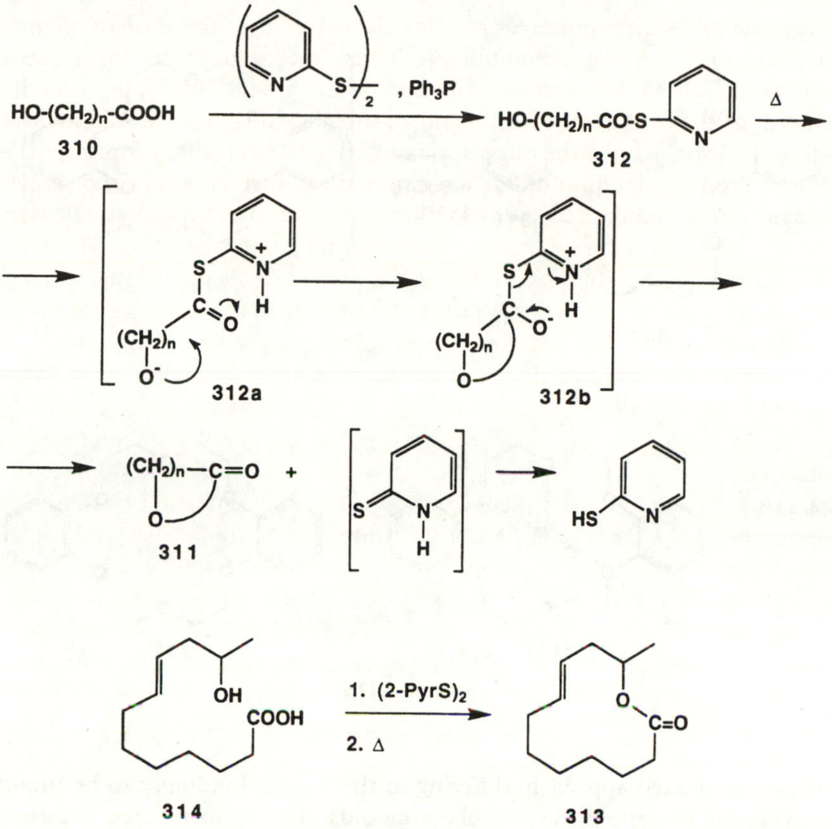

Scheme 2.114

macrocyclization proceeded efficiently without high-dilution conditions. In fact, up to 1 mole (360 g) of the product was prepared in a volume of 5 liters of solvent!

The striking efficiency of this reaction was correctly ascribed to the forced proximity of the reacting centers induced from the template effect of the sodium ion during the cyclization step. It was suggested that, owing to the coordination effect of the metal ion, the cyclization precursor **318** is formed with a set of donor sites (six ether oxygens) wrapped around the central ion. This pseudocyclic conformation is ideal for the geometry of the transition state required for the closure of the 18-membered ring and hence the formation of **315a** is greatly facilitated. Decomplexation of the latter produced crown ether **315**.

The same reaction performed in the presence of lithium or ammonium bases invariably led to the formation of linear oligomers. Further studies documented the universal validity in using multisite binding and an entry into exciting new areas of organic chemistry was opened. A more detailed discussion of the main trends and achievements in this field will be given in Chapter 4.

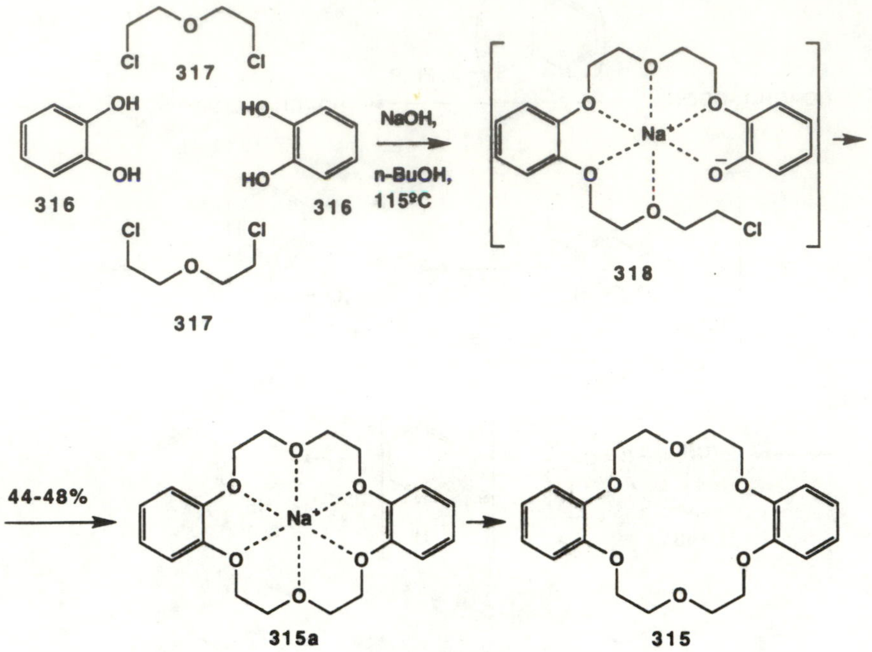

Scheme 2.115

A closely related approach, differing in the type of binding, can be found in the synthesis of certain macrocyclic alkaloids. The 13-membered macrocycle **319** (Scheme 2.116) constitutes the basic structural unit of the alkaloid celacinnine. The synthesis of **319** could have been envisaged *via* cyclization of the corresponding acyclic triamine **320**, which contained the required number of atoms in the chain and the appropriate functionalities at the ends of the chain.[30g] This route required, however, a regioselective intramolecular interaction between the ethoxycarbonyl group and the terminal amino group at the complete exclusion of the more favorable intermolecular reaction or competing intramolecular cyclization to the nine-membered product. The problem is actually unresolvable without auxiliary components to secure the opportunity to form the folded conformation of **320**. In this case, advantage was taken of the well-known ability of boron to form three covalent **B–N** bonds capable of hydrolytic splitting in mildly acidic conditions. In particular, earlier it had been shown that the interaction of triamine **321** with the boron reagent **322** proceeded smoothly with the selective formation of the bicyclic boron-containing heterocyclic compound **323** (see reference in ref. 30g).

A similar reaction with precursor **320** proceeded in a truly spectacular way and the required compound **319** was formed in 77% yield.[30g] The reaction route obviously involved the initial formation of the boraza heterocycle **320a**, an analog of compound **323**. The rigid structure of **320a** and the resultant proximity of the ethoxycarbonyl and amino functions secured the ease of the

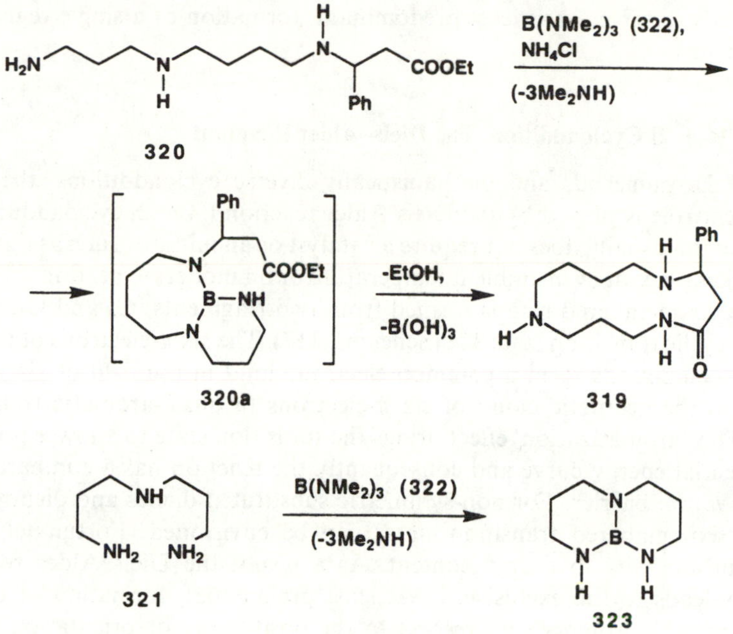

Scheme 2.116

final reaction step, namely formation of the lactam bond with concomitant protolysis of the **B–N** bond.

In this reaction there was also no necessity to resort to a high-dilution technique. The temporary covalent binding of the multidentate ligand around the central boron atom served as a scaffold to create an intermediate structure with the geometry required to form selectively the desired product. Later we will see examples that illustrate the breadth to which coordination around a central atom or ion is applied as a powerful tool to ensure pre-organization of the substrates.

2.19 CYCLOADDITIONS: METHODS SPECIFICALLY DESIGNED FOR THE FORMATION OF CYCLIC FRAMEWORKS

It is not difficult to recognize a common feature shared by the above-mentioned methods of ring formation. The ring-closure step is carried out as an intramolecular single bond formation from an acyclic precursor. There exist, however, many synthetic methods that are based upon an entirely different principle. In these methods the ring closure takes place either intra- or intermolecularly, but proceeds with the formation of two or more bonds in a single reaction event. This reaction type is known as cycloaddition. The intrinsic feature of these reactions is the increased selectivity exhibited in ring formation. The very mechanism of the interaction precludes the possibility of such side reactions as oligomerization or the formation of rings of other sizes. Cycloaddition also favors the exclusive, or at least predominant, formation of a single regio- and stereoisomer.

2.19.1 [4 + 2] Cycloaddition. The Diels–Alder Reaction

Among the numerous and mechanistically diverse cycloadditions, the most prominent role is played by the Diels-Alder reaction ([4 + 2] cycloaddition).[2a] This reaction usually does not require a catalyst or an initiator such as light and often takes place at even ambient temperature or on moderate heating.[31a,b] As a result, a six-membered ring is formed from two fragments, C_4 and C_2, *via* an ordered cyclic transition state **324** (Scheme 2.117). The six π-electrons of the two starting compounds form a common electron cloud in the field of six nuclei, similar to the aromatic cloud of six π-electrons (a quasi-aromatic transition state). This 'aromatization' effect brings the transition state to a lower point on the potential energy curve and consequently the reaction has a comparatively low activation barrier. For non-symmetric substituted dienes and dienophiles, several six-membered transition states can be envisioned. Fortunately they differ substantially in energy content. As a result, the Diels–Alder reaction typically leads to the exclusive, or at least preferential, formation of one of several possible isomers with respect to the positioning or orientation of the substituents. It is important to emphasize that the reaction course is well-accounted for in terms of the Woodward–Hoffmann concept of conservation of orbital symmetry.[31c] Its final outcome is generally predictable even for complicated cases.[31d]

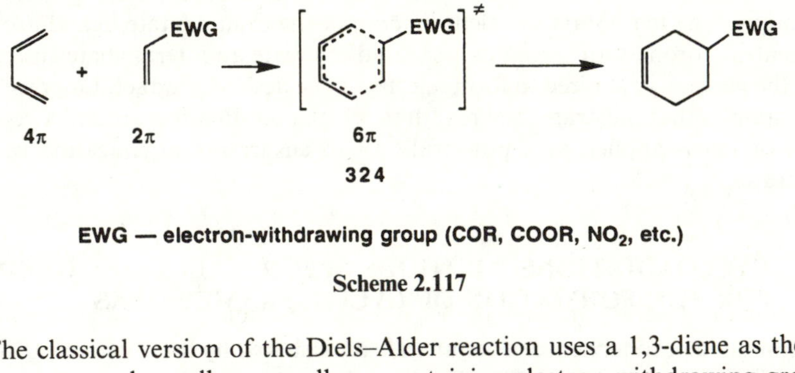

EWG EWG ≠ EWG

4π + 2π → 6π **324** →

EWG — electron-withdrawing group (COR, COOR, NO_2, etc.)

Scheme 2.117

The classical version of the Diels–Alder reaction uses a 1,3-diene as the 4π component and an alkene or alkyne containing electron-withdrawing groups (conjugated aldehydes, ketones, acids and their derivatives, nitroalkenes, *etc.*) as the 2π reactant. With appropriately chosen dienes and/or dienophiles, a wide array of purely carbocyclic or heterocyclic compounds can be easily synthesized. A representative set of simple examples is given in Scheme 2.118.

To assess properly the exclusive synthetic merits of the Diels–Alder reaction, it is necessary to consider the main peculiarities of its regio- and stereocourse. First of all, it must be emphasized that the very mechanism of this concerted cycloaddition implies complete retention of configuration of the substituents of

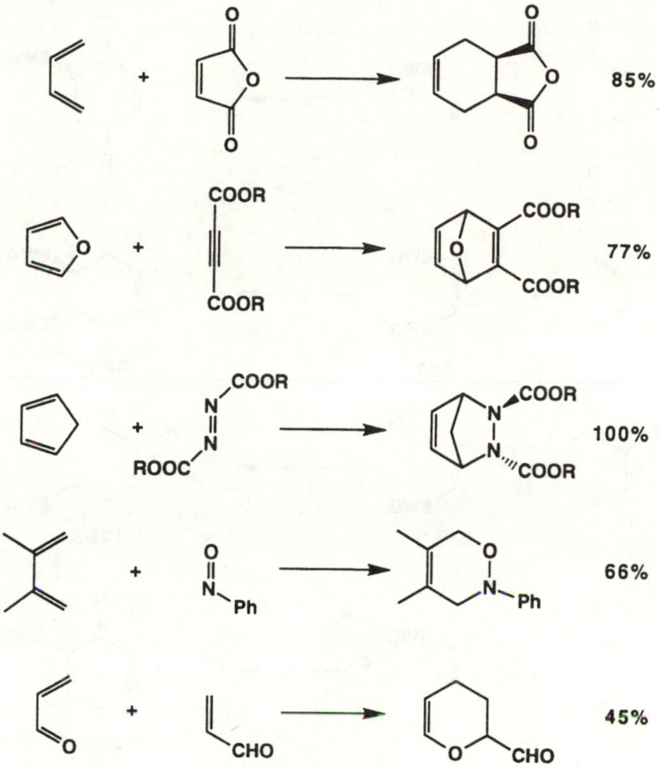

85%

COOR
COOR
COOR
COOR
77%

COOR
ROOC
COOR
COOR
100%

Ph
Ph
66%

CHO
CHO
45%

Scheme 2.118

the starting diene and dienophile moieties (**325** → **326**; **327** → **328**; Scheme 2.119). For non-symmetrical monosubstituted dienes and dienophiles the preferential formation of isomers, **329** or **330** correspondingly, is usually observed. The transition state for 1,4-disubstituted dienes may be formed *via* *endo* or *exo* orientation of the reactants and thus two isomeric products could be produced. Typically, the *endo* transition state is favored and the respective *endo* adducts produced, as is shown for the preparation of the nearly individual *endo* isomer **331a**.

Within the general scheme of the Diels–Alder reaction it is possible to achieve cycloadditions with the introduction of unexpected, at least at first glance, fragments due to the presence of easily removable or transformable substituents in either diene or dienophile counterparts. Consider, for example, a retro-Diels–Alder disconnection of the target molecules **332** or **333** (Scheme 2.120). Both compounds can be represented as enols **332a** and **333b**, respectively. The application of formal Diels–Alder disconnection to these compounds leads us to the dienes **334** and **335**. It might seem that this approach is not productive, as enol dienes **334** and **335** cannot be considered to be viable reagents. However, as we already know, various silyl enol ethers are stable, readily available compounds. With the use of the silyl ethers **336** and **337** (Scheme 1.121) the

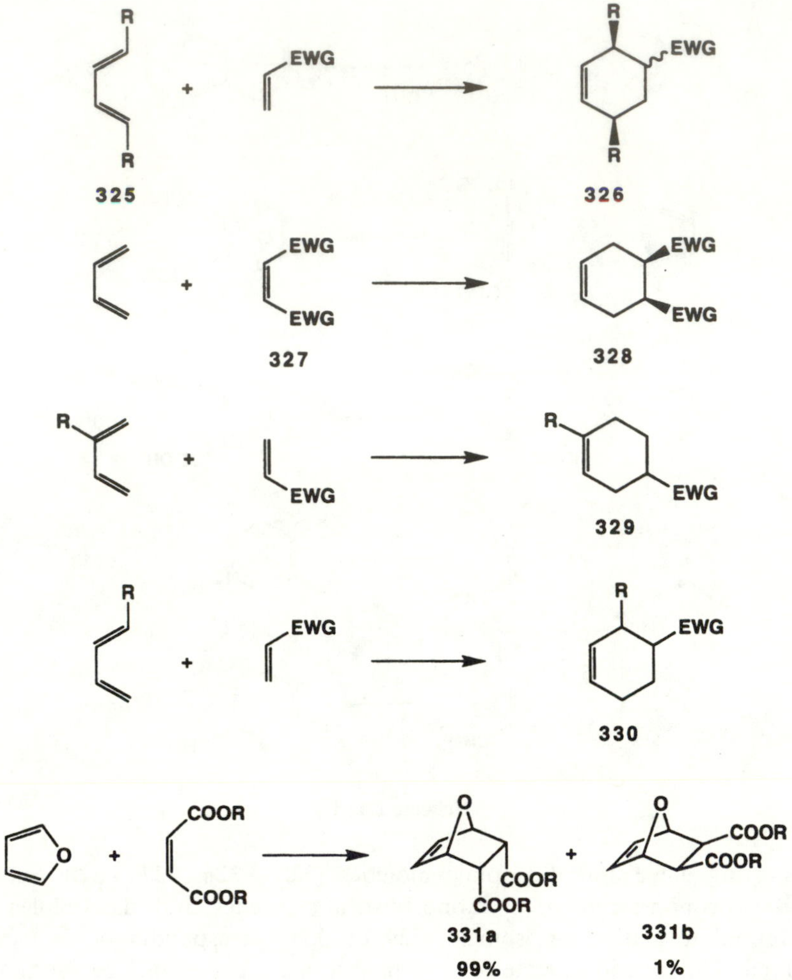

Scheme 2.119

desired transformations are easily accomplished in full accordance with the typical scheme of the Diels–Alder reaction followed by simple functional group transformations.[31e] In these preparations, dienes **336** and **337** with their latent carbonyl groups are employed as synthetic equivalents of the non-existing dienes **334** and **335**.

The examples given in Scheme 2.121 are typical of the general approach used to prepare six-membered rings bearing functional groups at any position *via* the Diels–Alder route from a diversified set of dienes and dienophiles. The flexibility of this protocol is further illustrated in Scheme 2.122. Utilization of 2-pyrone derivatives as diene components opens an entry toward the preparation of 1,3-cyclohexadienes *via* Diels–Alder reactions followed by a ready elimination of carbon dioxide from the initally formed adduct. For example, reaction of **338**

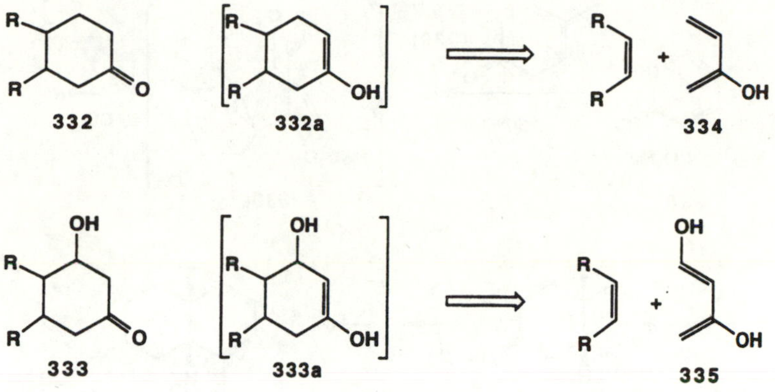

R R O 332 [R R OH] 332a ⟹ R R + OH 334

OH R R O 333 [OH R R OH] 333a ⟹ R R + OH OH 335

Scheme 2.120

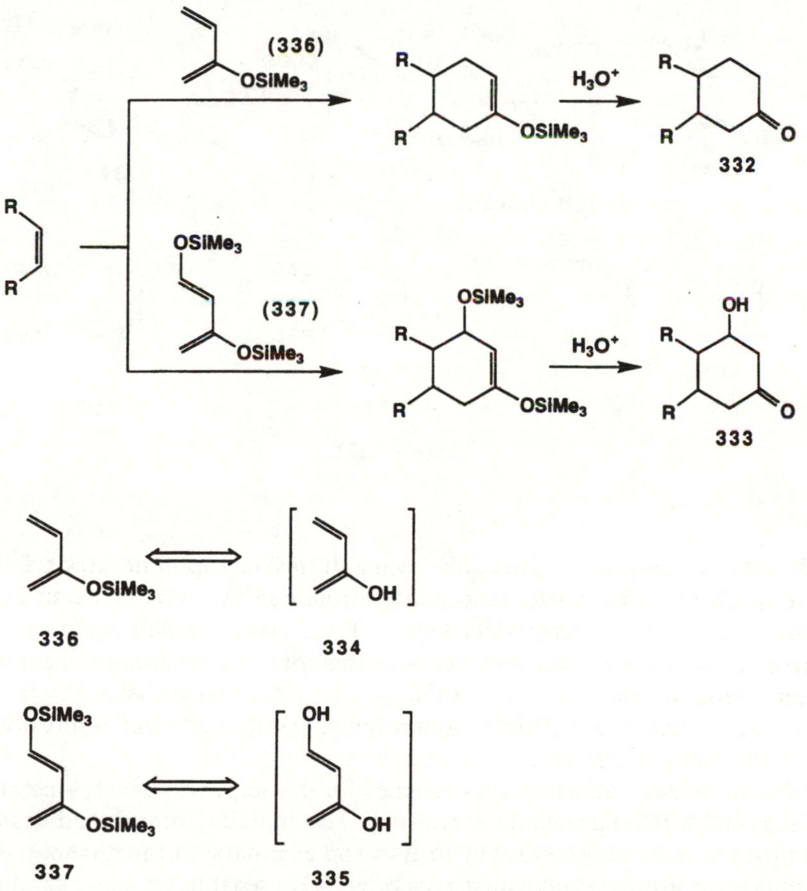

(336) OSiMe3 → R R OSiMe3 —H_3O^+→ R R O 332

R R — OSiMe3 (337) OSiMe3 → OSiMe3 R R OSiMe3 —H_3O^+→ OH R R O 333

OSiMe3 336 ⟺ [OH] 334

OSiMe3 OSiMe3 337 ⟺ [OH OH] 335

Scheme 2.121

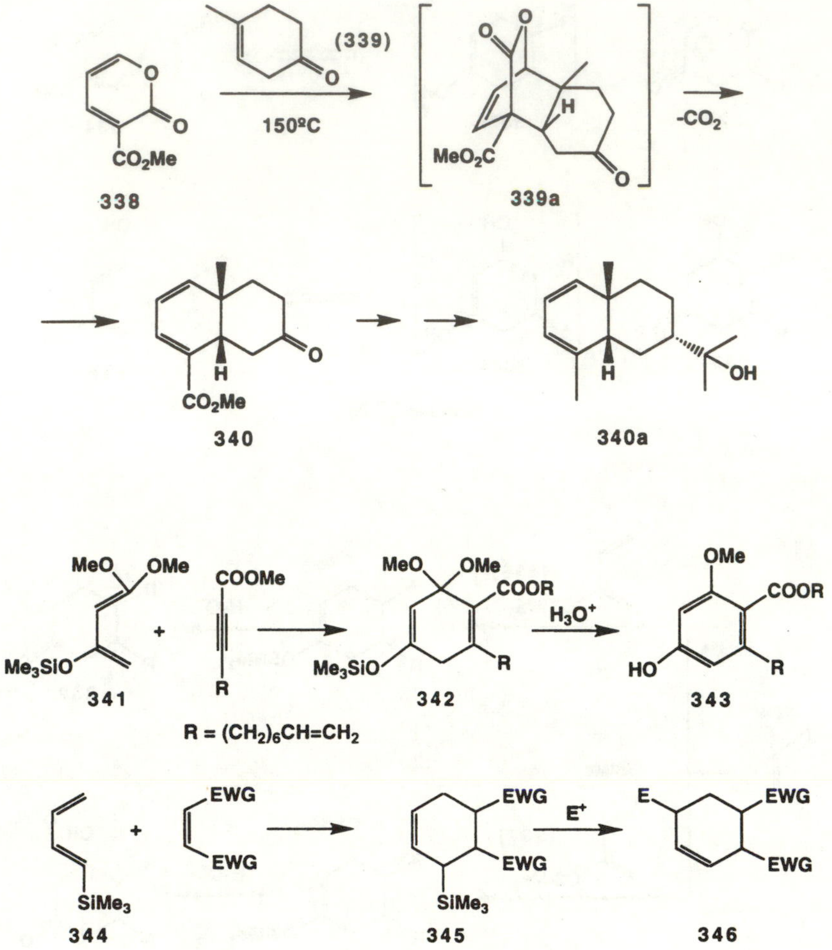

Scheme 2.122

with **339** gave unstable adduct **339a**, which furnished, upon decarboxylation, the bicycloketone **340**, an advanced intermediate in the total synthesis of the natural sesquiterpene occidentalol **340a**.[31f] The 2-pyrone residue could be easily incorporated into various polyfunctional compounds containing a dienophile moiety. Subsequent intramolecular Diels–Alder reaction (see below) with these substrates represents an efficient tool for the preparation of structurally diverse polycyclic natural compounds.[31g]

Siloxydiene **341** was specifically designed for the preparation of 1,3-methoxy-hydroxy substituted aromatic fragments. The initially formed adducts, for example **342**, are prone to easy hydrolysis and elimination of methanol to lead eventually to aromatic derivatives such as **343**, used further as a key intermediate in the synthesis of the plant growth inhibitor lasiodiplodine.[31h]

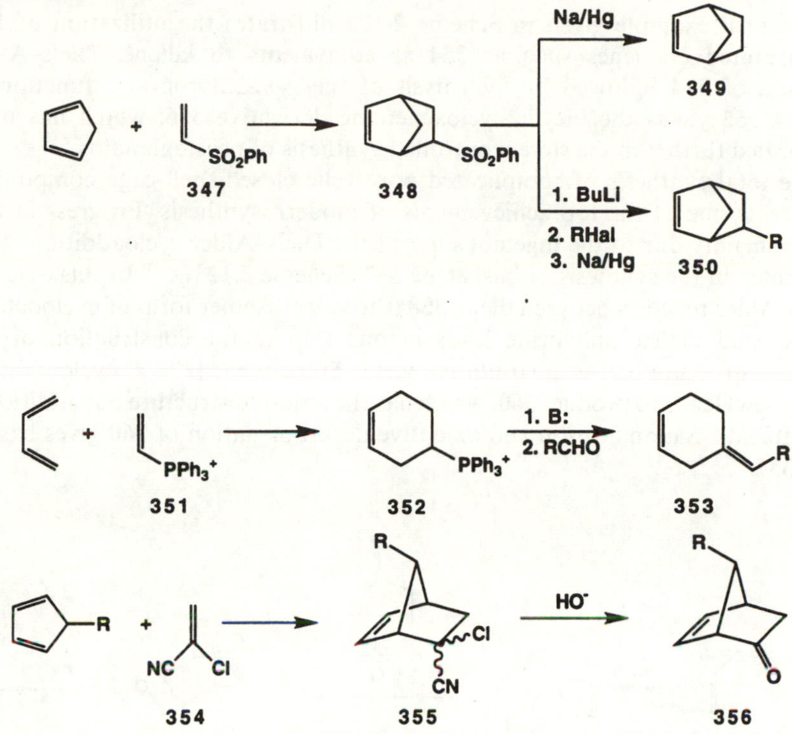

Scheme 2.122 (*continued*)

The presence of the trimethylsilyl group on diene **344** ensures the formation of the allylsilane moiety in cycloadducts such as in **345**. The ability of this moiety to react with electrophiles and eliminate the silyl substituent opens the entry into the preparation of a set of cyclohexene derivatives such as **346** with an unusual (for Diels–Alder adducts) substitution pattern.[31i]

Vinyl sulfone **347** easily forms adducts like **348**. The sulfone group can be reductively cleaved to give **349**. As an alternative, **348** can be alkylated first then cleaved to give **350**. In these cases, **348** plays the role of a synthetic equivalent to ethylene or terminal alkene dienophiles.[31j] The importance of the utilization of **347** stems from the fact that unactivated alkenes are poor dienophiles and their reactions with even the most active dienes, such as cyclopentadiene, require drastic conditions and are generally impractical.

The presence of the triphenylphosphonium substituent makes **351** (Scheme 2.122) an active dienophile. Resulting cycloadducts such as **352** are immediately recognized as precursors for the formation of ylides. Products with an exocyclic double bond such as **353** are easily synthesized *via* a tandem sequence of Diels–Alder and Wittig reactions.[31k] The formation of adduct **353** could have been achieved *via* the Diels–Alder reaction with the allene $CH_2=C=CHR$, but this direct route is generally unapplicable owing to the low activity of allenes as dienophiles.

The final example given in Scheme 2.122 illustrates the utilization of 1,1-disubstituted ethylenes such as **354** as equivalents to ketene. Diels–Alder reaction of **354** followed by hydrolysis of the *gem*-chlorocyano function in adduct **355** yields the bicylic cycloxenenone derivative **356**, which has been elaborated further in the stereocontrolled synthesis of prostaglandins.[31l]

The total synthesis of complicated polycyclic closed-shell cage compounds represents one of the top achievements of modern synthesis. Progress in this area is mainly due to the ingenious use of the Diels–Alder cycloaddition, as is illustrated in the synthesis of basketene **357** (Scheme 2.123).[31m] In this case the Diels–Alder reaction between diene **358** (the valent isomer form of cyclooctatetraene) and maleic anhydride leads in one step to the construction of the tricyclic structure **359** in quantitative yield. Subsequent [2 + 2] cycloaddition (see below) leads to product **360**, which has the required structure but additional substituents. Saponification and oxidative decarboxylation of **360** gives basketene **357**.

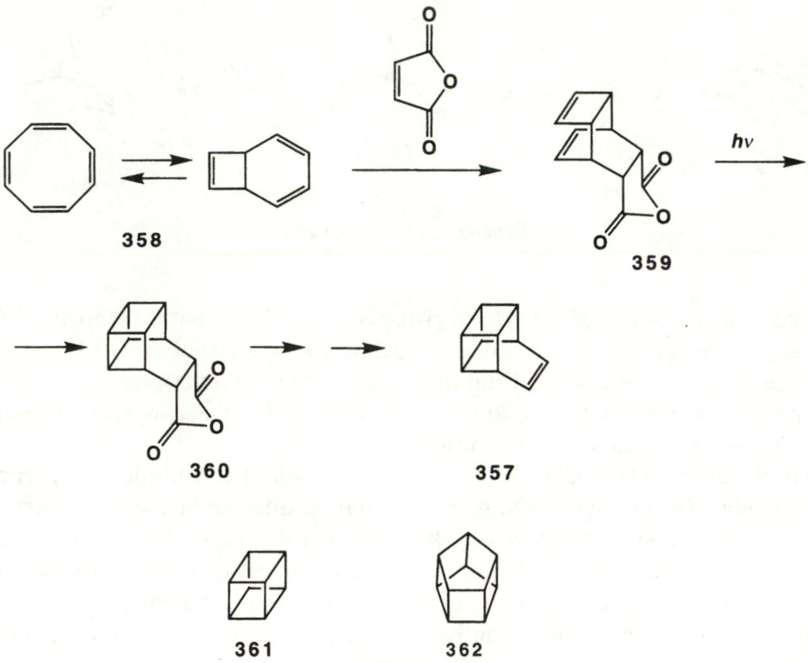

Scheme 2.123

The Diels–Alder reaction was used equally effectively as a key step in the preparation of cubane **361**, pentaprismane **362**,[31n] and many other exotic representatives of this artificially created class of organic compounds.

As we have already mentioned, intramolecular reactions generally proceed faster compared to their intermolecular versions because of the substantial reduction in the entropy barrier. Therefore it is not surprising that intramolecular Diels–Alder (IMDA) reactions, which emerged in the early 1960s, tremendously broadened the scope of the [4 + 2] cycloaddition in total syntheses.[31o] In Scheme 2.124 are representative cases, chosen from a list of several hundred examples, to illustrate the peculiarities of this reaction and the specific aspects of its synthetic utilization.

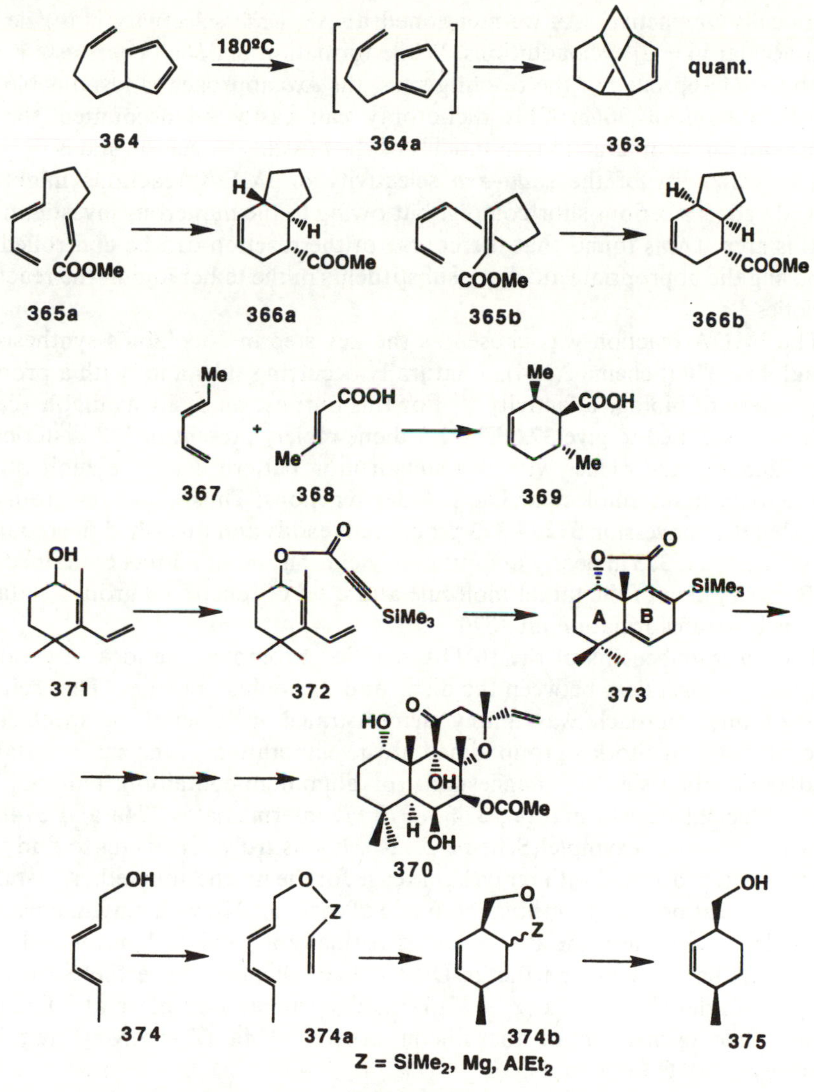
180ºC
quant.
364
364a
363
COOMe
H
365a
366a
365b
366b
Me
COOH
367
368
369
OH
O
SiMe3
A
B
371
372
373
HO
OH
OCOMe
370
374
374a
374b
375
Z
Z = SiMe2, Mg, AlEt2

Scheme 2.124

The unusual, strained structure of brex-4-ene **363** can be easily prepared in an almost quantitative yield from the simple precursor **364**,[31p] even though the latter contains an inactive alkenic double bond usually reluctant to participate as a dienophile in intermolecular [4 + 2] cycloadditions.

Transformations **365a** → **366a** and **365b** → **366b**[31o,r] are given as examples of the high regio- and stereoselectivity typical for IMDA reactions. It is not to be overlooked that these reactions proceed with a reversed regioselectivity compared to the course of the intermolecular reaction of (*E*)-1-methyl-1,3-butadiene **367** with (*E*)-crotonic acid **368** to give **369** [(*Z*)-1-methyl-1,3-butadiene is practically unreactive]. As we mentioned above, *endo* selectivity is a rule for bimolecular [4 + 2] cycloadditions. While formation of **366a** occurs as a result of the *endo* approach of the reacting sites, the *exo* approach is responsible for the formation of **366b**. This dichotomy can easily be accounted for by comparing steric interactions in models of the possible transition states.

The ambiguity of the *endo–exo* selectivity of IMDA reactions might be considered as a serious shortcoming, but owing to the numerous investigations in this area it was found that the course of the reaction can be controlled by choosing the appropriate auxiliary substituents in the tether and in the reacting moieties.[31o]

The IMDA reaction was chosen as the key step in Nicolaou's synthesis of forskioline **370** (Scheme 2.124), a naturally occurring substance with a promising pattern of biological activity.[31s] For this purpose an easily available diene, **371**, was esterified to give **372**. The 1,3-diene moiety present in **372** is sterically very hindered and dienes with this substitution pattern usually exhibit rather low activity in intermolecular Diels–Alder reactions. Thus it was rewarding to find that the conversion **372** → **373** proceeded readily and furnished the required tricyclic adduct **373** in nearly quantitative yield. The latter adduct contained the A–B ring system of the target molecule and a set of functional groups suitable for further transformation into **370**.

The obvious benefits of the IMDA reaction prompted the idea of creating *temporary connections* between the diene and dienophile moieties. The fruitfulness of this approach was amply demonstrated in a series of spectacular investigations by Stork's group aimed at the elaboration of the easily installed (and removable) silicon-, magnesium-, or aluminum-containing tethers.[31t] A highly efficient transformation, **374** → **375** *via* intermediates **374a** and **374b**, is shown as a model example (Scheme 2.124). It was truly surprising to find that the reaction proceeded with remarkable ease for the magnesium tether, installed upon the treatment of the lithium alkoxide of carbinol **374** with vinylmagnesium bromide. In this case, the dienophile is formally a vinyl carbanion and thus should not be active as a partner in Diels–Alder reactions. Nevertheless, for the magnesium derivative **374a** (Z = Mg) this conversion took place at 80 °C in 1 hour, while cyclization of the silicon-tethered **374a** (Z = $SiMe_2$) required heating at 160 °C for 3 hours.

The remarkable synthetic efficiency of IMDA reactions inspired a hypothesis about its possible involvement in the biogenesis of natural polycyclic compounds. It looks like Nature is well aware of the benefits of this approach and takes the advantage of it in cases of necessity. For example, convincing evidence was accumulated showing that the biogenesis of phytotoxins solanopyrone A and D, **376a** and **376b**, isolated from the pathogenic fungus *Alternaria solani*, involves an IMDA cyclization of the common acyclic precursor **377** (Scheme 2.125).[31u,v]

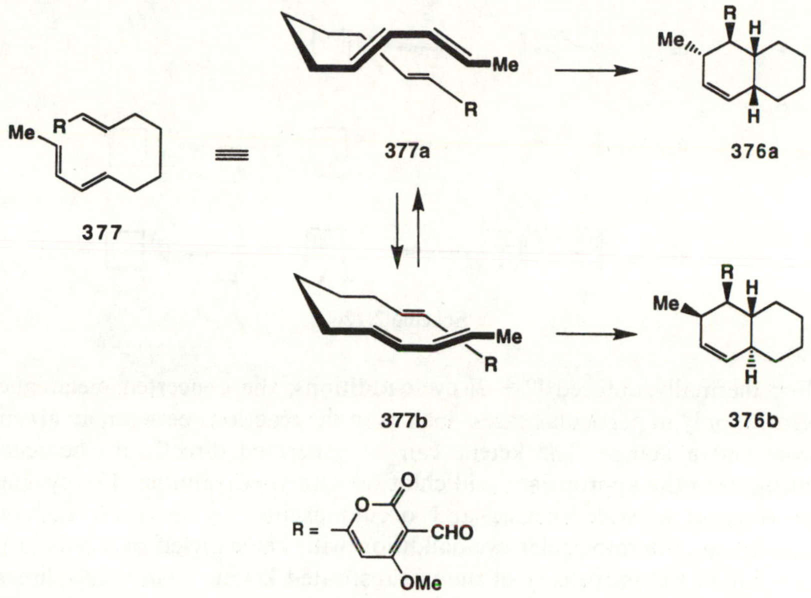

Scheme 2.125

A similar cyclization of **377** carried out under thermal conditions gave rise to two isomers, **376a** and **376b**, in comparable amounts.[31w] Notably, naturally occurring **376** is a mixture of the same diastereoisomers. Model studies revealed that this is an expected result for the cyclization of **377**, as its alternative reactive conformers **377a** and **377b** and the respective *endo* and *exo* transition states leading to **376a** and **376b** differ very little in their energy content.

In further sections we shall often refer to the Diels–Alder reaction as the most versatile and popular synthetic method. It is not an exaggeration to claim that, in the absence of this reliable tool, synthetic chemists would never have dared to pursue the syntheses of many challenging structures which eventually yielded to their efforts.

2.19.2 [2 + 2] Cycloaddition in the Synthesis of Cyclobutane Derivatives

The [2 + 2] cycloaddition represents the most general and direct pathway for the formation of a cyclobutane structure from two alkene moieties, as outlined in Scheme 2.126. This process may occur as a concerted reaction *via* a cyclic transition state (mechanism **a**), as a stepwise reaction involving the formation of an acyclic biradical (mechanism **b**), or through bipolar (mechanism **c**) intermediates. Depending upon the structure of the reactants, cycloaddition may occur by any of these mechanisms.

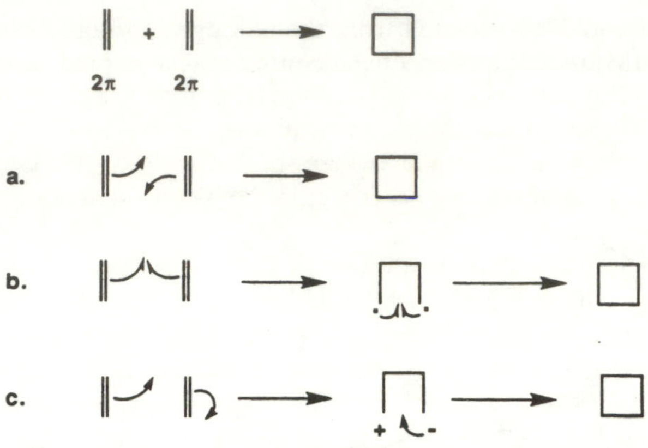

Scheme 2.126

For thermally induced [2 + 2] cycloadditions, the concerted mechanism is operative only in particular cases, such as in the reactions between an alkene or alkyne and a ketene. The ketene can be generated directly in the reaction mixture from the appropriate acid chloride with triethylamine. The cycloaddition reaction is stereospecific and occurs exclusively in a *cis* fashion.[32a] Although the intermolecular cycloaddition with ketene itself proceeds in poor yields due to the propensity of the unsubstituted ketene to undergo dimerization, it is quite an efficient reaction with ketenes containing electron-withdrawing substituents. Usually, α-chloro ketenes are employed as reagents formed *in situ* from the corresponding α-chloro acid chlorides. Typical examples are represented in the preparation of cycloadducts such as **378**[32b] and **379** (Scheme 2.127).[32c] The latter cycloadduct, prepared in modest yield (*ca.* 20%),

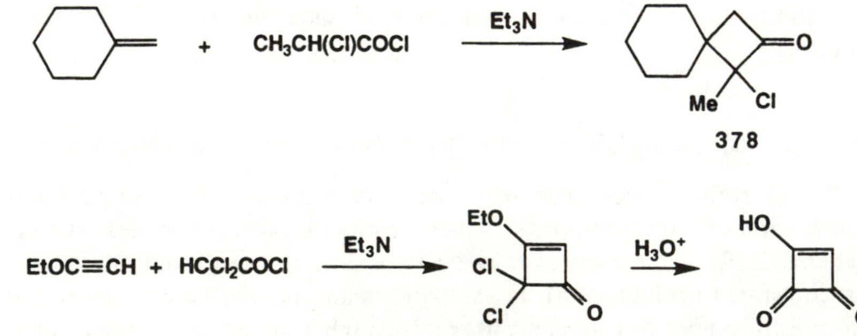

Scheme 2.127

was utilized as an immediate precursor in the synthesis of mycotoxin moniliformin **380**, a natural compound of surprising simplicity and unique structure.

α-Chlorocyclobutanones formed in these reactions are easily converted into their respective cyclobutanones *via* reductive dechlorination. Alternatively, they can be used as intermediates in subsequent synthetic manipulations owing to the high reactivity of the chlorine substituent as a good leaving group (see below).

Intramolecular [2 + 2] cycloaddition of ketenes to alkenes is also viable, but the efficiency of the reaction depends crucially upon the length and nature of the tether.[32d] A three-atom tether seems to be optimal and thus a 4,5-fused bicyclic system can be easily assembled with this method. The mild reaction conditions required to generate the ketene moiety in polyfunctional substrates, the ease of cycloaddition, and its highly selective course suggest numerous synthetic applications for this reaction. See, for example, the transformations of **381** → **382**[32e] and **383** → **384** (Scheme 2.128).[32f] The efficiency of the intramolecular option makes it possible to perform the reaction with *in situ* generated ketene intermediates not bearing halogen substituents, such as **381a** or **383a**, thus offering additional benefits in the synthetic utilization of [2 + 2] cycloadditions.

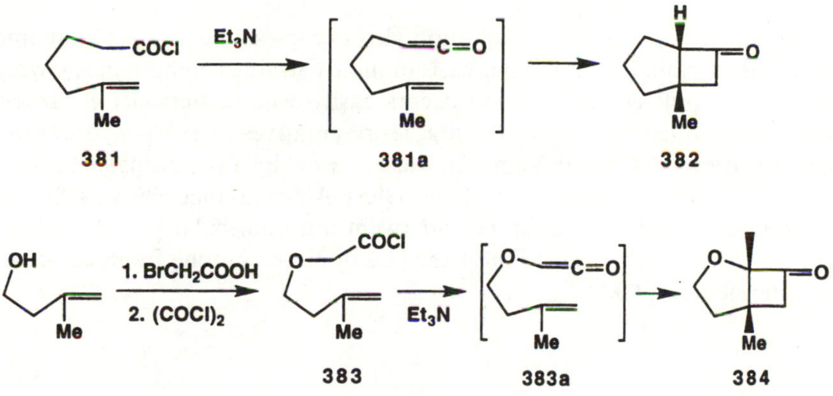

COCl Et$_3$N C=O H O Me Me Me

381 381a 382

OH 1. BrCH$_2$COOH 2. (COCl)$_2$ COCl O Et$_3$N O C=O O O Me Me Me Me

383 383a 384

Scheme 2.128

Photochemically induced [2 + 2] cycloaddition is of extraordinary importance in organic synthesis,[32g] as this is a method ideally suited for the preparation of sterically congested compounds. The reaction may occur by a concerted mechanism allowed by rules of orbital symmetry, or, more often, *via* a biradical pathway. For preparative purposes, the most widely exploited is the enone–alkene photochemical [2 + 2] cycloaddition. This reaction proceeds with high regioselectivity, although its stereoselectivity might be low. The first example of the utilization of this reaction for the synthesis of a natural compound, α-cariophyllene **385**, was described by Corey (Scheme 2.129).[32h] Adduct **386**, formed as a mixture of stereoisomers in high yield from simple precursors, was further transformed *via* the tricyclic intermediate **387** into the target compound **385** with the help of standard reactions. The broad applicability of enone–alkene photochemical [2 + 2] cycloaddition in the area of natural product synthesis is now well documented in numerous publications.[32i]

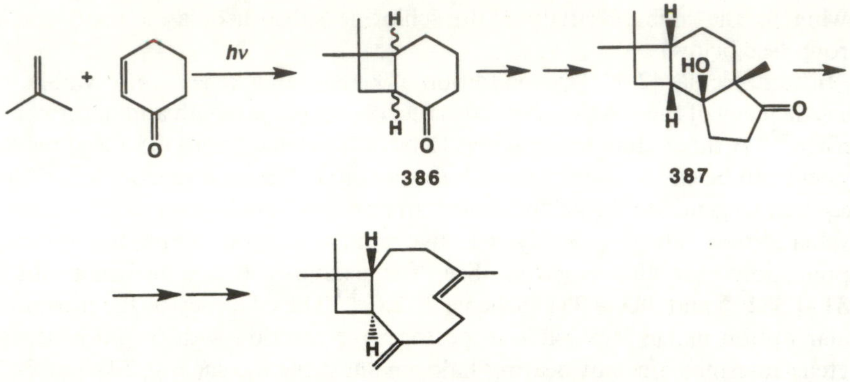

386

387

385

Scheme 2.129

Intramolecular [2 + 2] cycloaddition also occupies a pivotal position among the methods available for the synthesis of highly strained compounds. Owing to the proximity effect, this reaction occurs easily even in sterically encumbered cases. The synthesis of one of the first representatives of exotic hydrocarbons, Dewar benzene **388** by van Tamelen, was achieved by a surprisingly short route (Scheme 2.130). The readily available Diels–Alder adduct **389** was first converted into diene **389a**. The latter underwent intramolecular [2 + 2] cycloaddition which led to the formation of the [2.2.0] bicyclohexene framework of the key intermediate **390**.[1a]

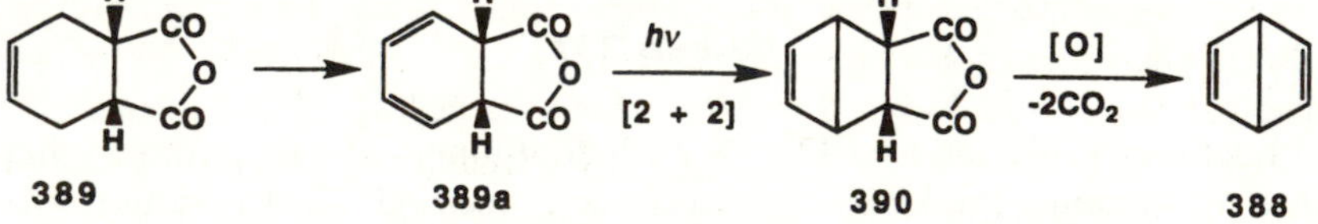

389 389a 390 388

Scheme 2.130

In the above-mentioned synthesis of basketene **357**, the intramolecular [2 + 2] cycloaddition was used to form the missing 'side' in the 'basket'. Intramolecular enone–alkene [2 + 2] cycloaddition[32j] is the method of choice in the total synthesis of many polycyclic natural compounds as it can be used as the key step, enabling a rapid escalation of the complexity of the assembling framework, as is shown in Scheme 2.131 .

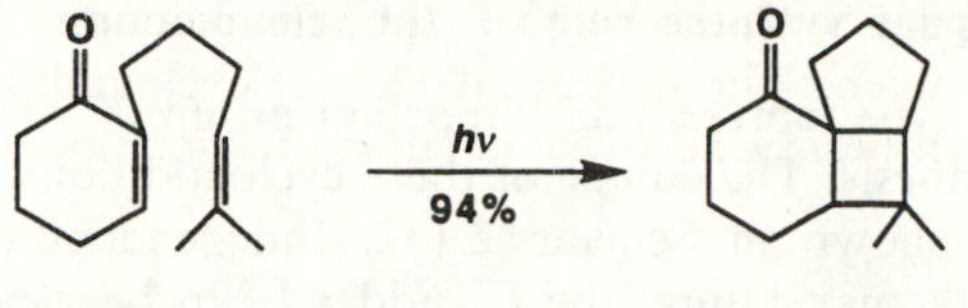

391 → 392 (hv, 94%)

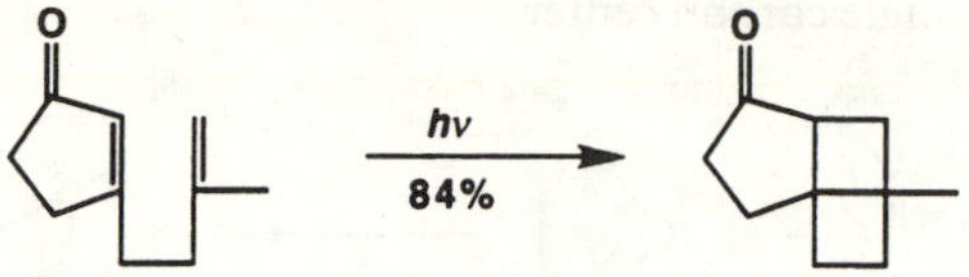

393 → 394 (hv, 84%)

Scheme 2.131

Transformations like **391** → **392**[32k] or **393** → **394**[32l] are generally performed in good to excellent yields.[32m] They are also remarkable for their nearly complete control over the regio- and stereoselectivity of cycloaddition, owing to the steric constraints of the emerging rigid frameworks which actually prohibit the formation of alternative isomers. Synthesis of the required dienone precursors, such as **391** or **393**, can be easily achieved with the help of well-established methods. Thus a fairly diverse carbon framework, consisting of multiply fused rings like those shown in Scheme 2.131, can be assembled in a versatile protocol based upon two reliable steps, namely, preparation of polyfunctional substrates and intramolecular [2 + 2] photocycloaddition.

Throughout the text, additional examples are provided of equally effective [2 + 2] cycloadditions involving a photoinduced interaction between two alkenes or alkene–enone moieties incorporated into various structures specifically designed for the particular synthetic goal. It is appropriate here to note that compounds like basketene **357**, cubane **361**, or Dewar benzene **388** represent typical examples of an absolutely rigid carbon framework, composed of a set of strained, and thus energy-rich, bonds. Their formation *via* photocyclization routes is a process of converting light energy into the energy of chemical bonds. Obviously the transformations leading to the opening of the strained fragments will be accompanied by a significant release of the accumulated energy. Therefore, photoinduced cycloaddition is being intensively studied as an approach to the creation of chemical devices capable of accumulating light energy, including solar energy, and storing it in the form of chemical energy.

The variety of methods for accomplishing the [2 + 2] cycloaddition and the reliability of this reaction secures its position as one of the most versatile tools for the creation of the cyclobutane fragment in a variety of structures. Additional attractive dimensions to the scope of this protocol come in the ease of further transformations through ring opening and skeletal rearrangements, typical for cyclobutane fragments[32m] (this aspect will be considered later in this chapter).

2.19.3 Cyclopropane Synthesis *via* [2 + 1] Cycloaddition

Reactions of this type represent an important group of methods employed in cyclopropane synthesis. The course of these cycloadditions is described in the formal equation shown in Scheme 2.132. Independent of the method of generation or its actual nature, the C_1 addend can be viewed as a synthetic equivalent to the *carbene* **395**, at least in terms of its ability to form two new **C–C** bonds at a single carbon center.

395

Scheme 2.132

Carbenes[33a] are intrinsically unstable species and their isolation is achieved only by entrapment in argon matrices at low temperature (77 K or less).[33b] The inherent reactivity pattern of carbenes as free species was thoroughly investigated for gas-phase reactions.[33c] The majority of conventional procedures for cyclopropane formation, formally described as carbene transfers, are based upon the interaction of an alkene with a carbene precursor at ambient temperature. These procedures might not involve the intermediacy of a true carbene species. Therefore, a neutral term, *carbenoid*, was suggested as a general, though not very specific, description of the nature of the intermediates possibly involved as carbene equivalents. Typical reactions leading to the *in situ* generation of these intermediates are shown in Scheme 2.133.

Among these pathways, reaction **A** became especially important following the elaboration of phase transfer catalysis, conditions which dramatically simplified the experimental procedure. In the adopted protocol, a chloroform solution of the substrate is treated with aqueous alkali in the presence of a phase transfer catalyst, for example a tetrabutylammonium salt. The generation of the dichlorocarbene takes place in the chloroform layer, exactly in the phase where the organic substrate is located. The entire process turns out to be efficient and practical. An alternative way of dichlorocarbene generation not involving the use of bases (or other catalysts) takes advantage of the ease of decarboxylation of sodium trichloracetate upon moderate heating (reaction $\mathbf{A^1}$). Method **B** is less general and most often utilized in intramolecular versions of [2 + 1] cycloadditions. Method **C** was employed mainly in mechanistic studies of carbene-mediated reactions. Chloro (or more generally halo) substituted cyclopropanes formed with the help of methods **A–C** could serve as versatile precursors for the synthesis of other cyclopropane derivatives.

Aliphatic diazo compounds bearing electron withdrawing groups are moderately stable and can be easily prepared from readily available precursors. Photolysis of these diazo compounds or thermolysis in the presence of a metal salt (Cu, Rh: reaction **D**[33d]) are among the most widely applicable protocols of cyclopropanations.

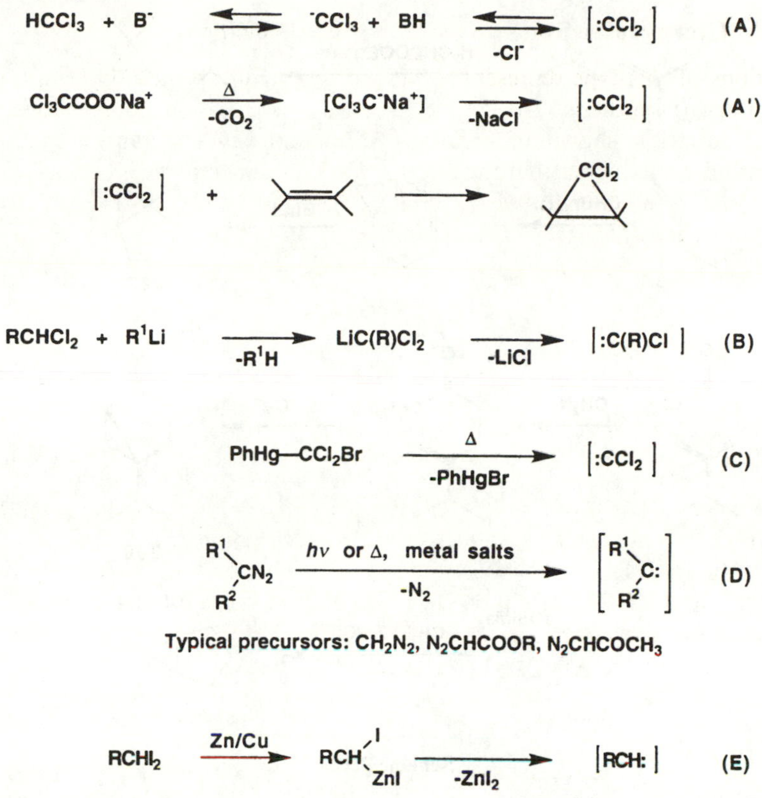

$HCCl_3 + B^- \rightleftharpoons {}^-CCl_3 + BH \xrightleftharpoons{-Cl^-} [:CCl_2]$ (A)

$Cl_3CCOO^-Na^+ \xrightarrow[-CO_2]{\Delta} [Cl_3C^-Na^+] \xrightarrow{-NaCl} [:CCl_2]$ (A')

$[:CCl_2]$ + alkene → CCl_2 cyclopropane

$RCHCl_2 + R^1Li \xrightarrow{-R^1H} LiC(R)Cl_2 \xrightarrow{-LiCl} [:C(R)Cl]$ (B)

$PhHg—CCl_2Br \xrightarrow[-PhHgBr]{\Delta} [:CCl_2]$ (C)

$R^1R^2CN_2 \xrightarrow[-N_2]{h\nu \text{ or } \Delta, \text{ metal salts}} [R^1R^2C:]$ (D)

Typical precursors: CH_2N_2, $N_2CHCOOR$, $N_2CHCOCH_3$

$RCHI_2 \xrightarrow{Zn/Cu} RCH(I)ZnI \xrightarrow{-ZnI_2} [RCH:]$ (E)

Scheme 2.133

Finally, a fairly convenient protocol for carbene transfer reactions utilizes species like ICH(R)ZnI (a carbenoid), generated *in situ* by the interaction of 1,1-di-iodides, $RCHI_2$, with a zinc–copper pair (Simmons–Smith procedure, reaction **E**[33e,f]).

A few specific examples of cyclopropanation using the above methods are shown in Scheme 2.134. The naturally occurring insecticide *trans*-chrisanthemic acid **396** served as an obvious target to check the viability of carbene addition as a preparative method. This compound was first synthesized (in the mixture with the *cis* isomer) by the monocyclopropanation of 2,5-dimethyl-2,4-hexadiene.[33g] Since then, numerous analogs of **396** were prepared by similar reactions. Some of the analogs are now widely used as efficient and ecologically safe pesticides. The formation of the tricyclic hydrocarbon **397** from 1,5-hexadiene proceeds as a sequence of inter- and intramolecular carbene transfer reactions. An initial carbene precursor, $CHBr_3$, is actually employed here as an equivalent of a unique tetradentate $\mathbf{C_1}$ synthon **:C:**.[33h] The preparation of **398**[33i] *via* intramolecular [2 + 1] photocycloaddition is a typical example of the efficiency of this route for the construction of the polycyclic framework frequently encountered in the structures of natural compounds.

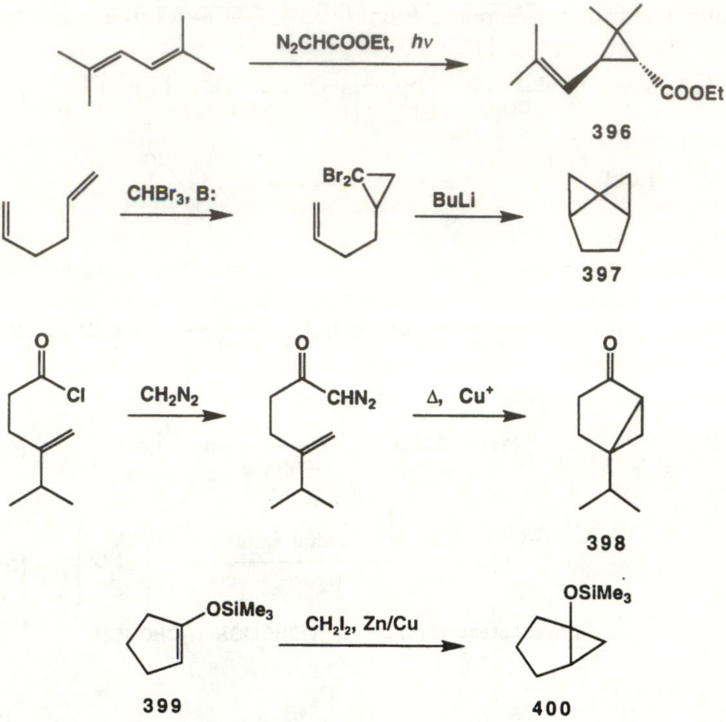

$N_2CHCOOEt$, *hv*

COOEt

396

$CHBr_3$, B:

Br_2C

BuLi

397

Cl

CH_2N_2

CHN_2

Δ, Cu^+

398

OSiMe$_3$

CH_2I_2, Zn/Cu

OSiMe$_3$

399

400

Scheme 2.134

Cyclopropanation of enol ethers, for example the conversion of **399** → **400**, proceeds especially easy under Simmons–Smith conditions.[33e] This reaction offers additional options for the synthetic utilization of carbonyl compounds capable of forming enol ethers. Some of them will be considered later in this chapter (see Section 2.23.2).

Here again it is not an exaggeration to claim that, in planning a synthesis, if it becomes expedient to resort to a cyclopropanation of the double bond, there should be no doubt about the possibility of selecting suitable conditions to achieve this conversion.

2.19.4 Cycloadditions Mediated by Coordination of the Substrate(s) around a Transition Metal

As we mentioned earlier (see Section 2.4), the first synthesis of cyclooctatetraene **137** was achieved by a rather laborious route. The multistep sequence involved functional group manipulations applied to a starting material already containing the eight-membered ring. This 10-step synthesis (overall yield 0.75%!) was repeated by other investigators only once, and for decades **137** was listed among the exotic and difficult to prepare compounds. As a result, no one was especially eager to investigate the chemistry of this highly unsaturated but very interesting compound. The situation changed dramatically in the 1950s, when **137** suddenly became a commercially available product, produced in one step from an abundant starting material, acetylene! This success, by Reppe's group,[34a] which evolved from the thoughtful development of a somewhat accidental observation, landmarked an appearance of a principally novel approach to the assemblage of cyclic frameworks.

The tendency of acetylene to undergo thermal oligo- and polymerization to give (actually in very low yields!) linear polyenes or benzene is a reaction known for almost a century. Formally, the same process is implied in the formation of **137** from four acetylene molecules. Yet this deceptively simple cyclotetramerization scheme turned to be a viable reaction only after Reppe's discovery that a simple catalyst, nickel(II) cyanide, could serve as a highly efficient device to control the course of acetylene oligomerization. It is the ability of this catalyst to form a complex, **401**, with four molecules of acetylene that ensured the required selectivity of cyclotetramer formation (Scheme 2.135).

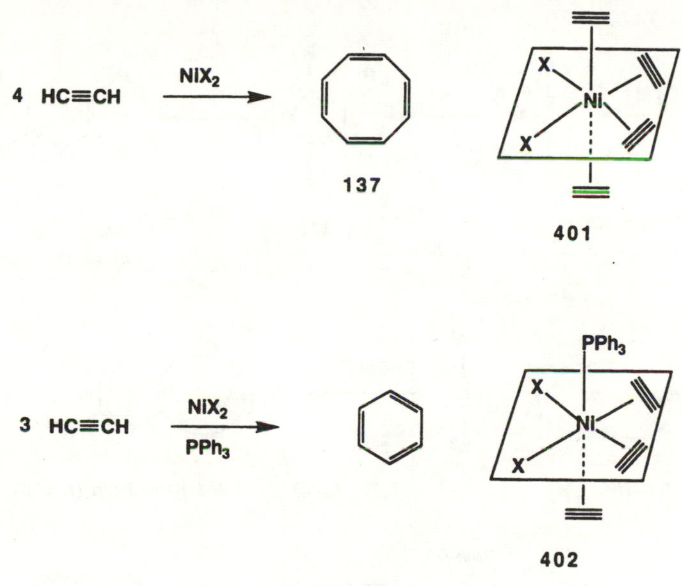

Scheme 2.135

The coordination of four acetylenic molecules as ligands around a central metal atom creates an arrangement ideally suited for selective intramolecular cyclooligomerization. In other words, an otherwise prohibitively high entropy barrier for the formation of a highly organized transition state from four acetylene molecules was substantially lowered owing to the propensity of nickel to form multidentate complexes like **401**. A similar reaction carried out in the presence of PPh_3, which is able to serve as a substitute for one of acetylenic ligands, proceeds as a cyclotrimerization (*via* formation of complex **402**) to give benzene.[34b]

Further studies revealed that the cyclotrimerization becomes a preferential pathway in the presence of $CpCo(CO)_2$. As was experimentally established, with this catalyst the reaction proceeds in a stepwise manner, first with the formation of the five-membered metallacycle **403**, which is converted further into the cyclotrimer **404**, as shown in Scheme 2.136. This process was originally applied to the preparation of a series of otherwise unaccessible, heavily substituted benzene derivatives (see, for example, hexaisopropylbenzene **404a**[34c]). An even richer area of preparative utilization emerged, however, with the development of a convenient procedure to cooligomerize alkynes with 1,*n*-diynes. Derivatives of indane **405a** or tetralin **405b** can be easily prepared from these precursors for $n = 3$ or 4, respectively. Far more important, though, was the finding that the reaction is equally effective even for precursors containing a bridge of only two methylene links. The reaction could thus be applied to the preparation of a wide set of substituted benzocyclobutenes, for example **406**.[34d]

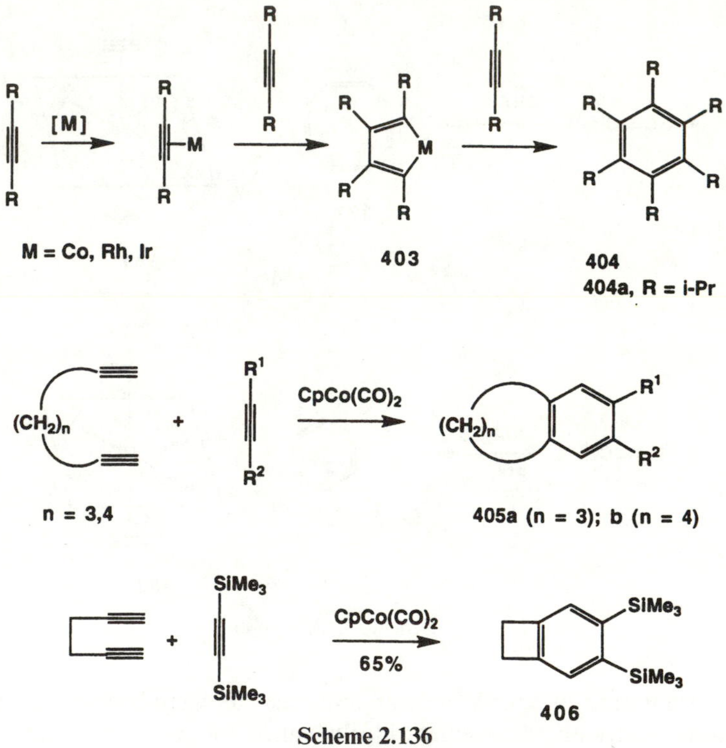

Scheme 2.136

Alkyne cyclooligomerizations such as those shown in Scheme 2.136 can be formally described as a [2 + 2 + 2] cycloaddition. They provide an especially important (and novel) approach to the synthesis of polycyclic compounds and will be referred to often in the next chapter.

Transition metal-mediated [2 + 2 + 1] cycloadditions are also viable and have numerous applications in total syntheses for the creation of five-membered ring systems. A fairly versatile protocol for the synthesis of cyclopentenones of general formula **407** was elaborated using cobalt carbonyl-mediated alkyne–alkene–carbonyl cycloaddition (the Pauson–Khand reaction[34e]). An intramolecular application of this process was elaborated by Shore's group.[34f] This versatile method is especially important for the formation of cyclopentenone moieties fused to a five- or six-membered ring system, as is represented in the general structure **408** (Scheme 2.137).

Initially, the Pauson–Khand reaction involved heating the substrate in a hydrocarbon solution at elevated temperatures and, as such, was not applicable to labile polyfunctional substrates. Later it was discovered that this cycloaddition could be greatly accelerated under the action of mild oxidants (morpholine

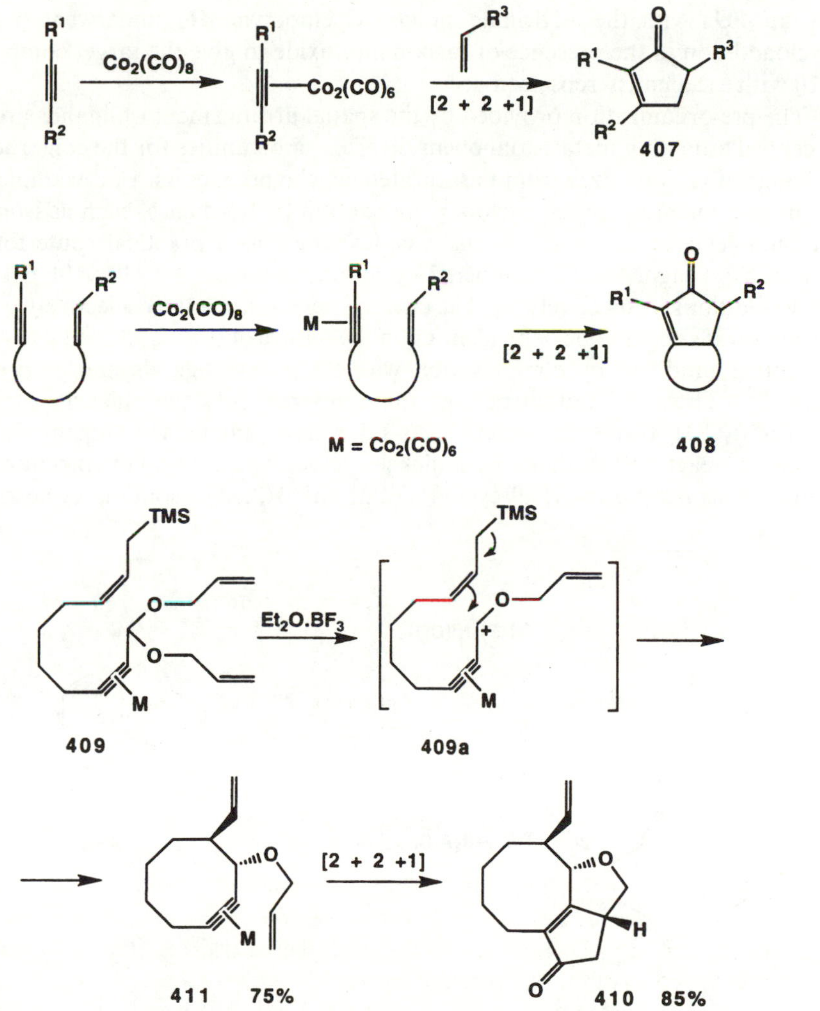

$Co_2(CO)_8$; $Co_2(CO)_6$; [2 + 2 +1] ; **407**

$Co_2(CO)_8$; M = $Co_2(CO)_6$; [2 + 2 +1] ; **408**

TMS ; $Et_2O.BF_3$; **409** ; **409a**

[2 + 2 +1] ; **411** 75% ; **410** 85%

Scheme 2.137

N-oxide[34g]) or by performing reaction on the surface of silica gel in the absence of any solvents.[34h] These conditions allowed the reaction to be carried out at much lower temperature and thus the scope of the reaction was greatly expanded.

An additional advantage of the intramolecular protocol stems from the opportunity to prepare easily the required polyfunctional precursors *via* cobalt carbonyl stabilized propargyl cations. The approach based on the tandem utilization of Co-mediated alkylation and Pauson–Khand annulation was developed in Schreiber's studies to elaborate short pathways for the synthesis of polycyclic compounds.[34i] An example of the efficiency of this protocol is the two-step transformation of the acyclic precursor **409** into the tricyclic derivative **410**. The cobalt-complexed acetal **409** was first transformed into the cyclooctyne derivative **411** *via* intramolecular reaction of the *in situ* generated propargyl cation **409a** with the allylsilane moiety. Cyclooctyne **411** underwent smooth cycloaddition in the presence of carbon monoxide to give the target compound **410** with excellent stereoselectivity.

The pre-organization provided by the spatial arrangement of ligands around a central transition metal atom opens diverse opportunities for the construction of rings of various sizes from unsaturated acyclic precursors. For example, the transition metal-catalysed cyclooligomerization of 1,3-dienes, such as isoprene or butadiene, at the 1,4-positions was developed as a practical route for the synthesis of eight- or 12-membered cyclic polyalkenes ([4 + 4] or [4 + 4 + 4] cycloadditions, respectively[34j]). The chemo-, regio-, and stereoselectively of this process was shown to be dependent upon the nature of the metal used and could be futher modified by complexation with additional ligands and/or promotors.[34j,k] Thus 1,3-butadiene in the presence of the nickel complex $(R_3P)_2Ni(CO)_2$ forms the dimer *cis,cis*-1,5-cyclooctadiene **412** (Scheme 2.138). A similar reaction, catalysed by a nickel–cyclooctene π-complex, produces the trimer *trans,trans,trans*-1,5,9-cyclododecatriene **413**. An isomeric cyclotrimer,

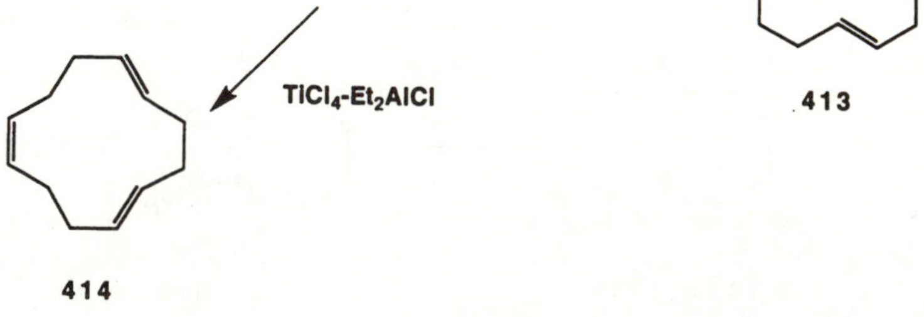

412

$(R_3P)_2Ni(CO)_2$

Ni

413

$TiCl_4$-Et_2AlCl

414

Scheme 2.138

trans,trans,cis-1,5,9-cyclododecatriene **414**, is formed with the catalytic system $TiCl_4$–Et_2AlCl.[34j,k]

Transition metal catalysts were also found to be useful for the mixed cyclo-co-oligomerization of two 1,3-dienes with an alkene or alkyne.[34l] This approach was further extended toward the preparation of heterocyclic systems.[34m]

In contrast to the conventional methods of cycloaddition, capable of forming only one type of product, the reactions catalysed by metal complexes are fairly flexible. This versatility offers numerous possibilities for elaborating a number of methods for the preparation of structurally diverse structures. However, there is still no consistent theory for transition metal complex catalysis which enables one to predict which catalyst and/or conditions are to be employed for the desired transformation. Results are often achieved through intuition and not reasoning. However, who would dare to negate the usefulness of tools like intuition and mere luck if they produce spectacular results like those met in the area of transition metal catalysis![34n]

2.20 RADICAL REACTIONS. NEWLY EMERGED TOOLS FOR THE SYNTHESIS OF CYCLIC COMPOUNDS

As we stated earlier, the majority of methods to form **C–C** bonds in a total synthesis are based on heterolytic processes involving the participation of carbanion- or carbocation-like species or imply the utilization of various cycloaddition reactions. The main reason radical reactions are generally less suited for this purpose can be easily understood if one takes into account the mechanisms involving homolytic scission/formation of covalent bonds. Typically these reactions proceed as a sequence of discrete steps: initiation, chain propagation, and termination, as is shown for the radical addition at the double bond in Scheme 2.139.

It is easy to grasp that a synthetically meaningful result of this sequence of events can be achieved only if intermediate radical **A**, formed upon the addition of **R·** at the double bond, is immediately trapped by a hydrogen donor to give the desired 1:1 adduct **B**. It is also possible for intermediate **A** to react with a second molecule of the unsaturated substrate to give the next radical intermediate, **A′**. This process may be repeated to yield eventually a mixture of oligomeric products. Therefore, methods based upon the radical additions in acyclic series usually utilize an excess of addition reactant to increase the ratio of the desired 1:1 addition to oligomerization. As a typical example, the preparation of levulinic aldehyde acetal **415** *via* radical addition of excess acetaldehyde to acrolein acetal is given in Scheme 2.139 (radical alkene hydroacylation reaction[35a]).

Despite these complications, the enormous and unique synthetic potential of radical reactions is also obvious. Radicals are highly reactive species and their addition to multiple bonds occurs easily, even with crowded substrates, under mild and essentially neutral conditions. Radical reactions are not generally sensitive to the influence of polar effects and tolerate the presence of functional groups otherwise incompatible with electrophilic and/or nucleophilic reagents.

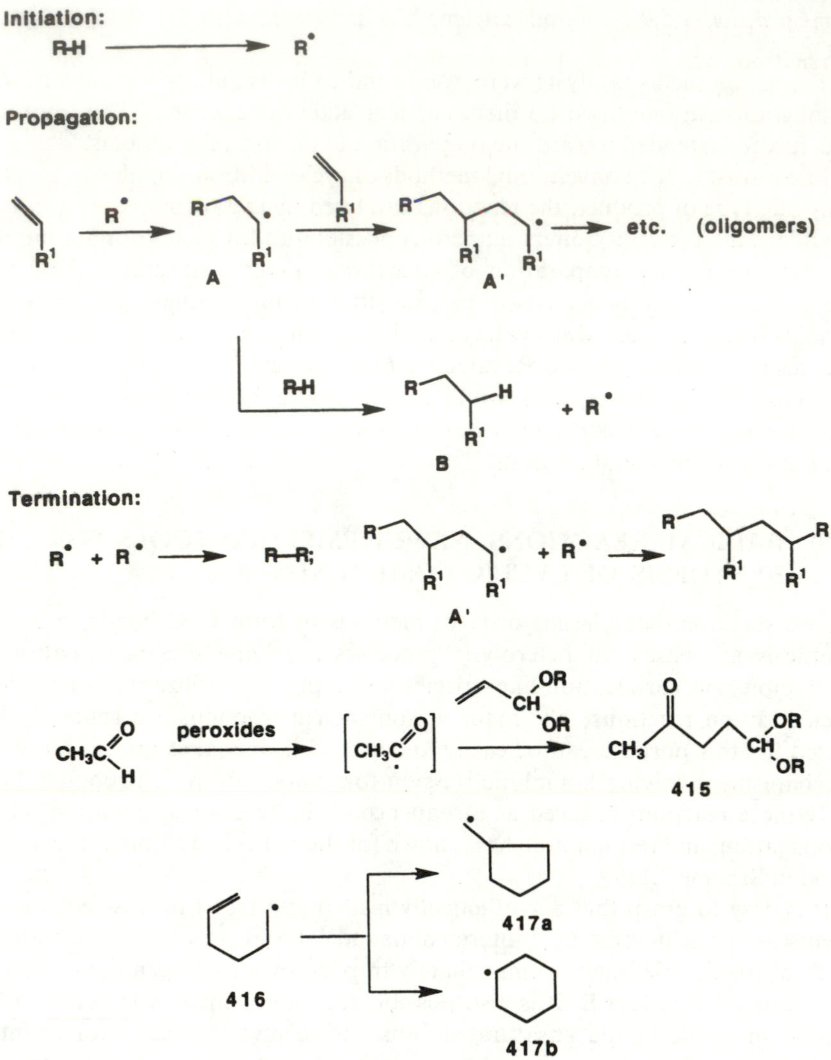

Initiation:

R-H ⟶ R•

Propagation:

Termination:

Scheme 2.139

It was the search for pathways to exploit these advantages in the early 1980s that revealed that the majority of the complications arising from oligomerization might be avoided, if the initially formed radical **A** is able to undergo an intramolecular cyclization to give a less active cyclic intermediate which can be quenched with an efficient radical scavenger. A thorough investigation of the cyclization course on simple models revealed that, for 5-hexenyl radicals such as **416**, the formation of a five-membered ring system as in **417a** is highly preferred (Scheme 2.139).[35b] The validity of this assertion was substantiated in further synthetic studies. Thus, radical-induced cyclizations emerged as an efficient and selective route to form cyclopentane moieties.[35b] Numerous total syntheses were carried out which took advantage of this approach. Some representative cases will be briefly highlighted in this section.

One of the earliest examples of the synthetic promise of radical reactions for preparing polycyclic products was provided by Corey's γ-lactone synthesis.[35c] This approach was actually based on a well-known reaction of α-carbonyl radicals, generated by manganese(III) oxidation of carboxylic acids, with unsaturated substrates. The mechanism of the basic steps shown for the preparation of lactone **418** (Scheme 2.140) involves initial addition of the α-carbonyl radical **419** to the double bond of styrene, followed by oxidation of the radical intermediate **419a** to carbocation **419b**, and subsequent intramolecular reaction with the carboxyl nucleophile to yield the lactone product.

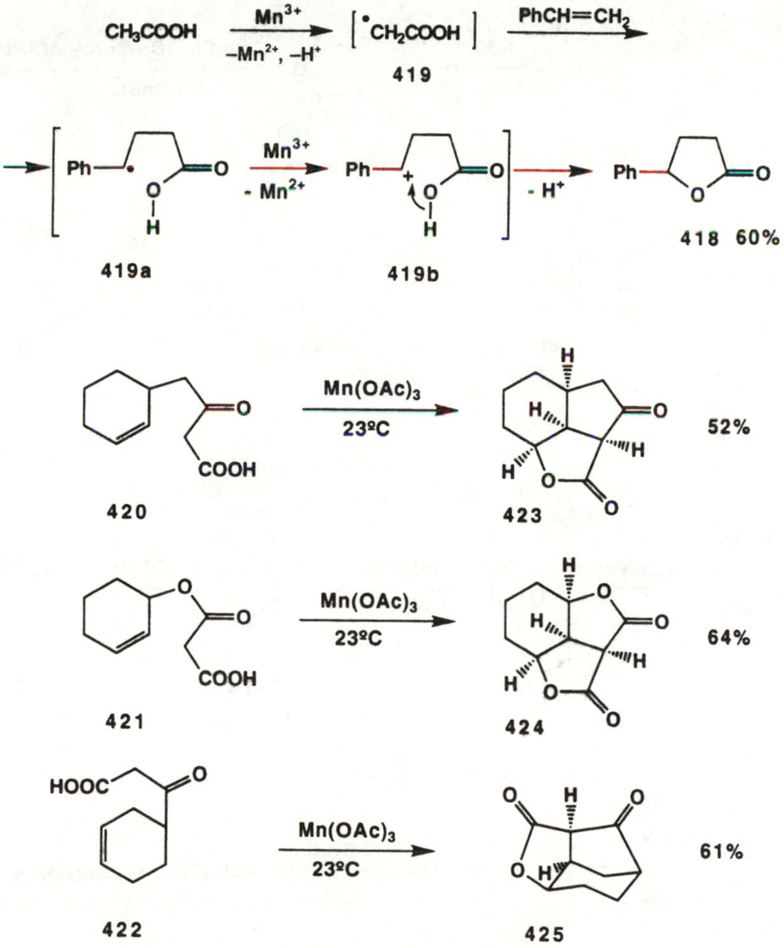

CH_3COOH —(Mn^{3+}; $-Mn^{2+}$, $-H^+$)→ [$^{\bullet}CH_2COOH$] **419** —($PhCH{=}CH_2$)→ [**419a** —(Mn^{3+}; $-Mn^{2+}$)→ **419b**] —($-H^+$)→ **418** 60%

420 —($Mn(OAc)_3$, 23ºC)→ **423** 52%

421 —($Mn(OAc)_3$, 23ºC)→ **424** 64%

422 —($Mn(OAc)_3$, 23ºC)→ **425** 61%

Scheme 2.140

An intramolecular version of this reaction was especially efficient with β-dicarbonyl derivatives such as **420–422**. In fact, the transformations of these substrates into the tricyclic lactones **423–425** occurred under mild conditions and gave the target compounds in good yields and with complete control over the regio- and stereochemistry of the ring junctions.[35c] It is also to be noted that the starting substrates were prepared by routine methods from readily available precursors.

The reactions shown in Scheme 2.140 involve the oxidation of radical species, for example **419a**, as the termination step in a radical addition. An alternative and highly useful protocol was elaborated with a system composed of a radical initiator and a hydrogen donor capable of reducing a similar radical intermediate. Utilization of azobisisobutyronitrile (AIBN) plus Bu_3SnH was shown to be especially efficient for this purpose. In this system, Bu_3SnH fulfills a dual

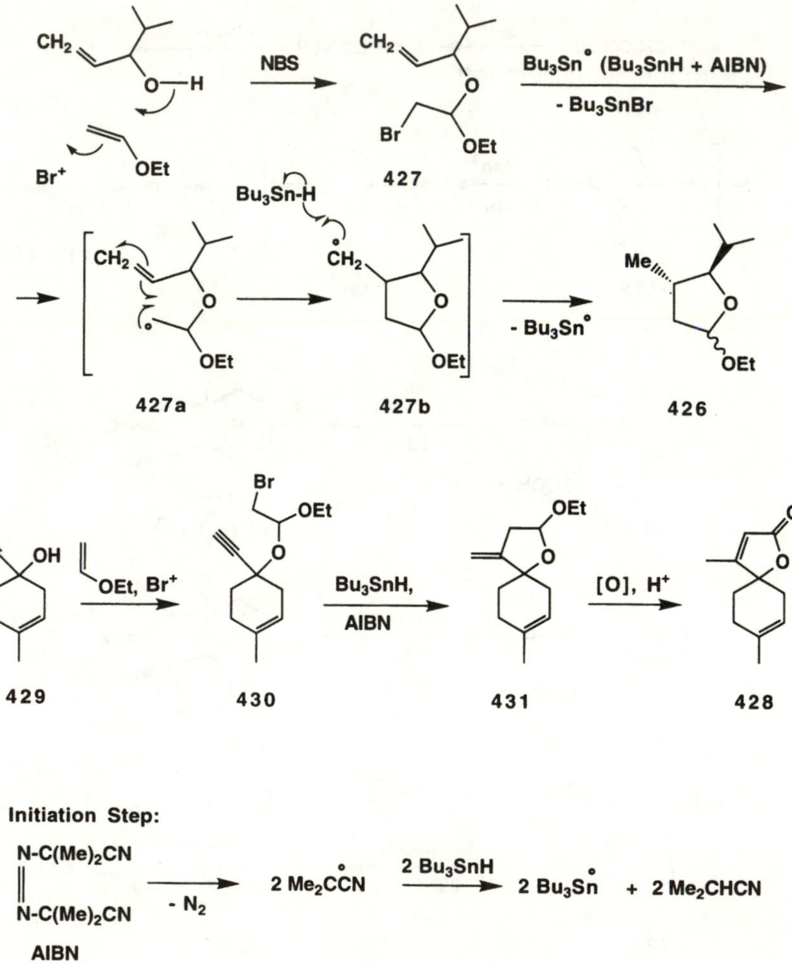

Scheme 2.141

role (Scheme 2.141). First of all it serves as the source of the $Bu_3\dot{S}n$ radicals generated under the action of AIBN (initiation step). These radicals tend to abstract halogen atoms and are thus able to generate site-selectively a radical center in a polyfunctional molecule capable of a subsequent intramolecular cyclization. At the same time, the Bu_3SnH is a powerful hydrogen donor and therefore can serve as an efficient quencher of the cyclized radical intermediate (propagation step), thus precluding the occurrence of side reactions.

The usefulness of this system in a total synthesis was first demonstrated by Stork's studies[35d] which ultimately led to the elaboration of a new and impressive approach to the synthesis of five-membered oxacycles from simple precursors (Scheme 2.141). A typical reaction sequence is illustrated in the preparation of **426** from a substituted allyl alcohol and a vinyl ether.[35d] Precursor **427** is made easily in a conjugate Ad_E reaction initiated by Br^+ (*N*-bromosuccinimide, NBS, as the reagent). Treatment of **427** with Bu_3SnH–AIBN results in bromine atom abstraction to give the radical intermediate **427a**. Cyclization of the latter to **427b** and a subsequent hydrogen atom transfer results in the formation of the target molecule **426**, and regeneration of the chain propagator, $Bu_3\dot{S}n$. The total synthesis of the natural terpenoid andirolactone **428** *via* steps **429** → **430** → **431** is shown as an example of strategy based on this protocol.[35e]

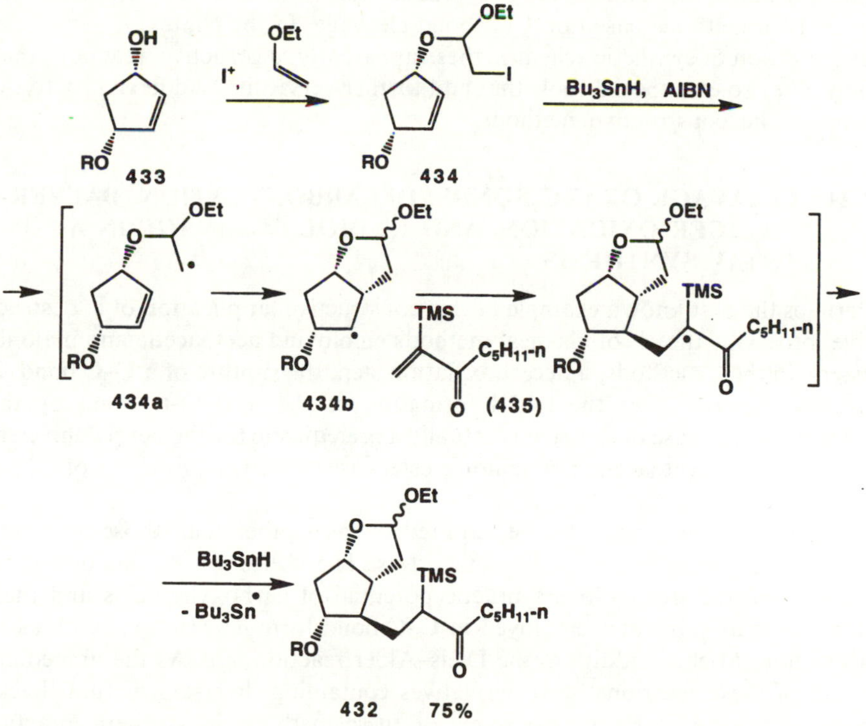

Scheme 2.142

The ingenious combination of a sequence of intra- and intermolecular radical reactions enabled Stork's group to achieve a short route preparation of compound **432** (Scheme 2.142), an advanced intermediate in the synthesis of prostaglandins.[35f] In this sequence an optically active diol **433** was converted into iodoacetal **434**, then treated with Bu_3SnH–AIBN in the presence of an excess of Michael acceptor **435**. The Michael acceptor intercepted the cyclized radical intermediate **434b** to give (after a final hydrogen atom transfer step) the desired adduct **432** in an impressive overall yield of approximately 75%. Further conversion of this adduct into the prostaglandin $PGF_{2\alpha}$ required only four conventional reactions.

In this section, we have presented some of the main approaches for the construction of cyclic systems, not a complete coverage of all possible routes. Our intention was to highlight general trends in the solutions of various problems in this area and to emphasize the flexibility of the elaborated pathways specifically adjusted for diverse synthetic tasks.

PART VII REMODELLING OF A CARBON SKELETON

In the preceding sections we examined basic types of **C–C** bond-forming reactions which are employed for the creation of the carbon skeleton of acyclic or cyclic molecules. This survey of synthetic tools should be now complemented by a set of methods based on **C–C** bond cleavage. In the context of a rational organization of synthetic schemes, these apparently 'destructive' reactions may play a key role as special tools that add further versatility and flexibility to the scope of the 'constructive' methods.

2.21 CLEAVAGE OF C–C BONDS. DECARBOXYLATION, BAEYER–VILLIGER OXIDATION, AND 1,2-DIOL CLEAVAGE IN A TOTAL SYNTHESIS

Perhaps the best known example of the 'constructive' application of a 'destructive' process is the set of classical methods employing acetoacetic and malonic esters. In these methods, a decarboxylation step, the rupture of a **C–C** bond, is usually required after the initial formation of the new **C–C** bond by the alkylation. The ease of this step is actually a prerequisite for the nearly universal application of acetoacetic and malonic esters as synthetic equivalents of C_3^- or C_2^- synthons.

Decarboxylation may also be required in cases other than those involving derivatives of acetoacetic or malonic esters. The usefulness of this operation stems from the tremendous synthetic potential of carboxylic acids and their derivatives as substrates employed in **C–C** bond-forming reactions such as α-alkylation, Michael addition, the Diels–Alder reaction, *etc.* As the immediate result of these reactions, acid derivatives containing diverse structural backbones are formed. Hence the scope of these methods in synthetic practice depends heavily upon the opportunity to remove the carboxyl group after it has fulfilled its auxiliary role. This goal can be achieved in a number of ways, but in essence all of them involve cleavage of the R–COOH bond and the release of CO_2. The above-mentioned decarboxylation of β-dicarbonyl compounds occurs with ease, usually upon moderate warming, owing to the presence of the second carbonyl group, as is shown for the model example **436** in Scheme 2.143.

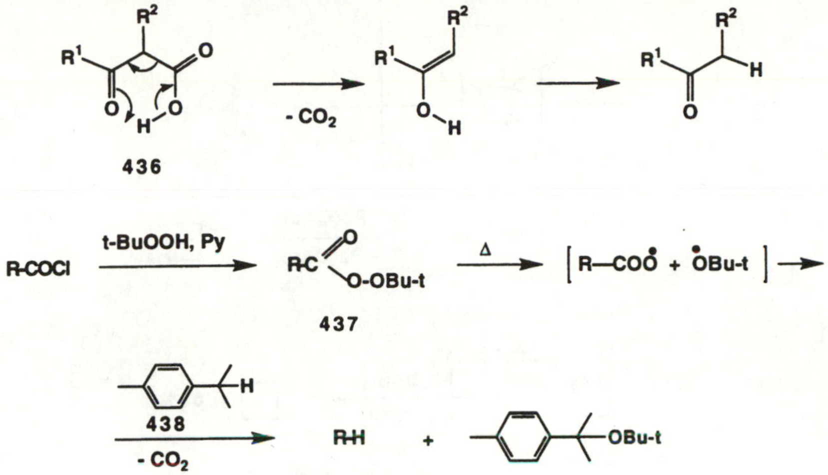

Scheme 2.143

In the absence of the activating second carbonyl functionality, it is necessary to use more ingenious methods to produce the same net effect. These procedures more often than not involve radical reactions. Among them is the thermolysis of *tert*-butyl esters of peroxyacids **437**,[36a] which are readily synthesized in a standard esterification of *tert*-butyl hydroperoxide with an acid chloride. Decarboxylation proceeds *via* an initial homolytic cleavage of the **O–O** bond, elimination of CO_2, and reduction of the incipient alkyl radical by an added hydrogen atom donor such as **438** (Scheme 2.143). Examples showing the exceptional synthetic importance of this decarboxylation procedure will be presented later.

Another widely used decarboxylation procedure involves the use of lead tetraacetate.[36b] Depending on the nature of the substrate and the reaction conditions, this reagent may transform a carboxylic acid into an alkane or alkene, or into the respective acetoxy derivative (Scheme 2.144). The most favorable conditions for alkane formation utilize a good hydrogen donor as the solvent. Usually this transformation is carried out as a photochemically induced oxidative decarboxylation in chloroform solution, as is exemplified in the conversion of cyclobutanecarboxylic acid in cyclobutane.[36c] In contrast, the predominant formation of alkenes occurs in the presence of co-oxidants such as copper acetate.[36d]

In the case of vicinal dicarboxylic acids, the interaction with lead tetraacetate in the presence of co-oxidants (O_2 or Cu^{2+}) invariably leads to the formation of an alkene.[36b] The decarboxylation of vicinal dicarboxylic acids is an especially

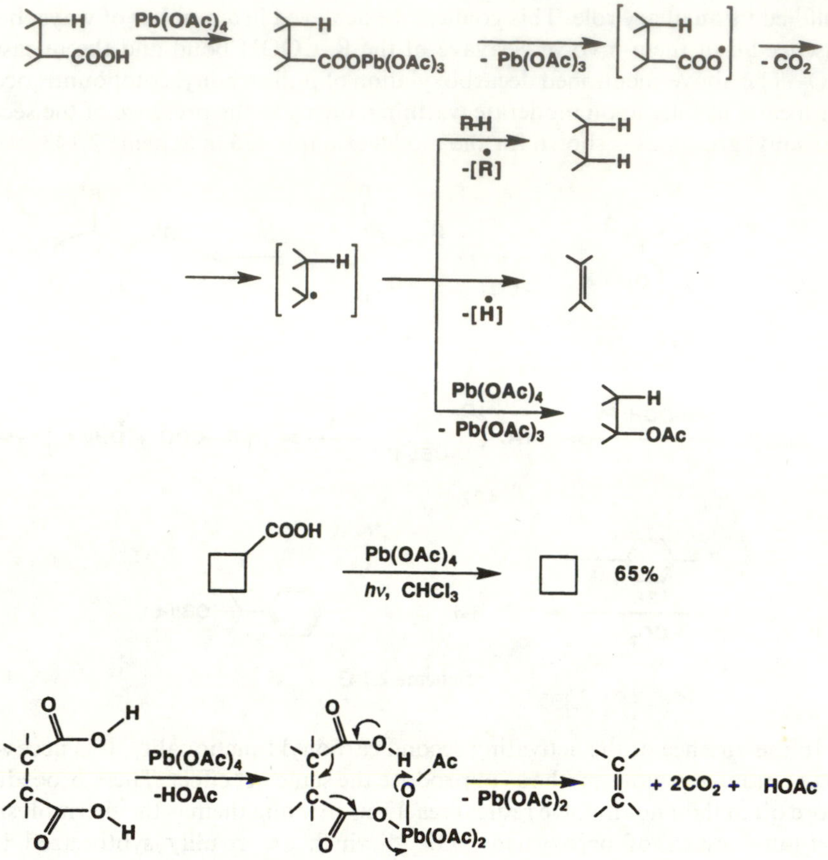

Scheme 2.144

important process for a large class of syntheses which utilize the Diels–Alder reaction as the key step in the generation of polycyclic structures (see, for example, the synthesis of basketene and Dewar benzene, Section 2.19.1). The Diels–Alder synthesis takes place easily with dienophiles such as maleic acid diesters or anhydride because of the presence of the two electron withdrawing groups. As the final goal in the cited examples was to obtain a hydrocarbon, it was imperative to be able to remove the carboxyl groups. This was successfully done in these cases with lead tetraacetate.[36e]

As has already been shown, numerous methods for the creation of carbon–carbon bonds are based on the use of carbonyl groups as an activating moiety to permit the introduction of various structural units at positions adjacent to the carbonyl function. The products of such reactions retain the carbonyl functional group and they can be transformed further without a change in the basic carbon skeleton, for example by nucleophilic addition at the carbonyl group. Alternatively, subsequent cleavage of the **C–CO** bond might well be of synthetic importance. The **C–CO** cleavage can be easily achieved with the Baeyer–Villiger reaction, an oxidation of ketones with peroxy acid (or with hydrogen peroxide in the presence of bases). This reaction proceeds initially as a nucleophilic addition to the carbonyl group to form peroxy intermediate **439**, which undergoes intramolecular rearrangement (**C** → **O** 1,2-alkyl shift) to form esters (for acyclic ketones) or lactones (for cycloalkanones), as is shown in Scheme 2.145.[37a]

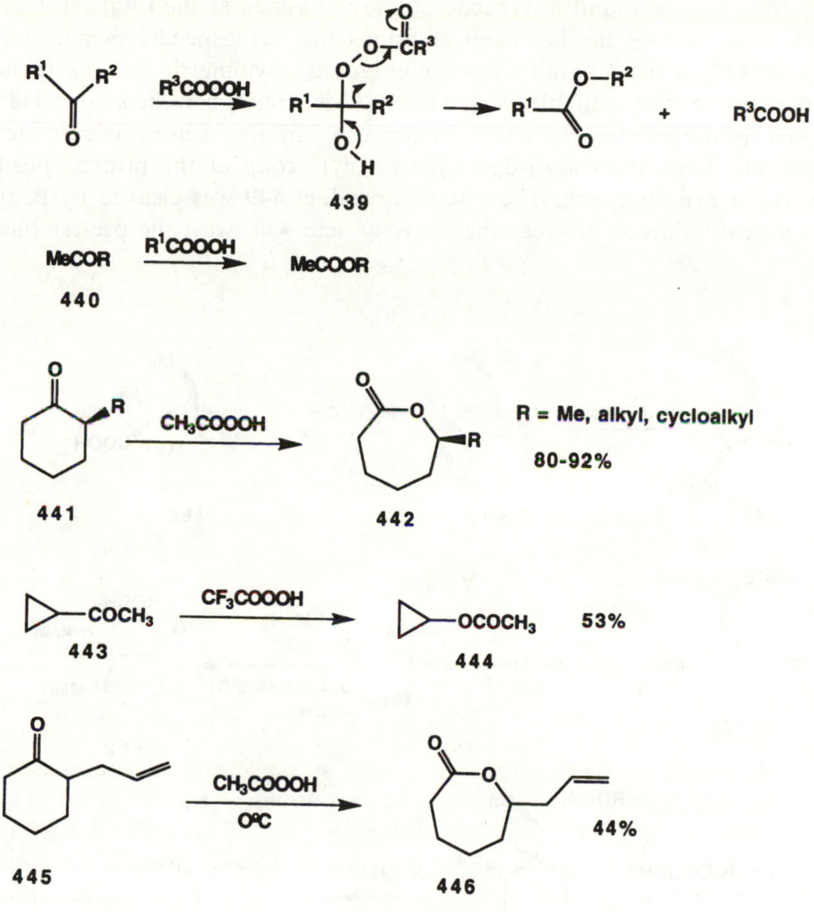

R¹COR² —(R^3COOOH)→ **439** → R¹COOR² + R^3COOH

MeCOR (**440**) —(R^1COOOH)→ MeCOOR

441 —(CH_3COOOH)→ **442** R = Me, alkyl, cycloalkyl 80-92%

443 (cyclopropyl–$COCH_3$) —(CF_3COOOH)→ **444** (cyclopropyl–$OCOCH_3$) 53%

445 —(CH_3COOOH, 0°C)→ **446** 44%

Scheme 2.145

An important feature of this reaction is its high regio- and stereoselectivity. In fact, the course of the reaction for nonsymmetrical ketones is governed by the migratory aptitude of the alkyl group, **R**, from carbon to oxygen, which is known to decrease in the order tertiary > secondary ≫ primary. Complete retention of configuration at the carbon center of the migrating fragment is also a well-established facet of the Baeyer–Villiger oxidation.

Thus methyl ketones **440** can be easily transformed into acetates, MeCOOR, regardless of the nature of the **R** residue (Scheme 2.145). 2-Substituted cycloalkanones like **441** are oxidized regio- and stereoselectively to lactones **442**.[37b] These reactions proceed under mild conditions and can thus be applied to transformations of acid-sensitive compounds, such as **443** → **444**.[37c] This reaction may also serve as a chemoselective method of oxidation for unsaturated ketones, for example **445** to the respective lactone **446**[37d] with the double bond being unaffected.

A spectacular example of the synthetic importance of the Baeyer–Villiger transformation is found in Grieco's studies,[37e] aimed at the total synthesis of calcimycin. One of the key steps envisions the stereospecific preparation of lactone **447**, which contains three contiguous asymmetric centers (Scheme 2.146). The readily available (*via* the Diels–Alder route) bicyclic ketone **448** was chosen as the starting material. Methylation of this ketone occurs stereospecifically from the *endo* side as the methyl group at the bridged position blocks the *exo* approach. The resulting product **449** was cleaved by Baeyer–Villiger oxidation to produce the hydroxy acid **450**, with the pre-established

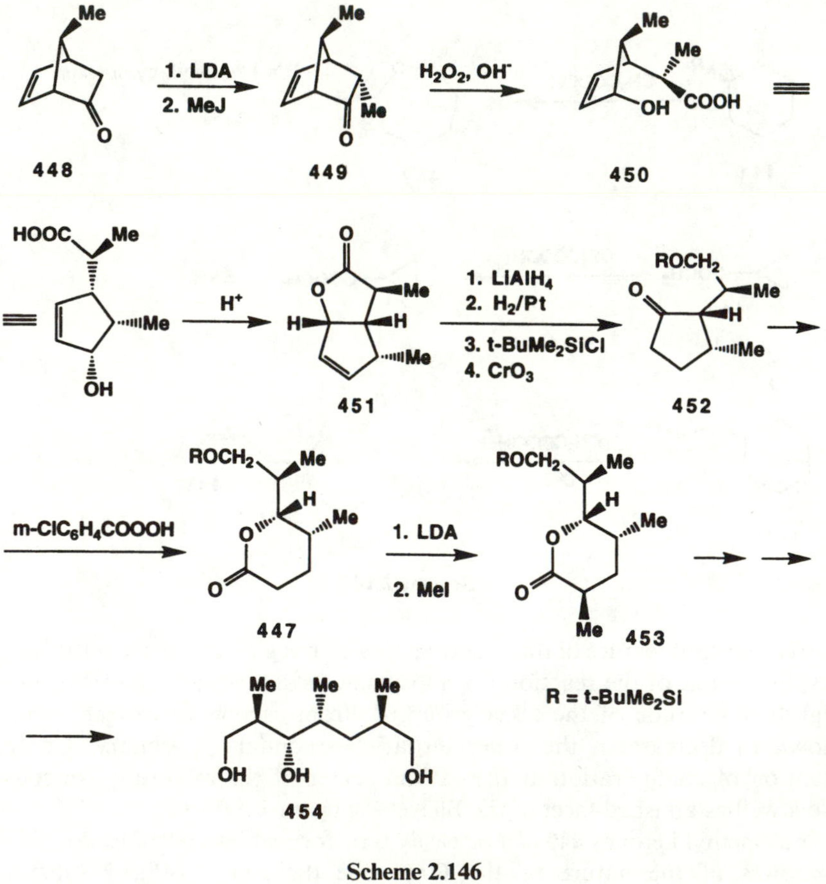

Scheme 2.146

configuration at all three stereocenters. The acid-catalysed lactonization of this product occurred with allylic rearrangement to give the bicyclic lactone **451**. The latter was transformed into the cyclopentanone derivative **452** in a series of trivial manipulations. A second Baeyer–Villiger reaction led to the formation of the desired key intermediate, lactone **447**, which was methylated to give **453**. The latter was further transformed, again in a relatively simple manner, into stereochemically pure **454**, a basic acyclic substrate used in the synthesis of the target compound. Thus a truly challenging stereochemical task of constructing an acyclic molecule with four chiral carbons was solved by the preparation of cyclic intermediates **449** and **452** with rigorously defined configurations. Two selective Baeyer–Villiger oxidations secured their stereospecific transformation into the required building blocks.

Of course, the cleavage of simple inactive **C–C** bonds can also be accomplished, but usually not selectively since more drastic conditions such as pyrolysis or combustion are required. Our interest, however, remains only with reactions that permit the *selective* cleavage of **C–C** bonds. Such reactions, generally speaking, are few in number. The most important involve the oxidative cleavage of vicinal glycols under the action of periodate in aqueous media, or lead tetraacetate in organic solvents, to form two carbonyl products.[38a,b] Oxidation proceeds *via* formation of cyclic intermediates such as **455**, as illustrated in the periodate cleavage of 1,2-diols (Scheme 2.147).

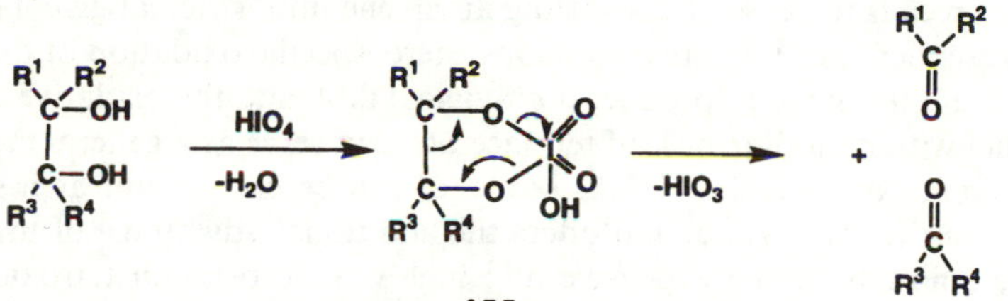

$$R^1R^2C(OH)\text{–}C(OH)R^3R^4 \xrightarrow[-H_2O]{HIO_4} \mathbf{455} \xrightarrow{-HIO_3} R^1R^2C{=}O + O{=}CR^3R^4$$

Scheme 2.147

These reactions historically have had a broad use in the chemistry of natural products, such as carbohydrates, both for structural elucidation and for preparative purposes. Numerous syntheses of chiral intermediates from carbohydrate precursors take the advantage of the selectivity and ease of *vic*-diol cleavage. The general route consists of modifying the chosen carbohydrate substrate in such a way to leave intact a single 1,2-diol fragment with subsequent cleavage in an oxidation procedure. An instructive example is provided by the preparation of a a key chiral building block, optically pure D-glyceraldehyde **456** (Scheme 2.148). Readily available D-mannitol **457** was first protected with two *O*-isopropylidene groups. The remaining vicinal glycol moiety in **458** was oxidized with sodium periodate, which acts as 'scissors' to cut the molecule in half at the central **C–C** bond.[38c] As the 'upper' and 'lower' halves of mannitol are identical, the result is the formation of a single product **459** (yield 70–80%), the protected derivative of optically pure **456**.

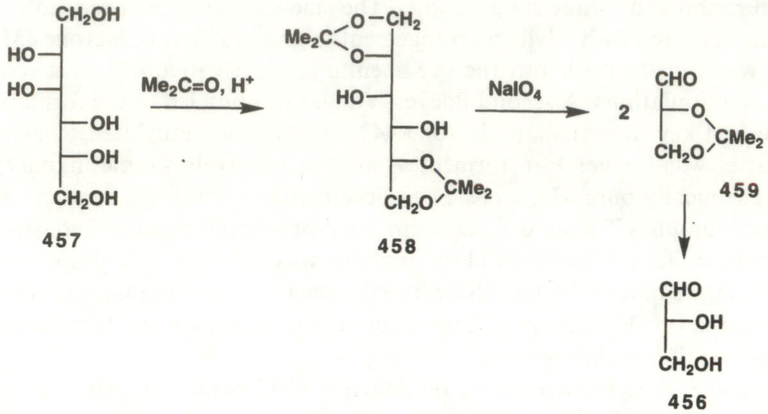

Scheme 2.148

2.22 SYNTHETIC UTILIZATION OF THE DOUBLE BOND CLEAVAGE REACTIONS

The creative potential of this **C–C** bond-cleaving reaction can be better assessed when one recalls the ease of converting an alkene into a vicinal glycol (Section 2.4). The sequence of these two reactions: stereospecific oxidation of the alkene with osmium tetroxide to produce a *cis*-vicinal diol, and the oxidative cleavage of the diol with periodate or lead tetraacetate, can serve as a general method to cleave alkenes oxidatively.[38d] This sequence can be carried out as a series of reactions in the same vessel and offers the additional advantage of using only catalytic amounts of the expensive and highly toxic osmium tetroxide. (The osmium tetroxide is regenerated in the reaction mixture owing to the presence of the excess of the second oxidant, periodate.)

One of the most popular ways of using the double bond cleavage sequence envisages its utilization for the preparation of 1,*n*-dicarbonyl compounds *via* oxidation of the respective cycloalkene derivatives. Thus oxidation of cyclohexene represents the easiest way to prepare the 1,6-dialdehyde **460**. Intramolecular aldol condensation of **460** proceeds with ease to give the respective cyclopentene derivatives **461** or **462** (Scheme 2.149).

Since both oxidative splitting of the double bond and aldol condensation represent reliable and general reactions, their sequence serves as an efficient route for the transformation of readily available cyclohexene systems (*e.g.* formed *via* the Diels–Alder reaction or Robinson annulation) into functionalized cyclopentene derivatives. This standard operational mode is extensively used in total syntheses. One of the numerous examples, the synthesis of helminthosporal **463**, the sesquiterpenoid toxin of fungi,[38e] is shown in Scheme 2.150. In the initial phases of the synthesis, commercially available (−)-carvomenthone **464** was transformed into **465** *via* Michael reaction with methyl vinyl ketone to give **466** and subsequent intramolecular aldol condensation.

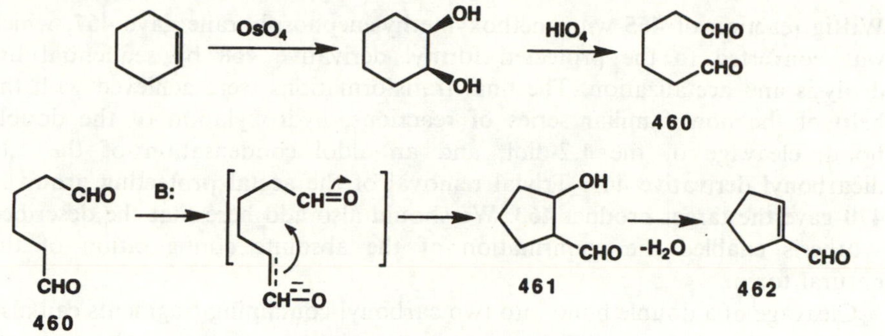

OsO_4 → OH, OH → HIO_4 → CHO, CHO **460**

CHO, CHO **460** → B: → [$CH{=}O$, $CH{=}O$] → OH, CHO **461** → $-H_2O$ → CHO **462**

Scheme 2.149

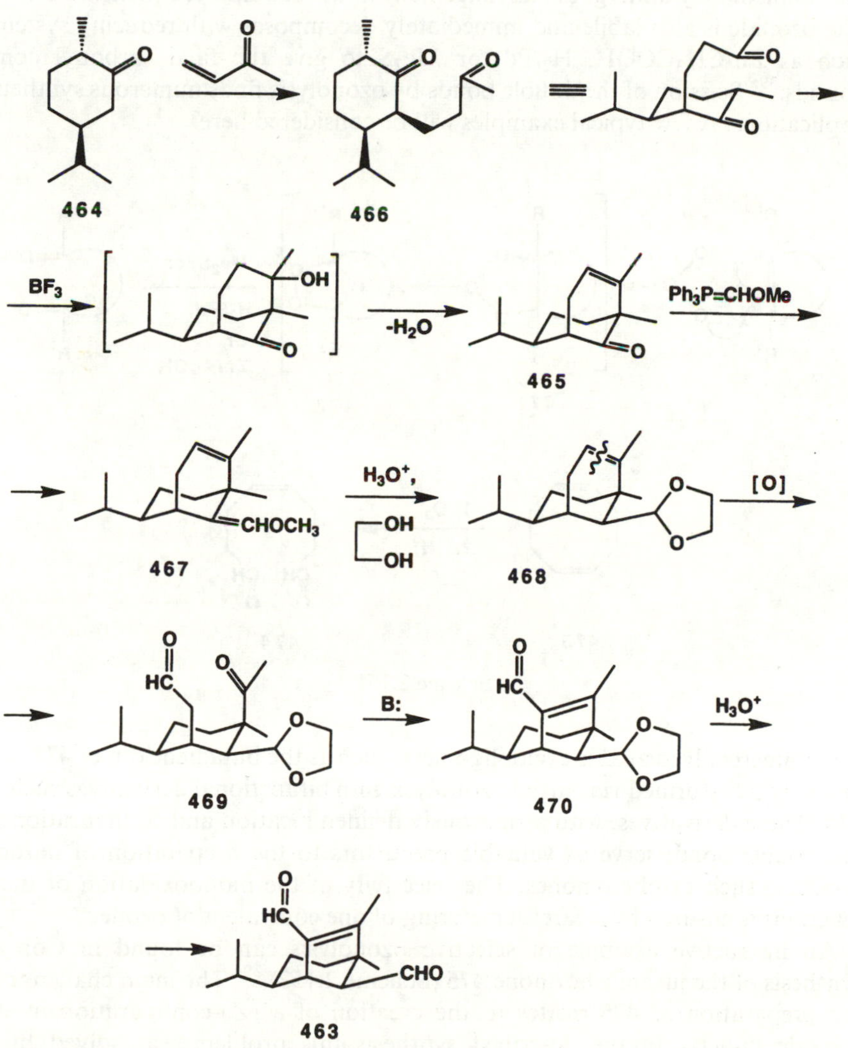

464 → **466** → BF_3 → [OH, O] → $-H_2O$ → **465** → $Ph_3P{=}CHOMe$ → **467** ($=CHOCH_3$) → H_3O^+, $HOCH_2CH_2OH$ → **468** → [O] → **469** → B: → **470** → H_3O^+ → **463** (HC=O, CHO)

Scheme 2.150

Wittig reaction of **465** with methoxymethylenephosphorane gave **467**, which was converted to the protected formyl derivative **468** by sequential hydrolysis and acetalization. The final transformations were achieved with the help of the now familiar series of reactions: hydroxylation of the double bond, cleavage of the 1,2-diol, and an aldol condensation of the 1,6-dicarbonyl derivative **469**. Trivial removal of the acetal protecting group in **470** gave the target product **463**. We should also add here that the described synthesis enabled the confirmation of the absolute configuration of the natural toxin.

Cleavage of a double bond into two carbonyl-containing fragments can also be accomplished using a different reaction: oxidation with ozone. Ozone reacts selectively and rapidly with double bonds to form an unstable cycloadduct **471** that immediately undergoes rearrangement to the ozonide **472** (Scheme 2.151). The ozonide is also labile and immediately decomposes with reducing systems such as Zn/CH_3COOH, H_2/Pd, or Me_2S to give the final carbonyl compounds.[39a] Scission of the double bonds by ozonolysis finds numerous synthetic applications. A few typical examples will be considered here.

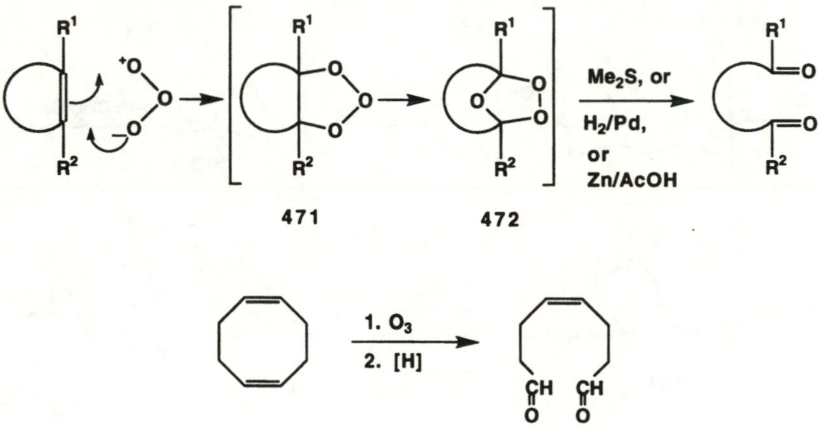

Me2S, or H2/Pd, or Zn/AcOH

471 472

1. O3 2. [H]

473 474

Scheme 2.151

Commercially available cyclooligomers, such as the butadiene dimer **473**, can be easily transformed *via* partial ozonolysis into bifunctional derivatives such as **474**. These derivatives, with a rigorously defined location and configuration of the double bond, serve as valuable precursors to the preparation of natural products such as pheromones. The selectivity of the monooxidation of these reactions is ensured by a careful metering of one equivalent of ozone.[39b]

An instructive example of selective ozonolysis can be found in Corey's synthesis of the juvenile hormone **475** (Scheme 2.152).[39c] The main challenge in the preparation of **475** relates to the creation of a (*Z*)-configuration in the epoxide moiety. In the described synthesis this problem was solved by a judicious choice of cyclic precursors and suitable methods for their transformation that preclude the formation of unwanted isomers. To this end, a readily available methyl ether **476** of *p*-cresol was first transformed into diene **477** (Birch reduction) and the latter was treated with ozone to give **478**. The selectivity of the monooxidation in this case was due to the enhanced reactivity of electron-rich double bond bearing the methoxy group. The efficient preparation of compound **478** actually resolved the key issue in the synthesis of **475**, as the correct configuration of the substituents around the double bond which was to be ultimately transformed into the epoxide moiety was already in place. Two sequential reductions were needed to convert product **478** into **479**. Subsequent chain elongation of **479** followed by functional group transformation at the terminal double bond led to the target molecule **475**. Thus, the Birch reduction, in conjunction with selective ozonolysis, enabled the use of the starting aromatic compound **476** as an equivalent of a functionalized acyclic C_7 synthon with a fixed (*Z*)-geometry at the triply substituted double bond.

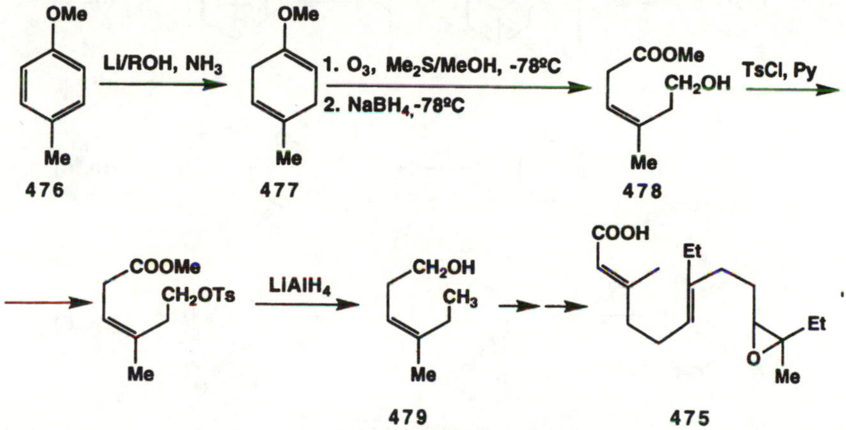

OMe, Me, **476**; Li/ROH, NH_3; OMe, Me, **477**; 1. O_3, Me_2S/MeOH, -78ºC 2. $NaBH_4$,-78ºC; COOMe, CH_2OH, Me, **478**; TsCl, Py; COOMe, CH_2OTs, Me; $LiAlH_4$; CH_2OH, CH_3, Me, **479**; COOH, Et, Et, O, Me, **475**

Scheme 2.152

2.23 REARRANGEMENTS OF THE CARBON SKELETON. SPECIFIC FEATURES AND SYNTHETIC BENEFITS

Almost all the reactions considered above, both constructive and destructive, have a common feature: the bond formation or cleavage occurs exclusively at the site of the reacting functional group and the rest of the molecule is left intact. There are, however, more complicated organic reactions which involve the participation of bonds further removed from the functional group and these may change the basic carbon framework. Classical examples of rearrangements are shown in Scheme 2.153. They include the pinacol rearrangement (rn. **A**), the closely related Wagner–Meerwein rearrangement (rn. **B**), the Claisen rearrangement of allyl–vinyl ethers (rn. **C**), and the oxy-Cope rearrangement of allylvinyl carbinols into unsaturated aldehydes or ketones (rn. **D**). In essence, a rearrangement also takes place in the Baeyer–Villiger oxidation of ketones, although this transformation is traditionally called a 'reaction' and not a 'rearrangement'.

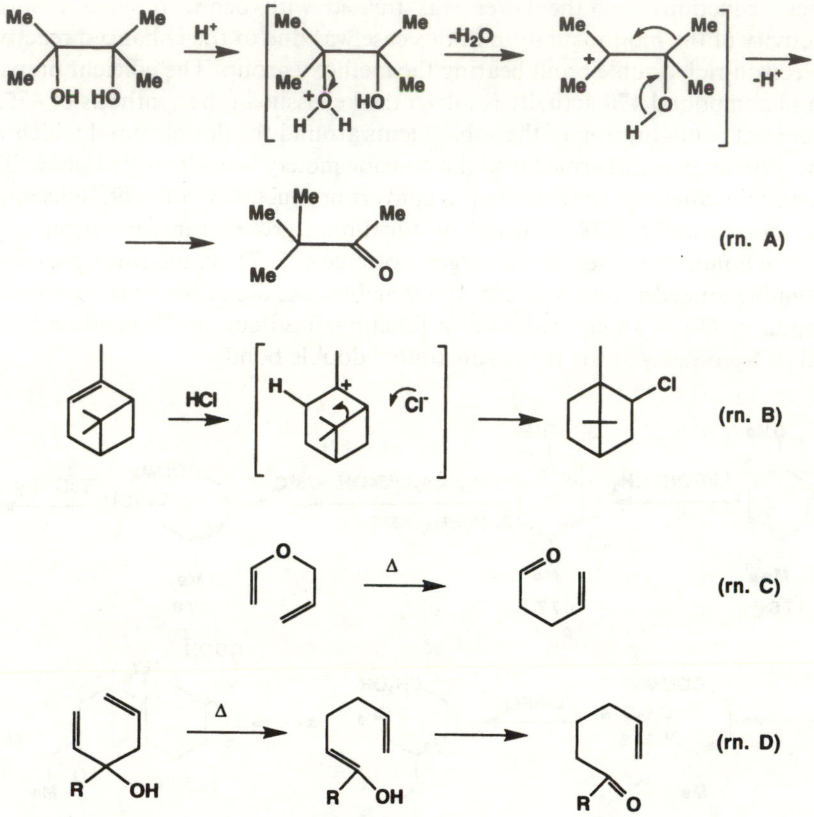

(rn. A)

(rn. B)

(rn. C)

(rn. D)

Scheme 2.153

The discovery of skeletal rearrangements produced a dramatic impact on the development of structural theory in the first half of this century. Numerous theoretical and experimental studies were carried out to formulate a consistent mechanistical description of the observed non-trivial transformations. Because of these efforts, the course of rearrangements in a diverse structural context can be now accurately predicted, or, at the very least, well accounted for. Skeletal rearrangements are therefore no longer a mere challenging curiosity of organic chemistry, but a new set of powerful and reliable synthetic tools.[40a]

2.23.1 Claisen–Johnson–Ireland and Oxy-Cope Rearrangements

In the area of acyclic and alicyclic compounds the Claisen and oxy-Cope rearrangements are especially important (rns. **C** and **D**, respectively, Scheme 2.153).

The Claisen rearrangement of allyl vinyl ethers **480** is a fairly general method for the preparation of γ,δ-unsaturated carbonyl compounds of the general formula **481** from simple precursors (Scheme 2.154).[40b] Synthetically, this transformation is equivalent to the well-known α-allylation of enolates, which gives ultimately the same product. However, the mechanisms and conditions of these two reactions differ and their synthetic potentials are complementary to each other.

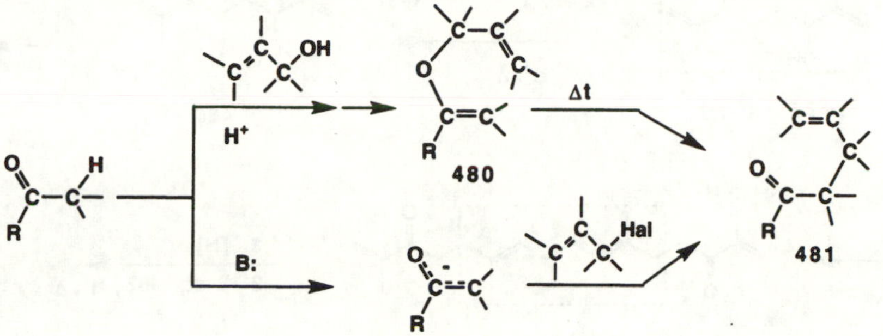

Scheme 2.154

The substrates for allyl vinyl Claisen rearrangements are conveniently obtained by transetherification of vinyl alkyl ethers with allylic alcohols. Typical examples of this rearrangement are represented in Scheme 2.155. The transformation of allyl vinyl ether **482** into aldehyde **483** illustrates the unique potential of the Claisen rearrangement as a method to prepare angularly substituted derivatives from readily available precursors such as **484**,[40c] a goal hardly achievable by other routes. Products of this type are used as key intermediates in the syntheses of many natural compounds.[40b]

The Claisen rearrangement belongs to a class of [3,3] pericyclic sigmatropic rearrangements proceeding *via* a quasi-cyclic transition state. Its course is sensitive to the presence of the substituents in both the vinyl and allyl moieties.[40b] This feature is especially useful when utilizing the Claisen rearrangement in stereocontrolled syntheses of acyclic compounds. The iterative protocol was designed by Johnson's group for the stereoselective preparation of regular isoprenoids, with the Claisen rearrangement of allyl vinyl ethers as a pivotal step (Scheme 2.155). In the sequence shown, a readily available bis-allylic alcohol, **485**, was converted into bis-ether **486** *via* transetherification with an excess of methoxyisoprene **487**. The Claisen rearrangement of **486** proceeded upon moderate heating and gave stereoselectively the bis-enone **488**, with the indicated geometry of the double bonds. Reduction of **488** yielded a bis-allylic alcohol and the entire chain-lengthening procedure was again repeated to give the required product **489**. The latter contained the carbon skeleton of squalene, a natural C_{30} hydrocarbon, thus prepared in a spectacularly short route involving symmetrical chain elongation in accordance with the general scheme $C_{10} + 2C_5 + 2C_5 = C_{30}$.[40d]

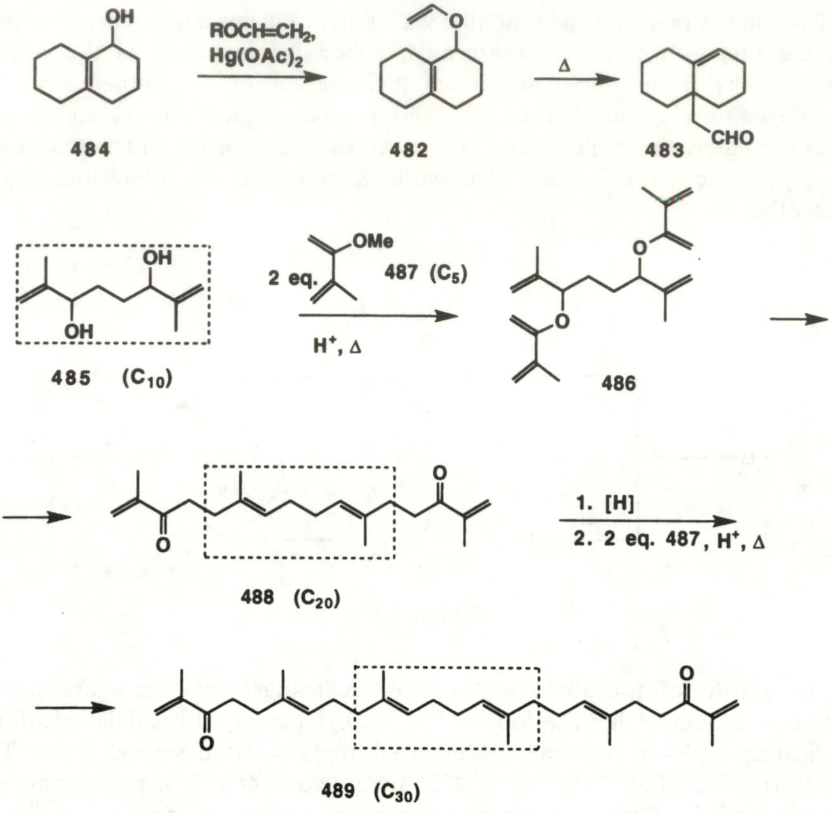

Scheme 2.155

Extremely useful ramifications of the Claisen rearrangement emerged with Johnson's discovery of the orthoester variant of this transformation. His approach (Scheme 2.156) involved the following sequence of steps, which were carried out in one reaction vessel: (i) transesterification of the orthoester with an allylic alcohol to give **490**; (ii) elimination to form the intermediate ketene acetal **491**; and (iii) [3,3] sigmatropic rearrangement to yield the γ,δ-unsaturated ester **492**. The Johnson–Claisen procedure is properly considered to be one of the most efficient methods available to prepare γ,δ-unsaturated esters such as **492**.[40d,e]

This method was further improved when it was found that readily available allyl esters of the general formula **493** could also be involved in Claisen rearrangements *via* intermediate formation of ketene derivatives such as lithium enolates **494** or trimethylsilyl ketene acetals **495** (the Ireland–Claisen variant[40f]). Moreover, rearrangement of these substrates into unsaturated acids **496** occurred easily at room temperature or below. This was in striking contrast to all previous versions of the Claisen rearrangement, which required heating at elevated temperatures (140–160 °C). The Ireland (silyl ketene acetal) variant of

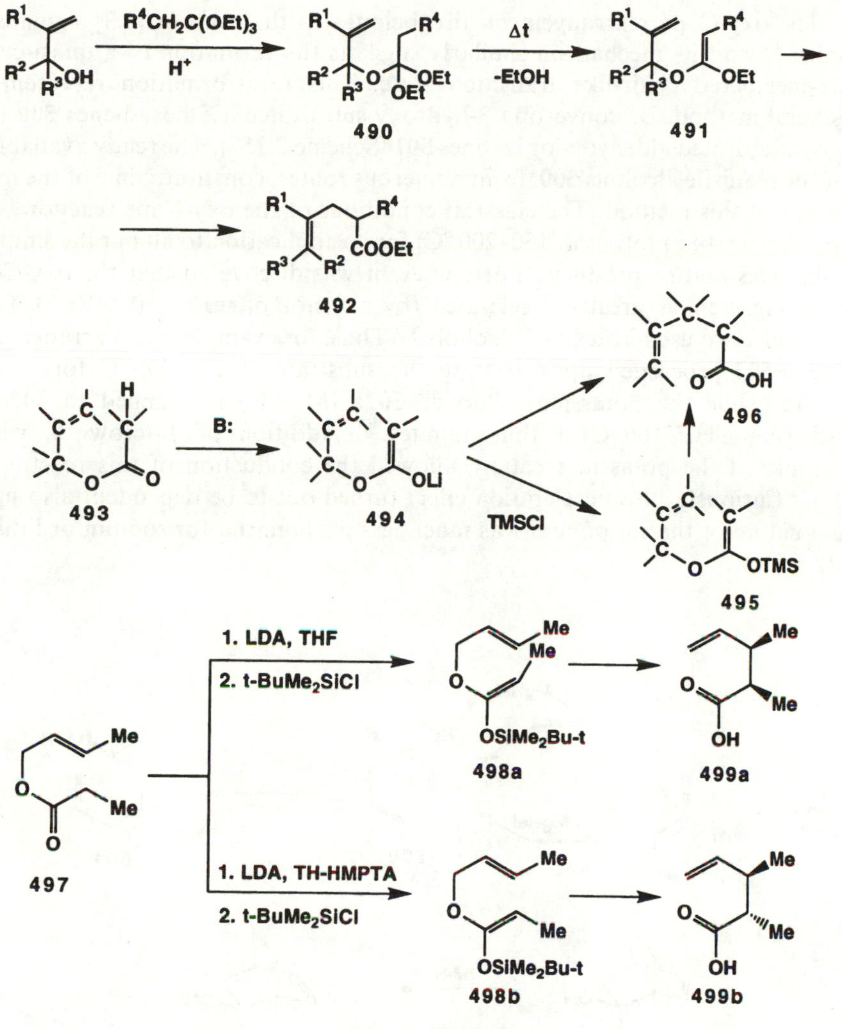

Scheme 2.156

the Claisen rearrangement proved to be especially suitable for the stereocontrolled synthesis of substituted acids of the general formula **496**. The relative stereochemistry of newly created sp^3 centers can be controlled by the proper choice of conditions required for the formation of the required (*Z*)- or (*E*)-enolates. Thus the treatment of *trans*-propenyl propionate **497** with LDA at $-70\,^{\circ}C$ in THF (kinetic conditions) followed by silylation with *t*-$BuMe_2SiCl$ afforded (*E*)-*O*-silyl ketene acetal **498a**, while deprotonation in THF–HMPA (thermodynamic conditions) led principally to (*Z*)-*O*-silyl ketene acetal **498b**. Rearrangement of both isomers proceeded cleanly to give stereospecifically the corresponding isomers **499a** or **499b**.[40b,g]

The oxy-Cope rearrangement also belongs to the class of [3,3] sigmatopic shifts[40h] and its mechanism similarly suggests the formation of a quasi-cyclic six-membered chair-like transition state. This transformation represents a general method for converting 3-hydroxy-substituted 1,5-hexadienes **500** into δ,ε-unsaturated aldehydes or ketones **501** (Scheme 2.157). The ready availability of the requisite alcohols **500**, from numerous routes, constitutes one of the main merits of this method. The classical conditions of the oxy-Cope reaction were too drastic (thermolysis at 150–200 °C) for its application to all but the simplest substrates and/or products. Fortunately it was discovered that the oxy-Cope rearrangement is greatly accelerated (by a factor of up to 10^{10}–10^{15}!) if the alkoxides are used instead of alcohols.[40i] Thus, for example, the rearrangement **502** → **503** proceeded upon heating the substrate at 170–200 °C for several hours, while the potassium alkoxide **502a** (M = K) rearranged to **503a** in refluxing THF (66 °C) within minutes.[40i] Addition of 18-crown-6, which complexed the potassium cation, allowed the conduction of this reaction at 0 °C. Curiously, this acceleration effect turned out to be dependent also upon the nature of the cation and was much less pronounced for sodium or lithium alkoxides.

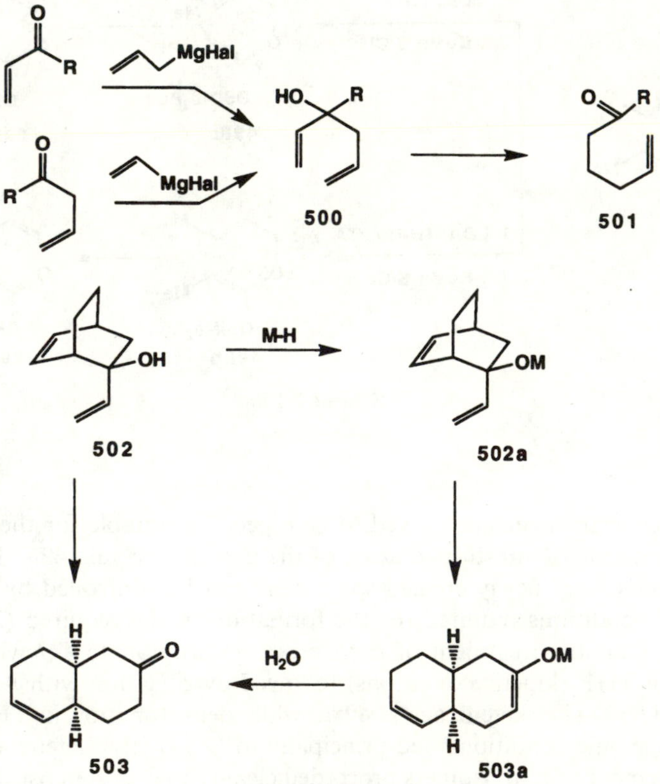

Scheme 2.157

With these improvements the oxy-Cope rearrangement acquired an almost universal applicability as a powerful transformation of strategic importance.[40j] Below are given some typical examples showing the unique synthetic opportunities offered by this method.

The major synthetic problem in the synthesis of the pheromone of the cockroach *Periplaneta americana*, periplanone **504**, is the construction of a 10-membered ring with the proper alignment of the functional groups (Scheme 2.158). Standard ring-forming reactions are rarely effective when applied to the formation of rings of that size. The oxy-Cope rearrangement affforded a general and rather simple solution to this synthetic task. Still's synthesis of periplanone[40k] commenced with the synthesis of allylvinyl carbinol **505**, from the precursor **506**, in a sequence of α-alkylation and carbonyl vinylation steps. Rearrangement of **505** proceeded smoothly and gave the 10-membered derivative **507**. With two double bonds in the right configuration and properly positioned substituents, compound **507** was further employed for the synthesis

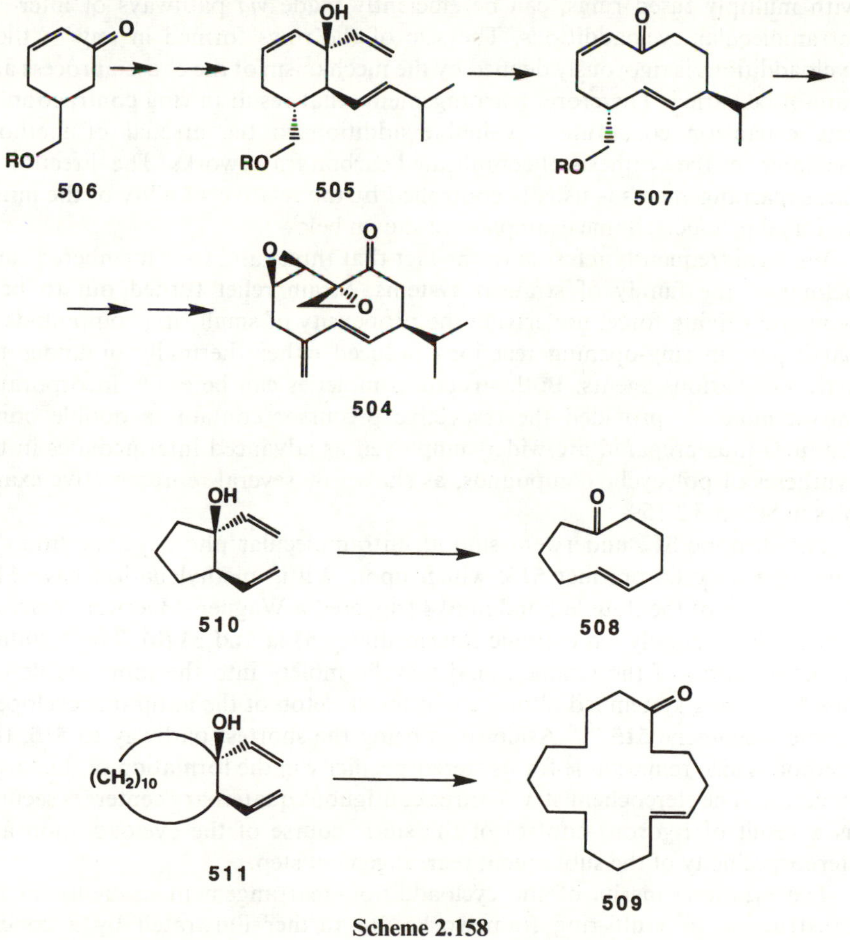

506 505 507

504

510 508

511 509

Scheme 2.158

of various stereoisomers of periplanone, including the naturally occurring isomer **504**. It is also worthwhile to note that alternative protocols for the preparation of **504**, based on entirely different starting materials, also envisioned the use of the oxy-Cope reaction as the key step.[40l]

The transformation **505** → **507** represents a specific case of a general protocol for ring expansion of 1,2-divinyl *n*-cycloalkanols into (n + 4)-cycloalkenones. This route is widely used in preparative practice and is exemplified in the preparation of cyclononenone **508** and hexadecenone **509** from the respective 1,2-divinyl cyclopentanol **510** and cyclododecanol **511** (Scheme 5.158).[40j]

2.23.2 Transformations of Small Ring Fragments and their Role in a Total Synthesis. Wagner–Meerwein Rearrangement, Fragmentation, Favorskii Rearrangement

As was shown in previous sections, various cyclic compounds, including those with multiply fused rings, can be efficiently made *via* pathways of inter- or intramolecular cycloadditions. The size of the rings formed in any of these cycloadditions is rigorously defined by the mechanism of the chosen process and cannot be varied. Therefore, rearrangements that result in ring contraction or ring expansion constitute a valuable addition to the arsenal of methods available for the synthesis of complicated carbon frameworks. The direction of these rearrangements is usually controlled by the relative stability of the initial and final products. Some examples are shown below.

We have frequently referred to the fact that three- and four-membered rings belong to the family of strained systems. Strain relief turned out to be a powerful driving force, underlying the propensity of small ring compounds to participate in ring-opening reactions induced either thermally or under the action of various agents. Both structural moieties can be easily incorporated into a molecule provided the respective precursor contains a double bond. Adducts thus prepared are widely employed as advanced intermediates in the syntheses of polycyclic compounds, as shown by several representative examples in Scheme 2.159.

The dienone **512** underwent smooth intramolecular photocycloaddition to form the tricyclic product **513**, which upon Wittig methylenation gave **514**. Protonation of the double bond in **514** triggered a Wagner–Meerwein rearrangement (presumably *via* cationic intermediates **514a** and **514b**). The resultant transformation of the strained [6.4] bicyclic moiety into the more stable *cis*-fused [5.5] ring system led ultimately to the skeleton of the natural tricyclopentanoid isocomene **515**.[40m] Aside from being the shortest pathway to **515**, this method is also remarkable for its stereospecificity in the formation of the target structure. The stereochemistry of three contiguous quaternary centers is secured as a result of rigorous control of the steric course of the cycloaddition and stereospecificity of the subsequent rearrangement step.

The synthetic merits of the cycloaddition–rearrangement sequence in the construction of multi-ring frameworks are further illustrated by a concise

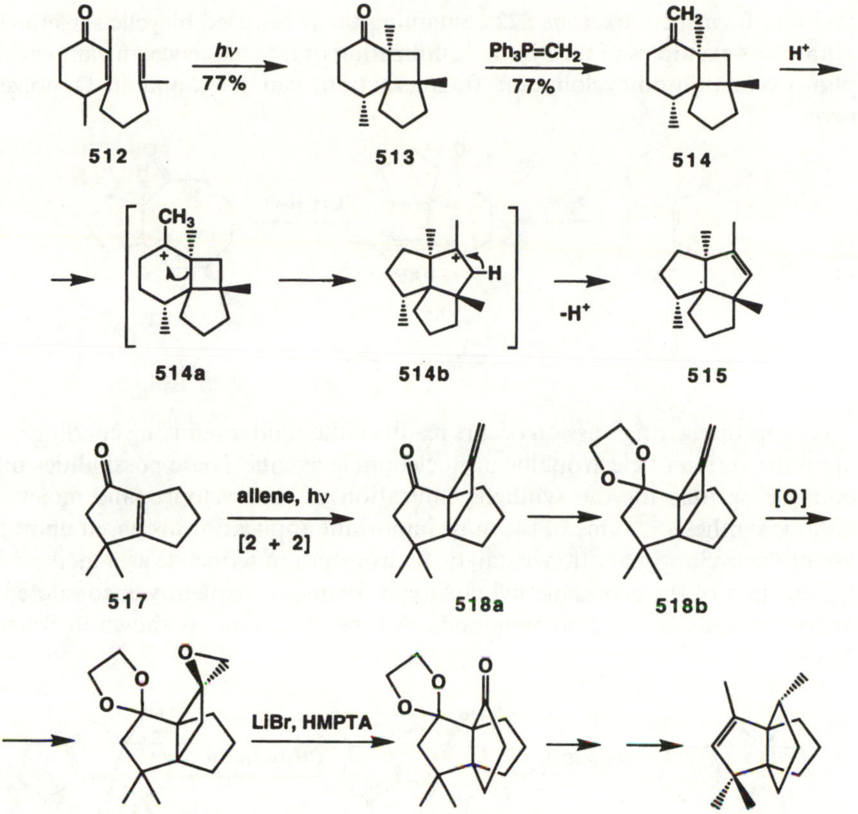

hν 77%
$Ph_3P{=}CH_2$ 77%
H^+
512 513 514
514a 514b 515
$-H^+$
allene, hν [2 + 2]
[O]
517 518a 518b
LiBr, HMPTA
518c 519 516

Scheme 2.159

synthesis of modheptene **516**,[40n] a natural triquinane with the unique [3.3.3] propellane ring system, from bicyclic enone **517** (Scheme 2.159). This route, elaborated by Tobe's group, takes advantage of the efficient enone–alkene photocycloaddition for the construction of the [3.3.2] propellane framework in adduct **518a**. This tricyclic ketone was further transformed into ketal **518b** and then into epoxide **518c**. Subsequent ring expansion to the [3.3.3] system was achieved by lithium bromide catalysed epoxide cleavage, which triggered rearrangement to the more stable ring system. With the basic carbon framework already created, further conversion of adduct **519** into modheptene **516** was not a difficult task.

In the presence of appropriate substituents, fused systems containing four-membered rings are susceptible to smooth fragmentation, a process that results in the formation of medium-sized ring moieties. This strategy is often employed for the preparation of compounds containing seven- or eight-membered ring fragments. Thus, for example, dienone **520** underwent smooth intramolecular [2 + 2] photocyclization to give the tricyclic product **521** (Scheme 2.160). Alkaline treatment of the latter adduct induced a retro-aldol fragmentation and thus furnished diketone **522** containing the [5.8]-fused bicyclic skeleton.[40o] Numerous examples of the synthetic utilization of this sequence, intramolecular photocycloaddition/cyclobutane fragmentation, can be found in Oppolzer's review.[32m]

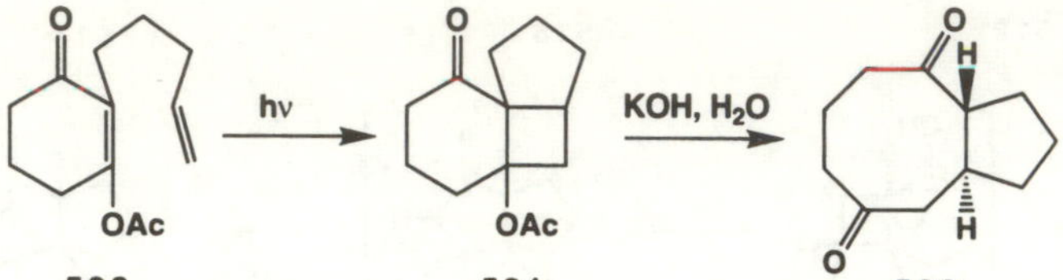

Scheme 2. 160

Cyclopropane ring scission occurs readily either under reducing conditions or upon the action of electrophilic or nucleophilic agents. These possibilities offer multiple options for the synthetic utilization of the cyclopropane moiety in organic synthesis.[40p] One of the most important applications is based upon the use of the cyclopropanation–catalytic hydrogenation sequence as a method for the creation of the *gem*-dimethyl moiety, a fragment frequently encountered in many naturally occurring compounds. A typical example is shown in Scheme 2.161.

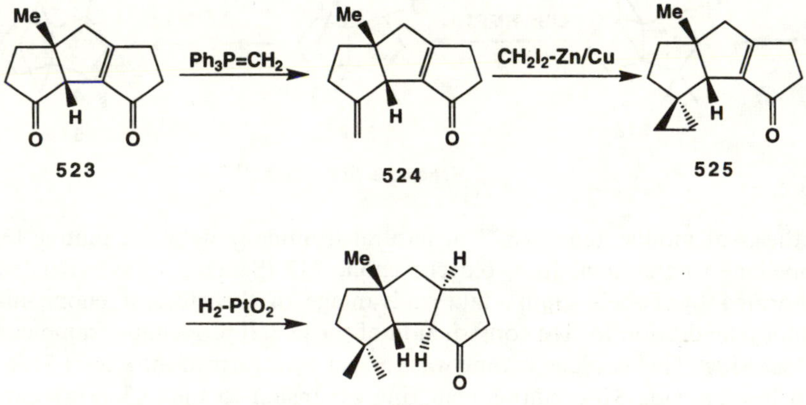

Scheme 2.161

The tricyclic compound **523** (for its preparation, see Chapter 4) served as the key intermediate in Mehta's synthesis of naturally occurring triquinanes. Its conversion into capnellenes[40r] required the introduction of a *gem*-dimethyl group on to ring A. To achieve this transformation, diketone **523** was first converted into the methylene derivative **524** by Wittig methylenation. Simmons–Smith cyclopropanation of the latter gave **525** and subsequent cyclopropane cleavage (accompanied by the reduction of the remaining double bond) proceeded readily to give the required product **526**, the immediate precursor of the target capnellene. Additional examples showing the synthetic usefulness of the cyclopropane moiety will be discussed in the following chapters.

Finally, we must also mention the Favorskii rearrangement, as it plays a key role in the preparation of many exotic strained systems. The net outcome of this rearrangement in the cycloalkane series is a ring contraction of α-halo (*n*) cycloalkanones **527** with the formation of (*n* – 1) cycloalkanecarboxylic acids **528**, as is shown in Scheme 2.162. The mechanism of this reaction involves the generation of carbanion **527a**, which affords the strained cyclopropanone intermediate **527b** *via* an intramolecular displacement reaction. Subsequent intermolecular nucleophilic attack induces ring opening of the strained fragment.[40s]

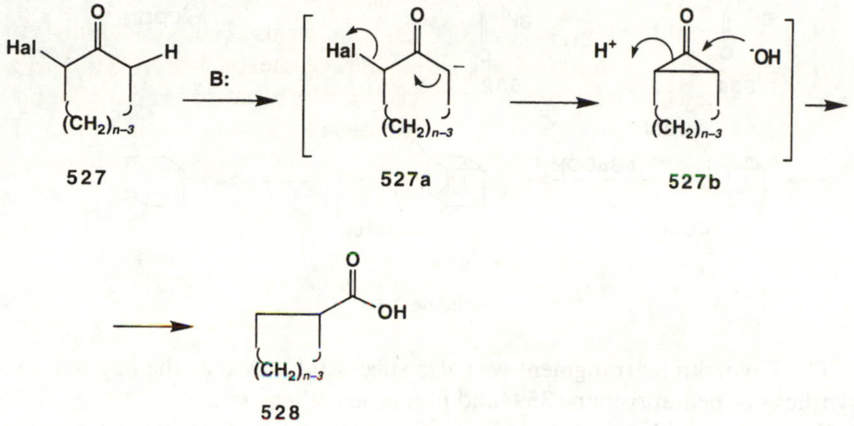

Hal, O, H, $(CH_2)_{n-3}$ **527** —B:→ [Hal, O, $(CH_2)_{n-3}$ **527a** → H^+, O, ^-OH, $(CH_2)_{n-3}$ **527b**] → O, OH, $(CH_2)_{n-3}$ **528**

Scheme 2.162

The tremendous potential of this skeletal rearrangement can be easily appreciated in the examination of a single example, the synthesis of cubane **361**,[40t] the first representative of Plato hydrocarbons (Scheme 2.163). Pettit's elegant synthesis starts with a Diels–Alder reaction between 2,5-dibromobenzoquinone **529** and an extremely unstable diene, cyclobutadiene, generated *in situ* from a stable complex with iron carbonyl, **530**. The resulting adduct **531** already contains two edges of the future cubane. A third edge is formed in a photochemical [2 + 2] cycloaddition to produce the Favorskii precursor **532**. This structure differs from the target hydrocarbon by the presence of two additional carbonyl bridges. Treatment of dibromide **532** with excess KOH at 100 °C effects a double Favorskii rearrangement which closes the missing edges to form the 1,3-cubanedicarboxylic acid **533**. The extraneous carboxyls are removed *via* conversion to *tert*-butyl peroxyacid esters **534** and thermal decarboxylation (see above). It is easy to recognize that the successful accomplishment of the whole sequence lies in the utilization of the Favorskii rearrangement in conjunction with the decarboxylation procedure. This protocol ensures the feasibility of utilizing the starting dibromoquinone as a synthetic equivalent for cyclobutadiene. The net outcome of the reaction sequence, leading ultimately to the formation of cubane, can be formally described as a double [2 + 2] cycloaddition of two molecules of cyclobutadiene.

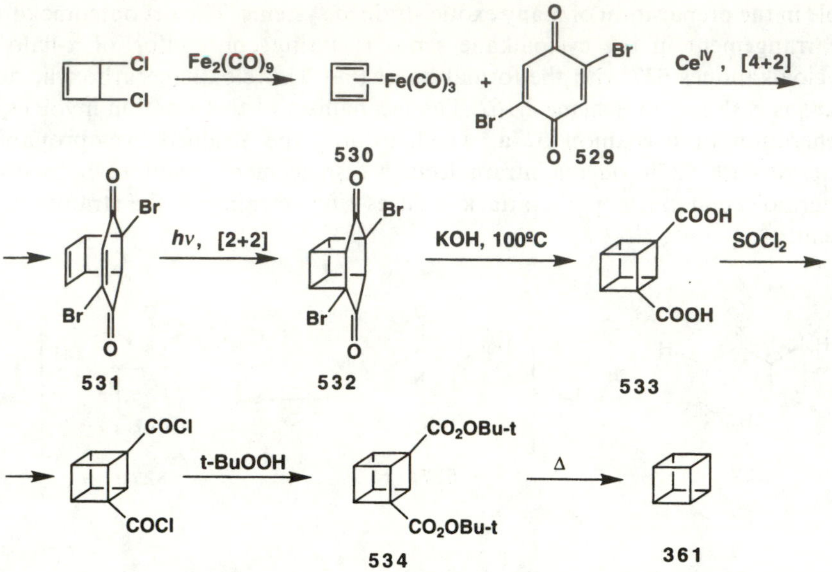

Scheme 2.163

The Favorskii rearrangment was also successfully used as the key step in the synthesis of pentaprismane **359**, and numerous other cases.

We have considered a limited set of examples that illustrate the enormous potential of reactions that involve bond breakage and skeletal rearrangements of various cyclic precursors. While the set given is limited, it is sufficient to illustrate the power and reliability of these reactions, reactions that should always be on the menu of methods in the course of planning a synthesis. Additional examples will be encountered in the following chapters.

CONCLUDING COMMENTS

The overview presented in this chapter cannot be considered as anything more than a sketchy perusal of the major tools employed in organic synthesis. Our main goal was to present, in a concise form, a set of underlying ideas for the elaboration of the principal methods applicable to solving the most diverse tasks encountered in the course of the construction of various organic structures. We hope that with a limited selection of the given factual material notwithstanding, the reader is able to evaluate the richness and versatility of the existing arsenal of synthetic tools available to the organic chemist. More in-depth and detailed information referring to the synthetic methods can be found in the already cited monographs, like Larock's *Comprehensive Organic Trans-*

formations or Smith's *Organic Synthesis* and references cited therein. The tremendous amount of data compiled in these sources may produce the impression that organic chemistry has achieved the status of a mature science capable of solving synthetic problems of almost any complexity. To a certain extent this is truly the case, and that is why from time to time we can hear the claims that, in fact, organic chemistry has lost its identity as a science. This issue is hotly debated in the review by Seebach entitled 'Organic Synthesis—Where now?' (*Angew. Chem., Int. Ed. Engl.*, **1990**, *29*, 1320). As a summary to his thorough and multi-faceted analysis of the present state of the art of organic synthesis, the author comes to the conclusion that 'organic synthesis continues to react forcefully and with vitality to new challenges, still ready to pursue old dreams' and thus organic chemistry and organic synthesis 'are neither stagnating nor are they on the decline'. Needless to say, we share, most enthusiastically, this attitude toward our discipline.

REFERENCES

[1] (a) Van Tamelen, E. E.; Pappas, S. P. *J. Am. Chem. Soc.*, **1963**, *85*, 3297; (b) Schleyer, P. von R.; Donaldson, M. M. *J. Am. Chem. Soc.*, **1960**, *82*, 4645.

[2] (a) The history of this discovery is highlighted in: Berson, J. A. *Tetrahedron*, **1992**, *48*, 3; (b) Normant, H. *Adv. Org. Chem.*, **1960**, *2*, 1; (c) Pedersen, C. J. *J. Am. Chem. Soc.*, **1967**, *89*, 2495; (d) for a review, see: Weber, W. P.; Gokel, G. W. *Phase Transfer Catalysis in Organic Synthesis*, Springer, New York, **1977**.

[3] The basic data related to the structure and properties of various carbocationic derivatives can be found in a treatise: Olah, G.; Schleyer, P. von R. *Carbonium Ions*, Wiley, New York, **1968–1976**, vols. 1–5; for a consise review, see: Bethell, G. *Reactive Intermediates*, Wiley, New York, **1978**, vol. 1.

[4] Bates, R. B.; Ogle, C. A. *Carbanion Chemistry*, Springer, New York, **1983**; see also: Stowell, J. C. *Carbanions in Organic Synthesis*, Wiley, New York, **1979**.

[5] Davies, S. G. *Organotransition Metal Chemistry: Applications to Organic Synthesis*, Baldwin, J. E., Ed., Pergamon, Oxford, **1982**; see also: Collman, J. P.; Hegedus, L. S.; Norton, J. R.; Finke, R. G. *Principles and Applications of Organotransition Metal Chemistry*, University Science Books, Mill Valley, CA, **1987**.

[6] For a review on synthetic utilization of allylic reagents, see: Magid, R. M. *Tetrahedron*, **1980**, *36*, 1901.

[7] Mori, K. *Tetrahedron*, **1974**, *30*, 3807.

[8] House, H. O. *Modern Synthetic Reactions*, 2nd edn., Benjamin/Cummings, Menlo Park, CA, **1972**, ch. 9, p. 492.

[9] (a) See: Eicher, T., in *The Chemistry of Carbonyl Group*, Patai, S., Ed., Interscience, New York, **1970**, pp. 621–693; (b) Imamoto, T.; Takiyama, N.; Nakamura, K. *Tetrahedron Lett.*, **1985**, *26*, 4763; Nagasawa, K.; Ito, K. *Heterocycles*, **1989**, *28*, 703.

[10] Ref. 8, ch. 10, p. 629.

[11] Masamune, S.; Choy, W. *Aldrichim. Acta*, **1982**, *15*, 47.

[12] (a) For reviews, see: Maercker, A. *Org. React.*, **1965**, *14*, 270; Bestmann, H. J.; Vostrowsky, O. *Top. Curr. Chem.*, **1983**, *109*, 85; (b) Wittig, G.; Geissler, G. *Ann. Chem.*, **1953**, *580*, 44; for historical background, see: Wittig, G. *Pure Appl. Chem.*, **1964**, *9*, 245; (c) Wadsworth, W. W., Jr. *Org. React.*, **1977**, *25*, 73; (d) Ramirez, F.; Desai, N. B.; McKelvie, N. *J. Am. Chem. Soc.*, **1962**, *84*, 1745; (e) Corey, E. J.; Fuchs,

P. L. *Tetrahedron Lett.*, **1972**, 3769; see also: Matsumoto, M.; Kuroda, K. *Tetrahedron Lett.*, **1980**, 4021.

[13] For a review, see: Bergmann, E. D.; Ginsburg, D.; Pappo, R. *Org. React.*, **1959**, *10*, 179.

[14] (a) Rapson, W. S.; Robinson, R. *J. Chem. Soc.*, **1935**, 1285; see also: du Feu, E. C.; McQuillin, F. J.; Robinson, R. *J. Chem. Soc.*, **1937**, 53; (b) for reviews, see: Jung, M. E. *Tetrahedron*, **1976**, *32*, 3; Gawley, R. E. *Synthesis*, **1976**, 777; (c) Nazarov, I. N.; Zav'yalov, S. I. *Izv. Akad. Nauk SSSR, Otd. Khim. Nauk*, **1952**, 300; (d) Tramontini, M. *Synthesis*, **1973**, 703; (e) see, for example, the list of more than a dozen of the modified Michael acceptors for Robinson-like annulations in Larock, R. C. *Comprehensive Organic Transformations*, VCH, New York, **1989**, p. 669.

[15] (a) Stork, G. *Pure Appl. Chem.*, **1968**, *17*, 383; (b) Taylor, R. J. K. *Synthesis*, **1985**, 364; (c) Posner, G. H. *An Introduction to Synthesis Using Organocopper Reagents*, Wiley, New York, **1980**; (d) 'Recent Developments in Organocopper Chemistry', *Tetrahedron*, **1989**, *45*, 349; (e) for a comprehensive summary, see: Perlmutter, P. *Conjugate Addition Reactions in Organic Synthesis*, Pergamon, Oxford, **1992**.

[16] (a) Normant, J. F.; Alexakis, A. *Synthesis*, **1981**, 841; see also: Ender, E. *Tetrahedron*, **1984**, *40*, 641; (b) Cahiez, G.; Alexakis, A.; Normant, J. F. *Tetrahedron Lett.*, **1980**, *21*, 1433; (c) Marfat, A.; McGuirk, P. R.; Helquist, P. *J. Org. Chem.*, **1979**, *44*, 3888; (d) Knight, D. W.; Ojhara, B. *J. Chem. Soc., Perkin Trans. 1*, **1983**, 955; see also: Baker, R.; Billington, D. C.; Ekanayake, N. *J. Chem. Soc., Perkin Trans. 1*, **1983**, 1387.

[17] (a) Smit, W. A. *Sov. Sci. Rev. B. Chem.*, **1985**, *7*, 155; (b) Fleming, I. *Chem Soc. Rev.*, **1981**, *10*, 83; (c) Weber, W. *Silicon Reagents for Organic Synthesis*, Springer, New York, **1983**, ch. 11, p. 173; (d) Mukaiyama, T. *Org. React.*, **1982**, *28*, 203; see also: Mukaiyama, T.; Murakami, M. *Synthesis*, **1987**, 1043; (e) Reetz, M. T.; Maier, W. F.; Chatziiosifidis, I.; Giannis, A.; Heimbach, H.; Löwe, U. *Chem. Ber.*, **1980**, *113*, 3741; for a review, see: Reetz, M. T. *Angew. Chem.*, **1982**, *94*, 97; (f) RajanBabu, T. V. *J. Org. Chem.*, **1984**, *49*, 2083; see also: Narasaka, K.; Soai, K.; Mukaiyama, T. *Chem. Lett.*, **1974**, 1223; (g) see reviews: Reetz, M. T. *Angew. Chem., Int. Ed., Engl.*, **1984**, *23*, 556; Reetz, M. T. *Acc. Chem. Res.*, **1993**, *26*, 462.

[18] (a) Nicholas, K. M. *Acc. Chem. Res.*, **1987**, *20*, 207; (b) Schegolev, A. A.; Smit, W. A.; Kalyan, Y. B.; Krimer, M. Z.; Caple, R. *Tetrahedron Lett.*, **1982**, *23*, 4419; see also: Caple, R., in *Organic Synthesis: Modern Trends*, Chizhov, O. S., Ed., Blackwell Scientific, London, **1987**, p. 119; (c) ramifications for the total synthesis are highlighted in the review: Smit, W. A.; Caple, R.; Smolyakova, I. P. *Chem Rev.*, **1994**, *94*, 2359.

[19] (a) Larock, R. C. *Comprehensive Organic Transformations*, VCH, New York, **1989**; (b) Soloveichik, S.; Krakauer, H. *J. Chem. Educ.*, **1966**, *43*, 532; see also the textbook: Mathieu, J.; Panico, P.; Weill-Reynal, J. *L'Aménagement Fonctionnel en Synthese Organique*, Herman, Paris, **1978**; (c) ref. 19(a), p. 354; (d) Murata, S.; Suzuki, M.; Noyori R. *J. Am. Chem. Soc.*, **1979**, *101*, 2738; (e) ref. 19(a), p. 604; (f) see, for references: Hudlicky, M. *Oxidations in Organic Chemistry*, ACS Monograph 186, Washington, DC, **1990**; (g) Epstein, W. W.; Sweat, F. W. *Chem. Rev.*, **1967**, *67*, 247; (h) Corey, E. J.; Gilman, N. W.; Ganem, B. E. *J. Am. Chem. Soc.*, **1968**, *90*, 5616; (i) ref. 19(a), p. 456; (j) Katsuki, T.; Sharpless, K. B. *J. Am. Chem. Soc.*, **1980**, *102*, 5974; see also: Hanson, R. M.; Sharpless, K. B. *J. Org. Chem.*, **1986**, *51*, 1922; (k) Kursanov, D. N.; Parnes, Z. N.; Loim, N. M. *Synthesis*, **1974**, 633; (l) Krishnamurthy, S.; Brown, H. C. *J. Org. Chem.*, **1982**, *47*, 276; employment of the most powerful complex hydride, lithium triethylborohydride (Super Hydride), enables one to reduce efficiently both alkyl halides and parent alcohols; see, for examples: Brown, H. C.; Krishnamurthy, S. *J. Am. Chem. Soc.*, **1973**, *95*, 1669; Brown, H. C.; Kim, S. C.; Krishnamurthy, S. *J.*

Org. Chem., **1980**, *45*, 1; (m) Corey, E. J.; Achiwa, K. *J. Org. Chem.*, **1969**, *34*, 3667; (n) Adlington, R. M.; Barrett, A. G. M. *Acc. Chem. Res.*, **1983**, *16*, 55.

[20] (a) Graham, J. C. *Tetrahedron Lett.*, **1973**, 3825; (b) Gilbert, K. E.; Borden, W. T. *J. Org. Chem.*, **1979**, *44*, 659; (c) Jacobson, R. M. *Tetrahedron Lett.*, **1974**, 3215; (d) Abdel-Magid, A. F.; Maryanoff, C. A.; Carson, K. G. *Tetrahedron Lett.*, **1990**, *31*, 5595; (e) Sharma, D. N.; Sharma, R. P. *Tetrahedron Lett.*, **1985**, *26*, 2561; (f) Jung, M. E.; Lyster, M. A. *J. Org. Chem.*, **1977**, *42*, 3761; see, for a review: Olah, G. A.; Nahang, S. C. *Tetrahedron*, **1982**, *38*, 2225.

[21] (a) Willstatter, R.; Waser, E. *Ber.*, **1911**, *44*, 3423; (b) Bochkov, A. F.; Zaikov, G. E. *Chemistry of the O-Glycosidic Bond: Formation and Cleavage*, Pergamon, Oxford, **1979**.

[22] (a) Bochkov, A. F. *Zh. Org. Khim.*, **1983**, *19*, 1654; (b) Izumi, Y; Tai, A. *Stereo-differentiating Rections*, Academic Press, New York, **1977**; Morrison, J. D. *Asymmetric Synthesis*, Academic Press, New York, **1983–84**, vols. 1–5.

[23] (a) For a general review on hydride reductions, see: Brown, H. C.; Krishnamurthy, S. *Tetrahedron*, **1979**, *35*, 567; (b) Hutchins, R. O.; Kandasamy, D.; Maryanoff, C. A.; Masilamani, D.; Maryanoff, B. E. *J. Org. Chem.*, **1977**, *42*, 82; (c) Adams, C. *Synth. Commun.*, **1984**, 1349; (d) see, for example: Crabbé, P.; Guzmán, A.; Vera, M. *Tetrahedron Lett.*, **1973**, 3021; (e) Sarkar, D. C.; Das, A. R.; Ranu, B. C. *J. Org. Chem.*, **1990**, *55*, 5799; (f) Fisher, G. B.; Fuller, J. C.; Harrison, J.; Alvarez, S. G.; Burkhardt, E. R.; Goralski, C. T.; Singaram, B. *J. Org. Chem.*, **1994**, *59*, 6378; (g) see, for example: Tsuda, T.; Yazawa, T.; Watanabe, K.; Fujii, T.; Saegusa, T. *J. Org. Chem.*, **1981**, *46*, 192; (h) Burgstahler, A. W.; Sanders, M. E. *Synthesis*, **1980**, 400; (i) Brown, H. C.; Krishnamurthy, S. *Aldrichim. Acta*, **1979**, *12*, 3; (j) Parnes, Z. N.; Loim, N. M.; Baranova, V. A.; Kursanov, D. N. *Zh. Org. Khim.*, **1971**, *7*, 2066.

[24] (a) See, for example: Hoehn, W. M.; Moffett, R. B. *J. Am. Chem. Soc.*, **1945**, *67*, 740; (b) Friour, G.; Alexakis, A.; Cahiez, G.; Normant, J. F. *Tetrahedron*, **1984**, *40*, 683; Friour, G.; Cahiez, G.; Normant, J. F. *Synthesis*, **1985**, 50; (c) Cahiez, G.; Rivas-Enterios, J.; Gragner-Veyron, H. *Tetrahedron Lett.*, **1986**, *27*, 4441; (d) Lipshutz, B. H.; *Synthesis*, **1987**, 325; (e) Gilman, H.; Straley, J. M. *Recl. Trav. Chim. Pays-Bas*, **1936**, *55*, 821 and refs. cited therein; (f) see, for example: Corey, E. J.; Posner, G. *J. Am. Chem. Soc.*, **1967**, *89*, 3911; (g) Corey, E. J.; Posner, G. *J. Am. Chem. Soc.*, **1968**, *90*, 5615; (h) Achyutha Rao, S., Knochel, P. *J. Org. Chem.*, **1991**, *56*, 4591; (i) Kotsuki, H.; Kadota, I.; Ochi, M. *J. Org. Chem.*, **1990**, *55*, 4417.

[25] (a) Brownbridge, P. *Synthesis*, **1983**, 1; *ibid.*, **1983**, 85; (b) Fleming, I.; Paterson, I. *Synthesis*, **1979**, 736; (c) see, for example, the method of regio- and streospecific alkylation of ketones *via* metallated *N,N*-dimethylhydrazone derivatives; Corey, E. J.; Enders, D. *Tetrahedron Lett.*, **1976**, 1; *ibid.*, **1976**, 11; (d) Schreiber, S. L.; Wang, Z. *J. Am. Chem. Soc.*, **1985**, *107*, 5303.

[26] (a) Odinokov, V. N.; Bakeeva, R. S.; Galeeva, R. I.; Akhunova, V. R.; Mukhtarov, Ya. G.; Tolstikov, G. A.; Khalilov, L. M.; Panasenko, A. A. *Zh. Org. Khim.*, **1979**, *15*, 2017; (b) see, for example: Bosch, M. P.; Camps, F.; Coll, J.; Guerrero, A.; Tatsuoka T.; Meinwald, J. *J. Org. Chem.*, **1986**, *51*, 773; (c) Greene, T. M.; Wuts, P. G. M. *Protective Groups in Organic Synthesis*, 2nd edn., Wiley, New York, **1991**; (d) Kocienski, P. J. *Protecting Groups*, Thieme, Stuttgart, **1994**; (e) Nicolaou, K. C.; Claremon, D. A.; Barnette, W. E. *J. Am. Chem. Soc.*, **1980**, *102*, 6611; (f) Jung, M. E.; Speltz, L. M. *J. Am. Chem. Soc.*, **1976**, *98*, 7882; (g) Keinan, E.; Sahai, M.; Shvily, R. *Synthesis*, **1991**, 641; (h) Mpango, G. B.; Mahalabanis, K. K.; Mahdavi-Damghani, Z.; Snieckus, V. *Tetrahedron Lett.*, **1980**, *21*, 4823; (i) Knapp, S.; Calienue, J. *Synth. Commun.*, **1980**, *10*, 837.

[27] (a) Corey, E. J. *Pure Appl. Chem.*, **1967**, *14*, 19; (b) Corey, E. J.; Cheng, X. M. *The Logic of Organic Synthesis*, Wiley, New York, **1989**; (c) March, J. *Advanced Organic Chemistry*, 3rd edn., Wiley, New York, **1985**, p. 422; (d) Casey, C. P.; Marten, D. F. *Synth. Commun.*, **1973**, *3*, 321; (e) for a concise exposure of the ideology of approach, see: Seebach, D. *Angew. Chem., Int. Ed. Engl.*, **1969**, *8*, 639; see also: Lever, O. W., Jr. *Tetrahedron*, **1971**, *32*, 1943; (f) Hase, T. A. *Umpoled Synthones*, Wiley, New York, **1987**; (g) for a spectacular utilization of this operation for a 'linch-pin' construction of unsymmetrical 1,4-diketones, see: Hermann, J. L.; Richman, J. E.; Schlessinger, R. H. *Tetrahedron Lett.*, **1973**, 3271; (h) for a review, see: Gröbel, B.-T.; Seebach, D. *Synthesis*, **1977**, 357; (i) Seebach, D.; Seuring, B.; Kalinowski, H.-O.; Lubosch, W.; Renger, B. *Angew. Chem.*, **1977**, *89*, 270; (j) Corey, E. J.; Kozikovsky, A. P. *Tetrahedron Lett.*, **1975**, 925; see also: Seebach, D.; Kolb, M. *Justus Liebigs Ann. Chem.*, **1977**, 811; Seebach, D.; Bürstinghaus, R.; Gröbel, B.-T.; Kolb, M. *Justus Liebigs Ann. Chem.*, **1977**, 830; (k) Chou, T-S.; Knochel, P. *J. Org. Chem.*, **1990**, *55*, 4791; (l) for a comprehensive review on heterosubstituted organometallics, see: Krief, A. *Tetrahedron*, **1980**, *36*, 2531; (m) Hase, T. A.; Koskimies, J. K. *Aldrichim. Acta*, **1981**, *14*, 73; (n) Stowell, J. C. *Chem. Rev.*, **1984**, *84*, 409.

[28] (a) Wiberg, K. B.; Lampman, G. M. *Tetrahedron Lett.*, **1963**, 2173; (b) Applequist, D. E.; Fanta, G. F.; Henrickson, B. W. *J. Org. Chem.*, **1958**, *23*, 1715; (c) see, for a review: Menchikov, L. G.; Nefedov, O. M. *Usp. Khim.*, **1994**, *63*, 471; (d) Misumi, A.; Iwanaga, K.; Furuta, K.; Yamamoto, H. *J. Am. Chem. Soc.*, **1985**, *107*, 3343; (e) ref. 8, p. 542; (f) Brewer, P. D.; Tagat, J.; Hergrueter, C. A.; Helquist, P. *Tetrahedron Lett.*, **1977**, 4573; (g) Whitesides, G. M.; Gutowski, F. D. *J. Org. Chem.*, **1976**, *41*, 2882.

[29] (a) Smith, M. B. *Organic Synthesis*, McGraw-Hill, New York, **1994**, p. 889; (b) for a review: Hudlicky, T.; Price, J. D. *Chem. Rev.*, **1989**, *89*, 1467; numerous applications of cyclopentanoannulation methods in total synthesis are shown in the review: Paquette, L. A. *Top. Curr. Chem.*, **1984**, *119*, 1; (c) Herrmann, J. L.; Richman, J. E.; Schlessinger, R. H. *Tetrahedron Lett.*, **1973**, 3275; (d) Mundy, B. P., Wilkening, D.; Lipkowitz, K. B. *J. Org. Chem.*, **1985**, *50*, 5727; see also: Furuta, K.; Misumi, A.; Mori, A.; Ikeda, N.; Yamamoto, H. *Tetrahedron Lett.*, **1984**, *25*, 669; (e) Piers, E.; Karunaratne, V. *J. Chem. Soc., Chem. Commun.*, **1983**, 935; (f) Bunce, R. A.; Wamsley, E. J.; Pierce, J. D.; Shellhammer, A. J., Jr.; Drumright, R. E. *J. Org. Chem.*, **1987**, *52*, 464; see also: Yamaguchi, M.; Tsukamoto, M.; Hirao, I. *Tetrahedron Lett.*, **1985**, *26*, 1723; (g) Nazarov, I. N.; Zaretskaya, I. I. *Izv. Akad. Nauk, Otd. Khim. Nauk*, **1944**, 65; (h) Santelli-Rouvier, K.; Santelli, M. *Synthesis*, **1983**, 429; (i) see, for a review: Trost, B. *Angew. Chem., Int. Ed. Engl.*, **1986**, *25*, 1.

[30] (a) The role of entropy and enthalpy factors in ring-closure reactions are thoroughly dicussed in: DeTar, D. F.; Luthra, N. P. *J. Am. Chem. Soc.*, **1980**, *102*, 4505; (b) March, J. *Advanced Organic Chemistry*, 3rd edn., Wiley, New York, **1985**, p. 184; (c) Ziegler, K., in *Methoden der Organischen Chemie* (*Houben-Weyl*), Müller, E., Ed.; Thieme, Stuttgart, **1955**, vol. 4/2; see also the discussion in ref. 29(a), p. 611; (d) for the reviews on the preparative chemistry of macrolides, see: Nicolaou, K. C. *Tetrahedron*, **1977**, *33*, 683; Masamune, S.; Bates, G. S.; Corcoran, J. W. *Angew. Chem., Int. Ed. Engl.*, **1977**, *16*, 585; (e) Corey, E. J.; Nicolaou, K. C. *J. Am. Chem. Soc.*, **1974**, *96*, 5614; for a review, see: Mukaiyama, T. *Angew. Chem., Int. Ed. Engl.*, **1979**, *18*, 707; see also highlights in: Mulzer, J. *Angew. Chem., Int. Ed. Engl.*, **1991**, *30*, 1452; (f) Corey, E. J.; Ulrich, P.; Fitzpatrick, J. M. *J. Am. Chem. Soc.*, **1976**, *98*, 222; for an example of the efficiency of an alternative approach involving the intermediate formation of the mixed anhydride of the cyclized ω-hydroxy acid and 2,4,6-trichlorobenzoic acid, see: Hikota, M.; Tone, H.; Horita, K.; Yonemitsu, O. *J. Org.*

Chem., **1990**, *55*, 7 and refs. cited therein; (g) Yamamoto, H.; Maruoka, K. *J. Am. Chem. Soc.*, **1981**, *103*, 6133.

[31] (a) Onischenko, A. S. *Diene Synthesis*, Davey, New York, **1964**; (b) Wollweber, H. *Diels–Alder Reaction*, Thieme, Stuttgart, **1972**; (c) Woodward, R. B.; Hoffmann, R. *Angew. Chem., Int. Ed. Engl.*, **1969**, *8*, 781; (d) Desimoni, G.; Tacconi, G.; Barco, A.; Pollini, G. P. *Natural Product Synthesis Through Pericyclic Reactions*, ACS Monograph 180, ACS, Washington, DC, **1983**, ch. 5, p. 119; (e) Danishefsky, S. *Acc. Chem. Res.*, **1981**, *14*, 400; (f) Watt, D. S.; Corey, E. J. *Tetrahedron Lett.*, **1972**, 4651; (g) see, for example, utilization in the synthesis of reserpine: Martin, S. F.; Rüeger, H.; Williamson, S. A.; Grzejszczak, S. *J. Am. Chem. Soc.*, **1987**, *109*, 6124; (h) Danishefsky, S.; Etheredge, S. J. *J. Org. Chem.*, **1979**, *44*, 4716; (i) Sadykh-Zade, S. I.; Petrov, A. D. *Zh. Obshch. Khim.*, **1958**, *28*, 1591; Carter, M. J.; Fleming, I. *J. Chem. Soc., Chem. Commun.*, **1976**, 679; (j) Carr, R. V. C.; Williams, R. V.; Paquette, L. A. *J. Org. Chem.*, **1983**, *48*, 4976; (k) Bonjouklian, R.; Ruden, R. A. *J. Org. Chem.*, **1977**, *42*, 4095; (l) Corey, E. J.; Weinshenker, N. M.; Shaaf, T. K.; Huber, W. *J. Am. Chem. Soc.*, **1969**, *91*, 5675; (m) Masamune, S.; Cuts, H.; Hogben, M. G. *Tetrahedron Lett.*, **1966**, 1017; (n) Eaton, P. E.; Or, Y. S.; Branca, S. J. *J. Am. Chem. Soc.*, **1981**, *103*, 2134; (o) for reviews, see: Oppolzer, W. *Angew. Chem., Int. Ed. Engl.*, **1977**, *16*, 10; Brieger, G.; Bennet, J. N. *Chem. Rev.*, **1980**, *80*, 63; (p) Brieger, G.; Anderson, D. R. *J. Org. Chem.*, **1971**, *36*, 243; (r) House, H. O.; Cronin, T. H. *J. Org. Chem.*, **1965**, *30*, 1061; (s) Nicolaou, K. C.; Li, W. S. *J. Chem. Soc., Chem. Commun.*, **1985**, 421; (t) Stork, G.; Chan, T. Y. *J. Am. Chem. Soc.*, **1995**, *117*, 6595 and references cited therein; (u) Oikawa, H.; Yokota, T.; Abe, T.; Ichihara, A.; Sakamura, S.; Yoshizawa, Y.; Vederas, J. C. *J. Chem. Soc., Chem. Commun.*, **1989**, 1282; Oikawa, H.; Yokota, T.; Ichihara, A.; Sakamura, S. *ibid.*, **1989**, 1284; (v) Oikawa, H.; Suzuki, Y.; Naya, A.; Katayama, K.; Ichihara, A. *J. Am. Chem. Soc.*, **1994**, *116*, 3605; (w) Ichihara, A.; Miki, M.; Tazaki, H.; Sakamura, S. *Tetrahedron Lett.*, **1987**, *28*, 1175.

[32] (a) See ref. 31(d), ch. 3, p. 33; (b) Brady, W. T.; Patel, A. D. *J. Org. Chem.*, **1973**, *38*, 4106; see, for a review: Brady, W. T. *Tetrahedron*, **1981**, *37*, 2949; (c) Springer, J. P.; Clardy, J.; Cole, R. J.; Kirksey, J. W.; Hill, R. K.; Carlson, R. M.; Isidor, J. L. *J. Am. Chem. Soc.*, **1974**, *96*, 2267; (d) Snider, B. B. *Chem. Rev.*, **1988**, *88*, 793; (e) Bisceglia, R. H.; Cheer, C. J. *J. Chem. Soc., Chem. Commun.*, **1973**, 165; (f) Snider, B. B.; Hui, R. A. H. F. *J. Org. Chem.*, **1985**, *50*, 5167; (g) reviews: Eaton, P. E. *Acc. Chem. Res.*, **1968**, *1*, 50; De Mayo, P. *Acc. Chem. Res.*, **1971**, *4*, 41; (h) Corey, E. J.; Mitra, R. B.; Uda, H. *J. Am. Chem. Soc.*, **1964**, *86*, 485; (i) see, for examples, the monograph of ref. 31(d), ch. 3, p. 42; (j) for a review, see: Crimmins, M. T. *Chem. Rev.*, **1988**, *88*, 1453; (k) Becker, D.; Haddad, N. *Tetrahedron Lett.*, **1986**, *27*, 6393; (l) Wolff, S.; Agosta, W. C. *J. Am. Chem. Soc.*, **1983**, *105*, 1292; (m) Oppolzer, W. *Acc. Chem. Res.*, **1982**, *15*, 135.

[33] (a) For a concise description of carbene structure, see: ref. 30(b), p. 170 and refs. cited therein; (b) Nefedov, O. M.; Maltsev, A. K.; Mikaelyan, R. G. *Tetrahedron Lett.*, **1971**, 4125; (c) Jones, M., Jr. *Acc. Chem. Res.*, **1974**, *7*, 415; (d) Tomilov, Y. V.; Dokichev, V. A.; Dzhemilev, U. M.; Nefedov, O. M. *Russ. Chem. Rev.*, **1993**, *62*, 799; Doyle, M. P. *Chem. Rev.*, **1986**, *86*, 919; (e) Simmons, H. E.; Cairns, T. L.; Vladuchick, S. A; Hoiness, C. M. *Org. React.*, **1973**, *20*, 1; (f) a general discussion of the generation and reactivity of carbenoids derived from α-halometalloorganic derivatives can be found in: Nefedov, O. M.; Dyachenko, A. I.; Prokofiev, A. K. *Russ. Chem. Rev.*, **1977**, *46*, 941; (g) Campbell, J. G. M.; Harper, S. H. *J. Chem. Soc.*, **1945**, 283; (h) Skattebøl, L. *J. Org. Chem.*, **1966**, *31*, 2789; (i) Vig, O. P.; Bhatia, M. S.; Gupta, K. C.; Matta, K. L. *J. Indian Chem. Soc.*, **1969**, *46*, 991.

[34] (a) Reppe, W.; Schlichting, O.; Klager, K.; Toepel, T. *Justus Liebigs Ann. Chem.*, **1948**, *560*, 1; (b) Schrauzer, G. N.; Glockner, P.; Eichler, S. *Angew. Chem., Int. Ed. Engl.*, **1964**, *3*, 185; for a discussion of the general problems of acetylene cyclooligomerization in the presence of transition metal catalysts, see: Maitlis, P. M. *J. Organomet. Chem.*, **1980**, *200*, 161; (c) Arnett, E. M.; Bollinger, J. M. *J. Am. Chem. Soc.*, **1964**, *86*, 4729; (d) Vollhardt, K. P. *Acc. Chem. Res.*, **1977**, *10*, 1; (e) for general reviews of the reaction, see: Pauson, P. L. *Tetrahedron*, **1985**, *41*, 5860; Schore, N. E. *Org. React.*, **1991**, *40*, 2; (f) Schore, N. E.; Croudace, M. C. *J. Org. Chem.*, **1981**, *46*, 5436; later data on the selectivity pattern and mechanism of the reaction can be found in: Krafft, M. E.; Scott, I. L.; Romero, R. H.; Feibelman, S.; Van Pelt, C. E. *J. Am. Chem. Soc.*, **1993**, *115*, 7199; (g) Shambayati, S.; Crowe, W. E.; Schreiber, S. L. *Tetrahedron Lett.*, **1990**, *31*, 5289; (h) Smit, W. A.; Gybin, A. S.; Strychkov, Y. T.; Mikaelian, G. S.; Caple, R.; Swanson, E. D. *Tetrahedron Lett.*, **1986**, *27*, 1241; (i) Schreiber, S. L.; Sammakia, T.; Crowe, W. E. *J. Am. Chem. Soc.*, **1986**, *108*, 3128; see also: Jamison, T. F.; Shambayati, S.; Crowe, W. E.; Schreiber, S. L. *J. Am. Chem. Soc.*, **1994**, *116*, 5505; (j) see, for a review: Wilke, G. *J. Organomet. Chem.*, **1980**, *200*, 349; (k) see, for example: Jemilev, U. M.; Ivanov, G. E.; Tolstikov, G. A. *Zh. Org. Khim.*, **1975**, *11*, 1636; (l) Brenner, W.; Heimbach, P.; Wilke, G. *Justus Liebigs Ann. Chem.*, **1969**, 727, 194; (m) for a review, see: Tolstikov, G. A.; Jemilev, U. M. *Khim. Geterotsikl. Soedin.*, **1980**, 147; (n) illustrious examples showing the achievements and promise of transition metal catalysis in the creation of the basic skeleton of various organic compounds are cited in the review: Trost, B. M. *Angew. Chem., Int. Ed. Engl.*, **1995**, *34*, 259.

[35] (a) For reviews, see: Walling, C.; Huyser, E. S. *Org. React.*, **1963**, *13*, 91; Vogel, H. H. *Synthesis*, **1970**, 99; preparation of levulinic aldehyde acetal is described in: Mondon, A. *Angew. Chem.*, **1952**, *64*, 224; (b) Jaspers, C. P.; Curran, D. P.; Fevig, T. L. *Chem. Rev.*, **1991**, *91*, 1237; (c) Corey, E. J.; Kang, M.-C. *J. Am. Chem. Soc.*, **1984**, *106*, 5384; (d) Stork, G.; Mook, R., Jr.; Biller, S. A.; Rychnovsky, S. D. *J. Am. Chem. Soc.*, **1983**, *105*, 3741; (e) Srikrishna, A.; Sharma, G. V. R. *Tetrahedron Lett.*, **1988**, *29*, 6487; (f) Stork, G.; Sher, P. M.; Chen, H.-L. *J. Am. Chem. Soc.*, **1986**, *108*, 6384; see also: Keck, G. E.; Burnett, D. A. *J. Org. Chem.*, **1987**, *52*, 2958.

[36] (a) See, for example: Wiberg, K. B.; Lowry, B. R.; Colby, T. H. *J. Am. Chem. Soc.*, **1961**, *83*, 3998; (b) Scheldon, R. A.; Kochi, J. K. *Org. React.*, **1972**, *19*, 279; see also: Serguchov, Y. A.; Beletskaya, I. P. *Russ. Chem. Rev.*, **1980**, *49*, 1119; (c) Kochi, J. K.; Bacha, J. D. *J. Org. Chem.*, **1968**, *33*, 2746; (d) Bacha, J. D.; Kochi, J. K. *Tetrahedron*, **1968**, *24*, 2215; (e) for representative examples, see the review: De Lucchi, O.; Modena, G. *Tetrahedron*, **1984**, *40*, 2585.

[37] (a) Hassal, C. H. *Org. React.*, **1957**, *9*, 73; (b) Starcher, P. S.; Phillips, B. *J. Am. Chem. Soc.*, **1958**, *80*, 4079; (c) Emmons, W. D.; Lucas, G. B. *J. Am. Chem. Soc.*, **1955**, *77*, 2287; (d) Payne, G. B. *Tetrahedron*, **1962**, *18*, 763; (e) Martinez, G. R.; Grieco, P. A.; Williams, E.; Ken-ishi, K.; Srinivasan, C. V. *J. Am. Chem. Soc.*, **1982**, *104*, 1436.

[38] (a) Dryhurst, G. *Periodate Oxidation of Diols and Other Functional Groups*, Pergamon, Oxford, **1970**, p. 191; (b) Perlin, A. S. *Adv. Carbohydr. Chem.*, **1959**, *14*, 9; (c) Schmid, C. R.; Bryant, J. D. *Org. Synth.*, **1995**, *72*, 6; (d) see, for a review: Schröder, M. *Chem. Rev.*, **1980**, *80*, 187; (e) Corey, E. J.; Nozoe, S. *J. Am. Chem. Soc.*, **1963**, *85*, 3527.

[39] (a) Bailey, P. S. *Ozonation in Organic Chemistry*, Academic Press, New York, **1978**, vol. 1, **1982**, vol. 2; see also: Odinokov, V. N.; Tolstikov, G. A. *Russ. Chem. Rev.*, **1981**, *50*, 636; (b) Odinokov, V. N.; Akhunova, V. R.; Bakeeva, R. S.; Galeeva, R. I.; Semenovsky, A. V.; Moiseenkov, A. M.; Tolstikov, G. A. *Zh. Org. Khim.*, **1977**, *13*, 532; (c) Corey, E. J.; Katzenellenbogen, J. A.; Gilman, N. W.; Roman, S. A.; Erickson, B. W. *J. Am. Chem. Soc.*, **1968**, *90*, 5618.

[40] (a) Ref. 31(d), ch. 7, p. 267; some specific aspects of the synthetic utilization of rearrangements are amply demonstrated in the review: Martin, S. F. *Tetrahedron*, **1980**, *36*, 419; (b) Ziegler, F. E. *Chem. Rev.*, **1988**, *88*, 1423; (c) Burgstahler, A. W.; Nordin, I. C. *J. Am. Chem. Soc.*, **1961**, *83*, 198; (d) Johnson, W. S.; Werthemann, L.; Bartlett, W. R.; Brocksom, T. J.; Lee, T.-T.; Faulkner, D. J.; Petersen, M. R. *J. Am. Chem. Soc.*, **1970**, *92*, 741; (e) see, for example: Johnson, W. S.; Gravestock, M. B.; McCarry, B. E. *J. Am. Chem. Soc.*, **1971**, *93*, 4332; (f) Ireland, R. E.; Mueller, R. H. *J. Am. Chem. Soc.*, **1972**, *94*, 5897; (g) Ireland, R. E.; Mueller, R. H.; Willard, A. K. *J. Am. Chem. Soc.*, **1976**, *98*, 2868; see also: Wilson, S. R.; Myers, R. S. *J. Org. Chem.*, **1975**, *40*, 3309; (h) see ref. 31(d), ch. 7, p. 279; (i) Evans, D. A.; Golub, A. M. *J. Am. Chem. Soc.*, **1975**, *97*, 4765; see also: Wilson, S. R.; Mao, D. T. *Tetrahedron Lett.*, **1977**, 2559; (j) Paquette, L. A. *Angew. Chem., Int. Ed. Engl.*, **1990**, *29*, 609; (k) Still, W. C. *J. Am. Chem. Soc.*, **1979**, *101*, 2493; (l) Schreiber, S. L.; Santini, C. *J. Am. Chem. Soc.*, **1984**, *106*, 4038; (m) Pirrung, M. C. *J. Am. Chem. Soc.*, **1981**, *103*, 82; (n) Tobe, Y.; Yamashita, T.; Kakiuchi, K.; Odaira, Y. *J. Chem. Soc., Chem. Commun.*, **1984**, 1259; see also a review: Bellus, D.; Ernst, B. *Angew. Chem., Int. Ed. Engl.*, **1988**, *27*, 797; (o) Begley, M. J.; Pattenden, G.; Robertson, G. M. *J. Chem. Soc., Perkin Trans. 1*, **1988**, 1085 and references cited therein; (p) Wong, H. N. C.; Hon, M.-Y.; Tse, C.-W.; Yip, Y.-C.; Tanko, J.; Hudlicky, T. *Chem. Rev.*, **1989**, *89*, 165; (r) Mehta, G.; Reddy, D. S.; Murty, A. N. *J. Chem. Soc., Chem. Commun.*, **1983**, 824; (s) Kende, A. S. *Org. React.*, **1960**, *11*, 261; (t) Barborak, J. C.; Watts, L.; Pettit, R. *J. Am. Chem. Soc.*, **1966**, *88*, 1328; see also: Eaton, P. E.; Cole, T. W., Jr. *J. Am. Chem. Soc.*, **1964**, *86*, 962; *ibid.*, **1964**, *86*, 3157.

CHAPTER 3

Strategy of Synthesis

3.1 IMPORTANCE OF PLANNING IN A SYNTHESIS

We now find ourselves familiar with methods of constructing a carbon skeleton for an organic molecule, how to introduce and transform functional groups, and how to achieve the required selectivity. One could almost expect the rich arsenal of contemporary methods to enable the chemist to solve practically any problem of synthetic chemistry. Yet it is not quite so simple. It is imperative to be skillful in planning a synthesis, to be able not only to master its tactics, but its overall strategy as well.

Woodward's synthesis of steroids,[1a] described in Scheme 3.1, illustrates the value of a carefully thought out general plan for a synthesis. This synthesis targeted the preparation of the tetracyclic keto aldehyde **1**, which can be used as an advanced intermediate to be converted into a set of natural steroids that includes progesterone **2**, deoxycorticosterone **3**, androsterone **4**, testosterone **5**, cholesterol **6**, and cortisone **7** *via* well-known routes. The synthetic plan for **1** required solving the following key tasks:

1. How to construct the tetracyclic ring system, **ABCD**.
2. How to ensure the natural *trans-anti-trans* configuration of all ring junctions.
3. How to introduce angular methyl groups at positions 10 and 13.
4. How to create the functionality pattern required for the preparation of the final targets, **2**–**7**.

We will follow the basic idiosyncracies of this synthesis that led to the solution of these tasks.

The tetracyclic skeleton was assembled by constructing the rings in sequence, **C** + **D** + **B** + **A**, with 5-methoxy-2-methylbenzoquinone **8** as the precursor for ring **C**. The infallible Diels–Alder reaction was used as a pathway leading eventually to the construction of the five-membered ring **D**. Actually, the immediate result was the attachment of a six-membered '**D**' ring, but the transformation into the required five-membered ring was a trivial task and left for the final steps. The bicyclic intermediate **9** thus prepared had the 'wrong' *cis* junction at the **C** and **'D'** rings. However, the presence of an adjacent carbonyl group secured an easy isomerization of **9** into the more stable product **10**, with the *trans*-fused system as is present in **1**. Thus, a sequence of fairly simple reactions led to the assemblage of the **'C–D'** portion of the target framework, with the correct ring junction and a methyl group properly placed at C-13.

A series of further transformations **10** → → **11** were intended to create the functionality needed to form ring **B**. To achieve the Robinson annulation (see Section 2.3.3), **11** was formylated at C-8 and then reacted with ethyl vinyl ketone. This operation produced not only ring **B**, but introduced as well the

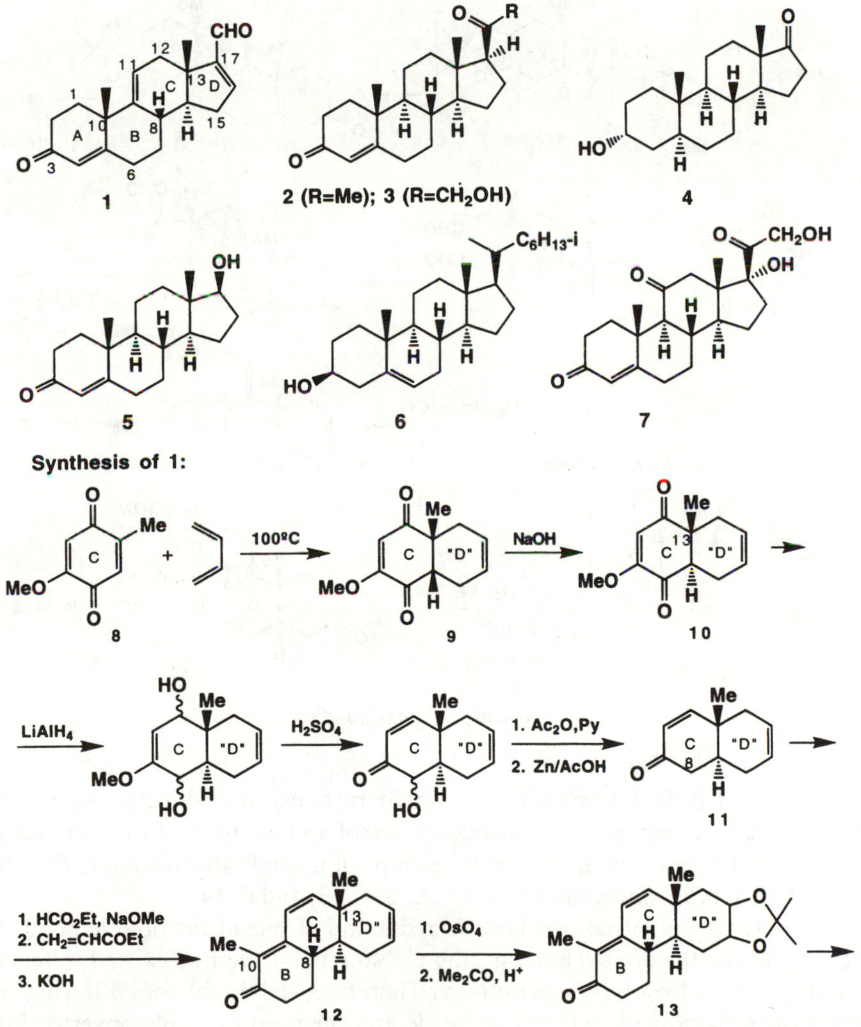

Synthesis of 1:

Scheme 3.1

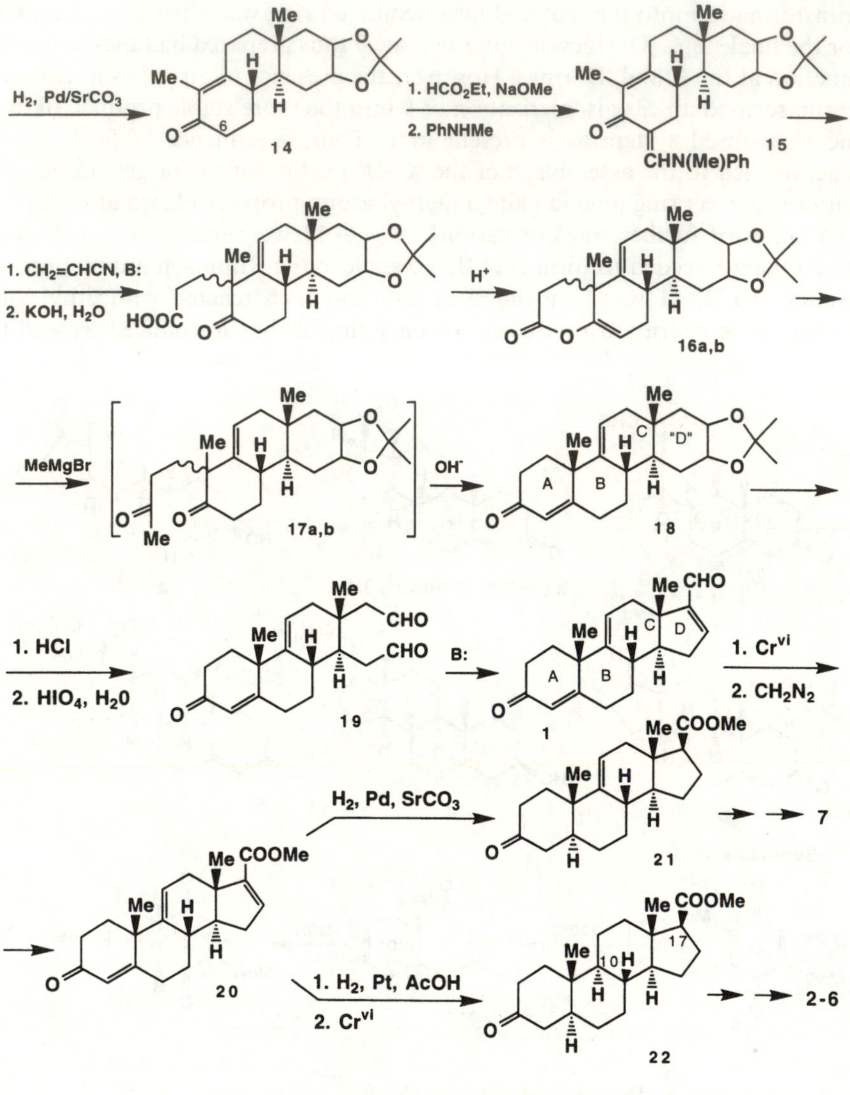

Scheme 3.1(*continued*)

methyl group at the future C-10 position. Steric constraints directed the closure of ring **B** with complete stereochemical control and led to the formation (after the hydrolytic cleavage of the formyl group) of a single stereoisomer, **12**, with the required *anti-trans* configuration at C-8, C-13, and C-14.

The next step involved the selective reduction of one of the double bonds in **12**, specifically the double bond in ring **C**. No reliable method existed to secure the required hydrogenation selectivity. Therefore, the double bond in ring **'D'** was hydroxylated with osmium tetroxide and the resulting diol converted into the corresponding acetal **13**. In addition to simplifying the selectivity problem (protecting the double bond in ring '**D**'), these steps were also useful to elaborate the five-membered ring in later steps. Now, the selective reduction of the less substituted double bond in diene **13**, the one in ring **C**, could be easily achieved *via* catalytic hydrogenation. To elaborate ring **A**, the following key steps were involved: (i) the C-6 nucleophilic site was protected by transforming **14** into **15**; (ii) the double bond was shifted concomitant with a Michael reaction of acrylonitrile at the only nucleophilic site present, C-10; (iii) lactone **16** was formed after hydrolysis; and (iv) *via* a Grignard reaction and subsequent intramolecular crotonic condensation of the diketone **17**, the tetracyclic product **18** was achieved. This overall transformation turned out to be rather troublesome. In fact, while cyanoethylation of **15** proceeded exclusively at C-10, it produced a mixture of stereoisomers at this center. Both isomers were converted into lactones **16a** and **16b**, but fortunately only the isomer with the desired configuration reacted smoothly with the Grignard reagent to give enone **18**.

Ring '**D**' was then ready for transformation into its proper shape. The diol formed from removing the protecting group in **18** was oxidized with periodic acid. The resultant 1,6-dialdehyde **19** readily underwent a regioselective intramolecular crotonic condensation to form the five-membered ring **D** bearing a formyl group at C-17.

Thus the first totally synthetic non-aromatic steroid, $(\pm)$-$\Delta^{9(11),16}$-bisdehydro-20-norprogesterone **1** was prepared in 20 steps from **2** with an overall yield of about 1% (average yield approximately 79% per step[1]).

Product **1** was transformed into ester **20**, with conversion of the latter into the target compounds achieved by way of two separate routes, both involving the reduction of double bonds as the first step. Selective reduction of both conjugated double bonds to give **21** was achieved by hydrogenation over a Pd catalyst. The presence of the double bond in ring **B** of product **21** provided the opportunity to install an oxygen-containing substituent at C-11. This series of reactions elaborated a successful synthesis of cortical steroids, including cortisone **7**. The route to prepare steroids **2–6** employed the complete reduction of all three double bonds *via* hydrogenation over a Pt catalyst and led eventually to the saturated ester **22**. This reaction established the required stereochemistry at C-10 and C-17. Transformation of **22** into steroids **2–5** was described earlier and, consequently, Woodward's synthesis of **22** encompassed the total synthesis of these steroids as well. Several additional and rather obvious steps (such as the elaboration of the aliphatic substituent at C-17 and functional group transformations) were employed to prepare cholesterol **6** from **22**.

In this synthesis, it is worth recognizing that all of the functional groups and structural elements play an active role only upon demand and only at the very moment they are required. For example, the methoxy group present in the starting quinone **8** is necessary to secure the regioselectivity of the Diels–Alder reaction. At the same time it serves as a 'masked' carbonyl group, to be easily unmasked in later steps. The double bond in ring '**D**' appears in the first step of the synthesis and stays undisturbed until the tenth step, when it is involved in the critical construction of ring **D**. The double bonds in rings **C** and **D** appear as 'side products' of the cyclization reactions, to be removed later. This apparent complication, however, plays an important role in the overall scheme, as it is the hydrogenation of these double bonds that solves the strategic task of creating the desired stereochemistry of the ring junctions.

The successful synthesis of such enormous complexity (for the 1950s!) was exhilarating. It resembled a symphony played by an orchestra of the finest professionals. In much the same way that the score and the arrangement for the instruments introduces each at the right time and in the right combination, the synthetic plan and the set of finely tuned methods were composed to achieve a true *harmony* of functional group interactions and interconversions.

The described total synthesis of sterols, as well as numerous other preparations of various natural products performed by Woodward's group,[1b] manifested in the most spectacular way the power and the beauty of organic synthesis. It is not an exaggeration to claim that the perception of organic synthesis as an art gained a firm foothold in the minds of organic chemists,[1c] primarily due to the benchmark achievements of its recognized Grand Master, R. B. Woodward. Since that time, hundreds of total syntheses of comparable caliber have been accomplished in many laboratories throughout the world. The artistry of the outstanding individuals of the past has been transformed nowadays into a skill which can be mastered by every qualified organic chemist.[1c] Undoubtedly the prerequisites for this transformation were created by elaboration of novel, versatile, and efficient synthetic methods. However, of equal importance was the recognition that there is a science behind the art of organic synthesis, the science with its unique and well-defined logical attributes, elaborated to solve both general and specific problems. It was perhaps inevitable that the role of the scientific component in this area of intellectual creativity has greatly increased over the years, although there is still plenty of room in the design of complex organic syntheses for excercising one's ingenuity and imagination.[1d]

In 1966, Corey presented a plenary lecture entitled, 'General Methods for the Construction of Complex Molecules' at the IVth International IUPAC Symposium on the Chemistry of Natural Products in Stockholm.[2a] As was stated in this lecture, 'The synthetic chemist is more than a logician and strategist; (the chemist) is an explorer strongly influenced to speculate, imagine and even to create. These added elements provide the touch of artistry which can hardly be included in cataloging of basic principles of synthesis, but they are very real and important... The proposition can be advanced that many of the most distinguished synthetic studies have entailed a balance between two different research philosophies, one embodying the ideal of a deductive analysis based on known methodology and current theory, and the other emphasizing innovation and even speculation'.[2a,b]

In 1990, the Nobel Prize in Chemistry was awarded to E. J. Corey for the 'development of the theory and methodology of organic synthesis'. As was summarized in Corey's Nobel lecture, the essence of his research was in 'advancing the level of synthetic science by an approach consisting of three integral components: the development of more general and powerful ways of thinking about synthetic problems, the invention of new general reactions and reagents for organic synthesis, and the design and execution of efficient multi-step syntheses of complex molecules at the limits of contemporary synthetic science'.[2c]

Not much could be added to the above quotations with regard to the role of both imagination and knowledge as complementary components of the true creativity in organic synthesis (or, for that matter, in any other area of both science and art). In the context of this chapter, however, it is essential to take a closer look at the rational background of modern organic synthesis, and its logistics as especially important components of the problem solutions in this area.

Theoretical analysis of the strategic problems in a total synthesis began to receive serious attention in the 1960s.[2a–d] Below we will present an outline of the general principles of synthetic strategy and its application to solve specific preparative tasks. The examples chosen illustrate guidelines essential to the development of the optimal synthetic plans. A more comprehensive treatment of this subject can be found in Corey's monograph[3a] as well as in several textbooks (see, for example, ref. 36).

3.2 STRATEGIC OPTIONS

In developing a strategy for a specific synthesis it is possible to encounter two extreme situations: (a) the starting compound is given and it is necessary to elaborate a route for its conversion into the target structure; (b) no specific starting compound is indicated, so the structure of the target compound must be analysed to identify pathways for its synthesis from simple precursors. Both approaches are usually encountered in the course of synthetic design. For ease of examination, however, we will treat them separately.

3.2.1 Planning 'from the Starting Material'

Planning a laboratory synthesis based on the choice of well-defined starting compounds is warranted in those cases where it is easy to identify structural fragments in the target molecules that are present in available materials. The most clear-cut examples of this approach can be found in the synthesis of biopolymers. Biopolymers, like proteins, polysaccharides, and nucleic acids, are constructed from relatively small monomeric blocks bound by heteroatomic bridges. The monomers in polypeptides and proteins are amino acids. An amidic bond serves as the bridge. Monosaccharides are the monomeric units of polysaccharides and these units are joined through an oxygen by glycosidic bonds. In nucleic acids the individual units are nucleotides connected *via* phosphodiester bonds.

Inter-monomer bonds are the easiest to 'break' in such destructive reactions as chemical or enzymatic hydrolysis. As easily as these bonds are broken, they can be formed by common functional group transformations from readily available monomeric units. In these cases, retrosynthetic analysis of the target molecule consists merely in identifying the units that should be combined. This almost 'automatic' identification of the starting materials by no means implies, however, that there exits an 'automatic' plotting of the strategy and tactics of a specific synthesis. Hurdles arise, a few of which are instructive to examine.

The majority of polysaccharides have a regular structure in the polymeric chain which suggests that such a chain can be constructed from repeating mono- or oligosaccharide links. The general strategy for their synthesis, therefore, consists of the polymerization or polycondensation of suitable monomers to form glycosidic bonds with the correct stereo- and regiochemistry.[4a] This was actually the strategy employed in one of the first syntheses of the polysaccharide **23** (Scheme 3.2), an analog of the bacterial polysaccharide dextrans.[4b] This biopolymer is composed of α-D-glucopyranosyl units linked by **1** → **6** bonds. D-Glucose **24** was properly identified as the obvious starting material. The main concern was to find a way to achieve the required regio- and stereospecificity in

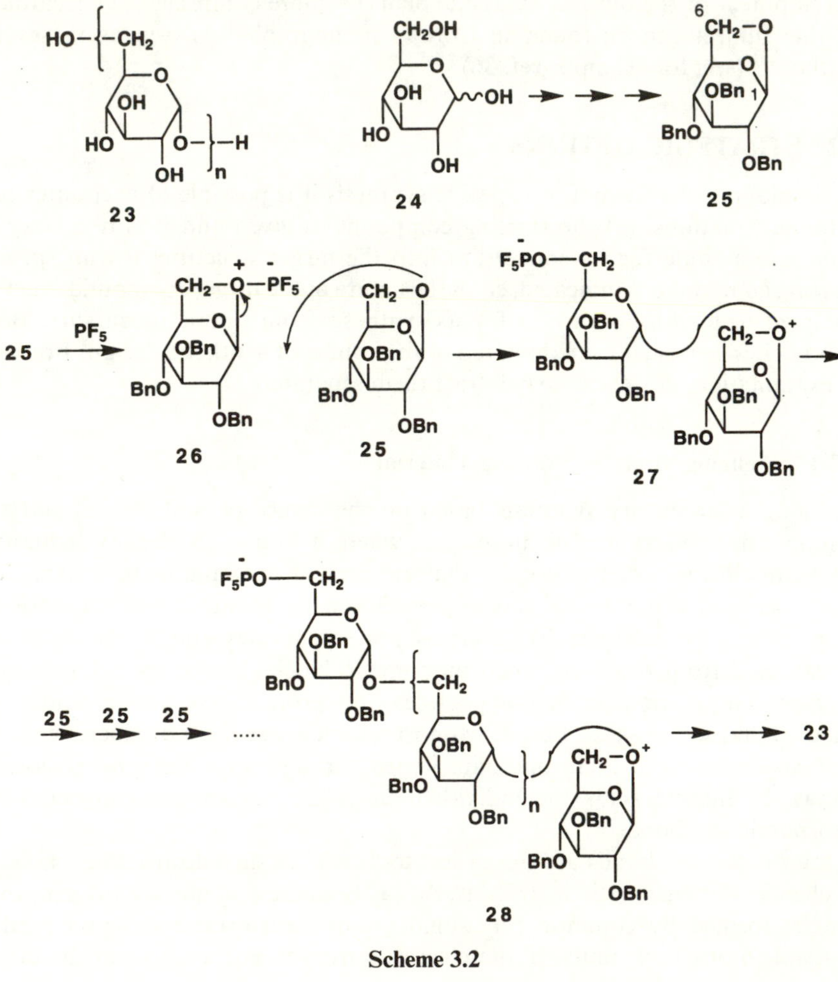

Scheme 3.2

the formation of the glycosidic bonds. The synthetic plan devised to meet these requirements involved the cationic polymerization of the 1,6-anhyro-β-D-glucopyranose derivative **25**, having benzyl-protected hydroxyl groups.

The cationic polymerization of monomer **25** occurs *via* the initial coordination the electrophilic initiator, PF_5, at the indicated ring oxygen atom to produce the oxonium ion **26**. The latter is attacked by another molecule of **25** at its electrophilic site, C-1, to form a new oxonium ion **27**. Repetition of this process leads to the consecutive formation of a set of glycosidic bonds in a stereospecific fashion as defined by inversion of configuration in every single attack at C-1. It is clear that the nature of the reaction and the structure of the monomer ensures both the desired stereo- and regiospecificity. Debenzylation of the final polymer **28** readily affords the target product **23**.

A different approach was used in the synthesis of another biopolymer, **1 → 6**-glucan **29**, possessing the β, rather than α, configuration of the intermolecular glycosidic bonds.[4c] The basic reaction employed in this approach is shown in Scheme 3.3. Here the necessary stereospecificity of formation of the β-glucosidic bond, as shown in a model glucoside **30**, is achieved by utilizing a triphenylmethyl cation catalysed reaction of cyclic ketals of type **31** with trityl ethers **32**, and proceeds *via* initial formation of the cyclic ion **33**.[4d]

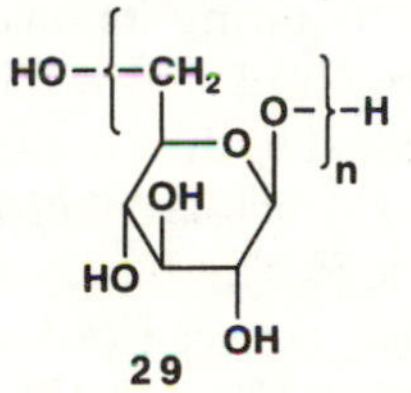

29

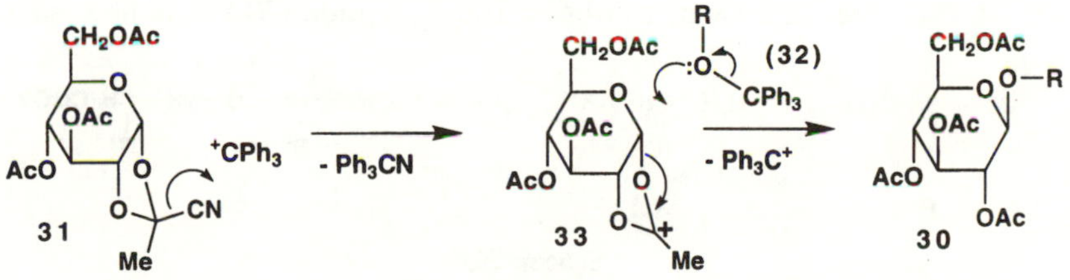

Scheme 3.3

The mechanism of this reaction implies that to achieve the stereospecific fastening of the β-glucosidic bond in the course of a polysaccharide synthesis, it is necessary to utilize a monomer with a 1,2-cyclic ketal moiety similar to that in **31**. As a necessary condition for the regiospecific formation of the **1 → 6** glucosidic bond, this monomer should contain, as well, the tritylated hydroxyl at C-6. These obvious considerations suggest that the protected glucose derivative **34** is the most promising monomer (Scheme 3.4). In fact, the trityl cation-initiated polycondensation of **34** proceeds smoothly and leads to the stereospecific formation of polysaccharide derivative **35**. Routine deprotection produces the desired glucan **29**.

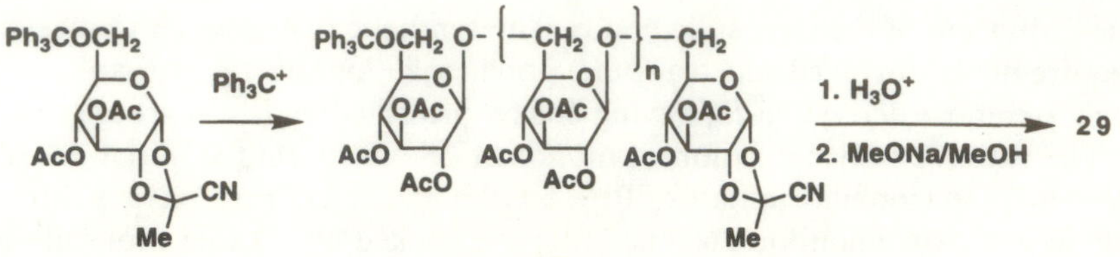

Scheme 3.4

The strategic tasks of polypeptide and polynucleotide syntheses take on an entirely different character. These syntheses also involve a stepwise formation of the appropriate inter-monomeric bonds and hence require the utilization of efficient and general methods to form amidic or phosphodiester bonds correspondingly. These biopolymers are composed of a linear, but irregular, sequence of non-identical monomeric units. This very sequence determines the unique chemical, physical, and physiological properties of a given biopolymer. Therefore, the key strategic problem consists of securing the required sequence of monomers in the growing polypeptide or polynucleotide chains. Obviously reactions like polymerizations or polycondensations are not applicable here and, in contrast to the homopolysaccharides, the formation of every link becomes an individual operation requiring its own set of reagents and conditions. Such a synthesis is necessarily broken down into a multitude of steps, the total number of which is at least equal to the number of links in the entire chain.

Let us briefly outline the general principles of problem solving involved in the task of synthesizing a polypeptide.[5a] The basic reaction in this synthesis is a trivial formation of an amide bond between two α-amino acids, one of which must have a protected amino group, **B**, and the other a protected carboxyl group, **A** (Scheme 3.5). Next, to obtain the tripeptide **CBA** it is necessary to

Scheme 3.5

remove the protecting group **Z** from the amino terminal of the dipeptide **BA** and to carry out an aminoacylation with the N-protected amino acid derivative **C**, **ZNHCHR^{3}COX**. Repetition of this sequence of operations (protecting group removal and condensation with an N-protected derivative of the next amino acid) leads to the sequential formation of a tetra-, penta-, and, eventually, *n*-polypeptide chain. Two simple operations for every step of the chain growth does not seem to be a very costly endeavor. It might have appeared that with readily available starting derivatives and sufficient patience there should be no problem in preparing a polypeptide chain of any given length, with a predetermined sequence of amino acid units. However (and certainly there is a 'however'!), this two-step cycle fails to be a realistic route because of the involvement of one additional, and purely technical, operation: the isolation of the intermediate oligopeptide derivative from the reaction mixture.

A separation procedure is the least standardized step in any reaction. The purification of an intermediate peptide, especially, can frequently be quite tedious. Let us suppose, for example, that in the sixth cycle we introduce the appropriate pentapeptide derivative of **EDCBA** containing a small quantity of the unreacted tetrapeptide **DCBA** as an impurity. In the course of preparing the desired hexapeptide **FEDCBA**, we will also obtain the by-product, pentapeptide **FDCBA**. These two products have the same termini, **F** and **A**, and nearly identical properties. Their separation may be virtually impossible. Hence the failure to achieve complete purification at a single step could lead to errors in the polypeptide 'text' being accumulated in every subsequent step, eventually leading to a nonsense 'text'. As the molecular size of the peptide increases, the difference in physical properties between the desired product and accompanying impurity, for example between oligopeptides containing 20 and 19 amino acid residues, will obviously be diminished. Correspondingly, purification will become increasingly difficult. Accumulation of several seemingly minor errors of this type can effectively sabotage an otherwise good synthetic plan.

There are generally two options available if a chemist is to avoid the complications outlined above. One must find conditions that ensure either quantitative yields at the chemical stages of condensation and deprotection or guarantee a 100% purification of the product formed at each step. Furthermore, these conditions should be applicable for a peptide of any structure. The first route, finding an errorless synthesis, is unrealistic since for all practical purposes there are no simple organic reactions that will secure a 100% yield of product in every case. As we mentioned above, 100% purification of the products in the course of peptide synthesis is also far from being an easy task. In fact it is a task of a formidable complexity. Therefore the first total synthesis of a peptide hormone, oxytocin (1953), consisting of only eight amino acids, was evaluated as an outstanding achievement. In 1955 it brought the Nobel prize to its originator, V. du Vigneaux.[5b] However, over the next two decades the synthesis of polypeptides of that complexity became routine, and at the present time the preparation of a polypeptide of more than 100 amino acid units is not considered a prohibitively difficult task.

What caused such a dramatic change in the area of protein synthesis? The answer is seemingly simple: in the early 1960s a novel approach was devised to resolve the isolation and purification problems in a peptide synthesis.[5c] How it happened was later related in a rather casual way by the discoverer of this approach, R. B. Merrifield, in his Nobel lecture: 'One day I had an idea how the goal of a more effective synthesis might be achieved. The plan was to assemble a peptide chain in a stepwise manner while it was attached at one end to a solid support'.[5d] This idea turned out to be truly brilliant in terms of both its simplicity and ingenuity. The 'trick' of Merrifield's approach consists of chemically binding the growing polypeptide chain to an insoluble and inert polymer support. As a result, the separation and purification procedure is reduced to a simple filtration of the polymer-bound product and a careful washing to remove excess reactants and by-products. Such a mechanical operation can be made completely quantitative, is easily standardized, and, as we will soon see, can even be automated. It is worth examining this procedure in more detail.

The polymer carrier in Merrifield's procedure is a granular cross-linked polystyrene (**P**) containing a certain number of chloromethyl groups on the aromatic rings. The latter group makes the polymer a functional analog of benzyl chloride and hence it will readily form an ester bond upon reaction with carboxylate anions. Condensation of this resin with an N-protected derivative of an amino acid will lead to the formation of the corresponding benzyl ester **36** (Scheme 3.6). Removal of the N-protecting group from the latter gives a protected derivative of the starting amino acid **37**, covalently bound to the polymer framework. Acylation of the amino group of **37** with an N-protected derivative of a second amino acid with subsequent deprotection gives the dipeptide **38**, again attached to the resin.

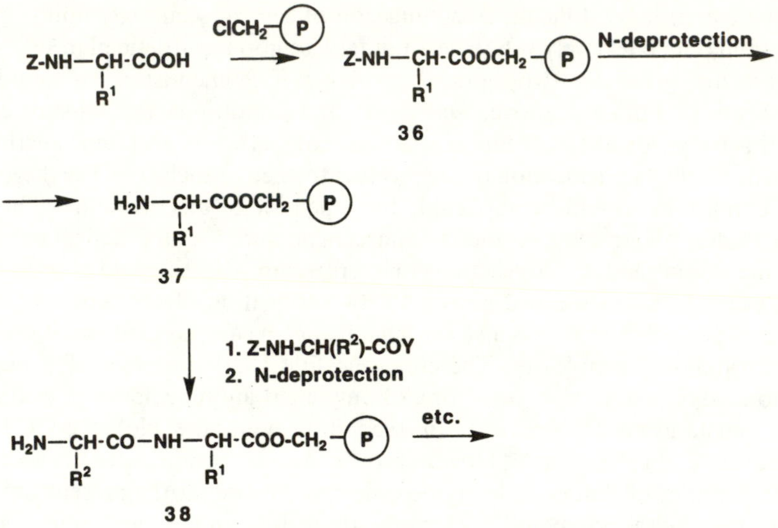

Z-NH—CH(R^1)-COOH —[$ClCH_2$–(P)]→ Z-NH—CH(R^1)-COOCH_2–(P) (**36**) —[N-deprotection]→ H_2N—CH(R^1)-COOCH_2–(P) (**37**)

37 —[1. Z-NH-CH(R^2)-COY; 2. N-deprotection]→ H_2N—CH(R^2)-CO—NH—CH(R^1)-COO-CH_2–(P) (**38**) —etc.→

Scheme 3.6

This two-step cycle can be repeated, in principle, as many times as desired to form the polypeptide chain of the required length. At the end, only a final work-up with acid (usually by a treatment with a strong anhydrous acid such as HF) is required to cleave the initial benzyl ester bond to the resin and liberate the free polypeptide.

The utilization of the solid support cannot by itself simplify the problem of separating the (n) polypeptide from the mixture of its immediate precursor, the ($n-1$) product, as both are bound to the resin. However, in this approach it is possible to use an excess of any reagent necessary to secure 100% conversion of the ($n-1$) precursor as this excess can easily be removed from the resin-bound target product.

It was quickly recognized that 'the ability to purify after each reaction by simple filtration and washing, and the fact that all reactions could be conducted within a single reaction vessel, appeared to lend themselves ideally to mechanized and automated processes.'[5d] In fact, it took only three years to develop automatic procedures and equipment to effect the programmed synthesis of polypeptides with predetermined sequences of amino acids. Initially both the equipment (bottles, valves, connecting tubes) as well as controlling devices (stepping drum programmer and a set of timers) were rather primitive. Nevertheless, the immense power of the general strategy was convincingly demonstrated by a number of polypeptide syntheses performed with the help of these almost 'caveman' tools. For example, insulin, a natural hormone containing two polypeptide chains composed of 30 and 21 amino acids connected by a disulfide bridge, was efficiently prepared by this semi-automatic procedure.[5e]

The utilization of the solid-phase technique has led to significant savings in the time and labor required to prepare peptides. For example, tremendous efforts were required for Hirshman and 22 co-workers to accomplish the remarkable synthesis of ribonuclease (an enzyme containing 124 amino acids) using the conventional solution technique.[5f] At almost the same time, however, this ribonuclease was prepared using the solid phase method.[5g] Using the automated procedure, the synthesis, involving 369 chemical reactions and 11 931 operations, was performed by Gutte and Merrifield in a matter of a few months (up to six amino acid residues were added per day to the growing polypeptide chain). Later improvement led to the creation of fully automated synthesizers. Thus, the challenging task of peptide synthesis, previously requiring an enormous expenditure of time and effort, can now be considered practically solved (at least for polypeptides of medium complexity).

The same principles found an even more spectacular application in the area of polynucleotide synthesis.[5h] Most importantly, it was recognized that the use of various types of solid support offered a number of promising strategic opportunities in other areas of organic synthesis.[5i,j] Thus the successful solution of what appeared to be initially a purely technical problem, of limited significance, led ultimately to results of enormous scientific importance.

For the biopolymers mentioned above, it is quite easy to identify the structure of the starting materials for their synthesis. In other cases the choice might not be so obvious. It is nevertheless desirable to start the analysis of a target molecule with an attempt to recognize structural fragments suitable for preparation from available starting materials. The significance of such an analysis can be illustrated with a few examples.

In 1982, microgram quantities of the pheromone of the female mosquito *Culex pipiens fatigans* were isolated. This mosquito carries the dreaded tropical disease elephantiasis. The insect used the pheromone to indicate the location of egg laying and to attract other females to deposit eggs at the same location. An examination of the structure **39** of the pheromone (Scheme 3.7)[6] immediately suggested that its synthesis could be achieved starting with a known, straight-chain C_{16} carboxylic acid **40**, containing a *cis* double bond at position 5. The conversion of **40** into the target material **39** was trivial.

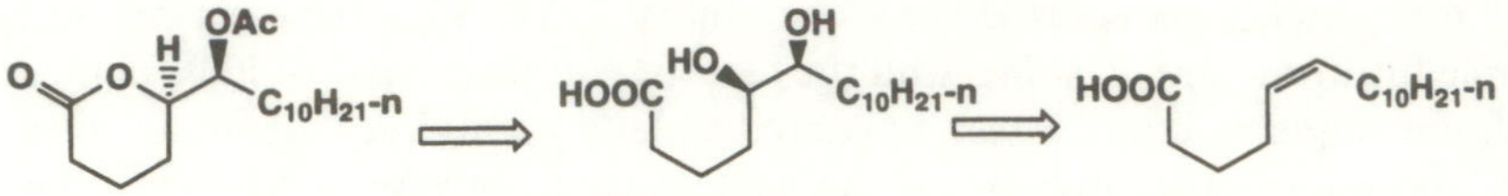

39 40

Scheme 3.7

The logic behind selecting the starting compound, however, is seldom so straightforward. Quite often, the *less than obvious* route turns out to be very rewarding. In this respect it is interesting to compare two solutions of the same classical synthetic task at the beginning of this century, the synthesis of tropinone **41** (Scheme 3.8), a key intermediate in the synthesis of the alkaloid atropine. The solution elaborated by Willstater[7a] was based on his perception of structure **41** as a derivative of cycloheptane. This obvious and tenable analysis identified cycloheptanone **42** as a starting material and dictated the sequence of simple transformations aimed first to create additional functionality and ultimately to introduce the methylamine bridge at the proper positions of the already present seven-membered ring. However, the simplicity of this approach to the synthesis of **41** suffered from drawbacks caused by the lengthy sequence of reactions involved. While all these reactions were routine and proceeded in good yields, the overall yield of **41** was frustratingly low.

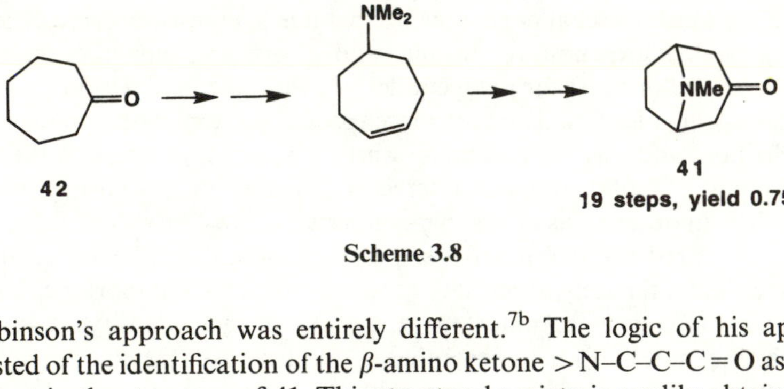

42 41

19 steps, yield 0.75%

Scheme 3.8

Robinson's approach was entirely different.[7b] The logic of his approach consisted of the identification of the β-amino ketone >N–C–C–C=O as the key fragment in the structure of **41**. This structural moiety is readily obtainable *via* the Mannich reaction[7c] between a carbonyl compound, a primary amine, and a carbon nucleophilic component, as shown in Scheme 3.9. From this reasoning, Robinson arrived at a rather unexpected set of starting compounds: succinic dialdehyde **43**, methylamine, and acetonedicarboxylic acid **44**. Interaction of these components proceeded as a double Mannich reaction and resulted in the immediate formation of the target molecular framework as present in **44a**. The latter product underwent smooth decarboxylation upon heating to give tropinone **41**. The yield of this product in Robinson's one-step preparation might appear to be rather modest, but it represented a spectacular improvement when compared to the overall efficiency of Willstatter's synthesis. Strategic advantages of Robinson's synthesis warranted further studies of this reaction to

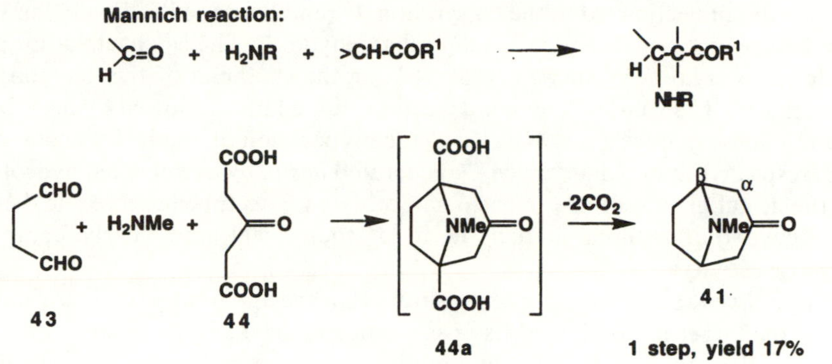

Mannich reaction:

Scheme 3.9

optimize its conditions. As a result of these studies, Schopf and co-workers were able to increase the yield to more than 90%.[7d] Thus a brilliant strategic idea led eventually to the creation of an entirely new and efficient synthetic procedure.

The advantages of Robinson's concept was confirmed in numerous total syntheses of alkaloids. One of the most illustrative examples is served by Stevens' synthesis of the tricyclic compound coccinelline **45** (Scheme 3.10), the pheromone of the ladybug (ladybird). As was acknowledged in Stevens' review article,[7e] 'in less time than it took to write this paragraph, we had developed on paper an attractive approach that relies on one of the oldest reactions known in alkaloid synthesis, namely the classical Robinson–Schopf condensation'. This approach involved the retrosynthetic transformation of **45** into the ketopiperidine derivative **46**, which is amenable to disconnection in accordance with the same logic as used by Robinson in the analysis of tropinone.

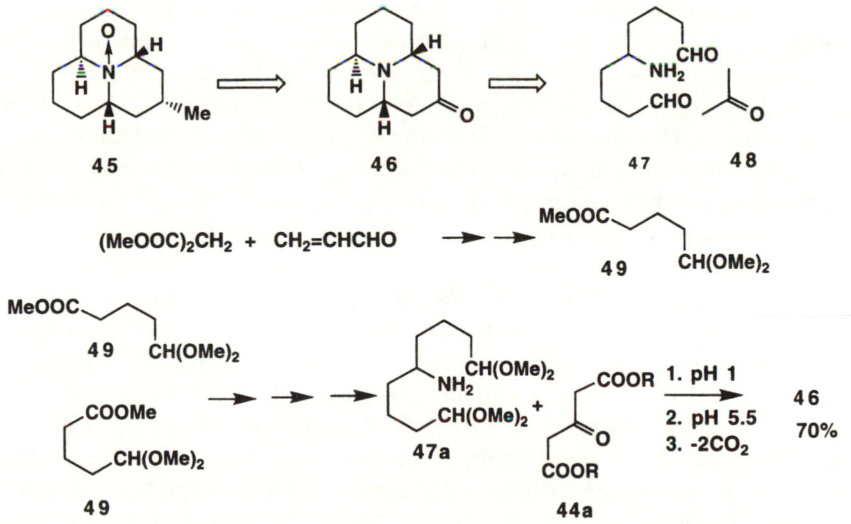

Scheme 3.10

This disconnection led to the C_3 synthon **48** (and hence to its already familiar synthetic equivalent **44**) and C_9 amino dialdehyde **47**. The Michael addition of malonic ester to acrolein was employed for the synthesis of the key starting material **49**. The Claisen ester condensation of the latter followed by decarboxylation and reductive aminolysis led to the preparation of amino-bis-acetal **47a**. The respective amino dialdehyde **47**, generated *in situ* by a controlled hydrolysis of the acetal groups of **47a**, reacted smoothly with acetonedicarboxylic diester and gave the required adduct **46** in a good yield and nearly complete stereoselectivity.

The above examples clearly demonstrate that the apparent structural similarity of the target molecule and its possible precursor is not necessarily the best lead in the elaboration of the optimal synthetic pathways. A thorough consideration of the functionality pattern of the target structure and its chemical ramifications might lead to the identification of non-trivial and much more efficient alternatives.

This is especially true in the analysis of the synthetic pathways to many natural compounds, where the most effective synthetic ideas can be often borrowed from the arsenal of the most skillful synthetic chemist of all times: *Mother Nature*. A good illustration of the fruitfulness of this so-called biomimetic approach is provided by the synthesis of morphine group of alkaloids (Scheme 3.11).

From the viewpoint of a 'pure' synthetic chemist it is not easy to identify a simple precursor for a short synthesis of the formidable pentacyclic skeleton structure of morphine **50**. The first total synthesis of **50**, accomplished by Gates in 1952,[8a] was based on the choice of bicyclic compound **51** as the starting material containing the **A**–**B** rings of the target framework. This precursor was converted into the *o*-quinone derivative **52**, which was used as a dienophile in the Diels–Alder reaction to form the tricyclic adduct **53**. Intramolecular reductive amination served as a key step for the formation of the **E** ring as is present in adduct **54**. An additional dozen steps were required for the transformation of **54** into **50**. While the overall yield of **50** prepared by this necessarily lengthy synthesis was negligible, this result was properly acknowledged as one of the outstanding achievements of synthesis in the 1950s.[8b]

In the 1960s, Barton, with an exquisite set of experiments, succeeded in showing that the key step in the biosynthesis of salutaridine **55a**, a common intermediate to **50** and its cogeners in plants, involves the formation of the central C–C bond *via* intramolecular oxidative coupling of two phenolic rings of the precursor, reticulin **56a**.[8c] This discovery prompted attempts to mimic this transformation, *in vitro*, by purely chemical methods not employing enzymes. After numerous fruitless experiments, it was finally found that the oxidation of **56a** into **55a** can be achieved under the action of potassium ferrocyanide. However, the yield of the product (0.03%) was disappointing, to say the least, and at best could serve as testimony to the futility of man's efforts to compete with Mother Nature. Nevertheless, this result was highly regarded as it gave unequivocal proof of 'the theorem of existence of the solution' for the problem of biogenetically patterned syntheses of morphine alkaloids. The strategic

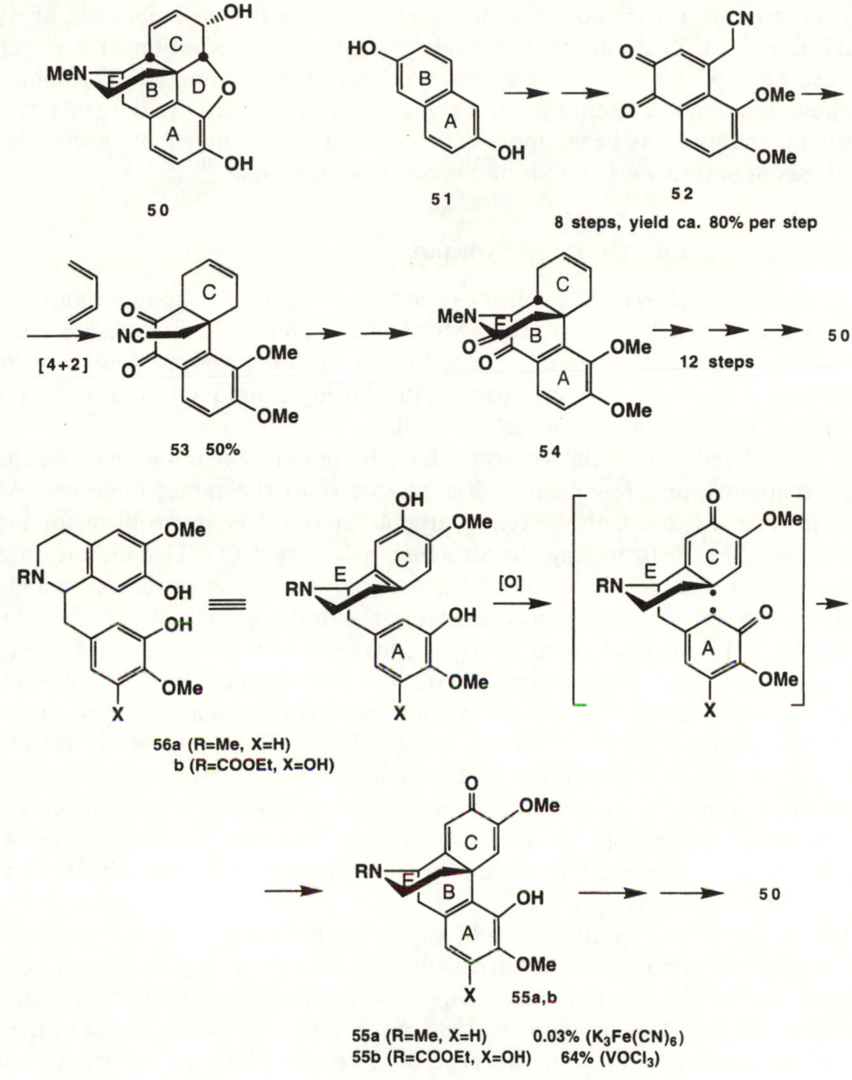

Scheme 3.11

gains of this approach were so obvious that numerous studies followed to develop the conditions optimal for this oxidative coupling. In the end, the determination of the synthetic chemists involved was rewarded with success. As was shown by Schwartz and Zoda,[8d] the conversion of the slightly modified analog **56b** into the respective derivative of **55b** proceeded smoothly under the action of $VOCl_3$ with 64% yield. The required precursor **56b** was prepared in six steps and 50% overall yield from readily available materials. The preparation of **55a** and **55b**[8e] constituted the formal synthesis of morphine **50**, as well, since the conversion **55** → **50** could be achieved easily by trivial transformations.

This example shows how a rather esoteric study in the biogenesis of **50** sparked a chain of events that ultimately led to the elaboration of a novel, concise strategy applied to the practical synthesis of this compound. During the last few decades, the usefulness and power of biogenetically patterned retrosynthetic analyses has been amply demonstrated by a number of spectacular syntheses of natural compounds of diverse structural types.[8f]

3.2.2 Planning 'from the Target Structure'

The benefits one gains from a judicious selection of starting compounds may not always be easy to achieve in practice. More often than not the analysis of a truly complicated molecule does not reveal the structural 'prompts' that provide immediate leads for identifying appropriate starting compounds and pathways for their conversion into the target material.

A general and more reliable approach (although more lengthy) involves the logistic application of sequential disconnections to the target molecule. As was formulated by Corey, 'retrosynthetic analysis is a problem-solving technique for transformating the structure of a target (TGT) molecule to a sequence of progressively simpler structures along a pathway that ultimately leads to simple or commercially available starting materials for a chemical synthesis'.[2a] The general methodology of this approach and its application to the solution of synthetic problems are summarized in an excellent monograph by Corey and Cheng.[3a] This methodology has already been employed in the preceding chapters of this book to rather simple examples. Below we are going to discuss problems of retrosynthetic analysis in a more systematic way. Rather obviously, it is absolutely impossible to present a comprehensive treatment of the whole subject in this text. Instead, we would like to emphasize only the most important features pertinent to the ideology of retrosynthetic analysis.

The retrosynthetic simplification of any target molecule involving a sequential rupture of bonds may be started, in principle, from any bond and thus carried out to the simplest precursors along a multitude of retrosynthetic pathways. If this dismantling is carried out in an unsystematic fashion, then the immense number of possibilities generated would make the approach useless for all practical purposes. On the other hand, a thoughtful retrosynthetic analysis, governed by chemical logic and conducted in accordance with carefully chosen criteria, represents a powerful approach to elaborate a sound synthetic strategy.

Retrosynthetic analysis involves several more or less distinct steps; probably the most difficult, and by far the most important, is the initial analysis of the synthetic target structure.

3.2.3 Debut

Before attempting the dissection of a TGT molecule it is useful to analyse the general synthetic task in order to identify and evaluate the complexity of a set of subtasks, and then to determine an optimized order of solving these subtasks. For example, in Woodward's synthesis of sterols, the key task was the assemblage of the tetracyclic system with functional groups at the C-3 and C-17 positions. The keto group at C-3 was needed to introduce the required functionalities in rings **A** and **B**; the functional group at C-17 was essential for the introduction of the required substituent at this center. In so much as we already know, the transformation of functional groups usually does not present a problem. The same refers to the attachment of alkyl pendants to the functionalized centers. Therefore, to a first approximation, the exact nature of functional groups introduced into specified positions of the intermediate products is irrelevant, since these groups could be modified or removed in order to match the functionality of the TGT molecule.

Woodward's synthesis illustrates the general idea of the initial steps of retrosynthetic analysis: start by splitting off 'pendants' (alkyl, aryl groups, *etc.*), replacing them with the functional groups at the appropriate center. Those 'stripped-off' groups in all likelihood can be introduced at the final steps of the synthesis — provided the latter is well-planned and Murphy's law fails.

In much the same way, if the structure contains heteroatoms that are not a part of the heteroaromatic system, it makes sense to start the analysis by rupturing a carbon–heteroatom bond as the reverse reaction represents, essentially, a trivial transformation of functional groups. The presence of small ring fragments such as cyclopropane or epoxide rings in the structure of the target molecule almost automatically dictates the retrosynthetic scission of these moieties in the initial steps of retrosynthetical analysis, as both these groups can be easily introduced with the help of very reliable methods.

The benefits of simplifying the target molecule are rather obvious. First of all, by doing this the final stages of the synthetic scheme relate to the more reliable and trouble-free reactions. The potentially risky steps are moved to the initial phases which, from the perspective of one's investment of time into a synthesis, carries obvious benefits. A second, but no less significant, advantage of this approach is that it obviates the need to drag along highly reactive and labile groups in a multistep sequence. Quite often this approach may also greatly simplify selectivity problems in the course of a real synthesis.

Thus, after splitting off side chains and removing or transforming 'extraneous' functional groups and other readily 'installable' moieties, the retrosynthetic analysis identifies the 'strategic core' of the original TGT molecule. The main focus of the planned synthesis, then, becomes the assemblage of this strategic core.

Here, it is appropriate to add certain proviso. The above recommendations by no means should be considered a set of rigid instructions. They are generally applicable to molecules of medium complexity and may be useless if applied to a target structure that contains an intricate set of interfering, polyfunctional groups at multiple chiral centers. A rather different approach should be used in such cases[3a] and its consideration definitely lies outside the limits and aims of our book.

3.2.4 Dissection of the Strategic Core of the Molecule

The complexity of the key steps of a retrosynthetic analysis may differ dramatically, depending upon the overall structure of the strategic core, which can be conventionally divided into three main groups: acyclic, monocyclic, and polycyclic.

The retrosynthetic analysis of acyclic systems usually does not require the formulation of special strategic concepts, as it usually can be based upon considering the positioning of functional groups and simple 'pendants'. This type of disconnection normally involves straightforward solutions derived from existing methods for creating C–C bonds. In fact we have already examined the strategic merits of various methods (see Section 2.4) and thus only a few additional comments seem to be appropriate here. As a rule, the dismantling of an acyclic chain can be carried out at almost any C–C bond and therefore, for even the simplest cases, one has to deal with a set of different retrosynthetic solutions. In fact, the number of options is even greater if one also takes into account that (i) more than one method may be feasible for the formation of a given C–C bond and (ii) within the limits of a given method, a set of fairly diverse reagents can be used. Rational selection among these options is determined by considering such factors as the availability of starting materials, the opportunity to exert rigorous control over the stereochemistry, the desire to minimize the number of steps, and the ultimate purpose of the projected synthesis. An impressive set of relevant examples can be found in studies targeting the synthesis of structurally uncomplicated acyclic compounds such as the juvenile hormone.[9a]

In regard to simple monocyclic systems, the basic principles of disconnection for the synthetic target molecule differ very little in essence from those of an acyclic system. In fact, here again the retrosynthetic analysis is more or less directly related to existing methods for the creation of rings (see Section 2.7). We will not regress to a second examination of these methods and will limit our discussion to a few specific points.

The presence of a six-membered ring in the synthetic target molecule does not necessarily imply that its formation should be carried out as a result of ring-forming reactions like, for example, the Diels–Alder reaction or Robinson annulation. It is useful to keep in mind alternative pathways based upon the transformations of aromatic six-membered systems such as the Birch reduction or catalytic hydrogenation. In fact this approach has two major strategic merits. First of all, it frees the chemist from creating the skeleton of the six-membered ring. Secondly, it may greatly simplify the problem of introducing the required functional substituents on the ring carbons, as the preparation of the respective derivatives in the benzene series may be a more or less routine task. It is definitely worthwhile to consider the plausibility of these routes at the very beginning of a retrosynthetic analysis, as it could negate our earlier recommendation of removing side chain 'pendants' and functional groups associated with a cyclohexane frame. In some cases it might actually be easier to introduce these groups at the aromatic precursor stage and then convert the intermediate product into a cyclohexane derivative rather than to devise options applicable for the introduction of these substituents into the assembled cyclohexane fragment. Hence a retrosynthetic dehydrogenation of a cyclohexane ring present in the target structure should be included in the list of initial steps to determine the optimal synthetic plan. A good model that illustrates the efficiency of such an approach is the synthesis of *cis*-4-*tert*-butylcyclohexane-carboxylic acid, as was discussed in Section 2.6. An example of the successful use of a Birch reduction as the key step for the preparation of polysubstituted cyclohexane derivatives will be considered below (see Scheme 3.24).

In the previous chapter we discussed some general methods specifically elaborated for the formation of a ring of a given size. Here it is appropriate to comment briefly about an additional method which is applicable to the formation of rings of almost any size and is especially applicable for the formation of medium and large ring systems. This method is the McMurry alkenation reaction, which involves an intramolecular reductive coupling of two carbonyl moieties to form an alkene fragment (Scheme 3.12).[9b] Owing to its reliability, the McMurry reaction represents a highly useful option to be considered in the course of retrosynthetic analyses of cyclic systems. Non-trivial strategic opportunities offered by this method can be illustrated in the synthesis of flexibilene **57**, a naturally occurring diterpene containing a 15-membered ring.[9c]

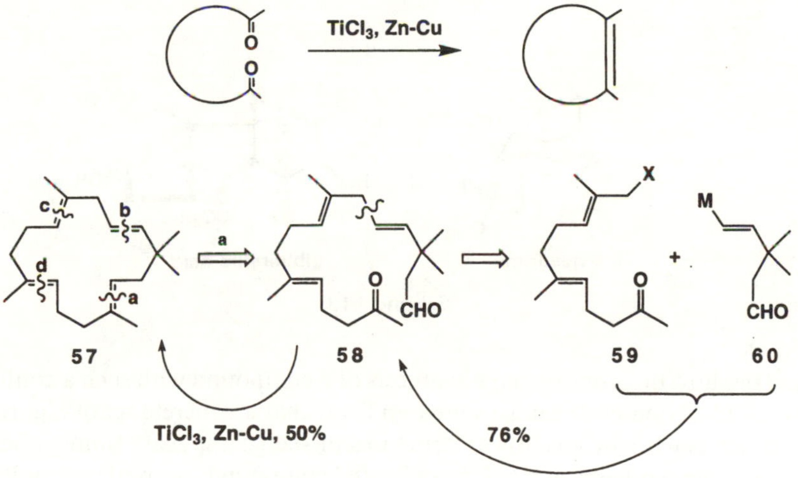

$TiCl_3$, Zn-Cu

57 ⟹(a) 58 ⟹ 59 + 60

$TiCl_3$, Zn-Cu, 50%

76%

Scheme 3.12

An opportunity to apply the McMurry coupling for the creation of C=C double bonds suggested four possible modes of retrosynthetic opening of the ring in **57**, affecting the double bonds (**a–d**). All modes are viable, but from the point of view of the accessibility of the acyclic precursor, route **a**, leading to the precursor **58**, seemed to be preferable. Further simplification of **58** could have been carried out by numerous routes in accordance with the general reasoning used in the retrosynthetic analysis of acyclic systems (see above). The retrosynthetic pathway shown in Scheme 3.12 was chosen by the authors as the shortest to arrive at the available materials **59** and **60**. Synthesis of **57** performed in accordance with the adopted plan turned out to be quite efficient.[9c]

No doubt the most difficult strategic tasks are encountered in the course of planning the synthesis of a polycyclic system. To appreciate the tremendous complexity of these problems, one only needs to glance at structures designed by organic chemists, such as cubane, asterane, and pentaprismane, or at those created by Mother Nature, like quadrone or gibberellic acid (Scheme 3.13). These are, by no means, the most complicated targets of modern synthetic pursuits! In these cases, examination of the structures does not readily suggest suitable starting materials or obvious pathways for the retrosynthetic simplification of the TGT molecule. It is not even clear where to begin dismantling these molecules. Something other than the simple applications introduced earlier are needed if appropriate precursors are going to be found for their synthesis.

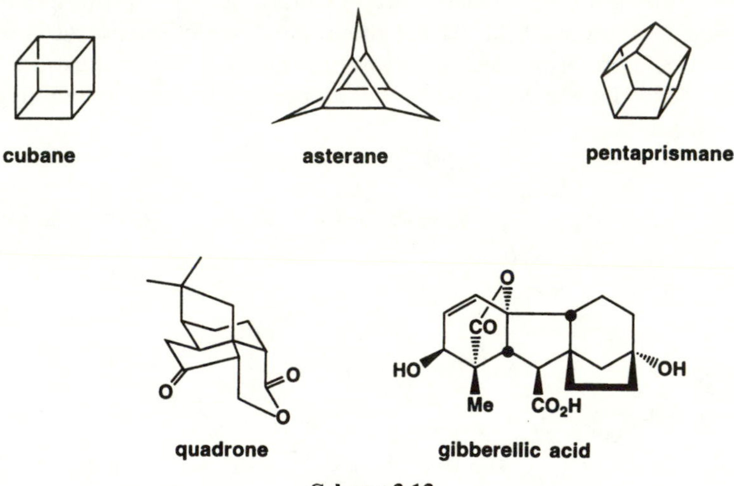

Scheme 3.13

In fact, how does one plan the synthesis of a compound with such a confusing array of C–C bonds? It can be stated up front that a concrete set of algorithms for the retrosynthetic analysis of structures of such complexity simply does not exist. Yet the syntheses of all of the indicated compounds, as well as a multitude of other even more complex structures, have been successfully accomplished. It thus follows that, despite the absence of rigorous rules, some guidelines must be available. The character of some of these guidelines can best be illustrated in the course of the analysis of the following representative examples.

The retrosynthetic analysis of polycyclic structures can proceed by one of two general routes: either break one bond at every step or disconnect, simultaneously, more than one bond. Both routes are valid and logical and both routes have been employed in contemporary syntheses. However, realization of these pathways requires entirely different arrangements of the synthetic plans and it therefore makes sense to examine them separately. We will begin by looking at the methodology of the first approach.

3.2.5 Selection of the 'Strategic Bond' in a Target Molecule

In 1978 the structure of the sesquiterpene quadrone **61** (Scheme 3.14) was established. Quadrone is a metabolite of the fungus *Aspergillus terreus* and exhibits a well-expressed inhibition activity toward several types of leukemia and carcinoma. Not surprisingly, there was an immediate surge of activity directed at the total synthesis of this rather complex tetracyclic compound. At present, more than a dozen total syntheses for quadrone have been described. It is instructive to compare the strategies of the most efficient syntheses. The first step in all of these approaches involves the rupture of the lactone ring **D**, as the reverse transformation represents a trivial task to be done at the final stage. In other words, the problem of the synthesis of **61** is actually reduced to the synthesis of its immediate precursor **62**. The strategic problem in the total synthesis of quadrone lies in assembling the tricyclic **ABC** core of **62**, and herein lies the differences between the pathways described in the reported approaches.

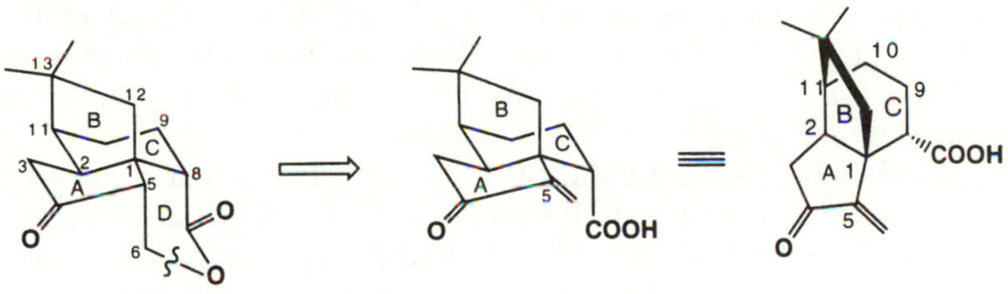

Scheme 3.14

In studies carried out by the groups of Danishefsky[10a] and Helquist,[10b] the scission of the C-8–C-9 bond was chosen as an initial step in the retrosynthetic breakdown of the TGT molecule. The strategic benefits of this approach seem to be obvious. In fact, as a result of this operation, the complicated bridged framework of **62** is transformed into a rather simple, linearly fused dicyclopentanoid structure of the precursor **63** (Scheme 3.15). Furthermore, the positioning of the substituents and the functionality pattern of **63** clearly dictates the logic of the subsequent retrosynthetic steps leading to the intermediate structures **64** and **65** and eventually to the set of simple starting materials **66–68**.

Before commenting on these steps it is appropriate to introduce a couple of useful new terms. The exact reverse of the synthetic operations applied for the retrosynthetic disconnection of the bonds in a synthetic target structure is called a *transform*. A structural unit or set of functional groups that should be present

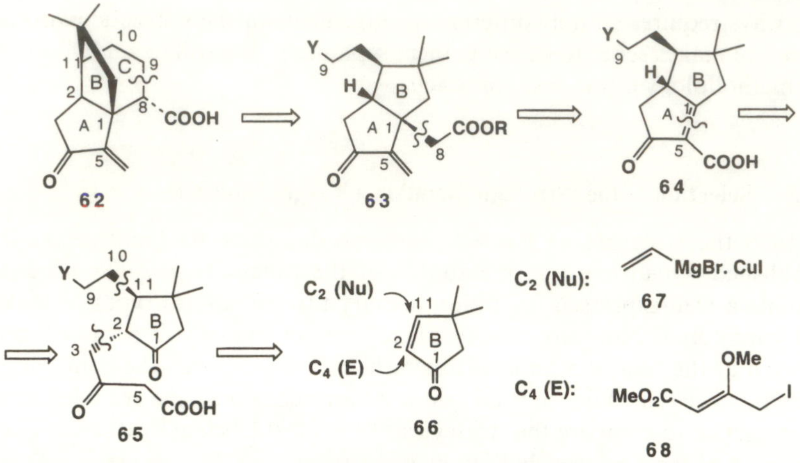

Scheme 3.15

in order to carry out a given synthetic reaction is called a *retron*.[3] Thus the presence of the required *retron* represents a mandatory prerequisite for the application of a given *transform* to simplify the target structure. In these terms the sequence shown in Scheme 3.15 is described as consecutive steps of transforms such as a carbonyl enolate alkylation (**62** ⇒ **63**), Michael addition (**63** ⇒ **64**), crotonic condensation (**64** ⇒ **65**), and tandem addition to the conjugated double bond (**65** ⇒ **66** + **67** + **68**).

The retrosynthetic sequence shown in Scheme 3.15 resulted in the identification of the constructive steps for the creating of the major framework and thus refers to only a general strategy for the synthetic approach. Additional retrosynthetic steps are required to install missing retrons, for functional group transformations, or to protect interfering functionalities. The actual synthesis carried out by Danishefsky's group, starting from **66**, required 19 steps and gave quadrone **61** in 3.1% overall yield.[10a] Helquist's synthesis[10b] was based on a similar retrosynthetic analysis, but it utilized different reagents and was accomplished *via* a shorter pathway.

From the point of view of a general strategy, these two approaches envisioned a gradual increase in the complexity of the system in the order of **B** → **BA** → **BAC**. This sequence had to surmount the problems associated with securing the required configuration at the stereocenters formed at the construction of each of these fragments.

These stereochemical problems were largely avoided in an entirely different approach in the retrosynthetic analysis of **62**, devised by Burke's group (Scheme 3.16).[10c] The authors identified the quaternary carbon C-1, common to all three rings, as the key element of the whole structure. The goal of their retrosynthetic analysis was to find a series of transforms which would eventually lead to the gross overall simplification of **62** with the C-1 center remaining intact. A set of retrosynthetic functional group transformations led to the subtarget structure

69. This subtarget was chosen as it contained the retron (enone moiety) that could presage the disconnection of the tricyclic framework into the bicyclic derivative **70** with the help of a crotonic condensation transform. A subsequent application of the Michael addition transform leads to the monocyclic precursor **71**. This precursor contains ring **A**, in which the quaternary carbon atom bears all the necessary fragments to assemble rings **B** and **C**. The selection of this 'not so obvious' strategy was determined by the results of previous studies by the same authors that elaborated reliable methods for the directed synthesis of spiro compounds such as **72**, which could be routinely transformed into derivatives bearing quaternary carbon similar to **71**.

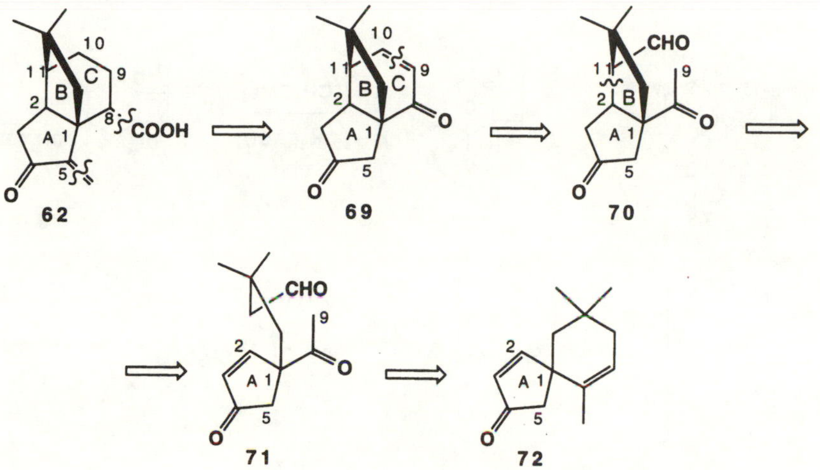

Scheme 3.16

Strategically, this route seems to be very appealing. In fact, the available starting structure **71** already contains a nearly complete set of the required carbon atoms and 'the only' remaining tasks needed to arrive at the target polycyclic system are to effect a couple of intramolecular conversions. Yet the viability of the suggested pathway was met with skepticism. In a later paper[10d] which described the real story behind this synthesis, Burke recalls: 'this idea, contained in my first research proposals, was described by one reviewer as "unlikely"; a less generous reviewer described it as "impossible". (The first comment was wounding, the second challenging.)' As an answer to this challenge, the author initially attempted to accomplish the transformation of **71** into the tricyclic enone **69** in a most spectacular way: as a single-flask sequence involving an intramolecular Michael addition and subsequent aldol cyclization. This one-stroke attack failed completely as the diketo aldehyde **71**, containing three electrophilic and three nucleophilic sites, reacted indiscriminately. Fortunately a judicious choice of conditions finally resulted in a stepwise protocol which involved the transformation of **71** into the bicyclo adduct **70** (yield 92%) followed by cyclization of **70** to **69** (yield 96%). Thus, the viability of the main strategic idea was spectacularly demonstrated; the basic framework of the synthetic target structure was assembled in only three steps from the spiro adduct **72** in an excellent overall yield (*ca.* 80%). More upsetting was the finding that the final, seemingly simple, transformations needed to elaborate **69** into **62** turned out to be rather troublesome (a case of Murphy's law!?). The completion of the synthesis required a dozen additional steps. Owing to these unexpected complications, the overall yield of quadrone amounted only to 6.2%.

An entirely different approach for the synthesis of quadrone was chosen by Yoshii *et al.*[10e] Yoshii's group started the assemblage of **62** from cyclohexenone **73** (Scheme 3.17), as a precursor of ring **C**. The synthetic sequence consisted of only 12 steps and was accomplished in an overall yield of 2.6%. The key steps in this synthesis, involving intermediates **74–78**, are shown in Scheme 3.17.

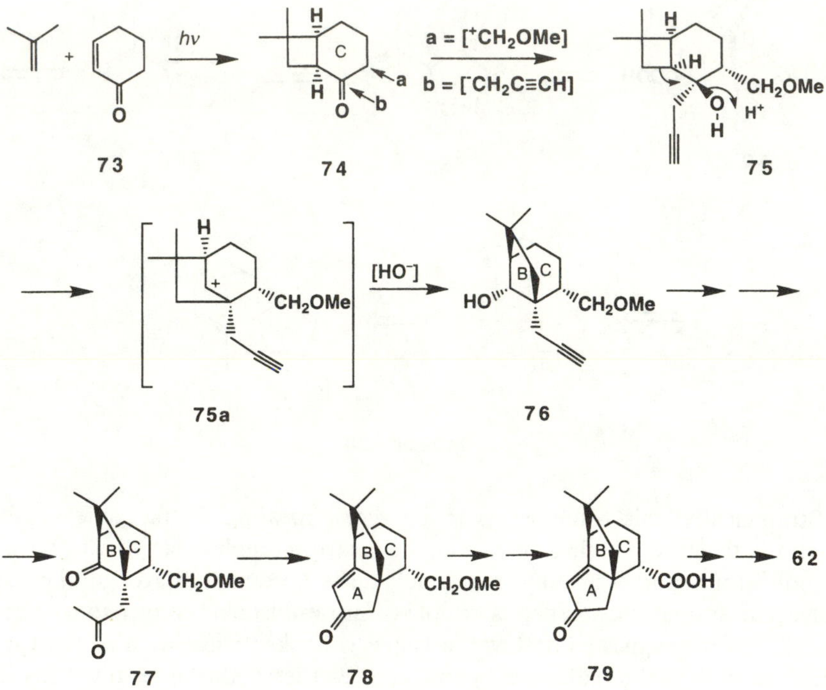

Scheme 3.17

The logic behind the choice of disconnecting steps used in this synthetic plan is rather straightforward (see Scheme 3.18). The authors chose the bicyclo[3.2.1]octane moiety formed by the **B–C** rings as a key fragment of the TGT molecule **62**. This choice might not have seemed to be very practical because of the apparent difficulty in synthesizing this fragment. Nonetheless, the authors reasoned that the well-known tendency of the strained ring system of bicyclo[4.2.0]octanes to rearrange into more stable bicyclo[3.2.1] systems might be employed in the present case. Therefore, the first steps of the retrosynthetic analysis of the tricyclic intermediate **80** were directed at the dissection of ring **A** at the C-2–C-3 bond to give, as a result of an aldol transform, the first subtarget **81**. An application of a carbocation rearrangement transform to **81** identified the next subtarget **82**. It was obvious, however, that the presence of the acetonyl and β-hydroxycarboxyl moieties in **82** virtually precluded its possible use as the substrate for a carbcation rearrangement. The authors were then forced to devise a synthetic equivalent of **82** that contained latent acetonyl and carboxyl groups. As a result, structure **75** was targeted as the key intermediate. The feasibility of this choice was demonstrated in the results of the acid-induced rearrangement of **75** → **76** *via* formation of the intermediate carbocation **75a** (Scheme 3.17). Retrosynthetic disconnections of the subtarget **82**, leading ultimately to **73**, are obvious.

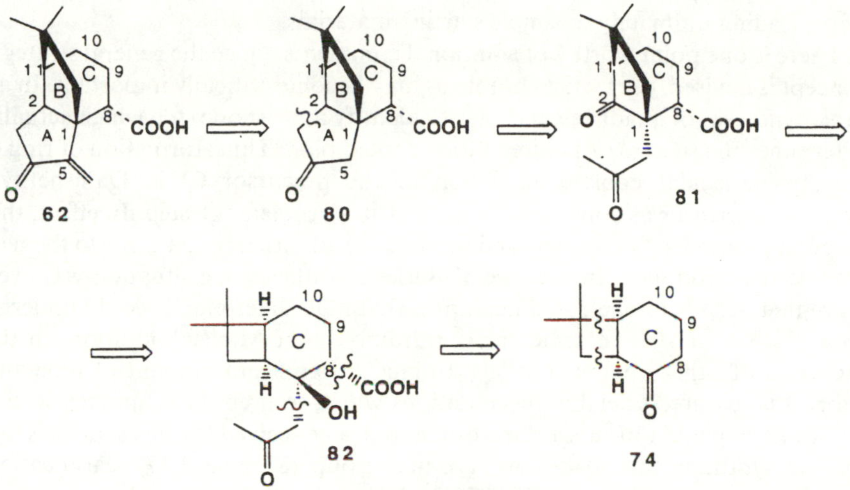

Scheme 3.18

Let us attempt to identify the general reasoning that guided the identification of the aforementioned different routes to dismantle the strategic core of quadrone. It is easy to notice that none of the authors attempted to disconnect the bonds to the two quaternary carbon atoms, C-1 and C-13, and in general did not try to disrupt ring **B** in the initial steps of retrosynthetic analysis. Why? First of all, the creation of quaternary carbon centers in an organic molecule is a rather difficult synthetic problem[10f] and for this reason it is better to place this risky operation in the beginning phases of a synthesis. This was especially true with regard to ring **B**, with four of its five bonds attached to quaternary carbon atoms, and therefore it was definitely preferable to avoid dismantling of this fragment at the initial steps of retrosynthetic disconnections.

Taking such considerations into account, it is not difficult to conclude that only a few among the available bonds in the synthetic target **62** could be considered as suitable candidates for disconnection. These include the C-8–C-9 and C-9–C-10 bonds in ring **C** and C-2–C-3, C-3–C-4, and C-4–C-5 bonds in ring **A**. Of the last three possibilities, bond C-2–C-3 appears preferable in view of the presence of the adjacent carbonyl group at C-4 which can be identified as a retron, securing the applicability of an enolate alkylation transform. As a result of this rather perfunctory analysis, it became clear that the search for the optimal pathway in the retrosynthetic simplification of **62** should be focused upon the scission of only three carbon–carbon bonds. In fact, each of these options was utilized in planning the syntheses of quadrone mentioned above.

So far we have emphasized the initial steps of a retrosynthetic analysis. The examples shown have also demonstrated that the very first bond selected for disconnection determines the strategy of the entire synthetic scheme. Therefore, this bond should be considered as a *strategic bond* (SB). Similar analysis aimed at the identification of the strategic bond could be also required for any intermediate structure generated in the course of the subsequent disconnection steps, leading ultimately to simple starting materials.

There is one point worthy of additional comments. Once the general strategic concept is devised, the tactical problems may become critically important. In all three syntheses of quadrone, it is easy to identify at least one step which actually determines the success or failure of the adopted plan. Thus, formation of ring **C** *via* intramolecular enolate alkylation of the precursor **63** in Danishefky's synthesis *a priori* was considered to be a rather problematic step. In effect, this reaction proceeded fairly easily and the authors admitted that it came to them as unexpected good luck. In the case of Burke's synthesis, the situation was even more tenuous as no one could guarantee that aldo diketone **71** would undergo an exclusive, or at least preferential, intramolecular Michael addition, to the exclusion of other obvious options. In fact, a significant amount of time and effort was required to elaborate conditions which secured the required selectivity of this step and thus to save the brilliant idea conceived for this synthesis. As for the synthesis carried out by Yoshii's group (Scheme 3.17), carbocation rearrangements of the type **75** → **76** had been studied earlier, but on much simpler examples. Here again a lot of additional experimentation was needed to establish the feasibility of this route for more complex substrates such as **75**.

An important feature of the philosophy of retrosynthetic analysis is clearly exposed in the described approaches to the syntheses of quadrone. In general, there are only a few bonds in the target molecule that can be reasonably considered as 'breakable' and even fewer of them as 'strategic'. The selection of the strategic bond will dictate the flavor of the rest of the dismantling down to suitable precursors and hence lead to synthetic plans as varied as those utilized for the the quadrone syntheses.

3.2.6 Analysis of the Structure as a Whole

The three variations of the retrosynthetic analysis of quadrone discussed in the preceding section were based upon one general concept: the sequential bond-after-bond disconnection of the core of the target molecule into a set of smaller fragments. Logistically it is an absolutely reliable approach; a consistent application of the retrosynthetic principles would inevitably produce a number of more- or less-reasonable pathways to synthesize the target molecule from available precursors. However, this reliable but overall rather lengthy procedure is not necessarily the most efficient mode of retrosynthetical analysis.

In contrast, the analysis of the TGT molecule as an *integral entity* can lead to the identification of peculiar structural characteristics that might prompt a much more economical route to its assembly. Such an in-depth examination is based upon both heuristic and logical considerations and, at times, the heuristic considerations might even be prevalent. This analysis of the molecule as a whole is aimed at the recognition of opportunities to put together the strategic core of the target molecule as a result of a single chemical operation (or a very few operations) and thus may be properly called a 'whole structure' strategy for the conception of a synthetic plan. In many cases this approach proved to be of paramount usefulness. Some illustrative examples will be considered below.

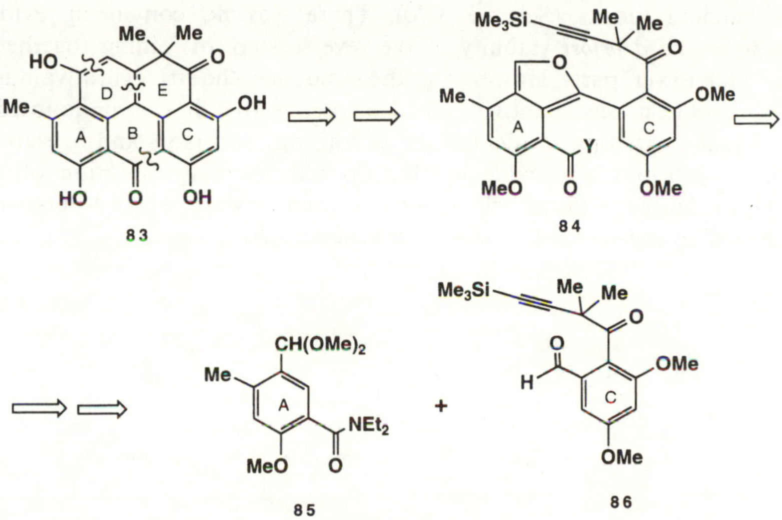

83 ⟹ ⟹ **84** ⟹ ⟹ ⟹ **85** + **86**

Scheme 3.19

It is hardly possible to identify a strategic core in the pentacyclic structure of the antibiotic resistomycin **83** (Scheme 3.19). The structure is also devoid of symmetry elements. Therefore, its synthesis through a stepwise protocol that envisioned a step-by-step elaboration of the **A**–**B**–**C**–**D**–**E** ring framework seemed to be the only likely option. This is exactly how the first synthesis of resistomycin was achieved by Keay and Rodrigo.[11a] The main steps of this synthesis are given in Scheme 3.19. Retrosynthetically, the structure of the target **83** was first simplified by the sequential application of two transforms: the Diels–Alder transform to dismantle the **D**–**E** ring system and the Friedel–Crafts acylation transform to disconnect ring **B**. Previous experience led the authors to suggest the utilization of an isobenzofuran moiety as the diene component. Accordingly, the subtarget **84** was identified as the immediate precursor for the preparation of **83**. Synthesis of **84** was envisioned *via* the intermolecular coupling of two polysubstituted aromatic derivatives, **85** and **86**. The preparation of these components from available aromatic compounds involved multi-step transformations but otherwise turned out to be rather efficient. This synthesis represents a fine example of the strategy based upon identifying a set of strategic bonds to assemble using reliable synthetic methods at key steps.

An entirely different approach in the retrosynthetic analysis of **83** was developed by Kelly and Ghoshal.[11b] Consideration of the ring system of **83** as a whole enabled the authors to recognize the close correlation between the substitution pattern of the bottom **ABC** rings of the TGT molecule and that present in an available anthraquinone, emodine **87a**. This striking structural resemblance prompted a retrosynthetic plan based upon severing the 'upper' and 'lower' parts of the structure, and led to structures **87a** and **88a** as the most likely building blocks (Scheme 3.20). There was no convincing evidence attesting to the *a priori* viability of the reverse step: of joining together the 'upper' and 'lower' parts. However, at the same time, the strategic advantage of such a dissection was so obvious and attractive that it was undoubtedly a worthy goal to attempt to identify the promising precursors and to elaborate reaction conditions applicable for the desired coupling. Solution of these problems required a lot of effort but as a final reward a truly outstanding synthesis of **83** was devised, as shown in Scheme 3.20.

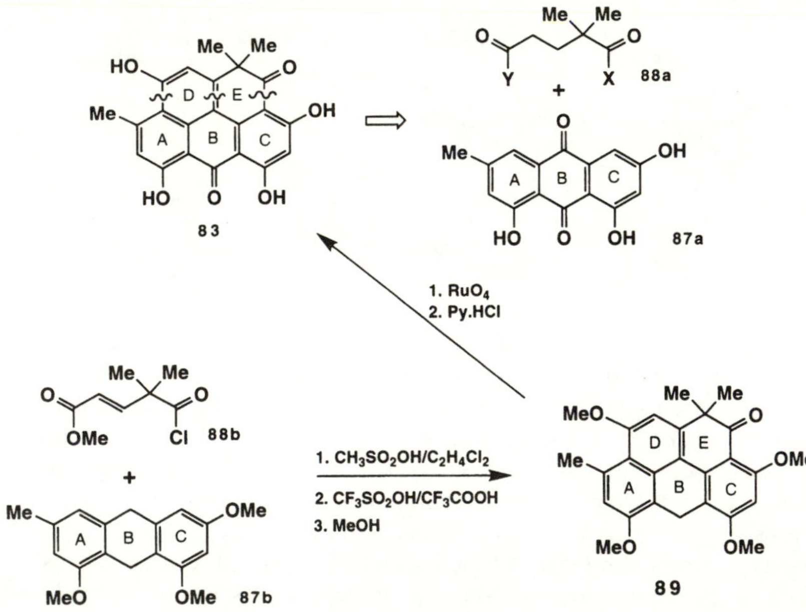

Scheme 3.20

The tricyclic precursor **87b** was easily obtained from **87a** in two steps in an overall 90% yield. The condensation of **87b** and **88b** *via* a Friedel–Crafts reaction, followed by an intramolecular Michael addition and oxidation, led to a one-pot formation of the intermediate pentacyclic product **89** in *ca*. 30% yield. The conversion of **89** into **83** was accomplished by standard procedures of oxidation and demethylation.

There is nothing wrong with the stepwise synthesis of **83** as shown in Scheme 3.19. At the same time, however, Kelly's approach (Scheme 3.20) seems incomparably more impressive owing to its laconic brevity and elegance. This example attests to the tremendous gains achieved in a retrosynthetic consideration of the synthetic target structure as an entity as a whole. Such an analysis may bring to one's attention a recognition of structural idiosyncracies which can sometimes lead to truly paradoxical and yet very efficient moves in the retrosynthetic planning. In particular, it may turn out that the greatest simplification of the synthetic task is achieved not *via* the cleavage of C–C bonds but by their retrosynthetic connection! A spectacular illustration of this claim can be found in the area of the synthesis of polycyclopentanoids (polyquinanes).

A number of natural sesquiterpenes like hirsutene **90** or corioline **91** have as their common structural unit a system of linearly fused five-membered rings (Scheme 3.21). The standard pathway of the retrosynthetic analysis of this system involves the search for strategic bonds in one of the rings, **A**, **B**, or **C**, disconnection of which would lead to the simplification of the target molecule and eventually to simple cyclopentane derivatives as available starting materials. As a result, diverse synthetic plans were devised and successfully employed in numerous synthetic studies in this field[12a] (see also the set of syntheses described in Section 2.23.2).

H H A B C H 90

O H A B C H COOH 91

Scheme 3.21

An absolutely different ideology in retrosynthetic analysis was suggested by Mehta.[12b] The author envisioned the linearly fused triquinane framework, present in the obvious subtarget **92** (Scheme 3.22), as a product of the scission of the cyclobutane ring in the pentacyclic caged compound **93**. In other words, the initial step of a retrosynthetic analysis suggested not *scission* but the *formation* of two novel C–C bonds between the opposite carbon atoms of the double bonds of **92**. One of rationales behind choosing the apparently more complex structure **93** as the subtarget was that the preparation of **93** can be readily accomplished. In fact, this caged structure is immediately related to the readily available compounds **95** and **96**, with the help of two reliable retro-synthetic steps *via* the tricyclic compound **94**.[12b]

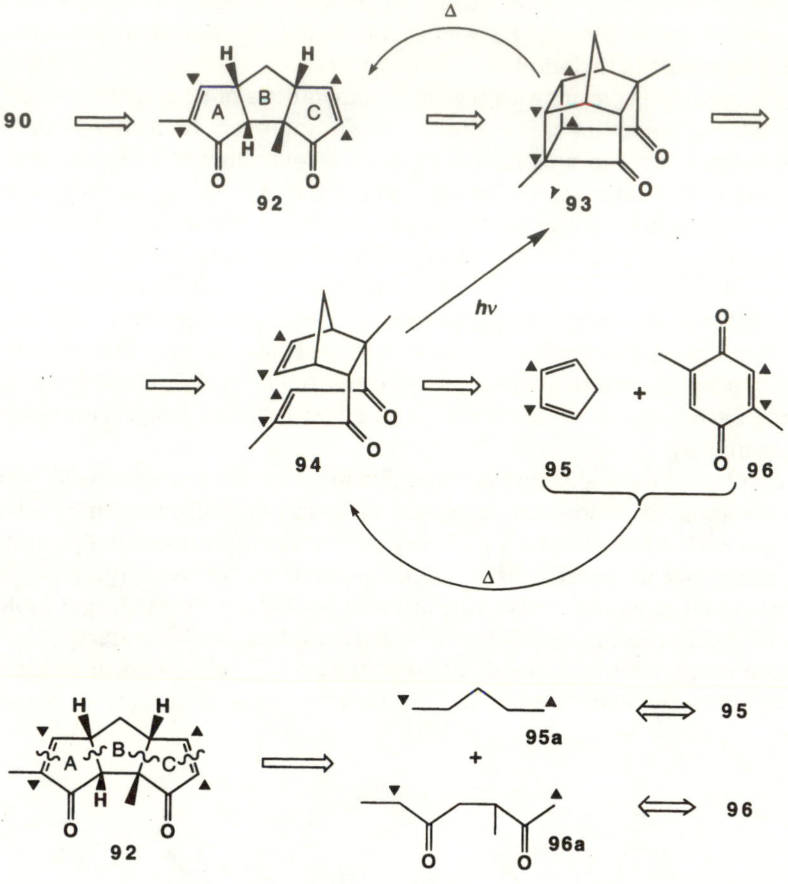

Scheme 3.22

In accordance with this plan, the target tricyclic system **92** was assembled in three reactions: a Diels–Alder cycloaddition of the 2,5-dimethylbenzoquinone **96** and cyclopentadiene **95** to give **94** (quantitative yield); an intramolecular [2 + 2] cycloaddition leading to **93** (85% yield); and finally, a quantitative conversion of the latter intermediate into the desired tricyclic system **92** by a [2 + 2] cycloreversion under thermolysis conditions.[12b,c] Thus the closure of the cyclobutane ring proceeds with the formation of two 'vertical' bonds, **C▲–C▲** and **C▼–C▼**, while its cleavage involves the breakage of its two 'horizontal' bonds, **C▲–C▼**. The net outcome of these two reactions corresponds to the most unusual 'isomerization' of **94** into **92**.

This retrosynthetic analysis corresponds to a somewhat bizarre disconnection of the target structure **92** into a pair of strange C_5 (**95a**) and C_6 (**96a**) synthons, as shown in Scheme 3.22. Yet rather trivial compounds, **95** and **96**, served as the adequate synthetic equivalents of these synthons. Only heat and sunlight were employed as additional 'reagents' in this simple, expedient synthesis of **92** from these components.

Of course, such an effective windfall upon retrosynthetic analysis does not happen very often. Nevertheless, it is generally recommended that the retrosynthetic analysis of polycyclic structures should be directed first to the search of pathways that lead to 'one stroke' framework assemblage. These possibilities can usually be accomplished with the help of cycloaddition reactions. In the course of the initial analysis, special emphasis should be given to identify structural features of the strategic core which might lead to the use of cycloaddition transforms. Let us examine a few more examples that illustrate the fruitfulness of such a retrosynthetic beginning.

The very structure of the natural sesquiterpene α-himachalene **97** (Scheme 3.23) suggests a rather obvious route of disconnection in only two steps: first by transformation into precursor **98** (generation of Diels–Alder retron) and then by applying a Diels–Alder transform into the available acyclic precursor **99**.[13a] The benefits of using a Diels–Alder reaction stem not only from its efficiency as a method of formation of a six-membered moiety within a diverse structural context, but also because of the predictability of its steric outcome. This is especially true for an intramolecular Diels–Alder reaction, which usually leads to a stereospecific formation of a single isomer (as was the case with the synthesis of **98**). It should also be emphasized that attempts to prepare this rather simple bicyclic compound *via* alternative stepwise pathways might prove to be rather cumbersome owing to the necessity to form the seven-membered ring and to control the stereochemistry of ring junctions.

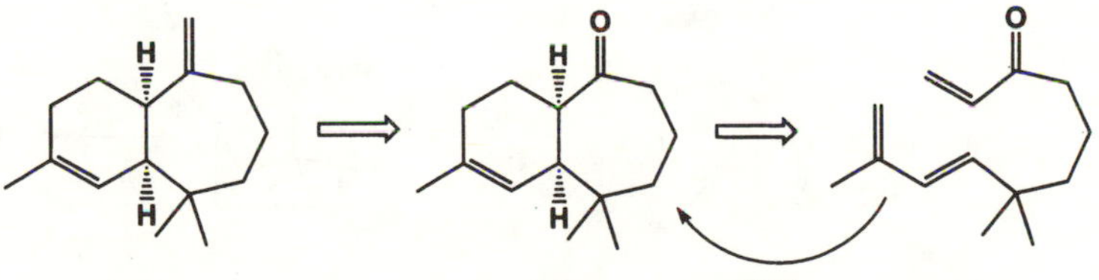

97 98 70% 99

Scheme 3.23

The tricyclic sesquiterpene 9-isocyanopupukaenone **100** (Scheme 3.24) is a toxic component isolated from the skin glands of the mollusk *Phyllida varicosa*. Its framework contains two bridged six-membered rings. There is more than one way to disconnect its core using the Diels–Alder transform. The ambiguity about which route to choose vanishes immediately if one first employs simplifying transforms (functional group transformations, pendant removal) to identify **101** as a strategic core. The latter contains a full retron for a Diels–Alder transform and application of the latter automatically leads to the monocyclic precursor **102**. In White's synthesis of **101**,[13b] shown in Scheme 3.24, the ether

102a was utilized as the substrate for the Diels–Alder cyclization. This key step, **102a** → **101a**, proceeded easily and furnished the desired product in nearly quantitative yield. Thus an efficient assembly of the entire polycyclic core of the target molecule **100** was achieved as the result of a single reaction. In preparing **102a**, advantage was taken of the accessibility of diene **103** in a Birch reduction of *p*-methylanisole. Sequential cyclopropanation of this diene afforded adduct **104**, which was readily converted under mild acid treatment into the diene **105**. Further transformation of the latter into the trienone precursor **102a** was carried out easily using several trivial reactions.

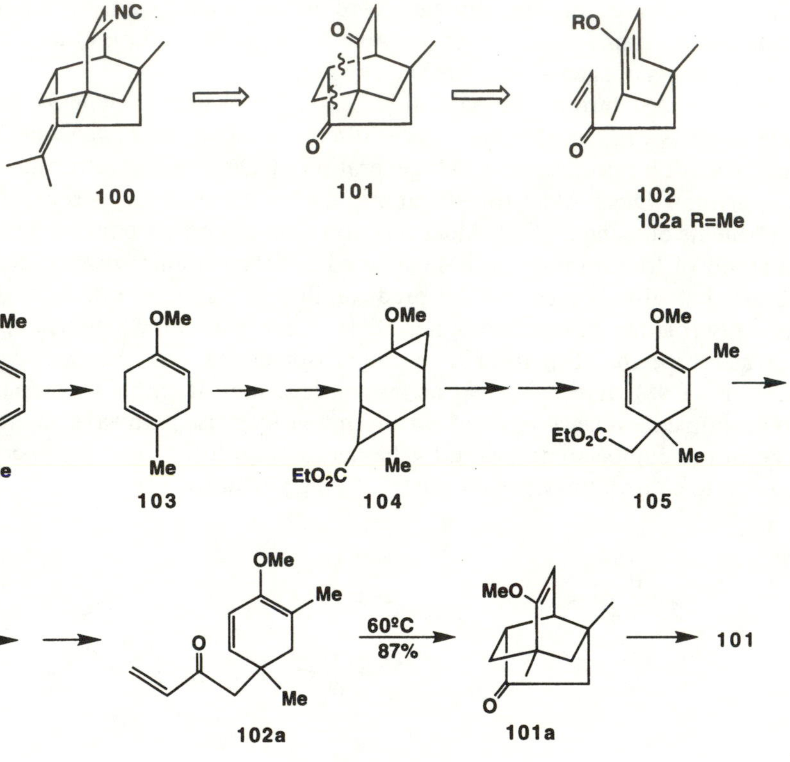

NC

100

101

RO

102
102a R=Me

OMe Me 103 EtO₂C 104 105

OMe Me

60ºC
87%

MeO

102a

101a

101

Scheme 3.24

Of course, not every cyclic structure is directly amenable to construction *via* cycloaddition reactions. Nevertheless, quite often the application of rearrangement transforms to the synthetic target structures may reveal welcome opportunities for the use of certain cycloaddition reactions to construct basic polycyclic frameworks. The efficiency of this approach is well-illustrated by Pirrung's synthesis of isocomene **106** (Scheme 3.25), a sesquiterpene with three angularly fused cyclopentane fragments.[14] At first glance there is nothing in this structure that would suggest any standard cycloaddition reaction as a route of assembly. It is well known, however, that highly strained fused systems of 4,6-membered rings are relatively easy to isomerize into a system of *cis*-pentalanes

(5,5-fused rings). Consequently, a transformation of the 5,5-fused system of the **A–B** or **B–C** rings present in the target molecule **106** into the 4,6-fused system is a viable retrosynthetical operation. The structure of the subgoal **107** generated as a result of this transform (as applied to the **A–B** pair of rings) suggests an appealing opportunity for its ready preparation from the monocyclic derivative **108** with the help of an intramolecular [2 + 2] enone–alkene cycloaddition reaction. These reactions, carbocation rearrangement and intramolecular photocycloaddition, are known to proceed with good stereocontrol. Thus the elaborated strategy provides not only for the creation of the required framework, but also secures the proper stereochemical relationships of the substituents in the target structure. In terms of Corey's retrosynthetical analysis, it meant that the stereocenters present in **106** could have been considered as *clearable* with the application of these transforms.

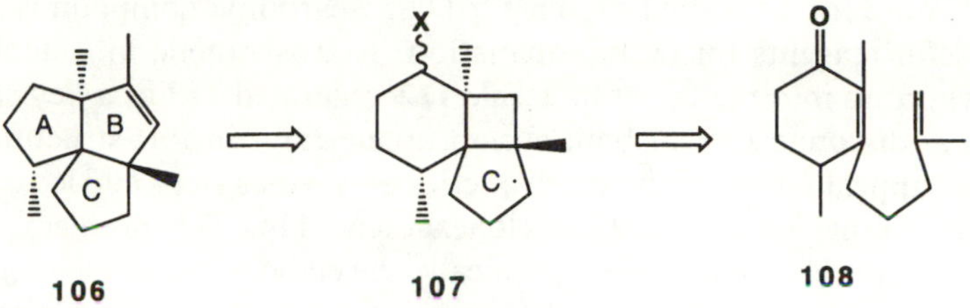

106 **107** **108**

Scheme 3.25

The validity of the conceived retrosynthetic plan was proven in its truly brilliant synthetic implementation. In fact, the synthesis of a rather complicated polycyclic framework containing three adjacent quaternary centers was achieved with record-breaking simplicity and efficiency (Scheme 3.26): in only

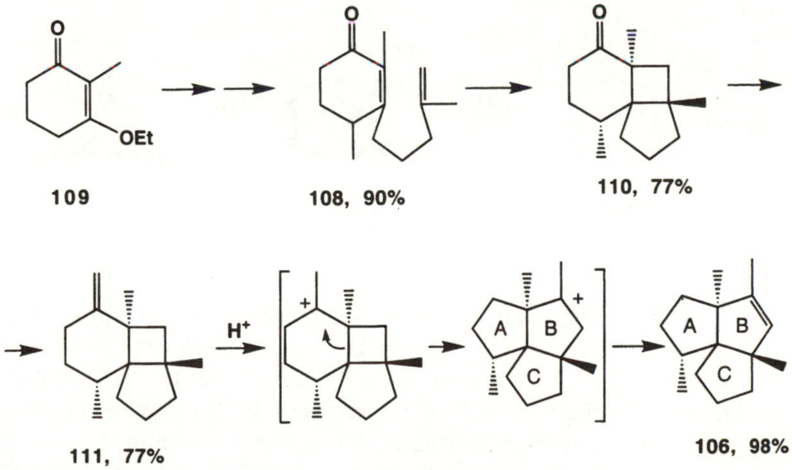

109 **108, 90%** **110, 77%**

111, 77% **106, 98%**

Scheme 3.26

five steps, a seemingly unrelated and rather trivial starting material, the enol ether of 2-methyldihydroresorcinol **109**, was transformed into isocomene **106** with an overall yield of 48%! In accordance with expectations, both transformations of **108** → **110** and **111** → **106** proceeded with high stereoselectivity.

The use of a variety of options of cycloaddition reactions makes possible the numerous efficient syntheses of such molecular structures as asterane, cubane, pentaprismane, basketane, dodecahedrane, *etc.* Some have been mentioned in the previous chapter. Here, it seems appropriate to examine the synthesis of asterane **112**, a rather peculiar molecular construction composed of three cyclohexane fragments locked in boat conformations (Scheme 3.27). It is fairly obvious that attempts to simplify this molecule through the disconnection of any single bond would complicate the synthetic task. On the other hand, the presence of a pair of cyclopropane fragments in the framework of **112** prompted the application of two consecutive [2 + 1] cycloaddition transforms as the most rational route for its simplification.[15] α-Diazo carbonyl compounds are especially useful reagents for cyclopropanation. Retrosynthetic introduction of a carbonyl group into the target molecule **112** generated **113** as a key subtarget. Sequential disconnection of both cyclopropane rings in this structure looked especially appealing since it involved a series of obvious steps and led eventually to a very simple precursor, 1,4-cyclohexadiene **114**. The problem was now reduced to a purely tactical task, specifically the elaboration of the reagents and conditions suitable for the sequential inter- and intramolecular cyclopropanations of **114** as outlined in Scheme 3.27. All reactions proceeded with acceptable yields.

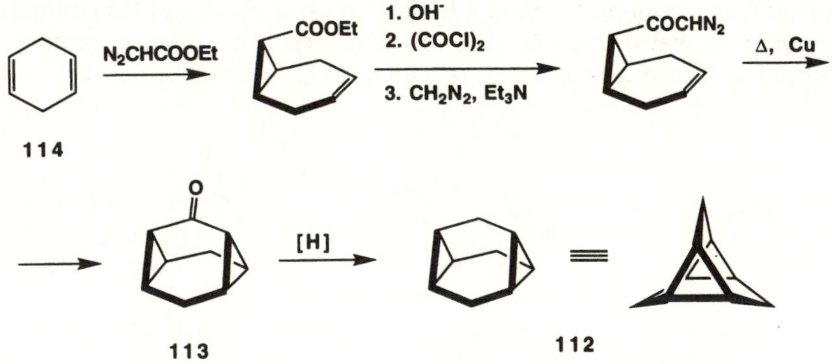

114 → ($N_2CHCOOEt$) → COOEt → (1. OH^-; 2. $(COCl)_2$; 3. CH_2N_2, Et_3N) → $COCHN_2$ → (Δ, Cu) → 113 → ([H]) → 112

Scheme 3.27

We do not want, however, to leave the reader with the impression that cycloadditions are some sort of ‘golden key’ that unlock the pathway to the creation of almost any cyclic system. Even reactions from this powerful arsenal can misfire and an otherwise brilliant retrosynthetic idea remains just that if it cannot be translated into a real synthesis. An instructive example is the exquisite proposal suggested by Woodward for the synthesis of dodecahedrane **115**.[16a] The idea of this synthesis corresponds to a symmetrical dissection of polyhedrane **115** to produce two identical units possessing the structure of the tricyclic triene triquinacene **116**. Woodward speculated that a concerted intermolecular cycloaddition of six double bonds might achieve a one-step preparation of **115** *via* the dimerization shown retrosynthetically in Scheme 3.28.

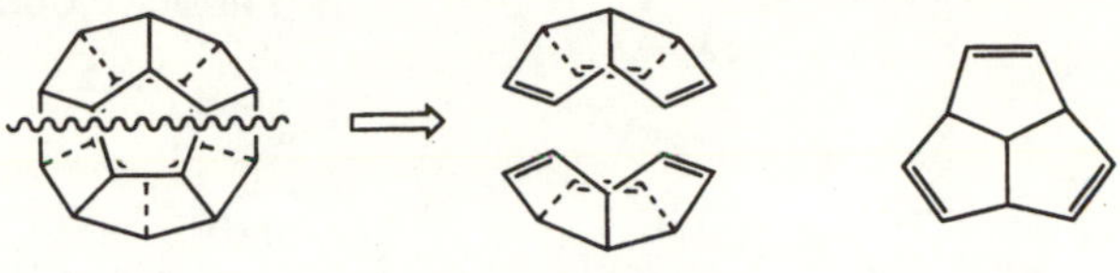

115 **116** **116**

Scheme 3.28

The concept seemed so rational and appealing that the rather difficult 17-step synthesis of **116** was undertaken. It is not difficult to imagine the utter disappointment of the investigators when all efforts to promote this seemingly viable cyclodimerization of **116** failed. While no complete paper was published describing these futile attempts, it is known that **116** did not reveal any tendency to undergo dimerization into **115** even under such forcing conditions as heating at 400 °C under a pressure up to 40 kbar.[16b] Thus the brilliantly conceived idea had to be abandoned. Such failures are often encountered in the practice of organic synthesis and we usually refer to them by saying that 'at times an organic reaction might exhibit a frustratingly low sense of responsibility'. More seriously, this case serves to demonstrate the hazards of ignoring a general strategic rule: never to place a risky, unchecked step at the end of a lengthy synthetic sequence.

Woodward's projected synthesis of dodecahedrane was actually based upon the recognition of the symmetry present in **115**. While this approach failed in this particular instance, considerations of this type might turn out to be extremely useful in the search for a short synthetic route even in cases involving rather complicated structures. The synthesis of tropinone by Robinson is probably the earliest example that illustrates the effectiveness of such an approach. This impressive accomplishment was actually achieved by utilizing symmetrical bifunctional reagents to secure the formation of a symmetrical bicyclic structure as a result of a single chemical operation (see Section 3.2.1).

This symmetry-guided approach was successfully employed by Cook *et al.* for the generation of a spectacularly concise pathway in the preparation of stauratetraen **117**, a tetracyclic hydrocarbon with the peculiar 'fenestrane' framework (Scheme 3.29).[17] The general strategy of this synthesis rests entirely on the high degree of symmetry in this compound. As an immediate precursor of the latter the symmetrical tetraketone **118** was identified. The choice of this subtarget was dictated by the obvious ease of its retrosynthetic conversion into the bicyclic intermediate **119**, which contains a pair of 'lower' rings with an angular substituent bearing two carboxyl groups needed to close the 'upper' rings. The proposed synthesis of the bicyclo[3.3.0]octa-1,5-dione moiety present

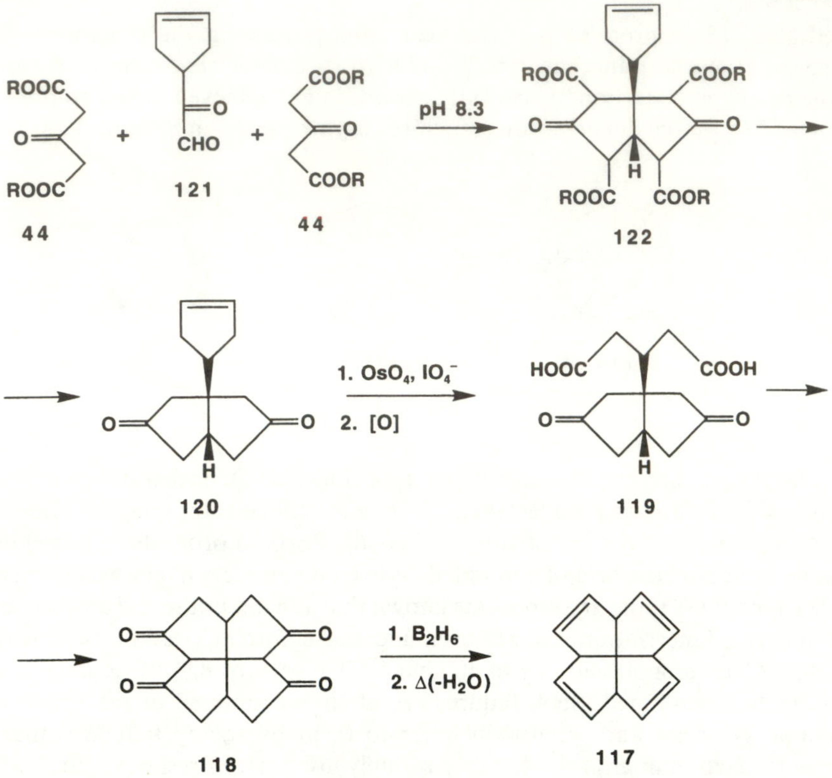

Scheme 3.29

in **119** seemed not to present any problems. However, the procedures initially elaborated for this purpose were based on a sequence of crotonic–Michael addition reactions and therefore could not have been directly utilized to prepare a heavily functionalized **119**. An additional retrosynthetic step (double bond oxidation transform) was required to arrive at the next precursor **120**. Now the stage was set for the swift disconnection of this structure, leading to a pair of available starting materials, namely acetonedicarboxylic ester **44** and keto aldehyde **121**. In fact, the base-catalysed condensation of **121** with two equivalents of **44** proceeded smoothly to give the tetraester **122** in 90% yield. Decarboxylation of the latter furnished the pivotal intermediate **120** in a nearly quantitative yield. Subsequent stages worked equally well and an efficient synthesis of **117** was accomplished.[17]

Taking advantage of the symmetry of the target structure often permits elaborating of the most economical pathways for assembling complicated frameworks. The direct application of a symmetry-guided retrosynthetic search is possible, though, only in the rare cases in which one has symmetrical target molecules. In any case, a thoughtful retrosynthetic analysis of even unsymmetrical structures may result in the disclosure of a 'masked' symmetry which might be made overt by a retrosynthetical introduction or removal of extraneous fragments, skeletal rearrangements, *etc.*

3.2.7 Organization of Synthetic Schemes: Linear *vs.* Convergent Mode

In the previous two sections we divided the principles of a retrosynthetic analysis into two conventional types: 'tactical', in which the molecule is viewed as a summation of bonds to be assembled, and 'strategic', in which the molecule is considered in its entirety. We will now try to examine the general principles for the organization of synthetic schemes.

The simplest and most obvious composition of a synthetic plan is a linear sequence of steps leading to the construction of a target product **P** from the appropriate starting materials, as is shown in a generalized way in Scheme 3.30.

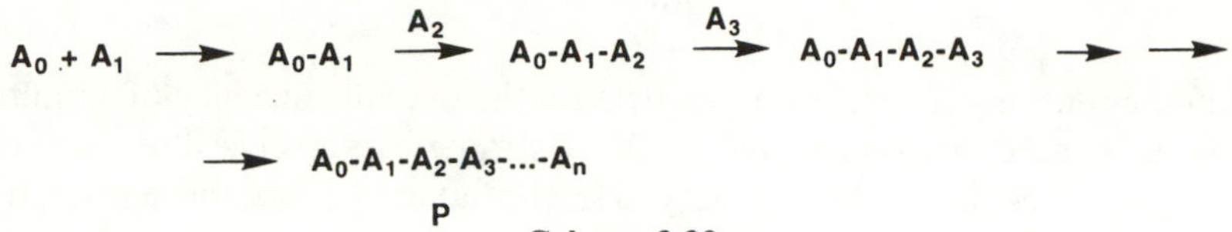

$$A_0 + A_1 \longrightarrow A_0\text{-}A_1 \xrightarrow{A_2} A_0\text{-}A_1\text{-}A_2 \xrightarrow{A_3} A_0\text{-}A_1\text{-}A_2\text{-}A_3 \longrightarrow \longrightarrow$$

$$\longrightarrow \underset{\mathbf{P}}{A_0\text{-}A_1\text{-}A_2\text{-}A_3\text{-}...\text{-}A_n}$$

Scheme 3.30

Almost all of previously discussed syntheses were based on this principle. The 'Achilles' heel' of this approach is the problem of achieving an acceptable overall yield. In consecutive reactions the final yield, of course, will depend upon the yields obtained in the intermediate stages. If the average yield per individual step is designated as Y, then the total yield of the product after the *n*th stage will be $Y_n = Y^n$. If Y is equal to 80%, a very acceptable yield, then the dependence of the overall yield on the number of steps would be as follows:

n	5	10	20	30	50	80
Y_n(%)	33	11	1.2	0.12	1.4×10^{-3}	2.0×10^{-6}

In real life, the overall efficiency of multistep syntheses is usually even worse. Woodward's synthesis of cortisone[1] involved 49 steps and gave the target product in 5×10^{-6}% yield.

Can we accept such a catastrophic drop in the yield? A yield of 2.0×10^{-6}% implies that in order to obtain 20 mg of the final product one must begin with one tonne of starting material. This is the direct and unavoidable result of the detrimental work of the 'arithmetic demon'. To alleviate its effects, all synthetic plans strive to maximize the yields at every stage, as well as to minimize the number of steps required. The first task is directly related to developing a reliable, high-yield synthetic procedure. Careful planning may also reduce the number of such auxiliary steps as protection and deprotection, functional group interconversions, *etc.*, and thus shorten the entire sequence. As a result of these combined efforts, even a rather lengthy linear synthesis may be accomplished with an acceptable total yield.

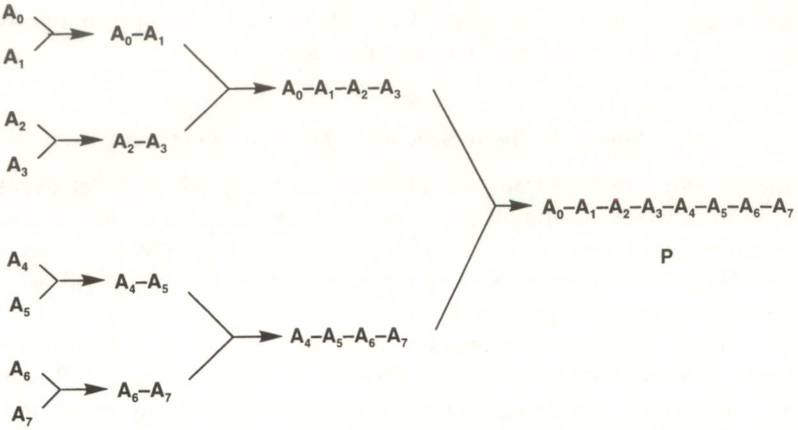

A_0 + A_1 → $A_0–A_1$

A_2 + A_3 → $A_2–A_3$

$A_0–A_1$ + $A_2–A_3$ → $A_0–A_1–A_2–A_3$

A_4 + A_5 → $A_4–A_5$

A_6 + A_7 → $A_6–A_7$

$A_4–A_5$ + $A_6–A_7$ → $A_4–A_5–A_6–A_7$

$A_0–A_1–A_2–A_3$ + $A_4–A_5–A_6–A_7$ → $A_0–A_1–A_2–A_3–A_4–A_5–A_6–A_7$ **P**

Scheme 3.31

Another entirely different way to increase the overall efficiency of a multistep synthesis is based on a convergent synthetic strategy; its essence illustrated in the model given in Scheme 3.31. It is easy to show that in this case the dependence of the overall yield of the final product upon the total number of steps is expressed by the equation $Y_n = Y^{\log_2 n}$. If we again assume Y = 80%, as the average yield per step, one can readily compare the relative effectiveness of the linear *versus* fully convergent routes for the assembling product **P** from the same fragments, A_i:

n	8	16	32	64	80
Y_n(%): linear scheme	16.8	2.8	0.08	6×10^{-5}	2×10^{-6}
convergent scheme	51.2	41	32.8	26	24.4

Assembling a target composed of 65 building blocks ($n = 64$) looks less than appealing if attempted *via* a linear route, as this route is virtually blocked by the 'arithmetic demon'. At the same time, though, the task does not look prohibitively complicated if a convergent route for its synthesis is feasible.

In addition to opening a way to 'defeat the arithmetic demon', the convergent route offers some additional benefits. First of all, convergent schemes are generally much more reliable. In fact, the failure of any single step of the convergent scheme does not invalidate the chosen pathway as a whole. It only indicates the necessity of surmounting this difficulty by finding a detour at some local point. On the other hand, the failure of one step of a linear approach may require a revision of the whole plan. Furthermore, in contrast to a linear approach, the question of compatibility between interfering functional groups is less likely to be a problem as the fragments bearing these groups can be treated on different branches of the synthetic tree which are joined in later steps. Owing to these features, the convergent scheme is also more suitable for the synthesis of a series of structural analogs. Finally, the very nature of these schemes provides an opportunity for simultaneous and independent studies aimed at elucidating the viability of all the entries ('branches') that lead to the point of convergency. Hence the progress of the whole project may be greatly enhanced.

The above comparison is, of course, idealized and one rarely encounters a total synthesis that is purely linear or purely convergent. The underlying principles, however, remain valid. In fact, there is a plethora of examples showing the enormous strategic advantages gained by the incorporation of even a single convergency point in a synthetic scheme. Several of these examples will be considered below.

The structure of prostaglandins, as exemplified by PGE_2 **123**, appears to be as if especially designed for the exploitation of convergent pathways for their synthesis. The presence of the cyclopentanone core bearing two acyclic appendages at C_α and C_β suggests the retrosynthetic removal of these groups to yield the respective cyclopentenone precursor, since the reverse coupling may be recognized as an example of the well-precedented tandem organocuprate–electrophile addition (path **A**, Scheme 3.32). Numerous syntheses of prostaglandins were carried out in accordance with this general strategic plan.[18a] However, in many cases, direct implementation was thwarted owing to the ease of equilibration of the intermediate enolate under the conditions required for the final step of the electrophilic alkylation (path **B**, Scheme 3.32). Several approaches were conceived to obviate this problem.[18b–d] One of the most efficient was elaborated by Johnson's group.[18d] The authors reasoned that the undesirable enolate equilibration and subsequent alkoxy group elimination could be largely suppressed by introducing an additional oxygen group into the cyclopentenone precursor. It was shown that enone **124** easily underwent a sequential addition of the stabilized cuprate nucleophile **125** and electrophile **126** to give the final product **127** in 64% yield. Further transformation of **127** into the target compound **123** required only two simple steps.

Besides its succinctness, there are additional advantages to utilizing such a convergent route. First of all, the convergent strategy allows one to dissect a rather complicated overall problem (such as that in the creation of the polyfunctional molecule **123**) into a set of smaller sub-problems. Each of the sub-problems, for example the preparation of the appropriate cyclopentenone derivative (*e.g.* **124**) and the selection of the reagents required for the introduction of two aliphatic pendants, in its own right are much simpler to solve. Moreover, method optimization studies can be conducted in non-interfering studies. Complications that may arise at the final point of convergency were early identified as the main obstacle toward the use of this strategy for PG synthesis. Yet these complications actually referred to a single step in an otherwise well-studied reaction sequence. The obvious strategic gains of overall synthetic plans warranted more thorough investigations of the factors affecting the course of the key reaction at the convergency point. These studies led, finally, to the elaboration of the optimal conditions for the desired tandem nucleophilic addition–electrophile trapping sequence.[18a–e] As a result, a general synthetic protocol was created which was shown to be applicable for the preparation of natural PGs and their analogs bearing fairly different functionalized substituents at C_α and/or C_β positions.

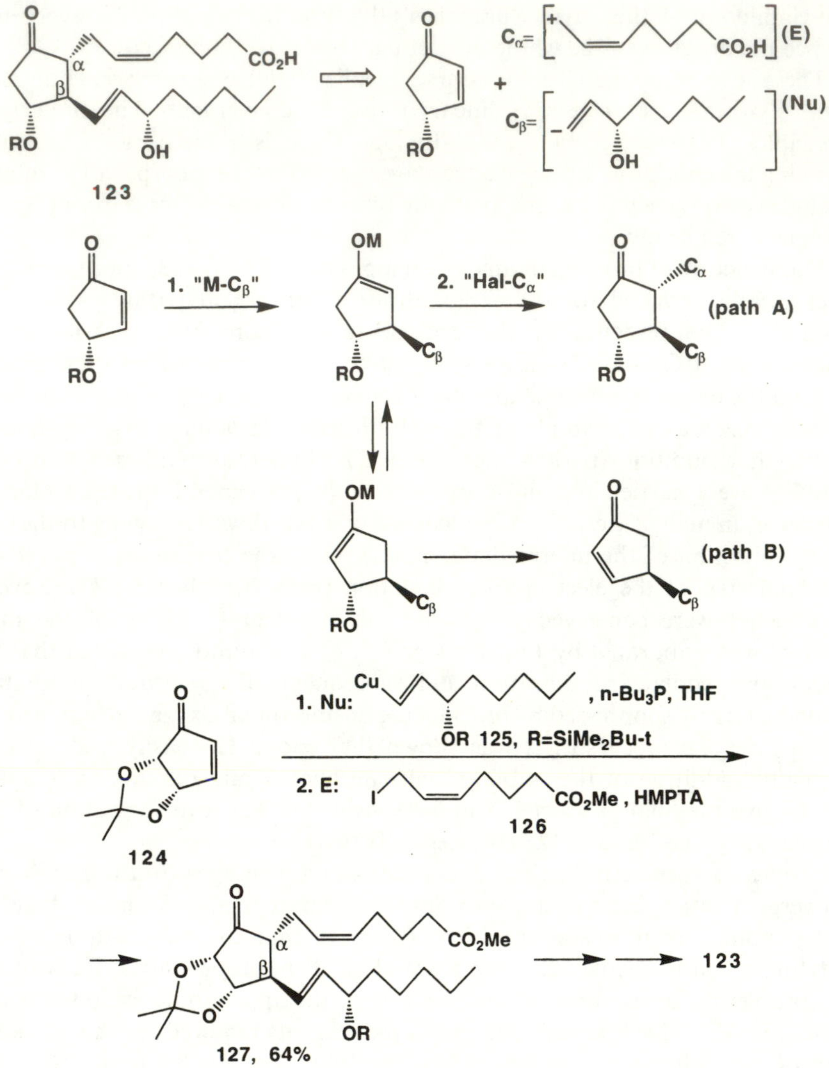

Scheme 3.32

The idea of convergency undoubtedly constitutes one of the basic strategic principles of contemporary organic synthesis. It can be stated without too great an exaggeration that the wide utilization of this principle was responsible, to a substantial extent, for the stunning successes achieved in the syntheses of various complex organic compounds during the last decades. Moreover, the need to execute convergent schemes served as a powerful motivation for the development of novel synthetic methods specifically designed to achieve multi-component one-pot coupling reactions.[19]

It is instructive to follow the evolution of the strategic principles with an example of the total synthesis of estrone **128** (Scheme 3.33), as it can be regarded historically as one of the 'trial cases' used to evaluate the effectiveness of various synthetic approaches.[20a] The first total synthesis of estrone, reported in 1948 by Anner and Mischner, involved 18 steps with an overall yield of 0.1% (based on the starting material, *m*-bromoanisole). In 1958, Johnson's group accomplished the synthesis in 10 steps with a 4.2% yield based on 2-methoxytetralone **129**. Finally, in 1965, Torgov and co-workers carried out a six-step synthesis, also based on **129**, but in 25% yield.

How was it possible to achieve such a remarkable improvement in the efficiency of estrone synthesis? Certainly an important role was played by the efficient solution of tactical problems such as increasing the yields at some steps, selecting optimal synthetic methods applicable for a given transformation, optimizing the choice of starting compounds, reducing the number of auxiliary steps, *etc*. However, the most important factor was undoubtedly the change from a linear strategy, the stepwise grafting of the tetracyclic skeleton **ABCD** of estrone in the sequence **A** → **AB** → **ABC** → **ABCD**, to the convergent assemblage of the fragments **AB** and **D** (**130** + **131** → **132**) with the subsequent formation of ring **C**. This, in fact, was the approach successfully employed by Torgov's group (Scheme 3.33).[20b]

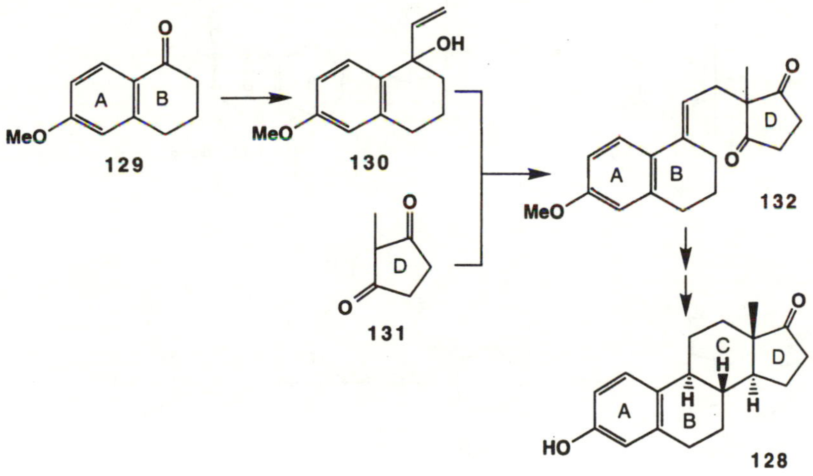

Scheme 3.33

In its day, Torgov's synthesis was recognized as being one of the most successful, as it had surpassed all previous achievements in estrone preparation. Further improvements in the solution of this task were deemed unlikely. Nevertheless, Torgov's record lasted only for about 15 years, after which no less than half a dozen highly efficient total syntheses of **128** and its analogs were reported. Why would there be this continued interest toward this seemingly 'solved' problem? The most general answer is that the newer syntheses are based on entirely different strategies and led to major improvements that are significant for the entire field. More specific comments will be given below upon examination of several representative examples.

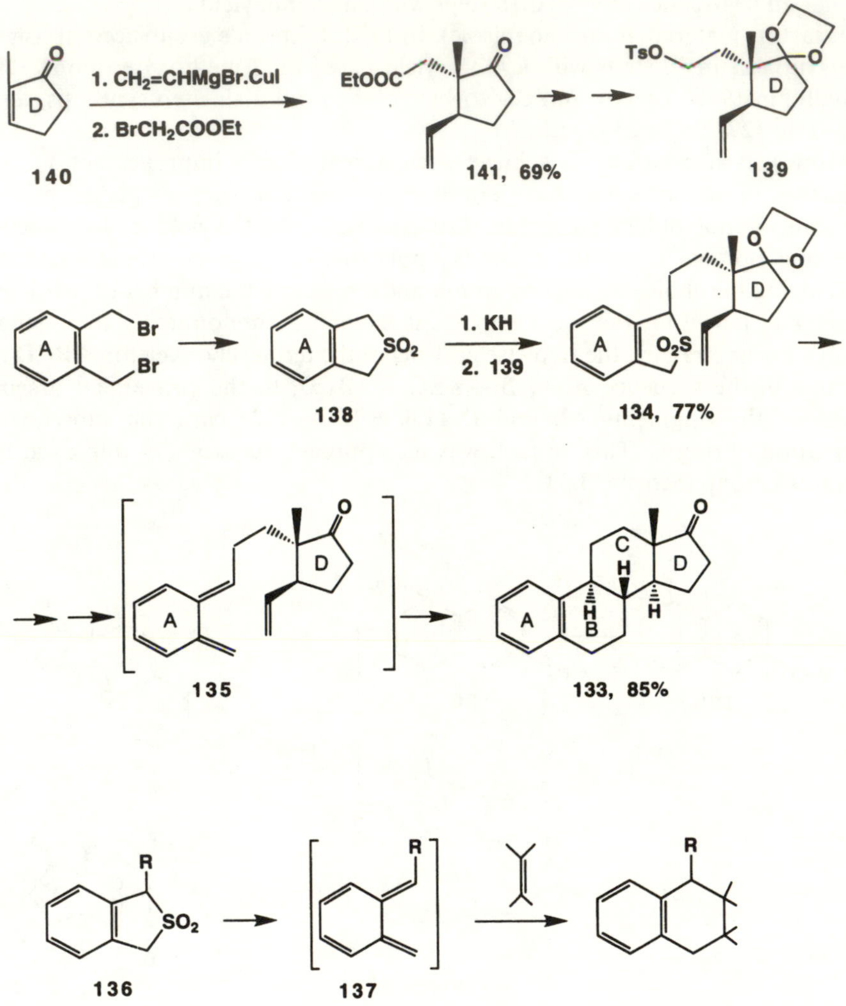

Scheme 3.34

In 1980, Nicolaou *et al.* completed the synthesis of estra-1,3,5(10)-trien-17-one **133**, using the approach outlined in Scheme 3.34.[20c] The unprecendented novelty of the strategic idea behind this synthesis is the simultaneous assemblage of the rings **B** and **C** from the precursor **134**, which contains the rings **A** and **D**. In a retrosynthetical sense it means that ring **B** in target **133** was considered as a cyclohexene unit in spite of the fact that its double bond is part of the aromatic ring **A**. Such a perception permits one to visualize the formation of ring **B** *via* an intramolecular Diels–Alder reaction. It might also be argued that the approach would be non-productive because the required diene fragment is found in the exotic form of the intrinsically unstable precursor *o*-quinodimethane **135**. This idea was basically sound, however, because it was well known that the thermal elimination of SO_2 from sulfones, such as **136**, leads to the formation of *o*-quinomethanes **137** which, in spite of their extreme instability, can be trapped at the moment of their formation as Diels–Alder adducts. In this manner, the completely stable compound, sulfone **134**, could be used as the synthetic equivalent to the Diels–Alder substrate **135**.

The very structure of the sulfone **134** dictated a convergent scheme for its preparation from the easily synthesized $\mathbf{C_8}$, **138**, and $\mathbf{C_{10}}$, **139**, building blocks. Furthermore, the structure of **139** also suggested the utilization of a convergent route for its preparation, with the help of a tandem Michael addition/electrophilic alkylation sequence from cyclopentenone **140** *via* adduct **141**.

Several of these tactical steps had to be carefully 'fine tuned' on model systems, but these additional efforts paid off when the entire synthesis was accomplished in seven steps in 29% overall yield. The general scheme of assemblage of the tetracyclic $\mathbf{C_{18}}$ framework from four small fragments is outlined in Scheme 3.35.

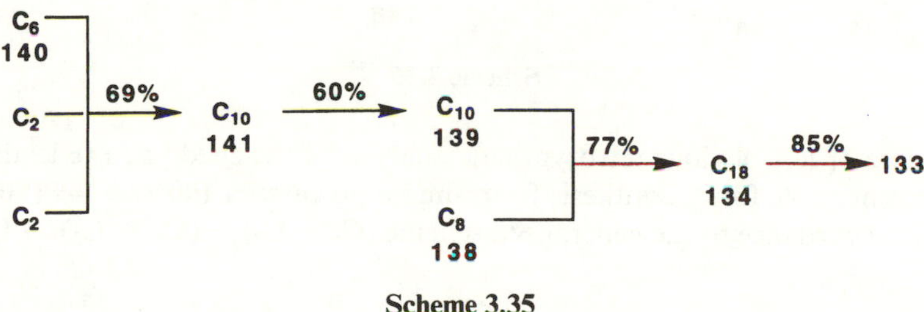

Scheme 3.35

Perhaps even more unusual is Vollhardt's strategic approach for the construction of the steroid tetracyclic framework.[20d] Here again the staring point in the retrosynthetic analysis implied the dismantling of rings **B** and **C** of the target structure **142** *via* a reverse Diels–Alder reaction to give **143** (Scheme 3.36). However, in the role of a stable compound synthetically equivalent to **143**, the investigators selected the benzocyclobutene derivative **144**, as it was well known that the thermolysis of such a system leads also to an *o*-quinodimethane. At first glance this selection might seem to be only a tactical change from the previous solution. It has, however, serious strategic ramifications as it opened the possibility of a non-trivial disconnection of ring **A** of **144** into the alkyne **145** and the enediyne **146**. The insight for this approach was based on previous studies of Vollhardt's group which led to the elaboration of a fairly general method of cobalt-catalysed cyclo-cotrimerization of alkynes as shown by the model **145** + **147** → **148**.

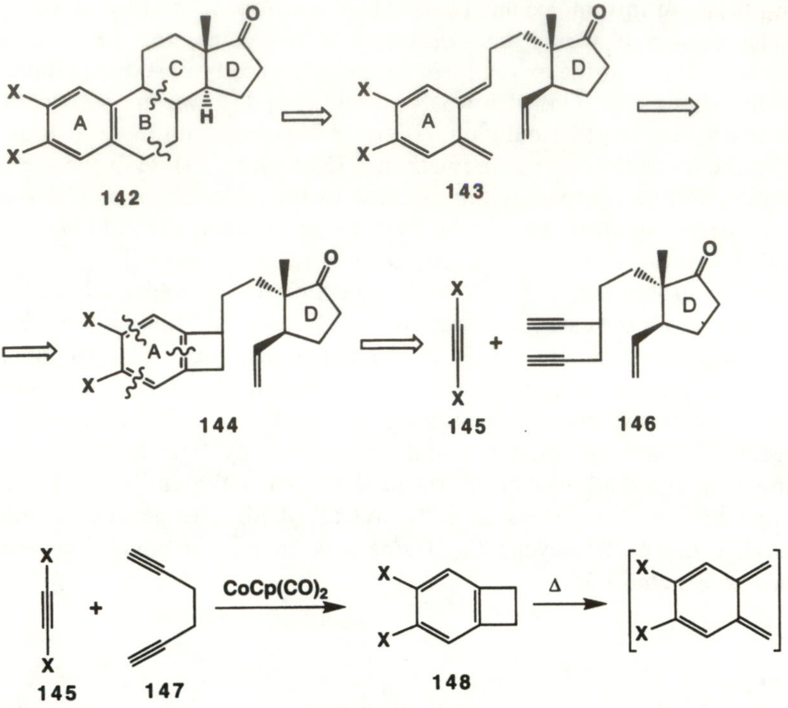

142 143

144 145 146

145 147 $CoCp(CO)_2$ 148 Δ

Scheme 3.36

A more or less obvious retrosynthetic analysis of the enediyne **146** leads to a convergent route for its synthesis from simple precursors **140** and **149** (Scheme 3.37) in accordance to the generalized scheme ($\mathbf{C_6} + \mathbf{C_2}$) + ($\mathbf{C_6} + \mathbf{C_2}$) → $\mathbf{C_{16}}$.

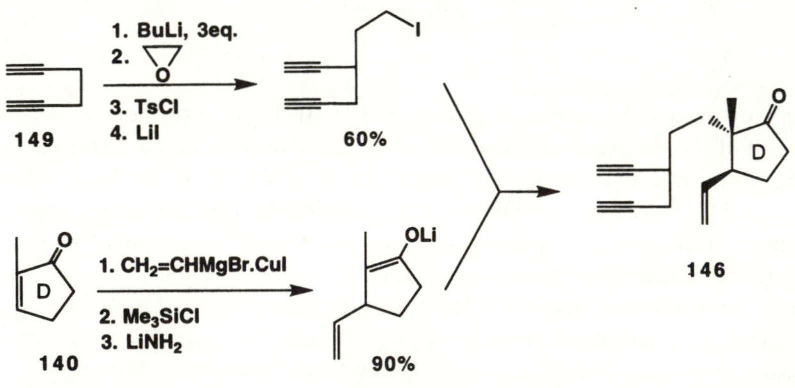

149 1. BuLi, 3eq. 2. (ethylene oxide) 3. TsCl 4. LiI 60%

140 1. CH_2=CHMgBr.CuI 2. Me_3SiCl 3. $LiNH_2$ 90%

146

Scheme 3.37

The next three reactions (Scheme 3.38), cotrimerization of enediyne **146** with bis(trimethylsilyl)acetylene **150**, thermolysis of the benzocyclobutene derivative **151**, and the intramolecular Diels–Alder reaction of the intermediate **152**, were accomplished in one operation upon warming components **146** and **150** in the presence of a cobalt catalyst, in a yield of 71%.

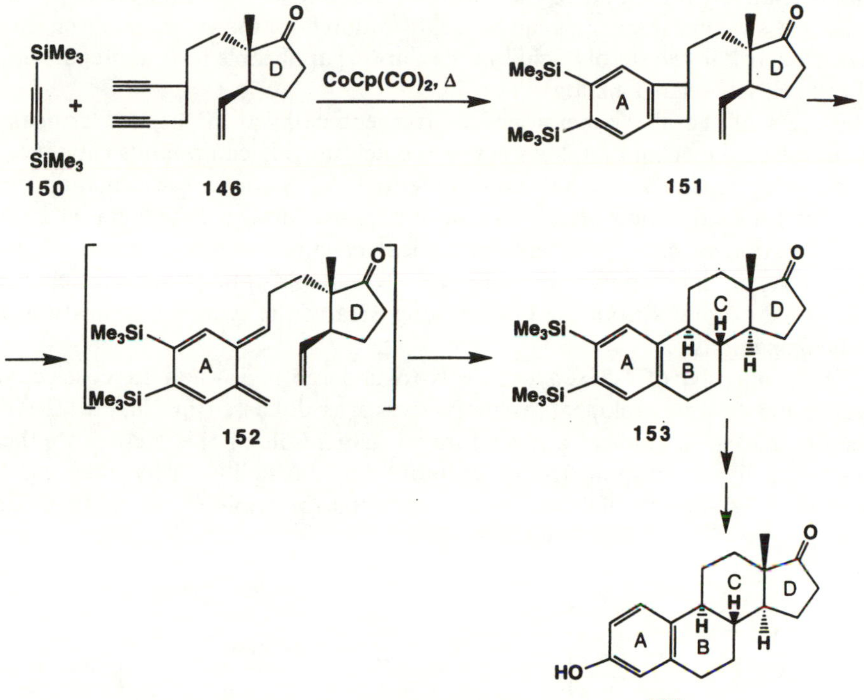

Scheme 3.38

The net result of this single operation is the assemblage of three rings (**A**, **B**, and **C**) in 'one stroke' with the formation of five new **C–C** bonds! The resulting product **153** was transformed *via* a sequence of trivial steps to estrone **128**, which was obtained in an overall yield of 21.5% in five steps (from **140**). This outstanding achievement can be attributed to both the brilliant retrosynthetic plan and the efficiency of its realization.

There is an exquisite elegance in the last two syntheses of estrone that we have examined. They can also be evaluated from a practical point of view since these syntheses differ sharply from the typical classical scheme for the syntheses of steroids. Thus, instead of the sequential growth of the rings (**C** + **D** + **B** + **A**) as in Woodward's synthesis of steroids or the (**AB** + **D** + **C**) route as used by Torgov in the synthesis of estrone, Nicolaou's scheme involves the simultaneous assemblage of the two rings **B** and **C**, while Vollhardt's synthesis can be visualized as a simultaneous formation of the three rings **A**, **B**, and **C**. In these syntheses, simpler starting materials can be used, and target compounds obtained in higher overall yields when compared to the results of classical linear routes. Of even greater importance is the enhanced generality of elaborated convergent schemes of sterol synthesis. This feature offers numerous advantages for constructing skeletons of a wide variety of steroid derivatives, owing to the fact that the key step in these pathways is a cycloaddition reaction that is relatively insensitive to the nature of attached substituents. On the other hand, classical linear schemes can be ideally tailored for the solution of a specific synthetic task in the steroid field but may not be applicable for the preparation of even closely related analogs.

Success of the above-mentioned convergent pathways depended critically upon the employment of 1,2-enones as the key starting compounds capable of undergoing controlled Michael addition/enolate alkylation. Additional opportunities for elaboration of convergent strategies can also be generated from consideration of entirely different chemical reactions of polyfunctional substrates. As an illustration of the diversity of available options, we have chosen additional examples from the fast-growing area of the synthetic utilization of radical reactions.[21a]

The propensity of 5-hexenyl radicals to undergo regio- and stereoselective cyclization to give cyclopentane derivatives was discussed in Chapter 2 (see Section 2.20). The radical adduct formed as a result of this step can either abstract a hydrogen atom from a suitable donor (usually tin hydride; eq. 1, Scheme 3.39) or interact with another double or triple bond to form an additional **C–C** bond (eq. 2, Scheme 3.39).

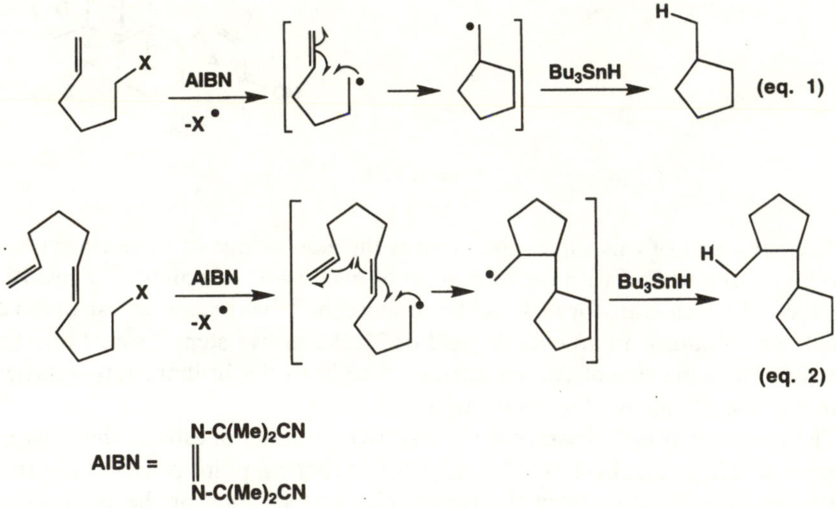

X AIBN $-X^{\bullet}$ Bu$_3$SnH H (eq. 1)

X AIBN $-X^{\bullet}$ Bu$_3$SnH H (eq. 2)

AIBN = N-C(Me)$_2$CN ‖ N-C(Me)$_2$CN

Scheme 3.39

The potential of sequential radical addition as a powerful method to achieve the formation of five-membered rings was fully realized in the 'tandem radical cyclization' strategy devised by Curran for the synthesis of triquinanes.[21a] In the case of linearly fused triquinanes, such as hirsutene **90** (Scheme 3.40), this strategy implies the retrosynthetic disconnection of the tricyclic framework by the application of two sequential radical cyclization transforms at rings **A** and

C, which leads to the biradical synthon **154** and as its equivalent sub-target **155**.[21b] Similar transforms applied to the structure of angularly fused triquinanes like silphiperfol-6-ene **156**, identified sub-target **157**.[21c] Both sub-targets are composed of the central cyclopentene ring **B** furnished with two appropriately positioned appendages, one containing the radical precursor (**C–Hal** bond) and the other a radical acceptor (an unsaturated fragment).

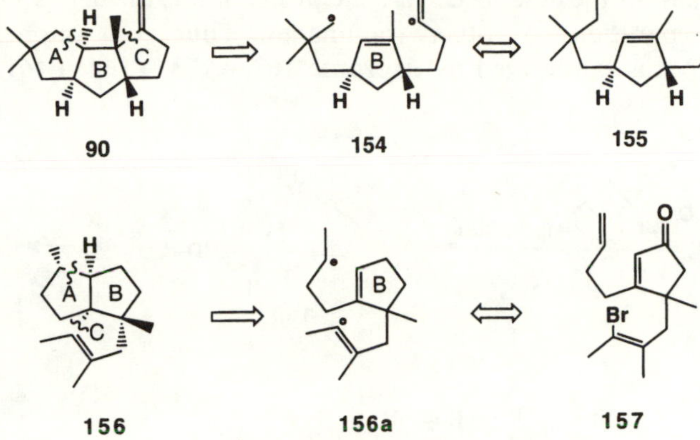

Scheme 3.40

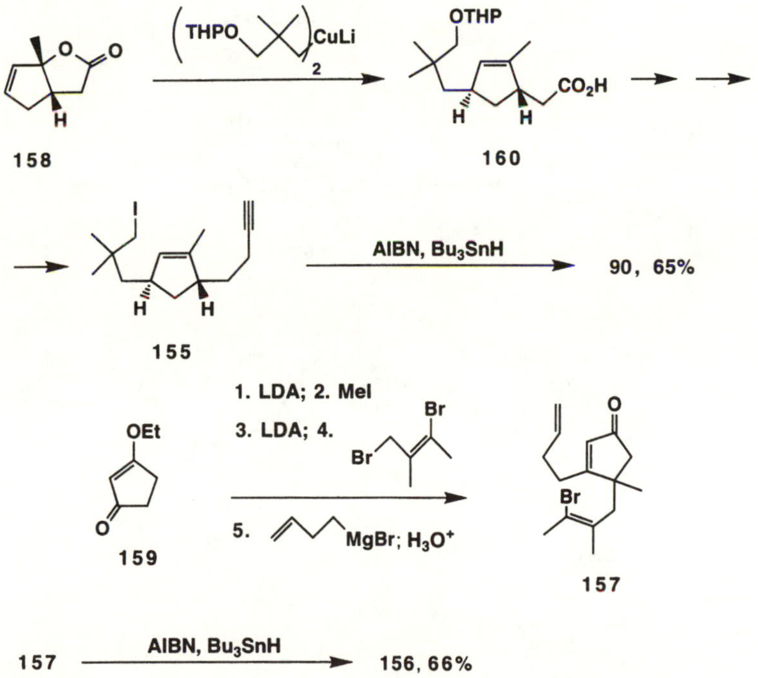

Scheme 3.41

The analysis of the structure of these sub-targets suggested the utilization of multifunctional cyclopentene derivatives such as the bicyclic lactone **158** and enone **159** (Scheme 3.41) as the most plausible precursors for the preparation of **155** and **157**, respectively. The latter preparations involved: (i) the synthesis of the acyclic reagents with a certain set of functionalities and (ii) utilization of these reagents to introduce the required appendage to the cyclopentane core of **158** or **159**, followed, when necessary, by a sequence of trivial transformations.

The usefulness of the described synthetic protocol was further ascertained by successful preparations of other triquinanes. Thus $\Delta^{9(12)}$-capnellene **161** (Scheme 3.42) was synthesized by the route **162** → **163** → **164** → **161**, which is

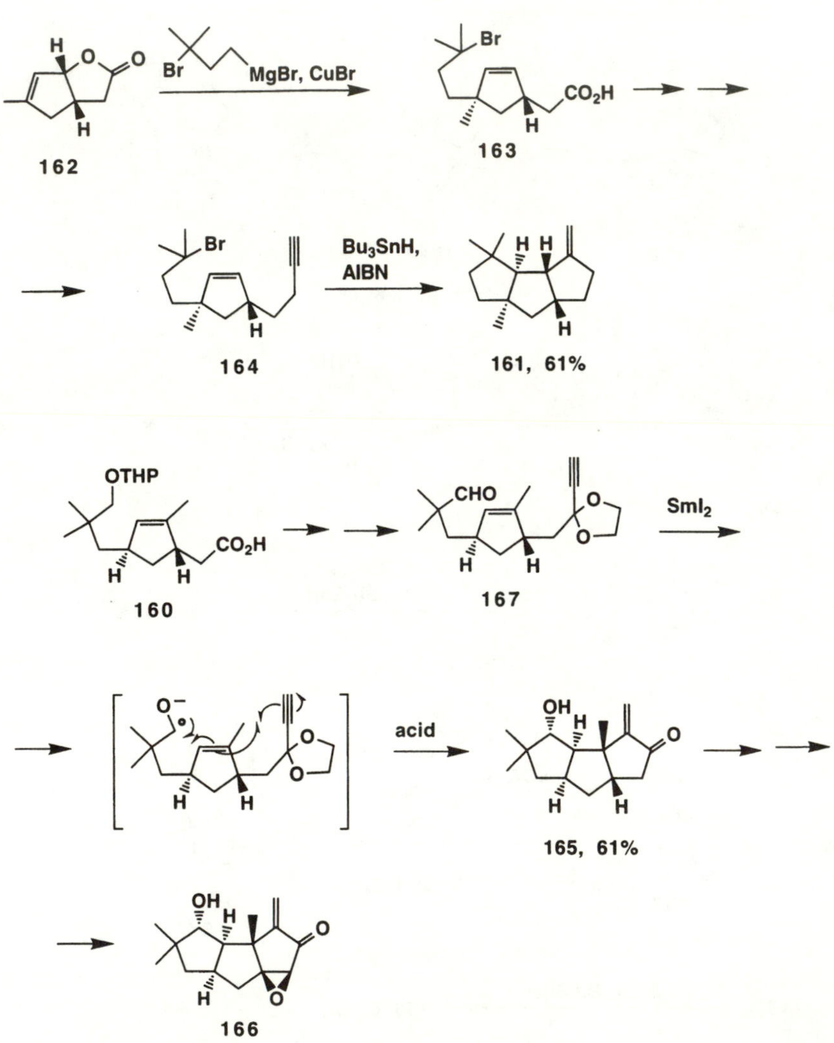

Br MgBr, CuBr

162

163

Bu3SnH, AIBN

164

161, 61%

160

167

SmI2

acid

165, 61%

166

Scheme 3.42

virtually identical with that described for **154** with the only difference being the utilization of the slightly modified starting material **162**.[21d]

More sophistication was required to elaborate the pathway applicable for the synthesis of triquinane **165**, a known precursor for the preparation of hypnofilin **166**.[21e] The presence of a hydroxyl group in ring **A** of **166** dictated the use of a modified substrate for the tandem radical cyclization and an entirely different method for its triggering. A one-electron reduction of the aldehyde carbonyl in **167** by SmI_2 proved to be the method of choice in this case. It is worthwhile to note that adduct **160**, already utilized in the synthesis of **154**, also turned out to be useful as an advanced intermediate for the preparation of **167**.

The synthetic protocol elaborated by Curran's group may serve as an especially instructive example of the benefits inherent to the convergent strategies. In fact, its versatility is secured due to: (i) a well-known low sensitivity of the key ring-forming radical reaction to the presence of functional groups and substitution pattern; (ii) the opportunity to achieve high efficiency by judicious choice of the radical-initiating system; and (iii) an easy preparation of the substrates through the help of reliable methods of coupling comparatively small and readily available fragments. No major changes are required to adjust the described protocol for the synthesis of various triquinanes. Of course, this strategy cannot be considered *universally* applicable, but it has a much broader scope of applicability for the synthesis of structurally diverse polyquinanes than the majority of alternative procedures.

3.3 A FEW GENERAL RECOMMENDATIONS

We have now illustrated, in a few typical examples, some of the actors that go into the planning and execution of a contemporary organic synthesis. One could well ask at this point, 'Is it possible to formulate a set of general rules that may direct the composition of an optimal synthetic plan?' Let us state up front that no such summary of rigorous rules exists that define a sequence of steps to be taken in the solution of some given synthetic task. It is possible, however, to list several *general recommendations* which, while not universal, are unquestionably useful in the planning a synthesis.

1. *Carry out a thorough examination of the target structure as a whole entity.* The main task at this initial stage of planning consists of the analysis of the general problem and sub-problems of the synthesis in order to identify the tasks of strategic importance.

The central problem in the synthesis of acyclic polyunsaturated compounds (for example, many pheromones) is the construction of an aliphatic chain possessing double bonds of the required configuration in specified positions. The strategy of synthesis in this area could be based either at the identification of unsaturated synthons with double bonds in the needed configuration (as was done in the synthesis of the juvenile hormone, see Section 2.8), or the employment of stereoselective reactions in the process of chain construction, such as carbometallation of alkynes (see Section 2.3.3).

In the area of prostaglandin synthesis, the general examination of the target structures immediately discloses the strategic task to be the creation of the cyclopentanone ring containing three or four properly positioned groups with the correct stereochemistry. Consideration of these peculiarities leads to the development of two alternative approaches. In the first of these, the proper orientation of substituents is provided for by a selection of suitable cyclic precursors with a well-defined stereochemistry for the functionalized substituents (see, for example, the use of a norbornene intermediate in Section 2.7.4). The second general route involves a stereoselective two-step addition to a cyclopentenone Michael acceptor (see the preceding Section 3.2.7).

In the case of quadrone (Section 3.2.5), one is pretty much stuck with the complex construction of the fused rings of **A**, **B**, and **C** and elaboration of an applicable strategy centered around this core. On the other hand, the lactone ring **D** is easy to formulate at the latter stages and therefore the presence of this moiety has no strategic ramifications.

Finally, by a thorough examination of the target structure as a *whole molecular construction* and not the sum of individual parts, it is sometimes possible to identify a very effective synthetic pathway such as the short entries into estrone exploited by Vollhardt and Nicolaou or Kelly's resistomycin synthesis.

2. *Select the strategic reaction.* We have already seen many examples that illustrate the use of a strategic reaction or a concerted sequence for the assemblage of the main core of the target structure. In the synthesis of 9-isocyanopupukeanone (Section 3.2.6) it was a Diels–Alder reaction, in helmintosporal (Section 2.80) it was an aldol condensation, in Torgov's estrone synthesis (Section 3.2.7) it was a sequence of reactions for assembling ring **C**, in periplanone (Section 2.8) it was an oxy-Cope rearrangement, in basketane (Section 2.7.4) and in cubane (Section 2.8) it was a combination of [4 + 2] and [2 + 2] cycloadditions, and in the estrone syntheses by Vollhardt and Nicolaou it was the use the *o*-quinodimethane as the 1,3-diene component in a Diels–Alder reaction.

The selection of a strategic reaction automatically dictates both the general composition and sequence of steps of the retrosynthetic analysis. The criteria for the selection of the optimal strategic reaction, generally speaking, can be very different, but convergency is most usually at the top of the priorities list. It is also good to remember that, with all other factors being equal, an intramolecular version of a reaction almost invariably is better than an intermolecular one. That is why this popular avenue is taken so often in contemporary organic syntheses.

3. *Select the strategic bond.* This recommendation is especially important in those numerous cases in which the initial analysis does not identify an efficient strategic reaction and, hence, a sequential bond-by-bond dismantling of the structure becomes obligatory. The first bond in this sequence (the strategic bond) defines all subsequent steps in the retrosynthetic analysis. This approach was illustrated in the syntheses of quadrone (Section 3.2.5).

There are no rigorous rules for the unambiguous selection of the strategic bond. Moreover, no generally applicable and strictly defined criteria for its selection can be formulated as any molecular structure is, in essence, an individual assemblage of atoms. However, several criteria do exist that unquestionably simplify the task when a stepwise disassemblage of the target structure is required.

Of course, it will be expedient to select in the role of a strategic bond the bond that leads to the greatest simplification of structure. For example, with polycyclic systems this usually will mean finding the bond which upon breakage will produce a structure with the least number of side 'pendants', bridged cyclic fragments as well as medium-sized rings. It follows that the most likely candidate for a strategic bond should be looked for among the bonds in the bridged cycles as well as among the bonds to the centers common to several cycles.

For an especially clear-cut example, let us look at how the strategic bond was identified for the structure of twistane **168**, composed of three locked twisted boat fragments. Analysis of possible pathways for the assemblage of this structure was presented in the thoughtful paper of Hamon and Young.[22a] Application of Corey's criteria for the identification of the strategic bond[2a] suggested four options for single bond disconnections affecting the common bonds of the ring system (routes **a–d**, Scheme 3.43). Among these options, the pathway **a** looks especially appealing since it involves the scission of the C–C bond between two common bridgehead atoms and thus immediately converts the doubly bridged framework of **168** into a trivial *cis*-decalin structure (type **A**).

In fact, the first synthesis of twistane, described by Whitlock, was carried out from the precursor having a bicyclo[2.2.2]octane structure (disconnection **B**).[22b] The ring closure leading to the tricyclic system of **168** proceeded efficiently but a rather lengthy sequence of reactions (more than 10 steps) needed for the preparation of the cyclization substrate made this synthesis too cumbersome. A number of other routes had been also employed for the preparation of the twistane system (see Hamon and Young[22a] for the references). Yet among these routes, the pathway corresponding to disconnection **a** turned out to be the most advantageous. Thus the initial studies of Deslongchamp's group revealed that the transformation of the readily available *cis*-decalindione **169** into 4-twistanone **170** can be easily achieved in only three steps *via* a precursor **169a**.[22c] This protocol was further simplified owing to later findings by the same group that the 8-acetoxy-4-twistanone **170a** can be obtained in one step from **169** upon treatment of the latter with boron trifluoride etherate in an AcOH–Ac_2O system.[22d]

Finally, the search for a strategic bond is always simplified when even a 'first glance' analysis leads to the identification of those bonds that clearly cannot be considered strategic. The latter include bonds in aromatic rings (see, however, an exception in Vollhardt's synthesis of estrone) or heteroaromatic rings, as well as bonds which are located in readily available fragments (such as monosaccharides, amino acids, natural fatty acids, *etc.*).

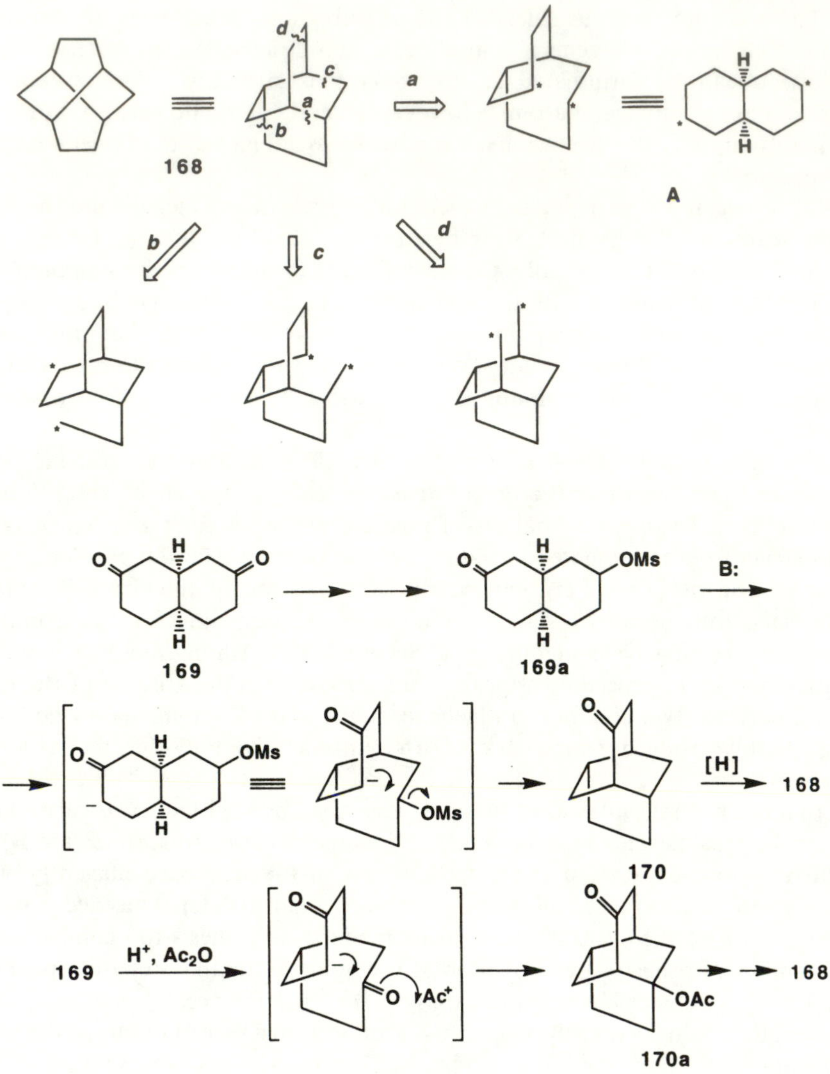

Scheme 3.43

4. *Carry out an initial retrosynthetic processing of the target structure*. This is also a very important step, essential to the above-mentioned procedures. It is mainly targeted at the identification of the shortest pathways for the retrosynthetic modification of the target structure in order to create the functionality pattern (retron) to secure the applicability of the chosen strategic transform. In the case of twistane the choice of enolate alkylation for the creation of bond **a** necessitated a retrosynthetic introduction of a keto group into target **168** to give the required retron in the immediate precursor **170**.

It is possible that preliminary retrosynthetic modifications of the target structure will implicate a few additional transformation steps that lengthen the whole scheme. However, the obvious gains provided by the opportunity to apply an efficient strategic reaction might well outweigh these expenses. It should also be kept in mind that, in the course of this step of the retrosynthetic analysis, not only the initial functional pattern but also the skeleton of the target itself need not be considered as fixed features. In fact, quite the contrary, the analysis of options suggested by various transformations of the basic framework might lead to truly imaginative solutions. An instructive example of a profitable application of this principle can be found in Mehta's synthesis of the tricyclopentanoids (Section 3.2.6). In this case the first stage of the retrosynthetic analysis was the paired crosslinking of the edges of cyclopentene rings that paved the way for a successful application of a well-elaborated sequence of [2 + 4] and [2 + 2] cycloadditions.

Finally, a significant aspect of most preliminary examinations of a target structure involves the retrosynthetic removal of every part of the target molecule that is not essential for the elaboration of the strategic concept. As a result of these pursuits, a strategic core could be identified. Various retrosynthetic operations may be utilized to achieve this goal, as was shown in a number of the examples given above.

The order in which we listed the above recommendations in no way should be considered as the preferential sequence of steps in a retrosynthetical analysis. For all practical purposes they must be considered simultaneously and, furthermore, throughout the entire sequence of dismantling down to the simple starting compounds. Such a systematic analysis will frequently lead to quite a number of possible routes. The selection among them can be dictated by very different reasons but, as we have already emphasized, the route that is tied together by a 'convergent knot' is almost invariably preferred. In general the minimization of the number of steps is a significant aspect of any synthetic plan. In an ideal sequence this implies that the intermediate product prepared at any stage is ready, without additional alterations, to serve as the precursor for the next constructive step.

It goes without saying that the optimal route employs the most effective synthetic methods that ensure the highest yields and greatest selectivity. This consideration includes simplicity of the required operation of separation and purification which, as we have mentioned, can turn out to be the 'Achilles' heel' of an otherwise excellent synthetic scheme.

No doubt the above recommendations are quite obvious and for that matter could have been formulated by a chemist in the 1920s. However, it was only the development of new synthetic methods during the last decades that provided these recommendations with real substance and routine reliability. It is worthwhile to illustrate this point by means of an example.

The Michael addition, a classical reaction in organic synthesis, was utilized for a long while almost exclusively for solving tactical tasks, usually the creation of a given **C–C** bond. The first example attesting to the extreme usefulness of this reaction as a truly strategic method was provided by the Robinson annulation protocol (Section 2.3.3). With further refinement this reaction was developed into a nearly universal method for sequentially introducing **C**-nucleophiles and **C**-electrophiles across the double bond of Michael acceptor. Thus a powerful tool to achieve the assembling of rather complex structures *via* a convergency pathway was created.

A similar metamorphosis was experienced by a number of other well-established tactical reactions, such as the Diels–Alder reaction, [2 + 2] cycloadditions, aldol condensations, *etc*. Of course it was both the inherent potential of these reactions and the requirements of synthetic practice that provided the stimulus to convert these reactions into the category of *strategically important methods*. At present the time lag between the discovery of a novel reaction and finding the way of its most productive exploitation in the field of total synthesis is dramatically shortened and the arsenal of synthetic methods is being permanently enriched.

Recommendations regarding the selection of the strategic bond likewise acquired significance primarily due to the existence of a rich selection of synthons. Their library contains synthons of different polarity with even the most unexpected functional group combinations. The analysis of options emerging due to the retrosynthetic cleavage of almost any bond, regardless of the structural context, might be considered viable if the respective synthons could be identified. It is also the judicious use of this arsenal that reduces the need to carry out functional group transformations and thus minimize the number of required synthetic steps.

The notion of a convergency as such did not become habitual until the mid-1960s.[23] Again this was not because the benefits of the convergency of synthetic pathways could not have been recognized earlier, but it was only at this time that the application of this principle became meaningful because of the development of the appropriate synthetic methods.

It seems obvious that the strategic setting of modern organic synthesis is rather flexible and reflects the state of art in the area of synthetic methods presently available. It should also be evident, therefore, that the utilization of the above recommendations is possible only on the basis of a broad synthetic erudition regardless of whether strategic reactions or functional group transformations are being considered. Likewise the use of the synthon concept requires the in-depth knowledge of main trends and achievements in this area. So this leads to the best recommendation to anyone wishing to be involved in organic synthesis — *read, analyse, and store data from current scientific literature!*

This need to tap a large database of chemical information (one that grows daily) brings us to the final consideration of this chapter. Certainly it would be strange in this day and age if an attempt was not made to utilize an artificial intellect as an aid in composing a synthetic plan. In the next section we will attempt to summarize only briefly the philosophy that is used in this approach.

3.4 THE COMPUTER AS A GUIDE AND ASSISTANT IN RETROSYNTHETIC ANALYSIS

We have already noted on several occasions that the methods of modern organic chemistry permit one 'to transform whatever you wish into anything you want'. Therefore, in principle, a retrosynthetic analysis can be initiated with any arbitrarily selected bond. Consequently, if the memory of a computer contains an exhaustive database on synthetic methods plus a program for the simplification of a structure by the sequential breaking of bonds, it should be possible to generate a multitude of retrosynthetic schemes leading eventually to simple starting materials. In fact the number of the formally correct schemes thus generated might be so large that it would be difficult or even impossible to select the few really sensible routes. A useful computer program, therefore, must have built in some chemical logic that will produce results which have a better chance of being rational. The program also should be able to recognize and terminate those pathways that proceed in an unrealistic direction. It is also obvious that there is no truly compelling reason to carry out the retrosynthetic analysis as a fully automated computer search without the assistance of human intelligence and experience. Several basically different concepts have been devised for the creation of programs for computer-aided elaboration of synthetic pathways.[24]

Initially the idea of computer utilization in the practice of organic synthesis was advanced in Vleduts's publication in 1963.[25] Yet the importance of studies in this area was recognized only after the famous work of Corey and Wipke published in 1969.[2d,26a] Below we will examine some aspects of the methodology elaborated in the studies of this group.

The initial and reasonable step in this development was to make it possible to introduce and retrieve chemical information in a language and format familiar to a chemist so that they can work with the computer in a dialogue regime. The so-called LHASA ('Logic and Heuristic Applied to Synthetic Analysis') program, developed in Corey's group, furthermore allows the chemist to be directly involved at all stages of retrosynthetic analysis. In reality it puts the chemist in the 'driver's seat' so that he/she is able first to specify the general direction of the retrosynthetic search and then to abort this or that route as being unpromising and direct the search along another more prospective direction ('pruning' of the retrosynthetic tree). Owing to this sort of 'computer–chemist interaction' their skills can complement each other. In fact, even an experienced synthetic chemist may overlook this or that not-very-trivial retrosynthetic pathway while the computer is specifically targeted at identifying all pathways satisfying the certain set of criteria. At the same time the chemist has an opportunity to exploit fully his/her experience, intuition, or merely rule-of-thumb to make the final choice of the most promising options for the subsequent retrosynthetic analysis.

Let us consider the general principles of operation of the LHASA program in finding a synthetic pathway to polycyclic system. The program can be aimed by the chemist along two modes of search. The first mode of operation implies that the computer analyses the structure of the strategic core of the target molecule with the goal of identifying a set of strategic bonds and suggesting methods suitable for their creation. This process may be repeated for each of the generated sub-structures all the way back to starting materials. In the second mode the program is oriented toward the search for the options to construct the target structure with the help of one of the most general and effective key synthetic methods (Diels–Alder reaction, Robinson annulation, *etc.*). We will start by examining the first mode.[26b]

The meaning of the term 'strategic bond', as well as the general criteria for the strategic bond selection in the target structure, have already been discussed in Section 3.2.5. To give credit where credit is due, the formalization of the concept of a strategic bond was developed exactly in connection with the use of the computer for planning a synthesis. In fact the LHASA program is much better prepared than a chemist to analyse exhaustively complex polycyclic molecules and to produce a complete selection of strategic bonds corresponding to certain pre-defined criteria. From this set of strategic bonds the LHASA program further proceeds with an analysis of the pathways applicable for disconnection (and hence connection) of each one of these strategic bonds. We will attempt to clarify the chemical logic and basic stages of this search.

To meet the needs of the LHASA program, all reactions leading to the formation of new **C–C** bonds are divided into two categories by formal structural criteria: *[A]*, reactions, the result of which is the formation of a fragment containing only one functional group, and *[B]*, reactions leading to bifunctional fragments. Examples of type *A* include the Grignard and Wittig reactions. Type *B* reactions are represented by the aldol condensation and the Michael reaction. This classification allows the program to select the proper tools of forming bonds based only on structural characteristics of a fragment to be assembled. In the LHASA program the retroreaction of type *A* is designated as 1-GRP and type *B* as 2-GRP operators.

The initial analysis of the distribution of functionality in the examined molecule is carried out with the goal of identifying the presence of one or two functionalities in the neighborhood of the strategic bond. The existence of suitable groups prompts the application of either a 1-GRP or 2-GRP operator that immediately provides a set of possible solutions to the task of disconnection at the given strategic bond. For example, the presence of a carbonyl group adjacent to the strategic bond in the tricyclic model **171** warrants the ready disconnection by a 1-GRP operator (enolate alkylation transform) to give the bicyclic precursor **172**. Similarly, a 2-GRP operator is immediately applicable for the retroaldol transformation of target **173** into **174** owing to the presence of the 1,3-hydroxycarbonyl fragment in **173** (Scheme 3.44).

The program also contains a set of algorithms for functional group interconversions and functional group additions (FGI and FGA operators, respectively). These operators are activated when there is no functionality in the neighborhood of the strategic bond. The application of various types of FGIs and/or FGAs enables the LHASA program to scan all possibilities for changing the functionality in the target structure, in order to identify the options suitable

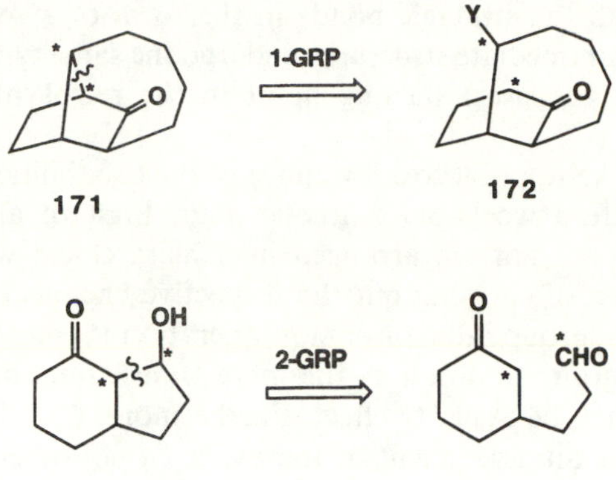

171 1-GRP ⟹ 172

173 2-GRP ⟹ 174

Scheme 3.44

for disconnection of a chosen strategic bond by 1-GRP or 2-GRP operators. For example, simplification of the bicyclic compound **175** (Scheme 3.45) leading to the readily available precursor **176** envisions an initial application of a FGI operator required for the formation of a β-hydroxycarbonyl moiety of the intermediate **177**. Retrosynthetic introduction of a carbonyl group into **178** (FGA operator) is required to secure the opportunity to cleave the indicated strategic bond in **179** to give structure **180** as a possible starting compound.

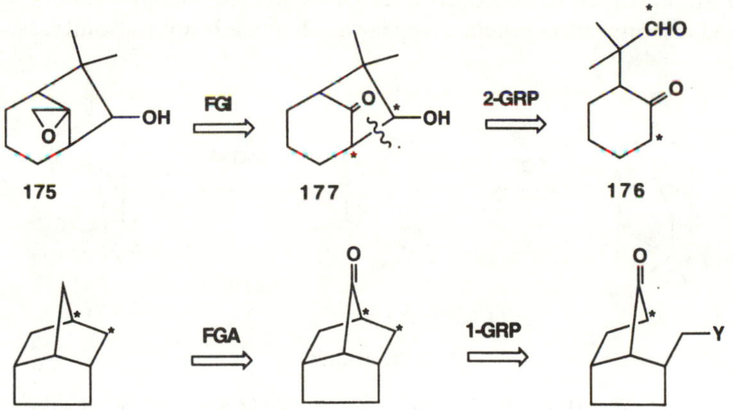

175 FGI ⟹ 177 2-GRP ⟹ 176

178 FGA ⟹ 179 1-GRP ⟹ 180

Scheme 3.45

The program is capable of applying sequentially several functional group interconversion and/or functional group addition operations (up to four steps) in the neighborhood of the selected strategic bond. If the strategic bond still will not yield to disconnection after this examination, the LHASA program switches to an analogous analysis on the next strategic bond. As a final result, the user obtains a complete selection of chemically meaningful variations for the disassemblage of all the strategic bonds in the form of a set of intermediate structures. These intermediate structures undergo the same type of analysis until an acceptable (for the user) starting point in the retrosynthetic network is achieved.

We would also like to point to a few more of the capabilities of this program in helping to create a workable synthetic plan. First of all, the operations employed by the program are arranged in a hierarchical way based on the relative effectiveness of carrying out the respective chemical transformations. Thus the functional group interconversion operation is considered preferential to the functional group addition as the mere transformation of a functional group should, in fact, be easier to effect. Furthermore, for all FGIs and FGAs the program makes an assessment of the expected effectiveness of the corresponding reverse reaction if used in the given structural context. Thus the retrosynthetic analysis of structure **181** along the route shown in Scheme 3.46 implies the use of a FGI operation to install the additional carbonyl group to prepare **182**, which is required for the disconnection of the chosen strategic bond by a 2-GRP operator (Michael transform) to yield, finally, **183**. In the course of this transformation (from intermediate product **182** into target molecule **181**), it is necessary to protect the upper carbonyl group of **182**. Since this entails the lengthening of the synthetic scheme by two steps (introduction and removal of the protection), the program automatically gives the sequence FGI plus 2-GRP a lower evaluation coefficient (rating). At moments like this the program might ask for the chemist's decision about the acceptability of this or that step. All of these control factors enhance the probability that the generated disconnection schemes appear as chemically sound solutions.

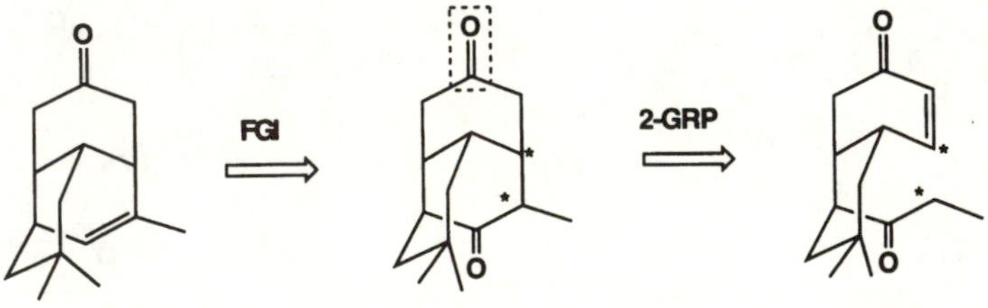

Scheme 3.46

As a check on the effectiveness of the LHASA program, the authors performed an analysis on 14 polycyclic compounds for which a synthesis was already reported in the literature. In 10 of these cases, the LHASA program included in the set of the generated solutions the very synthetic schemes that had been successfully realized.

As demonstrated above, the most efficient reactions in the construction of cyclic structures are those that lead to the formation of several bonds as the result of a single chemical operation. This consideration stimulated the development of an alternative version of the LHASA programs aimed at finding ways of dismantling target structures with the help of especially effective disconnections such as the Diels–Alder or Robinson annulation transforms. In this operational mode, chemists should first specify what particular strategic reaction is to be applied to the creation of the strategic core of the target structure and the program then proceeds to analyse the possible options for the solution of this problem.[26c]

The generic form of the Robinson annulation transform is presented in Scheme 3.47. It follows, therefore, that in order to be amenable to such a transformation the strategic core of the target molecule must contain a six-membered conjugated enone system. This structural moiety should contain at least one atom of hydrogen at **C-6** (for the [4 + 2] option) or at **C-4** (for the [3 + 3] option). In addition, this ring must not contain an electron withdrawing group at **C-5** since its presence would deactivate the respective Michael acceptor.

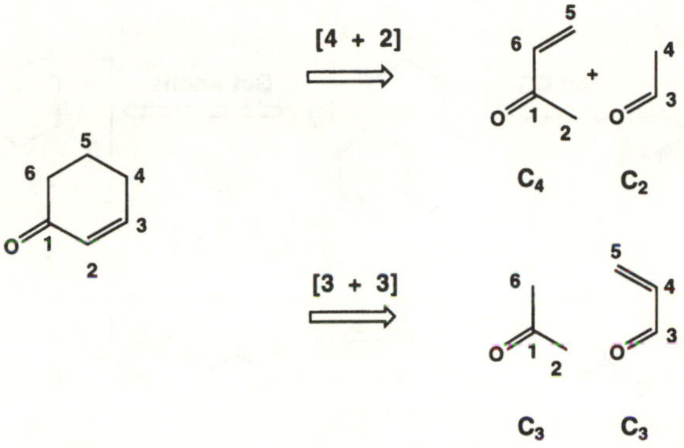

Scheme 3.47

The program starts its work with the arbitrary selection of a cyclohexane fragment and analyses the possibilities for its direct dismantling by a retro-Robinson annulation scheme. If such a disassemblage is feasible, the program proposes it as the first variant of an answer. If not, the program attempts to identify the structural peculiarities of the given ring that hinder application of the Robinson annulation transform. The program then proceeds to use a set of sub-programs (chemistry sub-routines) to remove retrosynthetically from the target structure elements interfering with the retro-Robinson annulation and to introduce fragments that facilitate its realization.

Among these sub-routines are those that dealkylate, introduce a double bond ('get DB'), introduce a carbonyl group (get C=O), *etc.* (Scheme 3.48). To achieve these results, a number of various methods or their sequence may be tried. For example, several dozen transforms are available and up to five consecutive chemical steps could be tried for the retrosynthetic installation of the ketone group ('get C=O').

The process of retrosynthetic matching of a target structure into a sub-goal containing the conjugated cyclohexenone fragment (thus amenable to retro-Robinson annulation simplification) involves the utilization of a combination of

Procedure 1:

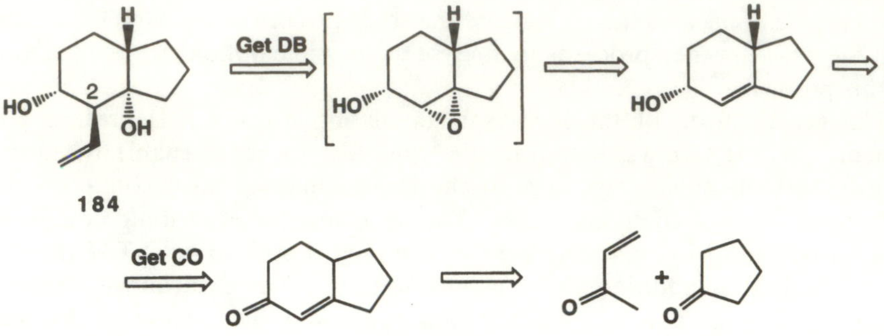
H
HO
2
OH
Get DB
O
184
Get CO
O
+

Procedure 2:

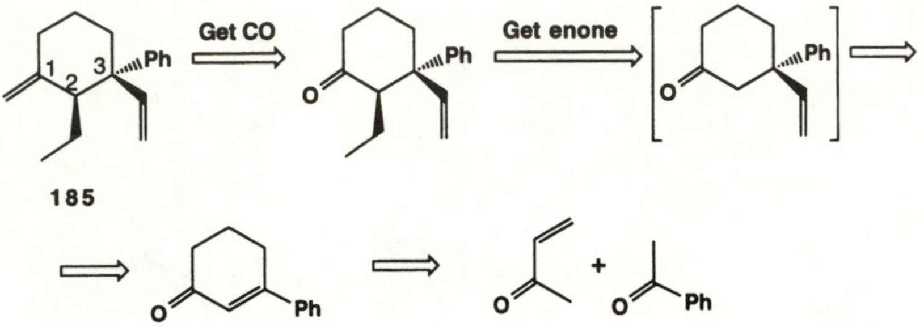
1
2
3
Ph
Get CO
O
Get enone
185
+
Ph

Procedure 4:

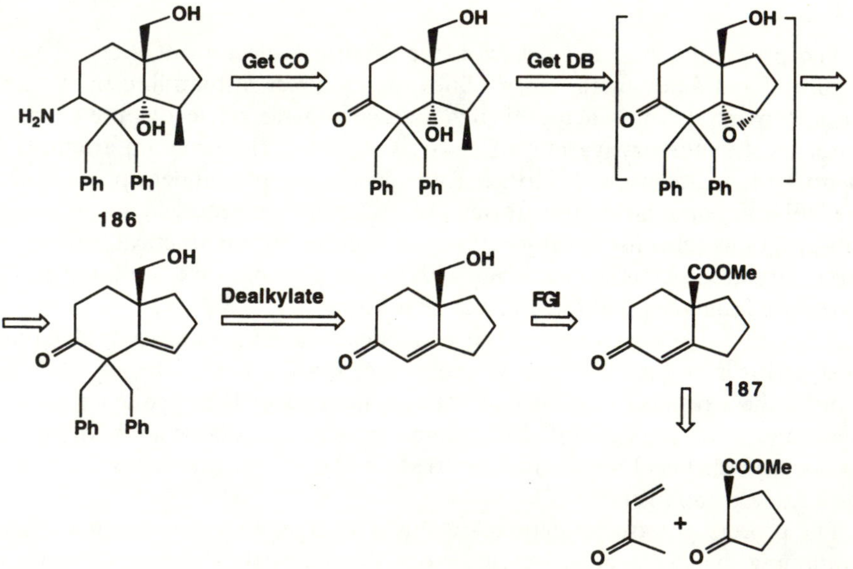
OH
H_2N
Ph
Get CO
O
Get DB
186
Dealkylate
FGI
COOMe
187
+

Scheme 3.48

the above-mentioned chemistry sub-routines (procedures). Seven such procedures have been selected, based on the various aspects of their reactivity pattern for α,β-unsaturated ketones and/or the related allylic alcohols. Scheme 3.48 illustrates the diversity of operation sets performed by some of these procedures.

For a target structure like **184**, Procedure 1 creates sequentially double bonds (*via* removal of the substituent at **C-2** and formation of an epoxide ring) and then a carbonyl group. Procedure 2 is able first to convert the 1-methylene substituent in target **185** into a 1-keto group and then to strip away both **C-2** and **C-3** substituents in a retro-Michael/enol alkylation sequence. The functionality pattern present in model **186** suggested the application of a series of retro reactions involving functional group interconversion and γ-extended enolate anion alkylation, which is aimed ultimately at the formation of the sub-goal **187** (Procedure 4).

For a complete analysis of a selected six-membered ring *via* a Robinson annulation transform approach, it is necessary to apply these Procedures sequentially to all six carbons, which leads to a total of 84 versions (six carbons, two directions for numbering, and seven procedures). Fortunately, the program is able to reduce the number of options due to the special module of preliminary evaluation (on an arbitrary scale) of the efficiency of the generated procedures. This evaluation accounts for the total number and ease of transformations implied by the application of a given procedure to the chosen structural fragment. The ratings developed by this module lead to the identification of the top ranking procedures, which are then presented to the chemist and from which the final strategic choice is made.

As an example, consider the analysis of the bicyclic terpenoid valerone **188**. LHASA identified 30 options for its construction *via* Robinson annulation pathways. Of these, only 15 were suggested as viable since the ratings of the others were too low. The three top choices that were identified by the computer as the most efficient are shown in Scheme 3.49.[26c] Route A was suggested as the most simple as it involved the least number of steps from readily available starting materials (although it did require the steps of selective protection–deprotection of one of the carbonyl groups, the one in the box). Route B represents an unexpected, but perhaps the most interesting, solution as it envisions a sequential formation of both rings as the result of an intramolecular Robinson annulation (two options). Route C implies the formation of ring **A** with the help of a Robinson annulation reaction. It is somewhat longer but leads to readily available precursors.

A similar computer-assisted approach was also elaborated envisioning the choice of the Diels–Alder transform as a strategic reaction. The task of the program in this case is to suggest the retrosynthetic transformations of the target structure leading to the creation of the cyclohexene moiety amenable to the retro-Diels–Alder disconnection. The general course of these pursuits is shown in Scheme 3.50.[26d]

The ideology of the program layout is similar to that described above for Robinson annulation. Here again it is targeted at the analysis of all cyclohexane fragments present in the target molecule, with the purpose of identifying all

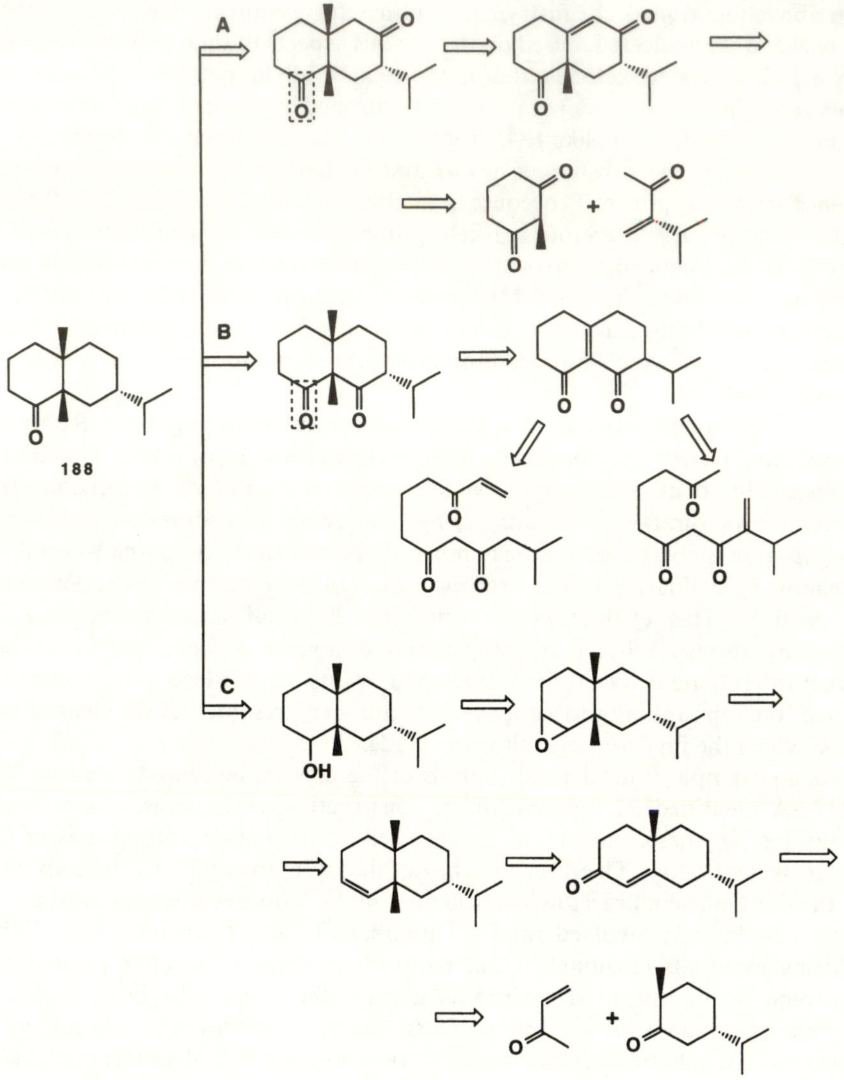

A

B

C

188

OH

EWG

Scheme 3.49

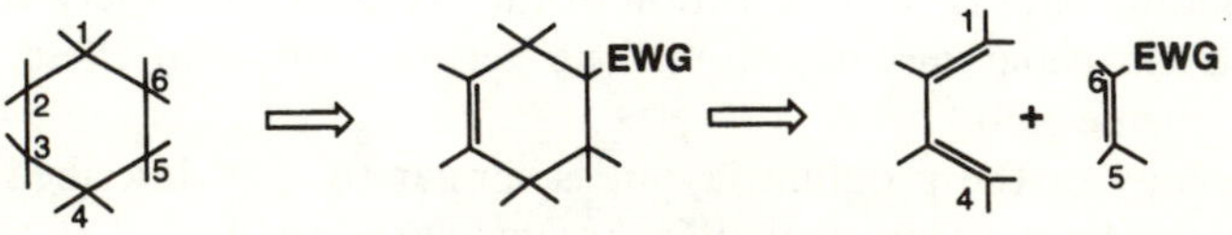

1 2 3 4 5 6 EWG

Scheme 3.50

possible routes of their dismantling *via* the Diels–Alder transform. This program also contains the necessary set of procedures and chemistry subroutines for retrosynthetic removal and/or introduction of the required structural elements. The program is capable of performing up to 15 sequential transformations in order to create the structural moiety corresponding to the Diels–Alder adduct, but in reality the depth of the searches is limited to no more than four steps, otherwise it will take too much time to analyse the result of these searches. Likewise, only those schemes that are considered feasible are presented to the chemists for their perusal. The application of this program for the analysis of the model example **189** is illustrated in Scheme 3.51.

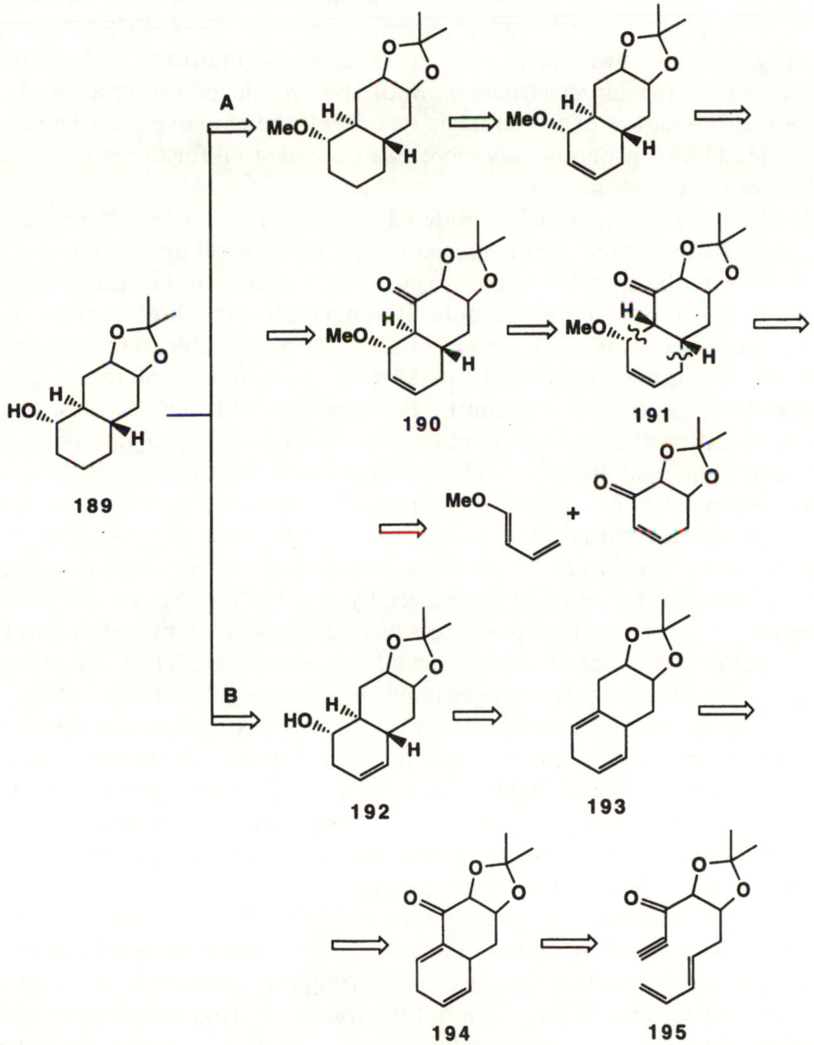

A

MeO

189

190

191

MeO

B

192

193

194

195

Scheme 3.51

It is interesting to note how the LHASA program surmounted the main 'road block' in this synthesis, namely assembling the needed *trans* fusion of the rings with the help of the Diels–Alder reaction. In route **A** this was accomplished by a set of retrosynthetic transformations into the sub-structure **190**, which contains a keto group. The presence of the latter secured the opportunity to apply an epimerization transform and thus to arrive to the next intermediate **191**, which is easily preparable by the standard Diels–Alder pathway. In route **B** the same stereochemical obstacle is removed by the transformation of sub-structure **192** into cyclohexadiene **193** and then to **194**. The creation of the 1,4-diene fragment in the target structure thus suggested an unanticipated intramolecular variant to the simultaneous assemblage of both rings from the precursor **195**.

The flexibility and wide scope of application are among the most important merits of the package of LHASA programs. It includes a well-elaborated set of sub-programs and procedures capable of effecting multistep retrosynthetic conversions of the target structure. Both the criteria of the strategic bond choice and the nature of the strategic reaction to be employed can be varied. Thus, the LHASA program may serve as a versatile tool for the retrosynthetic analysis of various structures.

The Corey–Wipke approach considered above actually represents a *backward computer-assisted search of synthetic pathways*, which is set up as an interactive procedure with the operator helping to guide the selection of the most promising routes. In this approach the computer is assigned to the job of accumulation, storage, and processing of all chemical information amenable to formalization, while the chemist is supposed to make a final evaluation of the suggested retrosynthetic solutions and to control the direction of further searches.

A rather different principle was utilized in the SYNGEN program developed in Hendrickson's laboratory.[27] This program was designed to be non-interactive, 'independent of user's preconceptions'. It is aimed at the generation of optimal synthetic routes for the *convergent assembly of the target molecule from available starting materials*. At the initial step the target structure is analysed in order to identify the set of bonds especially suitable for the two consecutive dissections into four (or less) precursors having the skeleton of materials found in its catalog (the latter contains over 6000 compounds). Then the program generates the functionality changes implicated in every bond connection and reveals the required functionality pattern of the starting material. If the latter is not found in the catalog, an attempt is made to elaborate its structure *via* a re-functionalization of the available compounds. The program operates efficiently without the user's interaction with not too complicated structures and it opens an opportunity for the exhaustive analysis of all possible simple pathways leading to the synthesis of the target structure.

As was noted in Bersohn–Esack's review,[24] 'The history of every computer application is marked by an initial over-optimism. The over-optimism results from the fact that people are successful in getting the computer to consider and solve simple problems in their area but the transition from simple problems to complicated ones is much more difficult than anticipated'. In fact, the general attitude toward the efficiency of computer assistance in the elaboration of synthetic plans is much more reserved than it used to be 20 years ago, following the landmarking publications of Corey's group. Yet there is a tremendous potential in the utilization of the power of modern computers in the solution of synthetical problems. The main area of their application were summarized in Hendrickson's review[27] as follows:

1. Complex calculations: molecular properties, reaction dynamics, and molecular modelling.
2. Information storage and retrieval: very large amounts of data stored, with retrieval completely dependent on efficient search system.
3. Artificial intelligence for intellectual problem solving: synthesis design and structure elucidation.
4. Automation of laboratory experiments: robotics to explore reaction optimization.

Not surprisingly, these tools are becoming a routine part of the repertoire used by human intellect in tackling the problem of constructing organic molecules of various complexity. No one can reasonably expect that computers of any capacity would ever be able to provide ultimate solutions to the problem of total synthesis. The ultimate goal of the computing chemistry is entirely different, namely to free the chemist from the necessity to spend mental effort at the solution of routine tasks of organic synthesis in order to exersize fully his/her imagination and intuition as the most valuable and unique features of human creativity.

Work with the computer required both loading its memory with an enormous amount of information regarding chemical structures and reactivity, reactions, and reagents, and the elaboration of the programs capable of processing this information. This necessity served as a powerful stimulus for a thorough analysis of empirical data accumulated by generations of synthetic chemists. As a result, it became possible to generalize this experience in a number of explicitly formulated concepts like, for example, those formulated in Corey's retrosynthetical analysis. Owing to these efforts to a substantial extent, organic synthesis is evolving into an independent discipline with its own methodology, approaches, and tools. The craftsmanship of organic synthesis has become as much a matter of learning as has any other professional skill.

REFERENCES

[1] (a) Woodward, R. B.; Sondheimer, F.; Taub, D.; Heusler, K.; McLamore, W. M. *J. Am. Chem. Soc.*, **1952**, *74*, 4223; (b) see, for the list of R. B. Woodward's outstanding contributions in total synthesis, *in memoria: Tetrahedron*, **1979**, *35*, No 19, p. iv; (c) for an assessment of the dramatic increase in complexity of synthetic goals during the last two decades, compare the list of objects in two editions of the monograph: Anand, N.; Bindra, J. S.; Ranganathan, S. *Art in Organic Synthesis*, Wiley, New York, 1st edn., **1970**; 2nd edn., **1988**; (d) a set of excellent papers, recounting the story of recent outstanding achievements is compiled in: *Strategies and Tactics in Organic Synthesis*,

Lindberg, T., Ed., Academic Press, New York, **1984**, vol. 1; **1989**, vol 2.; see also: Fleming, I. *Selected Organic Synthesis. A Guidebook for Organic Chemists*, Wiley, New York, **1973**.

[2] (a) Corey, E. J. *Pure Appl. Chem.*, **1967**, *14*, 19; (b) Corey, E. J. *Chem. Soc. Rev.*, **1988**, *17*, 111; (c) Corey, E. J. *Angew. Chem., Int. Ed. Engl.*, **1991**, *30*, 455; (d) Corey, E. J.; Wipke, W. T. *Science*, **1969**, *166*, 178.

[3] (a) Corey, E. J.; Cheng, X.-M. *The Logic of Chemical Synthesis*, Wiley, New York, **1989**; (b) for the step-by-step exposure of the basic principles of retrosynthetic analysis, see: Warren, S. *Organic Synthesis: The Disconnection Approach*, and *Workbook for Organic Synthesis: The Disconnection Approach*, Wiley, New York, **1980**; see also a textbook: Smith, M. B. *Organic Synthesis*, McGraw-Hill, New York, **1994**.

[4] (a) Bochkov, A. F.; Zaikov, G. E. *Chemistry of the O-Glycosidic Bond. Formation and Cleavage*, Pergamon, Oxford, **1979**, ch. 4; (b) Ruckel, E. R.; Schuerch, C. *J. Am. Chem. Soc.*, **1966**, *88*, 2605; see also: Ruckel, E. R.; Schuerch, C. *J. Org. Chem.*, **1966**, *31*, 2233; Ruckel, E. R.; Schuerch, C. *Biopolymers*, **1967**, *5*, 515; (c) Bochkov, A. F.; Obruchnikov, I. V.; Kalinevich, V. M.; Kochetkov, N. K. *Bioorg. Khim.*, **1976**, *2*, 1085; see also: Bochkov, A. F.; Obruchnikov, I. V.; Kalinevich, V. M.; Kochetkov, N. K. *Tetrahedron Lett.*, **1975**, 3403; (d) Bochkov, A. F., Kochetkov, N. K., *Carbohydrate Res.*, **1975**, *39*, 355.

[5] (a) Bodansky, M.; Ondetti, M. A. *Peptide Synthesis*, Interscience, New York, **1966**; (b) Vigneaud, V. D.; Ressler, C.; Swan, J. M.; Roberts, C. W.; Katsoyannis, P. G.; Gordon, S. *J. Am. Chem. Soc.*, **1953**, *75*, 4879; (c) Merrifield, R. B. *J. Am. Chem. Soc.*, **1963**, *85*, 2149; (d) Merrifield, R. B. *Angew. Chem., Int. Ed. Engl.*, **1985**, *24*, 799; (e) Marglin, A.; Merrifield, R. B. *J. Am. Chem. Soc.*, **1966**, *88*, 5051; (f) Denkewalter, R. G.; Veber, D. F.; Holly, F. W.; Hircshman, R. *J. Am. Chem. Soc.*, **1969**, *91*, 502; see also four subsequent papers; (g) Gutte, B.; Merrifield, R. B. *J. Am. Chem. Soc.*, **1969**, *91*, 501; (h) see, for example: Itakura, K.; Rossi, J. J.; Wallace, B. *Annu. Rev. Biochem.*, **1984**, *53*, 323; (i) review: J. M. J. Fréchet, *Tetrahedron*, **1981**, *37*, 663; (j) an overview of the scope of the utilization of reagents immobilized on the surface of polymer supports can be found in the monograph: Hodge, P.; Sherrington, D. C., Eds., *Polymer-supported Reactions in Organic Synthesis*, Wiley, New York, **1980**; an illustrative example of the utilization of the solid-phase multistep synthesis of a series of analogous organic compounds is given in: Chen, C.; Randall L. A. A.; Miller, R. B.; Jones, A. D.; Kurth, M. J. *J. Am. Chem. Soc.*, **1994**, *116*, 2661.

[6] Laurence, B. R.; Pickett, J. A. *J. Chem. Soc., Chem. Commun.*, **1982**, 59.

[7] (a) Willstatter, R. *Berichte*, **1901**, *34*, 129, 3163; (b) Robinson, R. *J. Chem. Soc.*, **1917**, 762; (c) for review on Mannich reaction see: Blicke, F. F., *Ozg. Reactions*, **1942**, *1*, 303; (d) Schopf, C. *Angew. Chem.*, **1937**, *50*, 779; (e) Stevens, R. in *Strategy and Tactics in Organic Synthesis*, Lindberg, T., Ed., Academic Press, New York, **1984**, ch. 10, p. 275.

[8] (a) Gates, M. *J. Am. Chem. Soc.*, **1950**, *72*, 228; Gates, M.; Tschudi, G. *J. Am. Chem. Soc.*, **1950**, *72*, 4839; **1952**, *74*, 1109; (b) see also an alternative synthesis of morphine: Elad, D.; Ginsburg, D. *J. Am. Chem. Soc.*, **1954**, *76*, 312; (c) Barton, D. H. R.; Bhakuni, D. S.; James, R.; Kirby, G. W. *J. Chem. Soc. (C)*, **1967**, 128; (d) Schwartz, M. A.; Zoda, M. F. *J. Org. Chem.*, **1981**, *46*, 4623; (e) see also: Szantay, C.; Blasko, G.; Barczai-Beke, M.; Pechy, P.; Dornyei, G. *Tetrahedron Lett.*, **1980**, *21*, 3509; for a review on the application of phenolic oxidation for the biomimetic synthesis of alkaloids, see: Kametani, T.; Fukumoto, K. *Synthesis*, **1972**, 657; (f) see, for example, the biogenetically patterned synthesis of progesterone: Johnson, W. S.; Gravestock, M. B.; McCarry, B. E. *J. Am. Chem. Soc.*, **1971**, *93*, 4332.

[9] (a) More than a dozen various strategies have been exploited for the preparation of this compound; for the leading references, see ref. 3a, pp. 146, 362; (b) McMurry, J. E.; Fleming, M. P.; Kees, K. L.; Krepsky, L. R. *J. Org. Chem.*, **1978**, *43*, 3255; for a review of the titanium-induced coupling reactions, see: McMurry, J. E. *Chem. Rev.*, **1989**, *89*, 1513; (c) McMurry, J. E.; Matz, J. R.; Kees, K. L.; Bock, P. A. *Tetrahedron Lett.*, **1982**, *23*, 1777.

[10] (a) Danishefsky, S.; Vaughan, K.; Gadwood, R.; Tsutsuki, K. *J. Am. Chem. Soc.*, **1981**, *103*, 4136; (b) Bornack, W. K.; Bhagwat, S. S.; Ponton, J.; Helquist, P. *J. Am. Chem. Soc.*, **1981**, *103*, 4647; (c) Burke, S. D.; Murtiashaw, C. W.; Saunders, J. O.; Oplinger, J. A.; Dike, M. S. *J. Am. Chem. Soc.*, **1984**, *106*, 4558; (d) Burke, S., in *Strategies and Tactics in Organic Synthesis*, Lindberg, T., Ed., Academic Press, New York, **1989**, vol. 2, ch. 2, p. 57; (e) Takeda, K.; Shimono, Y.; Yoshii, E. *J. Am. Chem. Soc.*, **1983**, *105*, 563; (f) for a review see: Martin, S. F., *Tetrahedron*, **1980**, *36*, 419.

[11] (a) Keay, B. A.; Rodrigo, R. *J. Am. Chem. Soc.*, **1982**, *104*, 4725; (b) Kelly, T. R.; Ghoshal, M. *J. Am. Chem. Soc.*, **1985**, *107*, 3879.

[12] (a) Paquette, L. A.; Doherty, A. M. *Polyquinane Chemistry: Synthesis and Reactions*, Springer, Berlin, **1987**; (b) Mehta, G.; Reddy, A. V.; Srikrishna, A., *Tetrahedron Lett.*, **1979**, 4863; (c) Mehta, G.; Reddy, A. V. *J. Chem. Soc., Chem. Commun.*, **1981**, 756.

[13] (a) Wenkert, E.; Naemura, K. *Synth. Commun.*, **1973**, *3*, 45; (b) Schiehser, G. A.; White, J. D. *J. Org. Chem.*, **1980**, *45*, 1864.

[14] Pirrung, M. C. *J. Am. Chem. Soc.*, **1981**, *103*, 82.

[15] Biethan, U.; Gizycki, U. V.; Musso, H. *Tetrahedron Lett.*, **1965**, 1477.

[16] (a) Woodward, R. B.; Fukunaga, T.; Kelly, R. C. *J. Am. Chem. Soc.*, **1964**, *86*, 3162; (b) Eaton, P. E. *Tetrahedron*, **1979**, *35*, 2189.

[17] Deshpande, M. N.; Jawdosiuk, M.; Kubiak, G.; Venkatachalam, M.; Weiss, U.; Cook, J. M. *J. Am. Chem. Soc.*, **1985**, *107*, 4786.

[18] (a) Noyori, R.; Suzuki, M. *Angew. Chem., Int. Ed. Engl.*, **1984**, *23*, 847; (b) Suzuki, M.; Yanagisawa, A.; Noyori, R. *J. Am. Chem. Soc.*, **1988**, *110*, 4718; (c) Corey, E. J.; Niimura, K.; Konishi, Y.; Hashimoto, S.; Hamada, Y. *Tetrahedron Lett.*, **1986**, *27*, 2199; (d) Johnson, C. R.; Penning, T. D. *J. Am. Chem. Soc.*, **1988**, *110*, 4726; (e) for later progress in the elaboration of a practical three-component coupling approach to PG-like structures, see: Lipshutz, B. H.; Wood, M. R. *J. Am. Chem. Soc.*, **1994**, *116*, 11689.

[19] Perlmuter, P. *Conjugate Addition Reactions in Organic Synthesis*, Pergamon, Oxford, **1992**.

[20] (a) For a review on the early steroid synthesis, see: Torgov, I. V. *Pure Appl. Chem.*, **1963**, *6*, 525; (b) Ananchenko, S. N.; Torgov, I. V. *Tetrahedron Lett.*, **1963**, 1553; see also: Koshoev, K. K.; Ananchenko, S. N.; Torgov, I. V. *Khim. Prir. Soedin.*, **1965**, 172; (c) Nicolaou, K. C.; Barnette, W. E.; Ma, P. *J. Org. Chem.*, **1980**, *45*, 1463; (d) Funk, R. L.; Vollhardt, K. P. C. *J. Am. Chem. Soc.*, **1980**, *102*, 5245, 5253.

[21] (a) Jasperse, C. P.; Curran, D. P.; Fevig, T. L. *Chem. Rev.*, **1991**, *91*, 1237; (b) Curran, D. P.; Raiewicz, D. M. *Tetrahedron*, **1985**, *41*, 3943; (c) Curran, D. P.; Kuo, S.-C. *Tetrahedron*, **1987**, *43*, 5653; (d) Curran, D. P.; Chen, M.-H. *Tetrahedron Lett.*, **1985**, *26*, 4991; (e) Fevig, T. L.; Elliot, R. L.; Curran, D. P. *J. Am. Chem. Soc.*, **1988**, *110*, 5064.

[22] (a) Hamon, D. P. G.; Young, R. N. *Aust. J. Chem.*, **1976**, *29*, 145; (b) Whitlock, H. W., Jr. *J. Am. Chem. Soc.*, **1962**, *84*, 3412; (c) Gauthier, J.; Deslongchamps, P. *Can. J. Chem.*, **1967**, *45*, 297; (d) Bélanger, A.; Lambert, Y.; Deslongchamps, P. *Can. J. Chem.*, **1969**, *47*, 795.

[23] Velluz, L.; Valis, J.; Nomine, G. *Angew. Chem.*, **1965**, *77*, 185.

[24] The basic principles of computer utilization in organic synthesis are thoroughly discussed in an excellent review: Bersohn, M.; Esack, A. *Chem. Rev.*, **1976**, *76*, 269.

[25] Vleduts, G. E. *Inf. Storage Retr.*, **1973**, *1*, 101.

[26] (a) Corey, E. J. *Q. Rev. Chem. Soc.*, **1971**, *25*, 455; (b) Corey, E. J.; Howe, W. J.; Orf, H. W.; Pensak, D. A.; Petersson, G. *J. Am. Chem. Soc.*, **1975**, *97*, 6116; (c) Corey, E. J.; Johnson, A. P.; Long, A. K. *J. Org. Chem.*, **1980**, *45*, 2051; (d) Corey, E. J.; Howe, W. J.; Pensak, D. A. *J. Am. Chem. Soc.*, **1974**, *96*, 7724.

[27] Hendrickson, J. B. *Top. Curr. Chem.*, **1976**, *62*, 49; see also a concise review: Hendrickson, J. B. *Angew. Chem., Int. Ed. Engl.*, **1990**, *29*, 1286.

CHAPTER 4

Molecular Design

> 'When you and I speak of design, we spontaneously think of it as an intellectual conceptualizing event, in which intellect first sorts out a plurality of elements and then interarranges them in a preferred way'.
>
> R. Buckminster Fuller, *Inventions*

INTRODUCTORY REMARKS

It can be claimed that, from the very beginning of its existence, organic synthesis has been based on molecular design. In fact, classical synthetic achievements, such as the synthesis of tropinone by Robinson or cyclooctatetraene by Willstater, was so carefully targeted, planned, and executed that the term 'molecular design' could have well been applied to these studies. Nevertheless, it is only recently that this term has become widely used and there is something more to it than just a fashionable catchword coined only for the sake of publicity. In this chapter, we are going to consider some main trends in modern molecular design, with special emphasis given to the peculiarities of the studies in this field. While there is no strict definition of this term, it is possible to delineate, mostly for the sake of convenience, two general types of problems dealt with in these studies that we would like to name as 'structure oriented design' and 'function oriented design'.

'**Structure oriented design**' refers to projects aimed at the creation of molecules with unusual structural characteristics not necessarily related to some useful property. This brings to mind a plethora of classical studies such as the previously discussed syntheses of cyclooctatetraene, Dewar benzene, or asterane. More recent examples include the syntheses of the dodecahedrane and tetrahedrane frameworks. The goal of the investigations in this area is first to invent and then to synthesize certain non-trivial molecules having unique structural features. This uniqueness very often refers to a novelty in the general shape of the molecules (as for dodecahedrane or catenanes), which are otherwise constructed in accordance with the classical concepts of structural theory. At the same time, quite a number of unprecedented structures have recently emerged as a result of studies conceived specifically to explore the limits of applicability of the classical descriptions of the structure of organic compounds. It may appear that the studies in this area are of those pursuing esoteric and narrowly focused goals, exercising the imagination and skill of the researchers in a contest for the best performance. If these were the only motivations, then these efforts could only be considered as additional examples of vanity and selfishness driving the scientists in their choice of targets for investigations. However, it must not be overlooked that the studies in the field of structure oriented design are actually carried out in a border zone between *knowns* and *unknowns* in organic chemistry. Therefore their immediate result is a tremendous expansion in the diversity of the objects studied by organic chemistry. Far more important is that this area represents an especially rich source of surprising discoveries with important ramifications for science, both pure and applied.

'**Function oriented design**' deals with the synthesis of compounds expected to possess a well-defined set of properties. Here the final goal of the study is to optimize the structure of the target compound in a way that maximizes its efficiency for the required function. The required functions may involve such valuable physical properties as electroconductivity (organic metals), the ability to form liquid crystals, catalytic activity similar to that of biocatalysts (enzymes), biological activity finely tuned for the treatment of a certain disease, or merely a peculiar pattern of reactivity specifically adjusted to meet certain synthetic demands. Here again, it is possible to claim that these were the most usual problems dealt with by organic chemistry for more than a century, long before the term 'molecular design' was introduced. The traditional search for useful compounds was based mainly on 'trial and error' and thus required an enormous amount of time and effort for the synthesis of thousands of analogs to arrive at just one suitable for the purpose. At present, there is a well-pronounced trend to elaborate less wasteful procedures. Quite often, at the beginning of such projects, various methods of molecular modelling are used with the aim to identify, with a certain probability, the set of specific structural parameters that would endow the target molecule with the capacity to serve the chosen goals. The results of initial experiments are further used to focus the studies more accurately. Eventually the search for the target structures is narrowed to a rather limited number of promising candidates. More often than not, 'function oriented design' gets its impetus from the results of 'structure oriented design', the discovery of new classes of structures with a set of novel and potentially useful properties.

Obviously no formal criteria exists that is useful to identify studies worthy of the label 'molecular design'. Nevertheless, we have chosen it as a title for this chapter. Below are considered a set of examples which illustrate the challenge and complexity of the goals as well as the elegance of the approach applied to the solution of the problems.

PART I STRUCTURE ORIENTED DESIGN

The synthesis of novel compounds is traditionally one of the main goals of organic chemistry. These syntheses are often aimed at the creation of fascinating molecules possessing intriguing and unprecedented structural features. At times, even the very possibility that these molecules can exist as stable or, at the very least, observable species seemed *a priori* questionable. In this area, imagination is probably at the top of the priority list of instruments required for both the identification of the goal and the execution of synthetic plans. Main trends and achievements in structure oriented design can be illustrated by numerous examples in almost any area of organic chemistry. The limited number considered below were chosen as a set that were representative of the diversity of motivations and ideas which provided the initial impetus for these target-oriented synthetic pursuits.

4.1 PLATO'S HYDROCARBONS AND RELATED STRUCTURES

We have already mentioned several examples of polycyclic molecules, the framework of which forms an enclosed spatial system or cage; hence the name 'cage systems' (see, for example, Section 2.19). Scores of compounds of this type were prepared and thoroughly studied. For decades, though, special attention has always been given to a very narrow group of polyhedral hydrocarbons (polyhedranes), known as Plato's hydrocarbons. The fascination with these shapes was kindled long ago, in antiquity, and it is interesting to view the evolution of ideas in this area.

As early as the 4th century BC, Plato recognized that there can exist five and only five regular polyhedrons: tetrahedron, cube, octahedron, dodecahedron, and icosahedron. Enchanted by the uniqueness of their geometry, he assigned four of these structures to the basic 'philosophical elements' of matter composing the whole world: fire (tetrahedron), earth (cube), air (octahedron), and water (icosahedron). During both the Medieval and the Renaissance periods the geometrical perfection and intrinsic beauty of these Platonic solids intrigued the minds of philosophers and scholars. Throughout these centuries, 'Perfection' and 'Harmony' persisted as the most important ideas intrinsic to God's created Universe. That is why so many efforts were spent to identify 'Elements of Perfection' in Nature and to find ways to connect the 'Perfection' of particular phenomena with the 'Laws of the Universe' as a whole. In this respect the very existence of Platonic polyhedrons has been considered as mysterious and deeply meaningful revelation. In view of that, it is not especially surprising to learn that such an outstanding astronomer as Johann Kepler (1571–1630) made serious attempts to utilize the geometrical pattern of these five Platonic solids as the basis to construct the orbits of the five planets known in his time.

Later, in more rationalistic times, these curious views were abandoned and nearly forgotten until, in the middle of the 20th century, the paradigm of perfection came to the minds of organic chemists (though in a very transformed shape). This time the problem was more modest in comparison with the previous attempts to apprehend the ideas of the Creator. Nevertheless, the very idea of reproducing the geometry of the perfect Platonic solids with artificially created hydrocarbons of the general formula C_nH_n represented a challenge. From valency rules it was obvious that only three of the listed regular Platonic polyhedrons could be reproduced in the shape of polyhedral molecules with the use of the CH fragment as the only building unit. These are C_4H_4 (tetrahedrane **1**), C_8H_8 (cubane **2**), and $C_{20}H_{20}$ (dodecahedrane **3**) (Scheme 4.1).

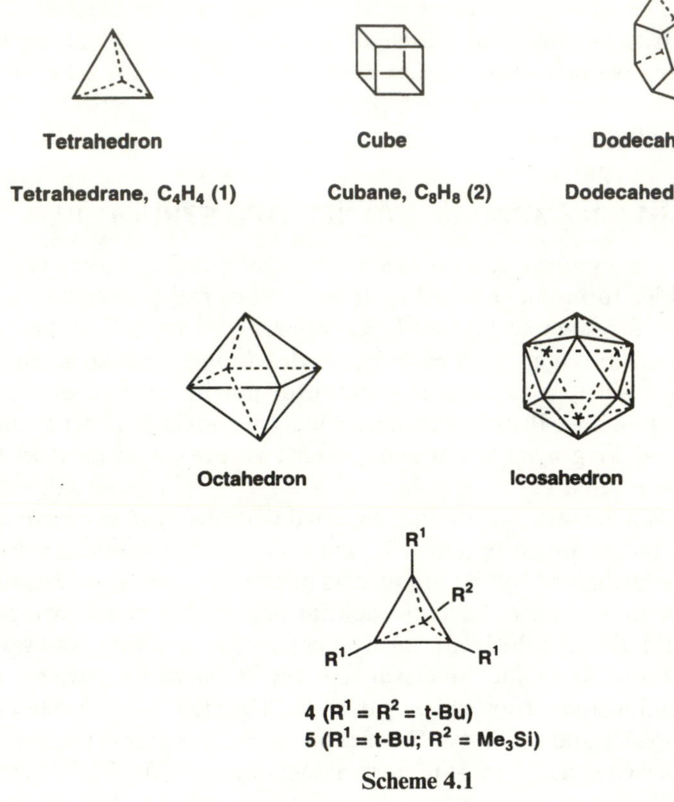

Scheme 4.1

The first synthesis of cubane **2** was reported in 1964 by Eaton.[1] It was considered to be a real breakthrough in this area. For about 20 years, several laboratories were engaged in attempts to solve the formidable problem of the dodecahedrane **3** synthesis. Dozens of approaches were tried and a great deal of interesting chemistry developed in connection with this project (see below). Finally, in 1982, a short communication in the August issue of the *Journal of the American Chemical Society*[2] announced that Paquette's team was the first to reach this goal. It was a spectacular achievement. Paquette's claim that they had conquered the Mount Everest of organic chemistry did not sound like too immodest an exaggeration.

While tetrahedrane **1** remains an elusive goal, some of its tetrasubstituted derivatives, namely tetra-*tert*-butyl- and (trimethylsilyl)-tri-*tert*-butyltetrahedranes (**4** and **5**, respectively, *vide infra*) have been synthesized by Maier's group.[3a,b]

As is usually the case with many of the most daring human endeavors (and for that matter the ascent of Mount Everest is not an exception!), there appears a fateful question to be answered: 'We finally reached the goal, so what?' Undoubtedly, the creation of molecules with the shape of Platonic solids *per se* constitutes an achievement that attests to the intellectual power and skill of modern science. This time, in addition to the usual claim about its unique ability to create its own object, chemistry can also claim that, like an art, it is able to create something really aesthetically appealing. Other aspects, however, are of no less importance. First, knowledge is acquired from the enormous efforts spent finding a viable pathway to the goal. The novelty of the target structure requires the elaboration of a non-trivial general strategy for the creation of unusual molecular frameworks and novel methods to perform even traditional transformations in an unusual structural context. Finally, and probably most importantly, the significant amount of data that was accumulated regarding unanticipated properties of the frameworks **1**–**3** (*vide infra*) enriched today's organic chemistry.

Let us now consider these issues more thoroughly, starting from cubane.[4a] *A priori* consideration of the cubane molecule prompted the conclusion that **2** must be a very unstable structure because of the severe angle strain caused by the geometry of the structure (all **C–C–C** angles in **2** are 90° instead of the regular 109.5° for sp^3 hybridized carbons). In fact, experimental data revealed an enormously high strain energy of 166 kcal mol^{-1} for this compound (approximately 14 kcal mol^{-1} per **C–C** bond). In striking contrast to expectations, **2** proved to be a kinetically stable molecule, withstanding heat without decomposition up to 200 °C! This unusual kinetic stability of a thermodynamically unstable (see the discussion of the general problem of kinetic *vs.* thermodynamic stability, Section 2.3) molecular construction can be probably best understood from an examination of the potential energy curve,[4b] calculated from the kinetic data on thermal rearrangements of **2** in the range 230–260 °C (Fig. 1).

Especially noteworthy is the huge activation energy (*ca.* 43 kcal mol^{-1}) for the rupture of the first two bonds. This means that an extremely unstable **2** (in thermodynamic sense!) lies actually within a rather deep energy well. It should be remembered, however, that curves like those shown in Fig. 1 represent only one cross-section of the multidimensional potential energy surface describing all transformations available for a given molecule. In fact, there exists more than one 'exit' available for the 'incarcerated tiger' **2** to get out of this well. It turned out that other routes for escape are not energetically as costly as the one shown.

Thus, a number of various reactions are known which proceed with surprising ease for **2** and its derivatives. For example, the stepwise catalytic hydrogenolysis of three **C–C** bonds in cubane occurs rapidly at room temperature (a rather unique transformation for a saturated hydrocarbon) with a release of 50,

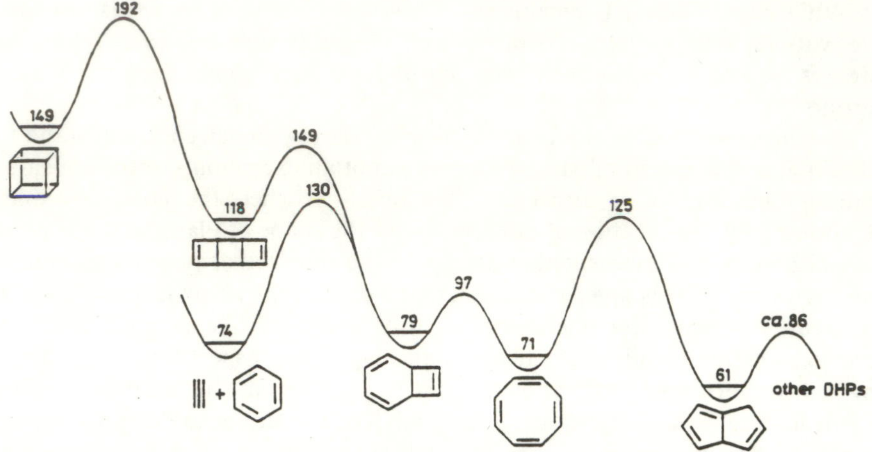

Figure 1 *Simplified potential energy hypersurface for cubane rearrangement (ΔH_f° values in kcal mol^{-1})*
(Reproduced from *J. Chem. Soc., Chem. Commun.*, **1985**, 964)

34, and 45 kcal mol^{-1} of the strain energy per each **C–C** bond broken in this sequence (Scheme 4.2).

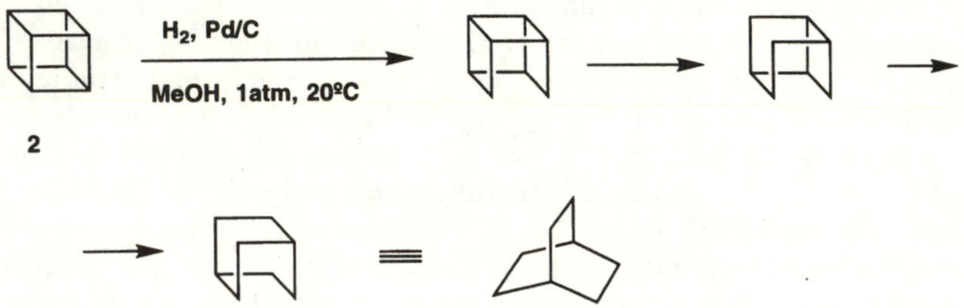

H_2, Pd/C

MeOH, 1atm, 20ºC

2

Scheme 4.2

Another unusual and highly useful reaction in the cubane series is based on the enhanced reactivity of **C–H** bonds in this framework toward strong bases. Cubane itself turns out to be rather inactive toward these reagents (owing to a rather low kinetic acidity of the **C–H** bond, even though it is 63 000 times higher than that of cyclohexane). However, some of its derivatives, like the carboxamides **6** or **7** (Scheme 4.3), can be directly metallated under the action of lithium amides. This reaction proceeds as a formal electrophilic substitution on an sp^3 carbon atom bearing no activating substituents in the α-position, a rather unusual pattern of reactivity. It is assumed that the presence of the carboxamide functionality in **6** or **7** ensures the necessary stabilization to the otherwise unstable lithiated derivatives. In fact this lithiation is a reversible process and only a small amount of the Li derivative is present in the equilibrium mixture. Nevertheless, the latter can be intercepted by transmetallation with mercury or (preferably) magnesium salts. The resulting cubylmercury or -magnesium derivatives could be further trapped by a number of regular electrophiles and thus a plethora of cubane derivatives like **8–10** are now accessible using this protocol (Scheme 4.3).[4a] In this respect it is worthwhile mentioning that the starting materials **6** and **7** are readily available compounds, as they are actually produced as advanced intermediates in the preparation of cubane. The problems encountered in the course of the metallation of cubane derivatives served as a motivation to search for new bases capable of doing this job. As a result of these efforts, a new set of strong and thermally stable bases, namely various magnesium amides like TMP–MgBr, emerged. They were shown to be the reagents of choice for the direct metallation of many other organic compounds besides cubane derivatives.[4a]

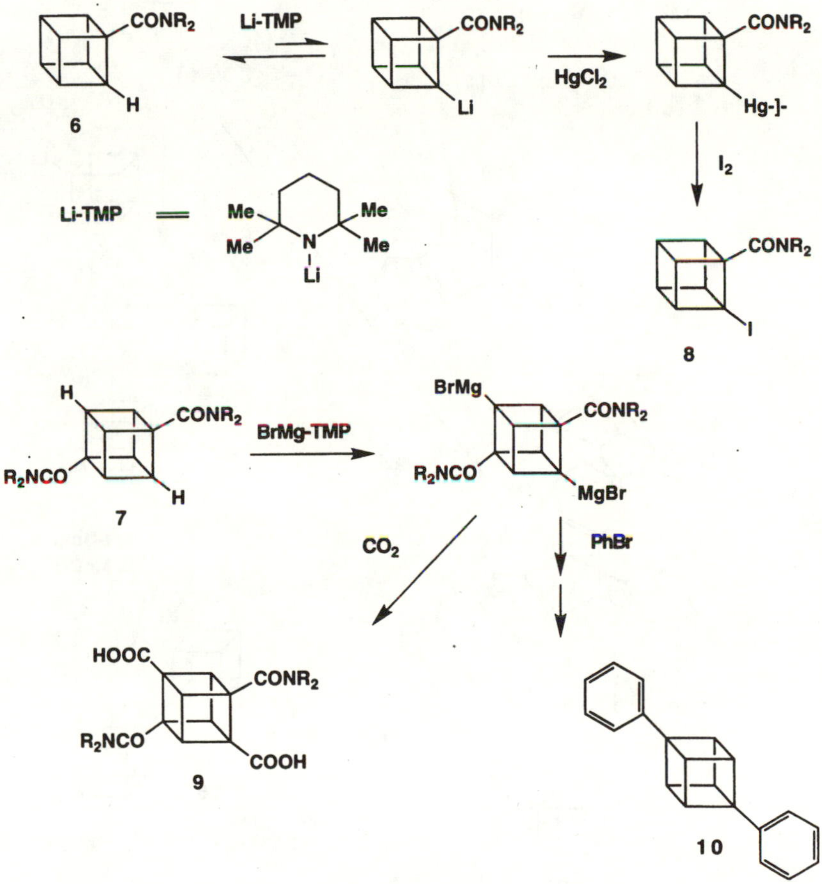

Scheme 4.3

Recent advances in the area of cubane chemistry are truly amazing. In fact, the high strain energy of cubane itself should have been taken as an unequivocal indication of the impossibility of introducing sp^2 center(s) into this framework. Nevertheless, the reactivity pattern for some cubane derivatives provides strong evidence in favor of the existence of an intermediate species like cubene **11**[4a,c] (Scheme 4.4) or the cubyl cation **12**[4a,d–g] (the problem of **12** will be considered later). In fact it was found that 2-iodo-1-lithio cubane **13**, generated *in situ*, is capable of undergoing facile 1,2-elimination, with the incipient **11** intercepted as the corresponding Diels–Alder adduct **14**.[4c] The same reaction performed in the absence of quenchers leads to the formation cubylcubane **15**, presumably *via* an interaction of **11** with the next molecule of **13**.

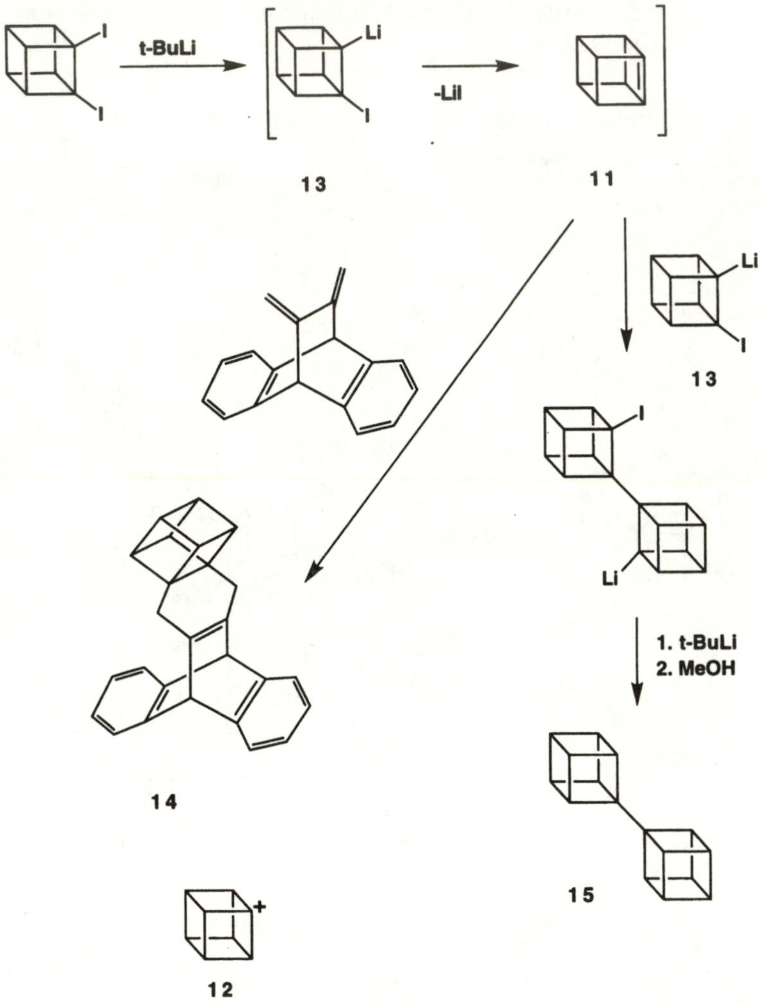

Scheme 4.4

Hydrocarbon **15** represents the first member of the new family of oligomers composed of cubyl units. A set of other representatives of this class was prepared along the route shown in Scheme 4.5. Here emerged another interesting aspect of cubane chemistry, namely the opportunity to generate the 1,4-cubadiyl **16** by an elimination route from the 1,4-diiodide **16a**. The interception of **16** by the cubyllithium reagent **17** (also generated *in situ* from **17a**) resulted in the formation of oligomeric intermediate **18** and, ultimately, oligomers like **19**.[4a] Compounds such as **19** belong to the recently emerged and sparsely populated family of rigid rod-like molecules. The disclosed opportunity to prepare a series of compounds of this type, with the distance between terminal groups being strictly defined by the number of cubyl units inserted, added yet one more dimension to the studies in the area of cubane derivatives. The search for novel and more versatile opportunities for the design and synthesis of rod-like molecules led to the elaboration of practical routes for the preparation of alkynyl cubanes. These compounds turned out to be suitable building blocks for directed syntheses of a set of structures with well-defined dimensionality and rigid-rod geometry.[4h] The efficiency of this approach was demonstrated by the preparation of 1,4-bis[(trimethylsilyl)ethynyl]cubyl-1,3-butadiyne **20**, one of the longest rods thus far prepared. As is stated in Eaton's paper,[4h] 'our ultimate goals are long (100–200 Å) rods of exactly defined structure and, further in the future, an extended three-dimensional network of dimensionally fixed,

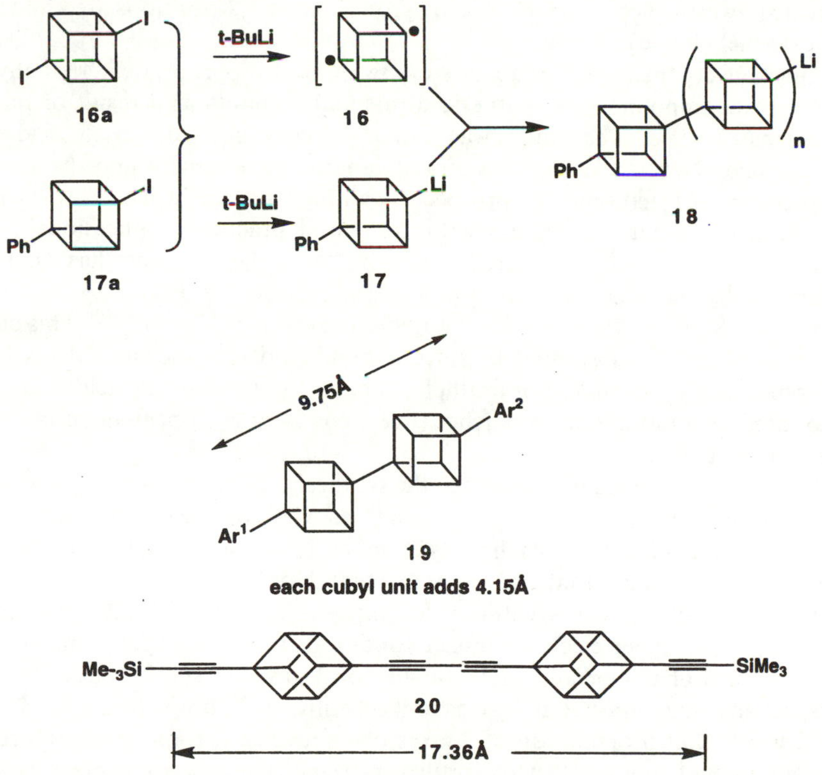

Scheme 4.5

electron-rich cavities ...'. Thus, this investigation, which initially stemmed from a purely academic interest toward cubane as an aesthetically appealing and exotic molecule, is being transformed into a molecular engineering project aimed at the design of novel structures potentially useful as materials for nanoarchitecture.

In addition to exciting chemistry, the renewal of interest in cubane and its derivatives is also due to the prospects of the utilization of cubane derivatives as high-energy propellants or explosives. In this respect, an especially appealing target is octanitrocubane.[4a,i] In fact, this compound combines, in a single molecule, both the property of an excellent fuel (high energy content and high density of cubane framework) and an oxidant (nitro groups) in conjunction with the optimal stoichiometry for an efficient explosion process in accordance with the formal equation: $C_8N_8O_{16} \rightarrow 8CO_2 + 4N_2$. It is not especially surprising that studies in the area of polynitrocubanes were strictly classified and for a long time very little was published about the progress achieved on these super-explosives. Only recently, Eaton's group reported an efficient syntheses of 1,3,5-trinitro- and 1,3,5,7-tetranitrocubanes, prepared in the course of studies aimed at preparing shock-sensitive, high-density explosives.[4j] It was surprising to find that both these derivatives were quite stable up to 250 °C. Unfortunately (?) it looks like octanitrocubane could not be prepared since cubane derivatives having powerful electron-withdrawing groups in the 1,2-positions are known to be extremely labile compounds.[4i,j]

One final comment to the cubane story seems quite appropriate. Originally, in 1964, this compound was synthesized in small amounts as a result of purely academic studies. The route was rather lengthy and the total yield not spectacular. Nevertheless, it was an outstanding achievement and the success appreciated. At that time, the prospects resulting from the broad investigations in the field of cubane, let alone the potential practical applications of its derivatives, were not considered seriously. Now the situation has changed dramatically. It was not vain for a recent review by Eaton to be titled: 'Cubanes: Starting Materials for the 1990s and the New Century'.[4a] This claim seems to be well substantiated, since a number of cubane derivatives have become readily available. For example, cubane-1,4-dicarboxylic acid is already produced at multi-kilogram batches (five steps from cyclopentenone in about 25% overall yield)!

The chemistry that came to light in the synthesis of dodecahedrane turned out also to be surprising in many respects.[5a] In view of the almost unstrained character of this molecule (each of its 20 sp^3 carbon atoms adopts a nearly ideal tetrahedral configuration), it was expected that **3** (Scheme 4.6) would be extremely stable. It melts without decomposition at 430 °C and produces a molecular ion under electron impact conditions in a mass spectrometer. The non-strained character and high stability of **3** gave no special indications to suspect anything unusual in regards to its chemistry. So it was rather surprising to discover that bromination of **3** takes place readily at room temperature (in the absence of any catalyst or irradiation!) and gives monobromide **21** in a quantitative yield. It should be noted that a similar reaction with a rather reactive adamantane (*vide infra*) requires heating with bromine. Derivative **21** can be easily converted into a number of other monosubstituted dodecahedranes *via* reaction pathways resembling those of the typical tertiary alkyl halides (Scheme 4.6). Formation of Friedel–Crafts alkylation products like **22** and then **23**[5b] seem to be especially noteworthy examples.

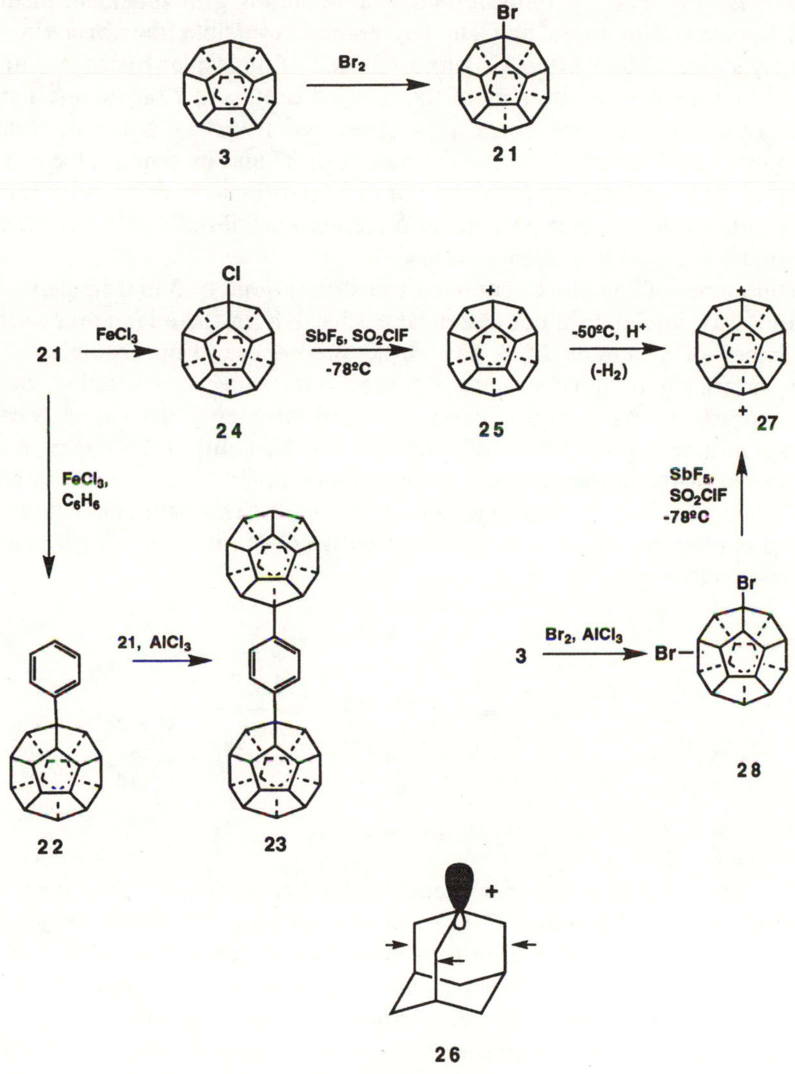

Br
Br_2
3
21
Cl
+
+
$FeCl_3$
SbF_5, SO_2ClF
-78ºC
-50ºC, H^+
($-H_2$)
21
24
25
+
27
$FeCl_3$,
C_6H_6
SbF_5,
SO_2ClF
-78ºC
Br
21, $AlCl_3$
Br_2, $AlCl_3$
3
Br–
28
22
23
+
26

Scheme 4.6

The treatment of monochlorododecahedrane **24** by SbF_5 in SO_2ClF at $-70\,°C$ leads to the formation of the stable monocation **25**. This fact, at first glance, seems to be an ordinary result of the well-known stability of tertiary carbenium ions. However, the corresponding acyclic ions acquire a planar sp^2 configuration which is unattainable in the framework of **25**. At the same time, a close relative to the latter among cage compounds and also non-planar, the 1-adamantyl cation **26**, presumably owes its stability to the presence of the adjacent **C–C** bonds oriented *trans* to the vacant orbital of the carbenium center (hyperconjugation). No such stabilizing effects can be operative in the case of **25** (*vide infra*). Monocation **25**, upon standing in superacid media at −78 °C or warming up to 20 °C, undergoes an irreversible transformation into the very stable 1,16-dication **27** (with evolution of molecular hydrogen). It was shown that the 1,16-positioning of the charged centers in **27** is the most stable arrangement, as the same ion can be generated from a mixture of isomeric dibromides **28**.[5c] Both the ease of formation of **27** and its remarkable stability are surprising and difficult to account for in conventional terms (a more detailed discussion of the evidence pertaining to the peculiarities of the caged-structure carbenium ions will be presented below).

While some of the aforementioned transformations of **3** and its derivatives are in a way similar to those described earlier (*e.g.* in the adamantane series), a number of reactions of **3** have little precedence. For example, it was truly surprising to discover that **3** turned out to be very reactive towards dichlorocarbene. As a result, the respective dichloromethyl derivative **29** can be formed in a very good yield under conventional conditions of phase transfer generation of dichlorocarbenes. Subsequent treatment of adduct **29** with *tert*-butyllithium leads to cyclopropadodecahedrane **30**. Thus, the formation of **30** from **3** implies the sequence of two consecutive **C–H** insertions of carbene-like species (Scheme 4.7).[5d]

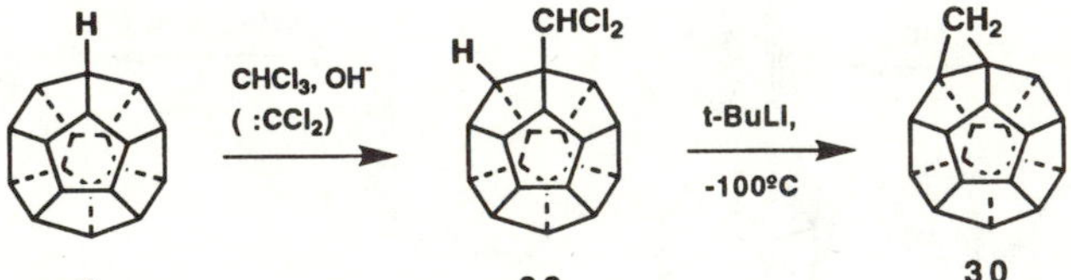

Scheme 4.7

Why are these reactions considered to be that peculiar? The insertion of carbene into a non-activated **C–H** bond of a saturated hydrocarbon is a well-known but not a very efficient process in the absence of catalysts. The fact that the insertion of carbene (both inter- and intramolecular) at the **C–H** bond of dodecahedrane frameworks proceeds with such an amazing ease has to be taken as a strong evidence for suggesting the operation of previously unknown and little-understood activating factors in this system.

While the story relating the design and synthesis of dodecahedrane sounds truly exciting, Paquette's synthesis of this compound required so much time and effort (23 steps starting from the cyclopentadienyl anion) that it seemed inconceivable to expect that any in-depth studies would be possible in that field. Fortunately, these concerns turned out to be not as serious as anticipated. Here it is appropriate to comment briefly on alternative strategic options elaborated for the synthesis of **3**. The original synthetic design of **3** was based on the consecutive additions of cyclopentane rings to the starting polycyclic precursor, prepared by the Diels–Alder route.[2,5a] It was essential to utilize photochemical cyclizations at key points late in this synthesis (Scheme 4.8) from heptaquinane **31** *via* octa-(**31a**) and nona-(**31b**) to decaquinane **32** (monoseco-**3**), since the non-bonded steric interactions increased progressively along this route. The final step in the synthesis, namely the conversion of rather strained **32** into unstrained target structure **3**, which required the removal of two hydrogen atoms from non-adjacent carbon atoms with the formation of a **C–C** bond, proceeded with unprecedented ease under the action of the usual catalysts for hydrogenation–dehydrogenation.

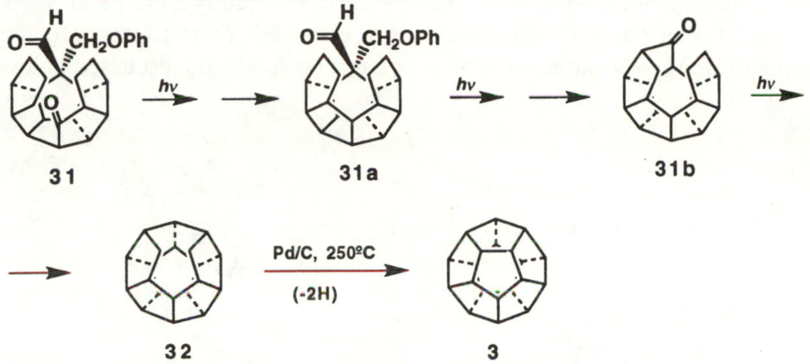

H
O= CH₂OPh
hν
31
H
O= CH₂OPh
hν
31a
O
hν
31b
32
Pd/C, 250ºC
(-2H)
3

Scheme 4.8

Molecular mechanics calculations on **3** revealed that this structure actually represents the global minimum of free energy for the $C_{20}H_{20}$ system. These data suggested an alternative solution to the problem of its synthesis based on the isomerization of various $C_{20}H_{20}$ hydrocarbons in a way similar to that used for the synthesis of adamantane (see Section 2.3). While these routes seem to be quite favorable thermodynamically, initial attempts in this field were not very rewarding. Although calculations suggested that an isomerization of the basketene dimer **33** into **3** (Scheme 4.9) must be accompanied by the release of 178 kcal mol^{-1} of strain energy, this reaction has never occurred under a variety of conditions tried.[5a]

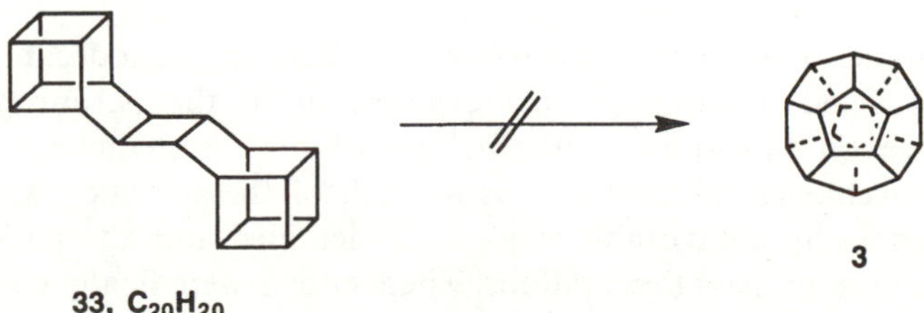

Scheme 4.9

Nevertheless, the viability of the thermodynamically driven isomerization approach has been brilliantly substantiated in the later investigations of Prinzbach's group in Germany.[6a] These studies started with the synthesis of another exotic $C_{20}H_{20}$ hydrocarbon, pagodane **34**, prepared from a commercial insecticide, isodrine, along the route outlined in Scheme 4.10 (a total of 15 steps, overall yield 15–18%). The authors reasoned that this compound, which also contains 12 five-membered rings (but otherwise bears no obvious structural similarity to **3**), could be converted into **3** under the action of hydrogenation–dehydrogenation catalysts. In effect, this conversion was achieved, but in a frustratingly low yield (Scheme 4.10). Transformation **34** → **3**, which can be depicted in a formal sense as a scission of two **C–C** bonds of the 'waist' cyclobutane ring ($+2H_2$) with the subsequent formation of two **C–C** bonds between the opposing four CH_2 groups ($-2H_2$), is actually a multistep sequence with a number of side reactions occurring at each step. (The attempts to achieve isomerization **34** → **3** under the action of superacids will be discussed below.)

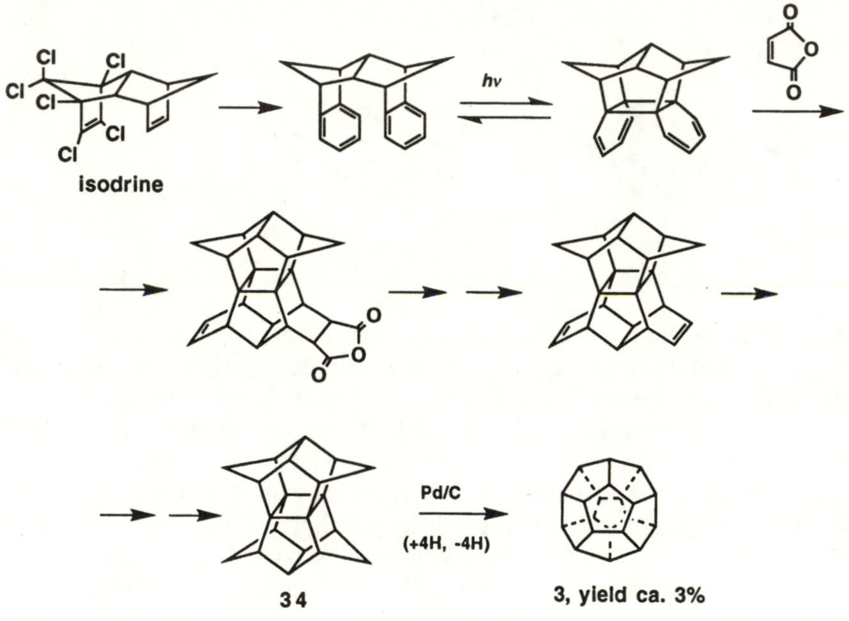

Cl Cl Cl Cl Cl Cl

isodrine

$h\nu$

Pd/C

(+4H, -4H)

34

3, yield ca. 3%

Scheme 4.10

The disclosed opportunity to transform pagodane into a dodecahedrane ring system warranted a more detailed exploration of the nature of possible intermediates on this route, with the use of both experimental data and molecular mechanics calculations. As a result of these studies, a number of thermodynamically unfavorable steps were identified and alternative options were designed to by-pass these pitfalls. These efforts were finally crowned by a spectacular demonstration of the promise of the suggested approach as a general route toward the preparation of **3** and a series of its variously substituted derivatives.[6b,c] As representative examples, the main steps in the syntheses of dodecahedradiene **35** and dodecahedrene **36** are shown in Scheme 4.11.

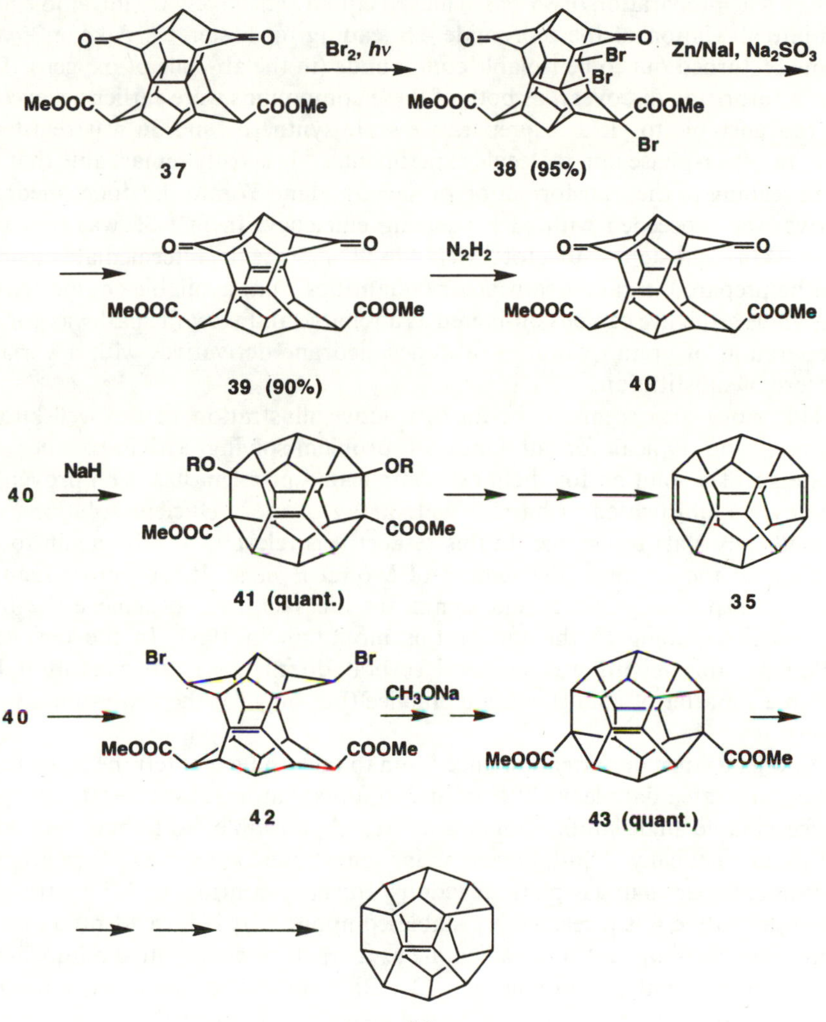

Scheme 4.11

The initial scission of cyclobutane ring of available pagodane derivative **37** is achieved as a result of homolytic bromination. The tribromide **38**, thus formed, underwent facile bromine elimination–fragmentation (and hydrogenolysis of the **C–Br** bond at the α-methoxycarbonyl center) under the action of metals to form diene **39**. Diimide reduction of **39** affected only one double bond and gave product **40**, which served as a common precursor for both **35** and **36**. The synthesis of **35** involved the formation of the two missing dodecahedrane bonds with the help of a double intramolecular nucleophilic addition at the ketone carbonyls. This yielded the dodecahedrene derivative **41**, which further underwent a standard set of transformations necessary to remove the extra functionalities. The preparation of **36** was achieved equally effectively *via* intramolecular enolate alkylation of the dibromide **42**, leading to diester **43**. Alkene **36** and diene **35** turned out to be isolable compounds (in the absence of oxygen). This was a surprising discovery, as both of these compounds were earlier believed to be too unstable to yield to preparative-scale synthesis, and **36** was registered only in vapor-phase ion–molecule experiments.[5a] It is truly remarkable that the steps leading to the transformation of the pagodane **37** into the dodecahedrane derivatives proceeded with such amazing efficiency. In fact **35** was prepared from **37** in eight steps with a total yield 51–54%! Since the intermediate diene **39** can be prepared in rather satisfactory quantities from available compounds,[6d] the whole sequence can be considered as a reliable and versatile pathway for the preparation of gram quantities of dodecahedrane derivatives with a variable pattern of substitution.

This story also seems to be an instructive illustration of the well-known phenomenon typical for all kinds of problem-solving activities: once the existence of a solution for the most complicated problem has been proven, in whatever sophisticated fashion, a wide array of more efficient solutions will immediately start to emerge. In this respect it is relevant to refer again to the analogy of the ascent of the summit of Mount Everest. It took more than 30 years of exhausting efforts and significant life sacrifices to achieve the final success in climbing to the top of this mountain in 1953. In the ten years following, this record was repeated several dozen times. At present it has become a matter of almost routine practice (but only for the professionals, for sure!).

Now let us turn our attention once again to the problem of tetrahedrane **1**. All the accumulated data leave little (if any) doubts that unsubstituted **1**, owing to its formidable angle strain, should be extremely unstable, both thermodynamically and kinetically.[7a] Still, there remains some hope to observe its formation as a transient species in gas phase reactions. In sharp contrast to **1**, its tetra-*tert*-butyl derivative **4** is a reasonably stable compound. It is formed rather easily from tetra-*tert*-butylcyclopentadienone as a result of consecutive photochemical reactions: firstly, intramolecular [2 + 2] cycloaddition and, secondly, CO extrusion. In fact, these reactions served as tools to convert the energy of light into the strain energy of the final product **4** (Scheme 4.12).[3a] Upon heating to melting (135 °C), **4** is transformed into its isomer, tetra-*tert*-butylcyclobutadiene **44**, with a partial release of the accumulated energy.

Thermochemical data substantiated the highly strained character of **4**. In fact, the strain energy per **C–C** bond of the tetrahedrane skeleton in **4** turned out to be 21.5 kcal mol^{-1}, the highest value known (*cf.* data for cubane, *vide supra*). Why then is it so stable? Obviously this is a purely kinetic stability, as was the case with cubane, but the origin of this stability of **4** is rather unusual. It is speculated that the presence of four *tert*-butyl groups creates a kind of 'corset'

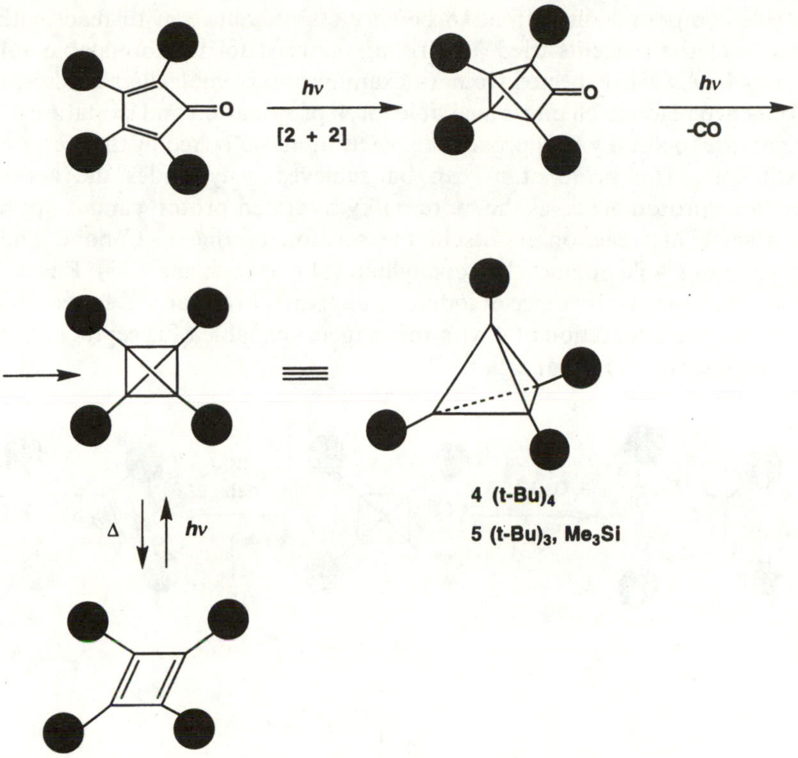

hv [2 + 2]

hv -CO

Δ hv

4 $(t\text{-}Bu)_4$

5 $(t\text{-}Bu)_3$, Me_3Si

44

Scheme 4.12

around the central tetrahedron core of **4**. Owing to the close packing of bulky substituents, any stretching of the **C–C** bonds of this core, which would lead to an increase of the distance between the two corresponding *tert*-butyl groups, would be opposed by increasing interactions with other pairs of the substituents. As a result, the activation barrier for the conversion **4** → **44** is comparatively high, while according to calculations it should be very low for the unsubstituted parent compound **1**. The operation of this stabilization effect requires the presence of four bulky substituents. Thus the presence of only three *tert*-butyl-group is not sufficient and all attempts to prepare the respective trisubstituted tetrahedrane ends with the formation of tri-*tert*-butylcyclobutadiene. A second representative of the tetrahedrane family, tri-*tert*-butyl(trimethylsilyl)tetrahedrane **5**, has also been prepared.[3b] In view of the aforementioned reasoning, it was anticipated that this compound might be stable, but it could not have been predicted that **5** would be noticeably more stable than **4** (no reaction occurred upon its heating to the melting point of 162 °C; conversion to the cyclobutadiene takes place only at 180 °C).

The chemical reactivity of **4** was both expected and, at the same time, rather surprising.[7a] What was surprising was that the thermodynamically very unstable compound turned out to be very stable, reluctant to react with the majority of the reagents tried (in striking contrast to its more stable valence isomer **44**). As was expected from an examination of molecular models, there are only two reacting channels available for **4**: protonation and oxidation. Other reagents are too bulky to approach the reaction sites blocked by the corset of the substituents. The protonation can be achieved only under the action of anhydrous proton acids, as the more bulky hydrated proton cannot approach the target. This reaction results in the scission of one **C–C** bond and the formation of stable homocyclopropenylium salts **45** (Scheme 4.13). Reaction of **4** with oxygen in the presence of iodine or copper(I) bromide yields peroxide **46**. Otherwise, the interaction of **4** with the reagents capable of accepting electrons produces the stable radical cation **47**.

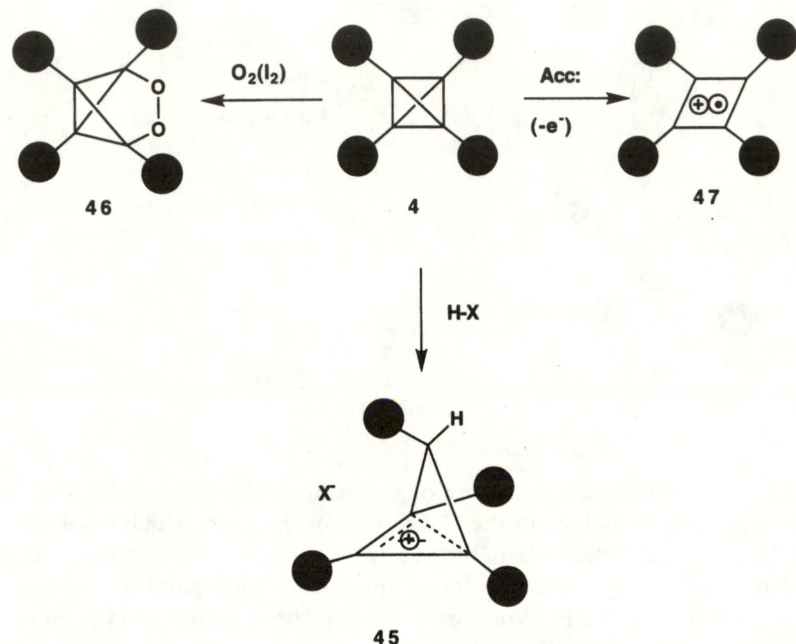

Scheme 4.13

These are only few of the representative examples showing the achievements and the problems related to tetrahedrane stability and reactivity. The importance of the studies in this seemingly narrow area is explicitly formulated by Maier in his review:[7a] 'The study of tetra-*tert*-butyltetrahedrane has: (a) sharpened our feeling for the subtle influence of sterically demanding groups (favoring of otherwise impractical synthetic routes; reversal of thermodynamic stabilities); (b) led to broadly applicable concepts (corset principle, silyl trick, formation of radical pairs); (c) opened up access to several unknown or at least difficultly accessible classes of substances; (d) allowed the isolation of crystalline derivatives of highly reactive systems; (e) allowed X-ray investigations with surprising results (extremely bent bonds, dependency of alternation on degree of substitution, gas-inclusion compounds, structural indicators of 'homoaromatic' character); (f) allowed detailed insight into the behavior of radical-cations; (g) made the discovery of new reaction mechanisms possible (addition reactions, autoxidation, and cationic rearrangements)'. It is really amazing to learn that the study originally aimed at the synthesis of a small and rather esoteric molecule resulted in so many ramifications for other areas of chemistry!

The above discussion touched only some aspects of the current activity in the area of synthesis of various exotic cage structures. Since no more Platonic hydrocarbons other than **1**, **2**, and **3** can exist, there is no room left for the structure oriented design in this particular field. Thus the present, and possibly most exciting, chapter dealing with this problem can be closed. However, there are no limits to the complexity of polyhedranes composed of various ring combinations. Hence, there will always exist a challenge to design a new organic molecule having the most intricate polyhedral structure and to develop synthetic methods enabling one to reach this goal. It seems appropriate to give here some examples showing the general trend of the design in this area. Irregular polyhedranes such as **48a** and **48b** composed of 15 and 16 rings respectively (see Scheme 4.14) were suggested in Paquette's review as worthy goals for 'those who welcome synthetic challenges of such enormous proportion'.[5a] Yet another series of hypothetical structures emerged from the analysis of constructions composed from cubane cells. Recent calculations with the use of *ab initio* molecular quantum mechanical methods predicted dicubane **49a** and its isomer dicubene **49b** (Scheme 4.14) to be *minima* on the $C_{12}H_8$ potential energy hypersurface and thus might represent 'achievable synthetic targets (?)'.[7b] Prohibitively high values of strain calculated for these compounds make chances for their successful synthesis marginal at best. In all likelihood the formidable complexity of the goal once again would serve as an additional and powerful motivation for ambitious explorers to undertake pursuits in this area.

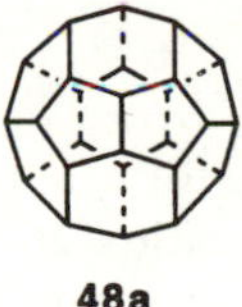
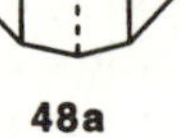
48a

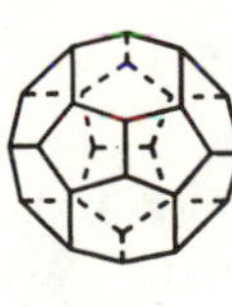
48b

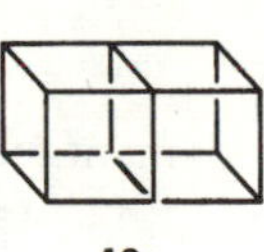
49a

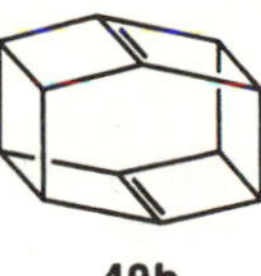
49b

Scheme 4.14

Hydrocarbons **48a** and **48b**, as well as **1–3**, belong to the same family of compounds of the general empirical formula $(CH)_n$. Much more diversity arises if **CH** fragments are not the only building blocks. In fact, a multitude of cage structures having different **C:H** ratios has been prepared and thoroughly studied during the last few decades. Several representatives of this type were mentioned earlier in this book (see, for example, Section 2.19.1). Among the set of various properties peculiar to the cage-like frameworks, probably the most noteworthy are those associated with the phenomenon of 'through space effects'. We feel it appropriate to discuss this effect briefly. Both chemical and physicochemical properties can be very sensitive to the operation of these effects, as shown in Scheme 4.15 for the classical example of adamantane derivatives.

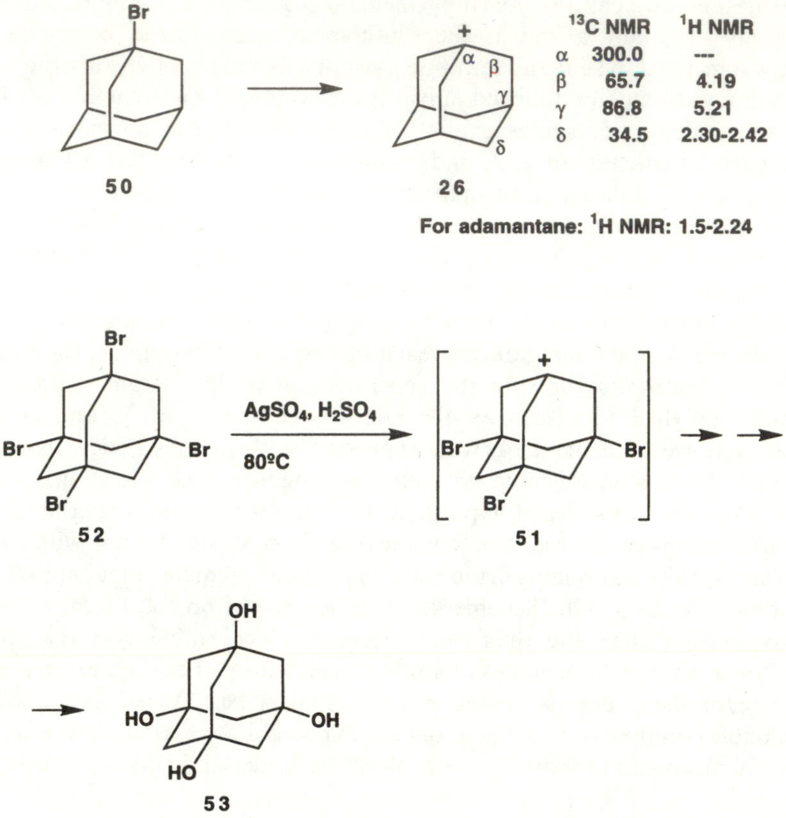

Br

50 → **26**

	^{13}C NMR	^{1}H NMR
α	300.0	---
β	65.7	4.19
γ	86.8	5.21
δ	34.5	2.30-2.42

For adamantane: ^{1}H NMR: 1.5-2.24

52 —($AgSO_4$, H_2SO_4, 80ºC)→ [**51**] → → → **53**

Scheme 4.15

It is well known that adamantane or its 1-halo derivatives, for example **50**, can be easily transformed into a stable tertiary 1-adamantyl cation **26**.[8a,b] As was already mentioned (*vide supra*), this cation owes its increased stability to the unique stabilization effects involving the participation of the remote centers in the charge delocalization. The pattern of NMR spectra may be used as a sensitive probe to the charge delocalization effects. Thus for the series of aliphatic tertiary carbenium ions, the presence of the positive charge induces a downfield shift of the ^{1}H NMR signals (relative to those of the parent covalent precursor) at the adjacent centers. The magnitude of this effect decreases in the order: $\beta \gg \gamma \approx \delta$. In the case of **26**, a substantial downfield shift of β-protons is also observed, but this effect is pronounced even stronger for the γ-protons (Scheme 4.15).[8b] This pattern of deshielding is even more pronounced in the ^{13}C NMR spectra of **26**. Here again, γ-carbons turn out to be shifted more downfield than β-carbons, which are positioned closer to the charged center. These unusual long-range effects were originally interpreted as the evidence of the specific interactions of the charged carbon orbitals with the inner lobes of the γ-carbon orbitals. Later calculations indicated that these interactions cannot be significant for the adamantyl system. The carbon–carbon hyperconjugation mechanism represents a more adequate explanation[8b] (see Scheme 4.6). It is certain that, for the adamantane system, these long-range interactions are more important than the usual inductive effects. Similar interactions leading to the destabilization of the tribromo-substituted 1-adamantyl cation **51** are actually responsible for the striking inertness of tetrabromide **52**[8c] toward nucleophilic substitution, in contrast to the ease of these reactions with monobromide **50**. In fact it has been found that the hydrolysis of **52** to give, eventually, **53** proceeds only upon treatment with silver sulfate in hot sulfuric acid. At the same time, **50** can be easily hydrolysed in aqueous acetone at room temperature,[8a] as is usually the case with typical *tert*-alkyl derivatives.

More recent examples also related to through-space effects can be found in the chemistry of cubane derivatives. We have already mentioned that the intermediacy of the cubyl cation **12** is inferred from the results of some transformations in this series. This issue deserves additional comments, since, as was specifically emphasized in Eaton's review, 'everything about the cubyl cation seems unfavorable'.[4d] In fact, owing to the peculiarity of the geometry of the cubane framework, the corresponding cation should have a geometry which is very far from the preferred planar sp^2 configuration (as is the case for the *tert*-butyl cation). In addition, no stabilizing effects similar to those mentioned earlier for the 1-adamantyl cation could be operative for **12**. This reasoning was supported by early molecular orbital calculations which suggested that cubyl trifluoromethanesulfonate **54** (X = H, Scheme 4.16) must be inert to solvolysis conditions.[4e] It came as an astonishing observation that **54** (X = H) is able to undergo ready methanolysis at least 10^{13} times faster than predicted by the previous calculations![4d,f] These findings prompted more accurate *ab initio* calculations of the same cation **12**. It was disclosed that **12** should be a relatively stable species due to the possibility of a partial delocalization of the positive charge to the α- and γ-carbons. In accordance with current views, stabilization

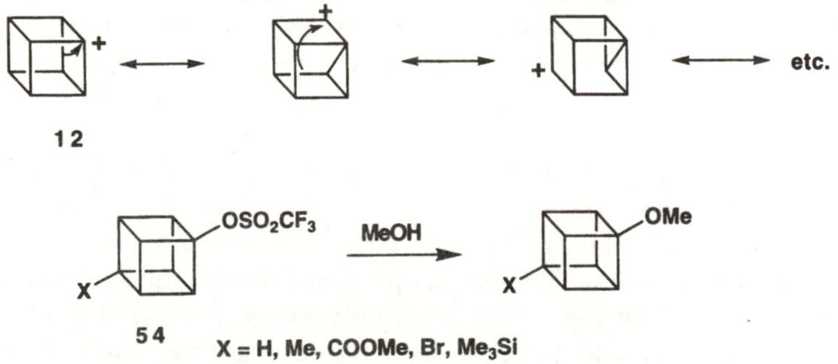

Scheme 4.16

of **12** can be depicted as the result of cross-bonding of the cationic center with three β-carbons, as is represented by the set of the resonance structures in Scheme 4.16.[4d,f,g] As an experimental probe to evaluate the significance of these effects, the solvolysis of substituted cubyl triflates **54** was examined. It was shown that introducing electron withdrawing substituents (*e.g.* X = CO_2Me or Br) at position 4 sharply decreased the rate of this reaction, whereas the presence of donating groups (*e.g.* trimethylsilyl) resulted in an acceleration of the process[4b,f] (Scheme 4.16). Moreover, the sensitivity of the solvolysis rate to changes in solvent polarity was found to correspond to the pattern typical for S_N1 reactions with the intermediacy of carbenium ions.

As we see, the pattern of non-bonded group interactions may change dramatically, depending upon the particular shape of the given cage system. Thus far it has been impossible to account for the observed phenomena of the through-space effects in terms of a unified description.

The multiplicity of these effects seems to be worthy of additional illustrations. As was already mentioned (*vide supra*), cationic species **25** and **27**, derived from dodecahedrane, are very stable. In striking contrast to the 1-adamantyl cation **26**, the positive charge in **25** and **27** is almost completely localized at the carbenium centers. In fact, it was observed that the magnitude of the downfield shift of ^{13}C signals of these centers has the highest value ever observed for carbocations, while the pattern of the ^1H and ^{13}C signals for other centers revealed no peculiar features compared to other tertiary carbenium ions (see Scheme 4.17).[9a] Therefore, the stability of these cations cannot be attributed to the charge delocalization effects mentioned earlier in discussing the peculiarities of cubane or adamantane derivatives. New concepts must be elaborated to interpret the stabilization effects in the dodecahedrane series.

Scheme 4.17 presents a couple of other strange-looking cationic species which were discovered in studies in a related field. In connection with the problem of dodecahedrane synthesis *via* the isomerization of pagodane **34** (*cf.* data in Schemes 4.10 and 4.11), Olah's and Prinzbach's groups engaged in studies of the behavior of pagodane derivatives under superacid conditions. Their hope was to force the cationic isomerization of **34** to **3**.[9b] Despite all attempts, this route was unworkable. As a reward for these apparently futile efforts they were able to observe the unexpected formation of a very stable cationic species, the pagodane dication **55** (Scheme 4.17). The pattern of its NMR spectra combined with the nature of its quenching adduct **56**, and the theoretical analysis of possible alternatives, enabled the authors to ascribe to this dication the unprecedented four-center/two-electron delocalized bis-homoaromatic structure.

While cation **55** was discovered almost by chance, the dicationic species **57** was prepared by deliberate efforts.[9c] Theoretical analysis of the specific configuration of the orbitals at the bridgehead carbons of the adamantane framework encouraged these studies. The analysis predicted a very substantial degree of stabilization for **57**, owing to the possibility of efficient overlap of all four orbitals of the bridgehead carbons inside the adamantane cavity. After some searching for an appropriate precursor, it was established that 1,3-dehydro-5,7-

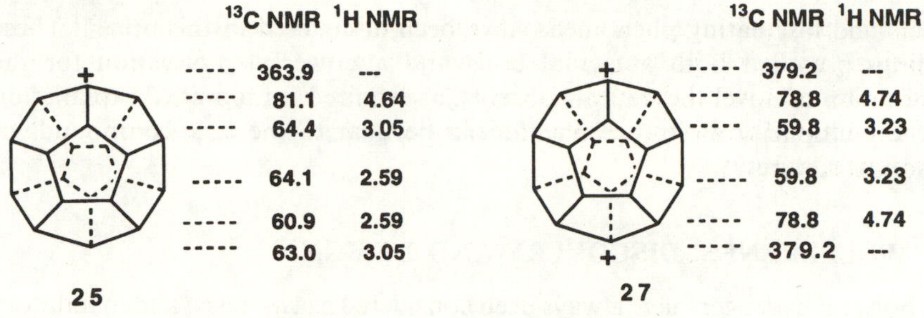

Scheme 4.17

difluoroadamantane **58** (a very unstable compound, which undergoes polymerization above 0 °C) when treated with SbF_5 at −80 °C produces the desired dication **57**. NMR parameters of this species revealed a tremendous shielding of the bridgehead carbons despite the presence of positive charges. This pattern is typical for hypercoordinate carbocationic centers as represented in formula **57a**. It is suggested that **57a** owes its stability to a very peculiar phenomena, so-called three-dimensional aromaticity.

The examples considered in these schemes clearly illustrate the promise of in-depth studies of various caged structures. Already a number of basically novel effects and fascinating phenomena have been discovered in this area.[9d] These findings provided both a factual basis and a powerful motivation for the elaboration of novel theoretical concepts, as required for a unified explanation of the multiplicity of strange phenomena peculiar to the cage compounds of various structures.

4.2 FULLERENES. DISCOVERY AND DESIGN

Carbon and hydrogen have always been considered as two basic and mandatory elements of organic compounds. Recent discoveries in the area of caged structures, however, reveal that a whole family of closed shell compounds composed of pure carbon with the general empirical formula C_n should be included as well in the list of objects to be studied by organic chemistry. At present, only two individual compounds, **59** (C_{60}) and **60** (C_{70}) (Scheme 4.18), have been prepared and unequivocally identified.

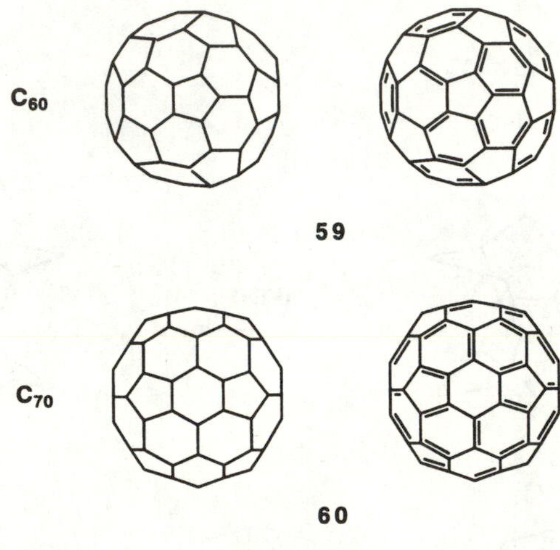

Scheme 4.18

'*Serendipity*, n. the faculty for making desirable discoveries by accident' (Webster's Encyclopedic Unabridged Dictionary of English Language, 1989). Probably the history of studies leading to the discovery of **59** and **60** can serve as an especially good example of the paramount role of *serendipity* in science.

As early as 1966, the British magazine *New Scientist* in the section 'Inventions of Daedalus' published an especially strange-looking project among other bizarre suggestions advanced in an equally quasi-serious way by David Jones ('Daedalus').[10a] The author speculated about the purely fantastic possibility of creating a substance with a density intermediate between that of gases (about 0.001 relative to water) and solids (0.5 to 25). Rather simple calculations led to the conclusion that hollow molecules with a diameter about 0.1 micrometer must have a bulk density of about 0.01. He further suggested that these closed shell structures could be constructed from flat sheets of graphite composed of benzene hexagons, provided some suitable impurities are introduced to warp these sheets. Later on, 'Daedalus' additionally specified that the required warping of these sheets can be achieved if 12 pentagons are included in the hexagon composed network.[10b] This idea actually originated from Euler's mathematical law describing the general prerequisites for the formation of closed polyhedrons from polygons, and in this particular case related to the use of hexagons as the main building blocks. It had been also pointed out by 'Daedalus' that somewhat similar hollow spheroidal structures are produced in quantities by microscopic sea-creatures, radiolaria *Aulonia hexagona*, whose beautifully symmetrical silica skeletons are composed of hundreds of hexagonal units in combination with non-hexagonal polygons.

Rather strangely, it was somehow overlooked at the time that a very similar design had also been suggested by the famous American architect and inventor Buckminster Fuller, who patented in 1951 an original framework for building

Buckminster Fuller
(Courtesy of Buckminster Fuller Institute, Santa Barbara, California, USA. Copyright (1960) Estate of Buckminster Fuller)

construction based upon the same principle of closing the spheroid-like forms ('geodesic dome') (Scheme 4.19).[11] Moreover, this design was extensively used in numerous constructions, including the fascinating US Pavilion at Expo-67 in Montreal.

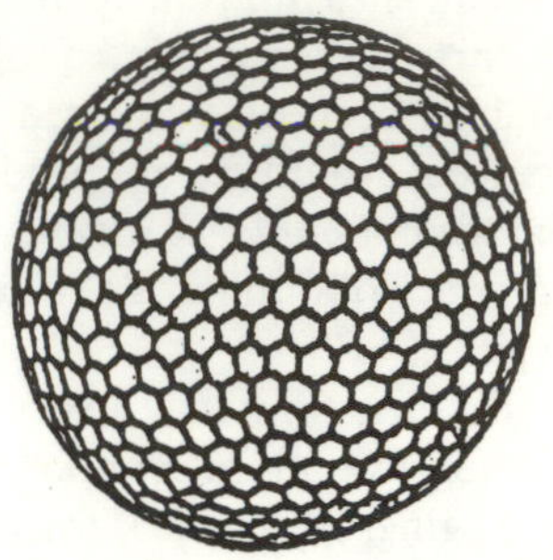

Aulonia hexagona Hkl. (c x 250)

Geodesic Dome design

Scheme 4.19

(*Aulonia hexagona* reprinted with permission from *On Growth and Form*, W. D'Arcy Thompson. Copyright (1992) Cambridge University Press)

Until recently, these ideas seemed to be too crazy to be sold to any 'truly serious' scientist. These speculations passed away almost unnoticed with other equally cute but hardly practical suggestions of 'Daedalus'. So it came later as a real surprise that, at least in this particular fantasy, the author had actually elaborated an original design of an unprecedented molecular shape and, moreover, had made the correct guess about the possibility of using graphite as a starting material for its construction!

The true chemical story of these structures started in the mid-1980s with completely unrelated studies aimed at elucidating the strange peculiarities of the absorption and emission spectra of interstellar matter believed to contain carbon-containing particles of unknown composition. In an attempt to solve this puzzle, model experiments were designed to elucidate the nature of the carbon species formed upon the laser-induced vaporization of graphite, conditions supposedly similar to that of interstellar space. Initial experiments[12a] indicated the apparently indiscriminate formation of a wide set of carbon species C_n within the whole range of $n = 1$–190. However, it was also noted that only even-numbered clusters, C_{2m}, were present in the range $10 < m < 45$ (Fig. 2, curve C). These data were interpreted as evidence of the formation of the linear $(\mathrm{C{=}C})_m$ species related to the previously suggested new form of carbon, carbyne. A year later, in 1985, these experiments were repeated using a slightly modified technique to provide better conditions for the clustering of the initially formed small carbon fragments. The results turned out to be quite astonishing. Instead of the more or less even distribution of carbon clusters that was reported previously,[12a] the mass spectrum registered the appearance of the C_{60} cluster peak about 40 times more intense than neighboring clusters (Fig. 2, curve A).[12b]

It became obvious that a linear carbyne structure could not be accepted as a viable explanation for the observed stability of such a huge molecule. An

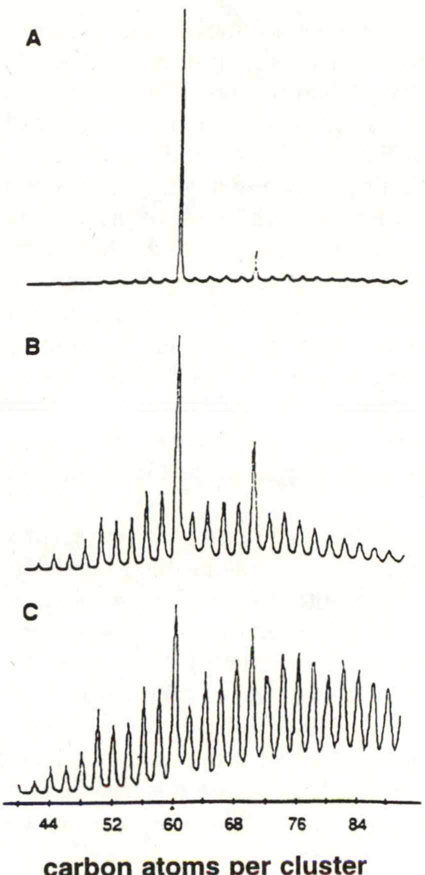

Figure 2 *Mass spectra of carbon cluster distributions in a supersonic beam produced by laser vaporization under conditions of increasing extent of clustering (C to A).* Reprinted (abstracted/excerpted) with permission from *Nature*, **1985**, *318* 162. Copyright (1985) American Association for the Advancement of Science)

alternate and absolutely non-traditional explanation was desperately needed to solve this puzzle. The solution was found almost immediately in a very daring suggestion: this C_{60} cluster represented a new stoichiometric allotropic form of elementary carbon possessing the structure of a closed shell, a polyhedron with 60 vertices, 32 faces, 12 pentagons, and 20 hexagons, *i.e.* **59**.[12b] In fact, at that moment there was no conclusive evidence supporting this suggestion other than its ability to account for the appearance of highly predominant signal of the C_{60} cluster in the mass spectrum (see above). However, as was pointed later by Harold Kroto, one of the authors of this discovery, 'after all, it (this suggestion) was surely too perfect a solution to be wrong'.[12c] In contrast to the similarity between the shapes of the previously known Plato's hydrocarbons or adamantane to well-known forms of crystals, no such counterparts could be found for **59**. The closest material analogy to the truncated icosahedron shape of **59** can be found among the geodesic domes of Buckminster Fuller. In fact, the knowledge of Fuller's 'constructs ... had been instrumental in arriving at a plausible structure'.[12d] While the original patents of Buckminster Fuller[11] had actually embraced all possible areas of application of the suggested principle in building construction (from an Eskimo hut in Alaska to a 150-foot-diameter dome in Hawaii, with the materials utilized ranging from plywood to concrete[11]), it took a rather long time to realize the fruitfulness of the same principle when applied to molecular constructions assembled from a different kind of building block, namely carbon atoms. As an appreciation of the impact of Fuller's ideas, the name 'buckminsterfullerene' was coined for the C_{60} species. Among the number of alternative names, 'soccerene' was also suggested for **59** and 'rugbyball' for **60** (Scheme 4.18), owing to the obvious similarity of these structures to the design used in making balls for European and American football, respectively. As a sort of compromise, the name 'buckyball' was finally adopted for **59**, with the generic name 'fullerenes' suggested for the whole family of C_n closed shell compounds.

The discovery of fullerenes came as a complete surprise for its authors and thus it seems to have nothing in common with rational design (and hence should not be considered in this chapter!). However, *post factum* exhaustive literature search revealed[12d] that the enhanced stability of polyhedrane C_{60} with the shape of buckyball had been already predicted in the early 1970s owing to careful calculations carried out independently by Japanese[12e] and Russian authors[12f] for a variety of potentially possible cage structures composed exclusively of carbon atoms. As a matter of fact then, buckyball had been *designed*, in the exact meaning of this word, long before this molecule was actually prepared. However, as happens too often, this pen-point discovery passed virtually unnoticed.

Not surprisingly, the initial attitude toward the buckyball discovery as a result of mass spectrometry studies was rather cautious. There was a great deal of argument regarding the validity of evidence in favor of the ascribed structure. At the same time, however, its appeal, both purely aesthetic and scientific, was so overwhelming that quite a number of laboratories joined in the race pursuing the ambitious goal of creating a new allotropic form of carbon.

Nothing particularly amazing appeared in this area until 1990. In that year it seemed as if an avalanche was triggered. At present, almost any issue of the leading chemical journals includes articles dealing with the chemistry of **59** and similar structures. The breakthrough was fostered by the elaboration of procedures[13a,b] which opened the way for the preparation of both **59** and **60** (albeit minute quantities were sufficient to determine unambiguously the spectral parameters of these species).[13a] The spherical shape of **59** was ascertained from direct images produced by scanning tunneling microscopy of the solid C_{60}.[13c] These data actually removed all doubts that persisted regarding the identity of this exotic compound. It became clear that the signal in the mass spectrum (shown in Fig. 2, curve A) was actually nothing else but the ID displayed to chemists by a newly emerged VIP member of the community of unusual structures.

Almost immediately, conditions for graphite vaporization were further refined and several publications appeared describing the preparation of **59**, first on hundreds of a milligram scale and then in gram quantities.[13d,e] The whole procedure looks incredibly simple ('cookbook-level recipe' as is stated by Smalley and co-workers[13d]) for a compound that emerged just a few years ago as nothing more than a peak in the mass spectrum of a mixture produced by a very sophisticated technique. In fact, simple evaporation of graphite rods by resistive heating (contact arc) under a partial helium atmosphere (0.3 bar) produces a soot like material which contains up to 14% of a mixture of **59** and **60** in an 85:15 ratio.[13d,e] Both compounds are extracted with boiling benzene and separated chromatographically over alumina.[13e,f] C_{60} is produced as a mustard-colored solid, while C_{70} is reddish brown. Strem Chemicals has now added to its catalog 'an exciting new product, buckyball', which can be now bought for a rather reasonable price (50 mg for $100). The stunning discovery of geochemists followed, as they were able to identify the presence of buckyballs in natural sources. Samples collected from the deposits in the 37-mile-long and 17-mile-wide Sudbury crater, formed 1.85 billion years ago in a collision with a meteorite, were shown to contain up to a few parts per million of C_{60}.[13g] In parallel and independent searches, fullerenes were also found in samples from Cretaceous–Tertiary boundary sites in New Zealand.[13h] The origin of fullerenes in this layer was traced to extensive combustion which occurred in the Earth's history some 65 million years ago, presumably due to a huge meteor impact.

It seems appropriate to make a few related comments. How many elementary compounds are soluble in organic solvents? Certainly halogens (in many solvents), oxygen (in perfluorinated ethers), sulfur and phosphorus (in carbon disulfide), and what else? As for carbon, both diamond and graphite, until recently the only known allotropic forms of this element, are totally insoluble in organic or inorganic solvents. It would not have been seriously considered possible to make any chemical experiments with carbon dissolved in organic solvents. Both C_{60} and C_{70}, however, are moderately soluble in usual organic solvents. It is now possible to manipulate a solution of elementary carbon in benzene or toluene. This unique property opens completely new and very exciting prospects into the various areas of science and technology based on carbon and its compounds. Studies originally based upon the idea of modeling the conditions of the interstellar space ended up with the elaboration of a preparative method for the synthesis of previously unknown spheroid allotropes of carbon, with tremendous ramifications for science as a whole. This story is reminiscent of many other stories of totally unexpected and outstanding discoveries.

How could it happen that, as was pointed out by Kroto,[12c] only 100 μs are required to create C_{60} 'when Paquette and co-workers had taken somewhat longer to make dodecahedrane, $C_{20}H_{20}$'? As we already mentioned, the thermodynamic stability of C_{60} was predicted long ago. At that time, though, no attempts were made to envision the possible routes toward creating this structure. The experimental discovery of the formation of this carbon cluster posed the problem of a reaction mechanism leading to the formation of a highly ordered arrangement of 60 atoms of **59** as a result of a spontaneous process without any 'scaffolds' to control its course. The problem was formulated by Kroto as follows: 'How could the entropy factor inherent in the spontaneous creation of so symmetric an object from chaotic plasma have been overcome?' As a working hypothesis, it has been suggested that, in the course of fragmentation of the graphite network, a set of very reactive C_n clusters with a number of 'dangling bonds' is formed.[12b,c] Among the numerous reaction routes open for these clusters, the one leading to the formation of fullerenes is supposed to start with the warping of the arising intermediate clusters owing to the formation of pentagonal fragments surrounded by five hexagons. A well-known stable compound, corannulene **61**[14] with its bowl-like shape, can be considered as a good model for this arrangement. The formation of fragments with corannulene-like structures provokes a further icospiral nucleation process (see Scheme 4.20), leading to structures like **62**, **63**, and **64**, and eventually to the

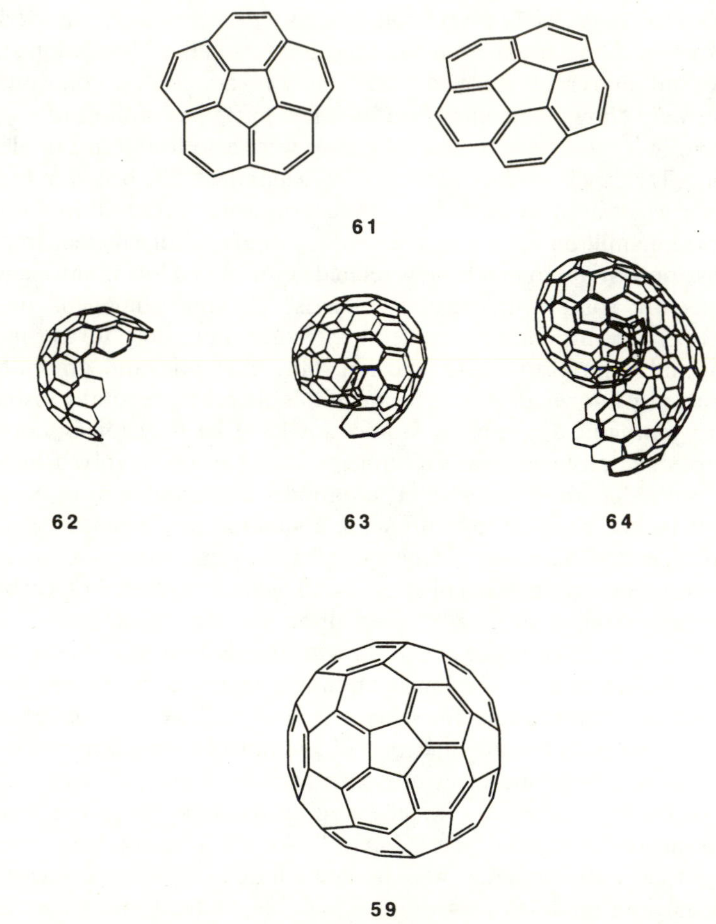

Scheme 4.20

(Reprinted (abstracted/excerpted) with permission from *Science*, **1988**, *242*, 1139. Copyright (1988) American Association for the Advancement of Science)

saturation of all the valencies owing to the irreversible formation of the closed shell structure of **59**.

From a standpoint of organic chemistry this representation cannot be really considered as a mechanism of the reaction, but is rather a description of a complicated stepwise process leading to the transformation of the initial chaos of C_n clusters into the highly ordered structures of C_{60} and C_{70} fullerenes. It is assumed that thermodynamic control serves as the driving force over the sequence of these reversible and kinetically plausible steps.

The sudden appearance of **59** on the chemical scene is likely to produce a truly dramatic effect on the further development of organic chemistry, comparable in its significance with the discovery of benzene in 1825 by Michael Faraday. The essential difference between these two discoveries is that it took nine years to establish the molecular formula of benzene (Mitcherlich, 1834), an additional 31 years to understand its structure (Kekulé, 1865), and several decades more to develop the chemistry in this area. The arsenal of modern science, on the other hand, made it possible to cover such a distance for the case of **59** and related compounds in a matter of just a few years.

From *a priori* considerations it was clear that **59**, consisting of sp^2 carbons comprising aromatic rings, should be a very reactive compound. In fact, **59** can undergo Birch reduction (with Li in liquid NH_3–*t*-BuOH) very smoothly to give hydrocarbon $C_{60}H_{36}$ **65** as a major product[13d] (Scheme 4.21). This reaction is supposed to proceed as a reduction of every cyclohexatriene fragment into a cyclohexene moiety, the latter being stable to Birch conditions. Treatment of **65** with a mild oxidant, 2,3-dichloro-5,6-dicyanobenzoquinone (DDQ), regenerates the starting compound **59**. Reaction of **59** with OsO_4 in the presence of 4-*tert*-butylpyridine gave the monoadduct $C_{60}OsO_4$(4-*tert*-butylpyridine)$_2$ **66**, formed as a result of attack at the double bond of the fused 6,6-ring fragment. This compound was utilized for the first exact determination of the structure of the whole framework by X-ray analysis.[15a] Spectacular results were obtained upon the interaction of **59** with $[(C_2H_5)_3P]_4Pt$.[15b] Here a single isomer containing six platinum atoms, $\{[(C_2H_5)_3P]_2Pt\}_6C_{60}$ **67**, was obtained in 88% yield. The enhanced electron affinity of **59** is evidenced by its easy electrochemical reduction sequentially into mono-, di-, and trianions.[15c] Reduction of **59** with lithium in THF yields a mixture of polyanions which could be quenched with electrophiles like MeI with the formation of a mixture of methylated buckminsterfullerenes, containing up to 24 methyl groups with hexa- and octamethylated derivatives as the predominant components.[15d] The derivatization of **59** is an area of intensive study due to both the obvious chemical interest and the prospects of obtaining new materials (*vide infra*). High temperature chlorination of **59** led to the formation of a polychlorinated product **68**, containing more than 24 atoms of chlorine.[15e] The interaction of **68** with methanol in the presence of KOH produced the respective polymethoxy derivative **69**, while its reaction with benzene in the presence of $AlCl_3$ follows the usual pattern of Friedel–Crafts substitution and gives the polyphenylated compound **70a** with $n \leq 22$. Remarkably, direct reaction of **59** with benzene also proceeds easily in the presence of $AlCl_3$ and yields polyphenylfullerene **70b** with $n \leq 12$.[15f]

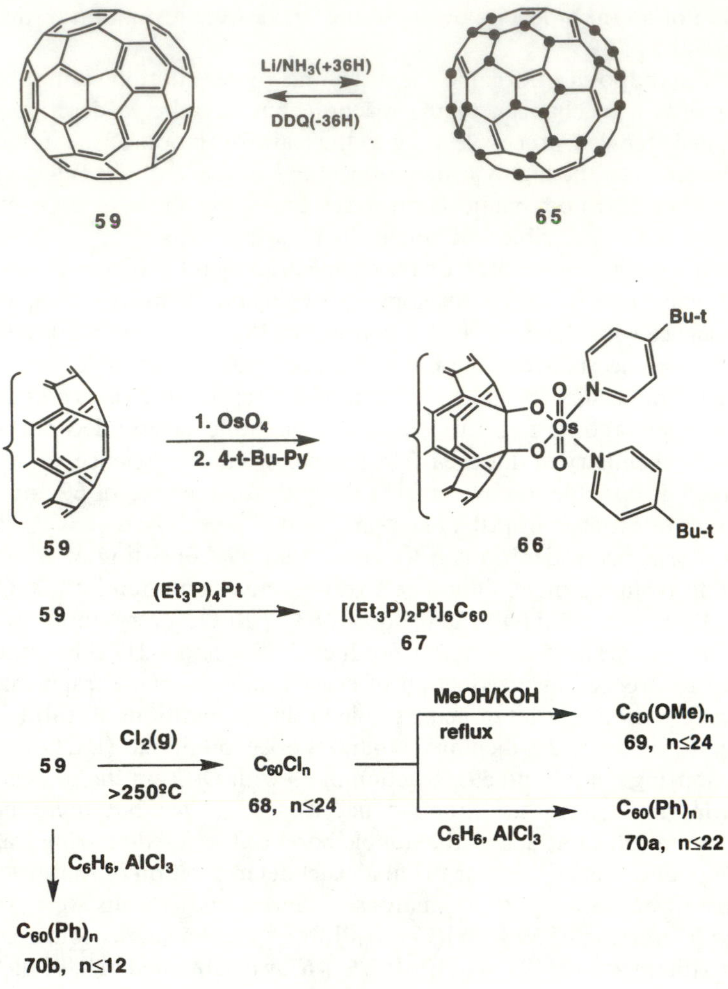

59 $\xrightarrow[\text{DDQ(-36H)}]{\text{Li/NH}_3\text{(+36H)}}$ **65**

59 $\xrightarrow[\text{2. 4-t-Bu-Py}]{\text{1. OsO}_4}$ **66**

59 $\xrightarrow{(Et_3P)_4Pt}$ $[(Et_3P)_2Pt]_6C_{60}$ **67**

59 $\xrightarrow[>250^{\circ}C]{Cl_2(g)}$ $C_{60}Cl_n$ **68, n≤24**

$C_{60}Cl_n$ $\xrightarrow[\text{reflux}]{\text{MeOH/KOH}}$ $C_{60}(OMe)_n$ **69, n≤24**

$C_{60}Cl_n$ $\xrightarrow{C_6H_6,\ AlCl_3}$ $C_{60}(Ph)_n$ **70a, n≤22**

59 $\xrightarrow{C_6H_6,\ AlCl_3}$ $C_{60}(Ph)_n$ **70b, n≤12**

Scheme 4.21

Buckyball **59** exhibits high reactivity toward a manifold of diverse reagents but, in many cases, the analysis of the reaction course is complicated owing to the formation of a mixture of the initial adducts and/or the occurrence of secondary reactions. Thus **59** was shown to be an active 2π component in reactions with regular dienes such as cyclopentadiene, furan, or anthracene, but the instability of the adducts precluded rigorous proof of their structure. This was finally achieved for the product formed in the reaction with **71** (Scheme 4.22), as in this case an initially formed Diels–Alder product easily underwent carbonyl elimination leading to the stabilized aromatic derivative **72**.[15g] Interaction of **59** with the rather sophisticated diene **73** also gave a stable adduct, **74**.[15h] The efficient formation of the products **72** and **74** opened an entry into the preparation of various other functionalized derivatives containing the fullerene

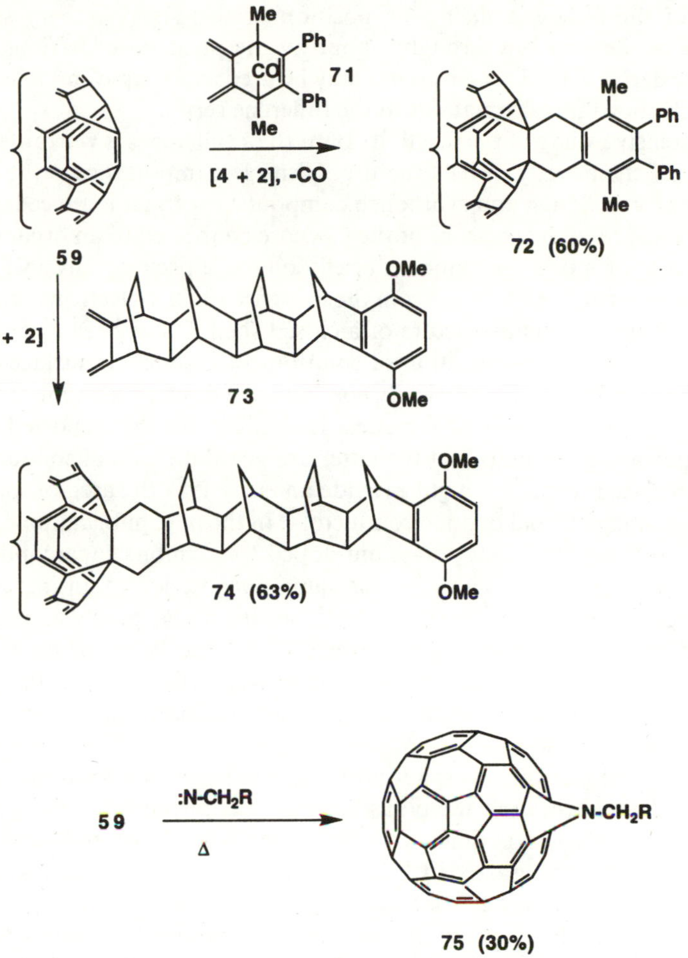

Me
Ph
CO 71
Ph
Me
[4 + 2], -CO
Me
Ph
Ph
Me
72 (60%)
59
OMe
[4 + 2]
73
OMe
OMe
74 (63%)
OMe
59
:N-CH$_2$R
Δ
N-CH$_2$R
75 (30%)

Scheme 4.22

moiety. While, in the reactions mentioned above, the fullerene framework stayed intact, it may be modified in the interaction with reagents like carbenes or nitrenes. Thus reaction of **59** with a nitrene generated *in situ* proceeded as nitrene insertion at the **C–C** bond of the 5,6-fused rings and produced adduct **75**.[15i]

The peculiarity of the fullerene structure suggested some unique opportunities for the utilization of NMR methods for structural studies of substituted fullerenes. Conventional application of ^{13}C NMR spectroscopy for this purpose is often too troublesome because of the appearance of the manifold of low-intensity signals. A spectacular and ingenious solution to this problem, suggested recently, takes advantage of the fullerene ability to form an endohedral complex with the noble gases.[15j] ^{3}He NMR of a helium inclusion complex revealed the presence of a single narrow signal with substantial chemical shifts (C_{60}, -6.3 ppm; C_{70}, -28 ppm from dissolved ^{3}He as reference) due to the effects of the fullerene shell. The position of this signal is very sensitive to changes in the structure brought about by the presence of various groups in fullerene derivatives. This sensitivity may be used as a simple and reliable tool to follow chemical transformations in the fullerene series.

An intensive study of the reactivity pattern of fullerenes is warranted not only because of the novelty of its structure. Of no less importance is the promising opportunity to design and synthesize compounds with an unprecedented set of properties (due to the presence of the C_{60} core connected to an organic moiety). Here it is appropriate to comment briefly on some peculiar physical properties of fullerene itself, which arose from the presence of 60 π-electrons arranged in a three-dimensional rigid structure of a closed shell. It has been established that films formed by the evaporation of solutions of C_{60} on the surface of various materials could be made electrically conductive if doped with a number of alkali metals.[15k] The magnitude of the effect is similar with that reported earlier for doped polyacetylene films, but the latter are one-dimensional conductors while those produced from C_{60} might provide an entry into the area of long-sought-after three-dimensional organic conductors. In the next publication by the same group it was reported that potassium-doped C_{60} exhibits superconductivity at 18 K.[15l] The use of rubidium for the same purpose led to an increase of this temperature, up 30 K.[15m] This is still far from the practical requirements (working at liquid nitrogen temperature, 77 K), but it can be considered as a significant leap toward this goal, especially when compared to the previously achieved record of 7 K (under normal pressure) for superconductors made from organic compounds.

The cavity inside the C_{60} spheroid is large enough to trap metal atoms. The first proof demonstrating the plausibility of this phenomena was provided by the initial gas-phase experiments. In fact it has been shown that laser vaporization of graphite impregnated by UCl_3 produces $C_{60}U$ as a positive ion.[13d] Later, several endohedral clusters like La@C_{60} were isolated in mg quantities and unambiguously identified. It is very likely that quite a number of useful applications can be envisioned owing to this unique property of molecules like **59**. According to some recent calculations, small molecules trapped inside this spheroid would 'exhibit properties radically different from those in the gas phase', owing to the very profound effects of this particular host cluster on the structure of the guest molecules.[15n] At the same time, it may also have ominous ramifications. It was mentioned by Kroto[12c] that clusters like metallofullerenes may have been formed in the Chernobyl disaster. If this really happened, it would not be easy to find a way of removing this extremely dangerous contaminant from the polluted area. The point is that ions encapsulated inside hydrophobic molecules differ sharply from 'free' ions in their solubility and diffusion properties in water. Hence the existing methods, specifically elaborated for monitoring and removing pollutants as hydrophilic inorganic ions, may fail completely if the ions are accumulated in the form of hydrophobic species incarcerated within spheroids **59**.

Above we mentioned just a few examples of the fascinating chemistry of the closed-shell, all-carbon compounds like **59**.[16a] The progress of investigations in this area is so rapid that it is not very wise to speculate about the nature of further discoveries both in the chemistry and in the applications of buckyballs. The best we can do is to refer to the same, less than serious, publication of 'Daedalus', who suggested that the designed hollow materials are able to entrap 100 times their own weight of smaller-size molecules (carbon balloons) and 'would have a host of uses, in novel shock adsorbers, thermometers, barometers, and. . ...in gas-bearings'.[10a,b] It must also not be overlooked that spheroids like **59** actually represent molecular size vacuum chambers. In principle, they can be used to carry out gas-phase reactions under conditions of absolute vacuum. Since Daedalus' main prediction about the existence of these spheroids was rather prophetic, it is hardly surprising to observe that his other speculations regarding their properties are now considered quite seriously. Again we come across a very instructive illustration that in fact 'Imagination is more important than knowledge. . ..' (Albert Einstein).

The story of the discovery of buckyballs originated from a study aimed at the clarification of a particular problem, which represented a certain interest of a select group of astronomers and astrophysicists. At that time, there was no one especially eager to provide financial aid for the investigations of such an esoteric problem as the composition of the cosmic carbon dust. Initial progress in this field was achieved mainly due to the use of leftovers from other, supposedly more useful, projects. However, in a matter of a few years the situation changed dramatically. All of a sudden, physicists and spectroscopists dealing with this project found themselves involved in the creation of a new area of organic chemistry, an area which might turn out to be a real gold mine for exciting discoveries in different branches of science. At present, it causes little surprise to see that several dozen famous laboratories all over the world are desperately trying to seize leadership in these studies, with their efforts generously supported by giants such as IBM or AT&T. The best indication of the fervor of these studies (in addition to an ever increasing flow of publications) can be found in the enormous number of authors (8–17) of even very short papers dealing with C_{60} (see refs. 13–15). With such a boom at hand, it seems very likely that the time span between scientific discovery in this area and its practical application will not be too lengthy.

It can be still argued that the material discussed in reference to C_{60} has little (if any) relevance to molecular design, since in reality this structure did not emerge as a result of a carefully planned experiment. However, design means not only planning and creativity, but insight and imagination as well. Moreover, a very challenging problem for molecular design can be formulated to elaborate the route for a directed synthesis of **59** and related compounds. The problem looks formidable and thus far no attempts have been made to achieve this goal. It seems obvious that a step-by-step protocol will not be especially promising for the total synthesis of **59**. Consideration of the symmetry suggests a number of pathways based upon the assemblege of the target molecule *via* the coupling of two hemispheres. One of the possible options leading to the C_{30} precursor **76** is shown in Scheme 4.23. Previous attempts to perform a similar dimerization on a simpler model (Woodward's design for the synthesis of dodecahedrane, see

Section 3.2.6) resulted in a very spectacular failure. Nature may be more lenient to human efforts this time, especially if one finds the way to achieve a preliminary coordination of two molecules of **76** as ligands in a transition metal complex. To the best of our knowledge, undecacyclic compound **76** has not been synthesized. Its synthesis, however, does not overtly look prohibitively complicated (see, though the preparation of the isomeric decacyclic C_{30} hydrocarbon, described recently[16b]). The dimerization approach as outlined in Scheme 4.23 can hardly be made competitive with the graphite evaporation procedures for the preparation of the parent fullerenes, but it might be useful for the synthesis of the hetero analogs of fullerenes with a predetermined pattern of distributed heteroatoms, as the latter goal cannot be achieved utilizing the conventional stochastic process of thermal synthesis.

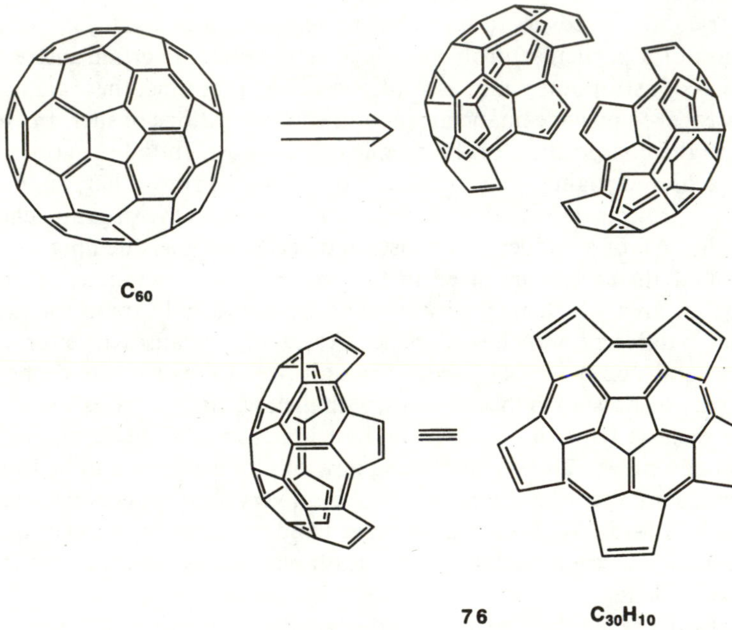

C_{60}

76 $C_{30}H_{10}$

Scheme 4.23

We started this section with a reference to the ancient perception about the elements of 'Perfection' and their role in the Universe. Here it seems timely to remember that the highest symbol of the 'Harmony' and 'Unity' in the World had been always represented in the shape of the Sphere. The repercussions of these beliefs can be found in all religions, very ancient and rather modern. Therefore it is hardly surprising to observe that the preparation of a nearly perfect spheroidal molecule, buckminsterfullerene **59**, besides its overwhelming scientific importance, elicits a special excitement due simply to the intrinsic beauty of this structure and its appeal to the common perception of Harmony.

4.3 TREE-LIKE SHAPED MOLECULES. STARBURST DENDRIMERS, ARBOROLS

By a strange, though probably meaningful, coincidence, during the last few years there appeared quite a number of publications in the area of spheroid-like structures which employed essentially classical approaches for the construction of this shape. The general strategic principle can be easily illustrated using an analogy with the formation of a tree crown *via* the growth and branching of its limbs.

The formation of the idealized ball-like tree crown (the lime tree can serve as pretty close approximation to this model) can be described as a process starting with formation of ***n*** branches from a common center (trunk). Each initially formed branch, after some period of growth, produces ***m*** secondary branches. This consecutive process is repeated until a nearly spherical crown of leaves is closed. If this principle is applied to the growth and branching of organic molecules, then two general patterns can be considered for $n = 3$ or 4 corresponding with $m = n - 1$ (Scheme 4.24).[17a] The distance between the branching points may vary, but the lower limit is determined by the minimal space necessary for the placement of the growing chains. By considering space-filling models it is obvious that, after several cycles of growth, one should arrive at a sphere-like structure ('stuffed ball') with the surface covered with terminal groups ('leaves') and the interior filled with the links ('branches') connected to the central core ('trunk'). As we see, the general design of these tree-like structures is rather straightforward.

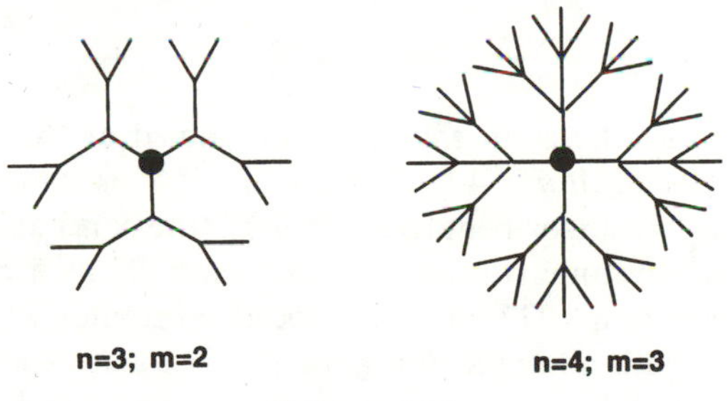

Scheme 4.24

In principle, the strategy of their synthesis is also very simple, since the periodicity of the target structure implies an iterative sequence of a small number of reactions applied to appropriately chosen polyfunctional substrates. In a very general form, it is outlined in Scheme 4.25 for the particular case of $n = 3$. In this scheme, **C-[X$_3$]** represents a trifunctional core of the system and **R-[Z$_2$Y]** is a trifunctional reagent. One functional group, **Y**, of the latter should be able to react with **C-[X$_3$]**, while two others, **Z**, are actually protected forms of function **X** present in the core. Coupling of one equivalent of **C-[3X$_3$]** with three equivalents of **R-[Z$_2$Y]** produces a molecule of the first generation, **Gen-1[Z$_6$]**.

Removal of the protecting groups liberates all six functional groups and the resulting **Gen-1[X_6]** is ready for further reactions with the same reagent.

Thus starting from only two precursors and repeating the coupling–deprotection sequence, one would have an opportunity, at least in principle, to arrive at a branched structure of any size. This prospect may seem to be a little too boring and unchallenging. Since the structures of the tree-like shape had never been prepared previously, it may have been anticipated that complications would inevitably arise in the course of a real synthesis. In fact, the real synthesis was not as smooth as shown in Scheme 4.25.

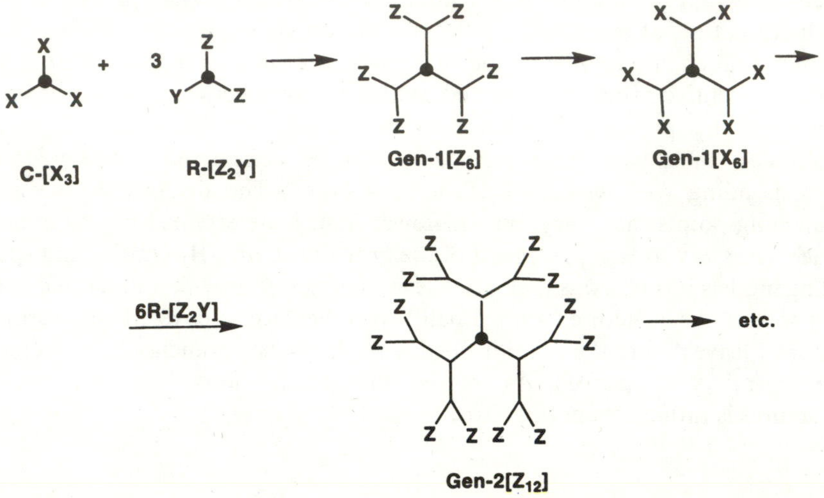

C-[X_3] + 3 R-[Z_2Y] → Gen-1[Z_6] → Gen-1[X_6] → $\xrightarrow{6\text{R-}[Z_2Y]}$ Gen-2[Z_{12}] → etc.

Scheme 4.25

One of the first syntheses in this area was aimed at the preparation of a 'starburst dendrimer' with $n = 4$ (Scheme 4.26).[17b] This synthesis started from the tetrafunctional precursor pentaerythritol **77** (the core) and utilized a triply protected form of the same compound, orthoester **78**, as a chain lengthening unit. The transformation of **77** into the respective bromide **77a** followed by the standard coupling with alkoxide **78a** gave the first-generation protected dendrimer **79**. Removal of the protecting groups from the latter produced dodecaol **80**, which underwent the same sequence of steps, leading eventually to the dendrimers of the second, **81**, and the third, **82**, generations. Thus, extremely simple chemistry (in essence, merely good old Williamson's method for the synthesis of ethers) turns out to be very efficient for the construction of rather complex structures. Further progress along these lines was virtually impossible, though, owing to a trivial reason. Analysis of molecular models of the dendrimer of the third generation, **82**, revealed that these ball-like molecules (about 22–24 Å in diameter) are so closely packed that it is unrealistic to expect that all of the 108 surface hydroxyls could be transformed into the corresponding polybromide, not to mention the opportunity to achieve complete conversion in coupling with the sterically demanding reagent **78a**.

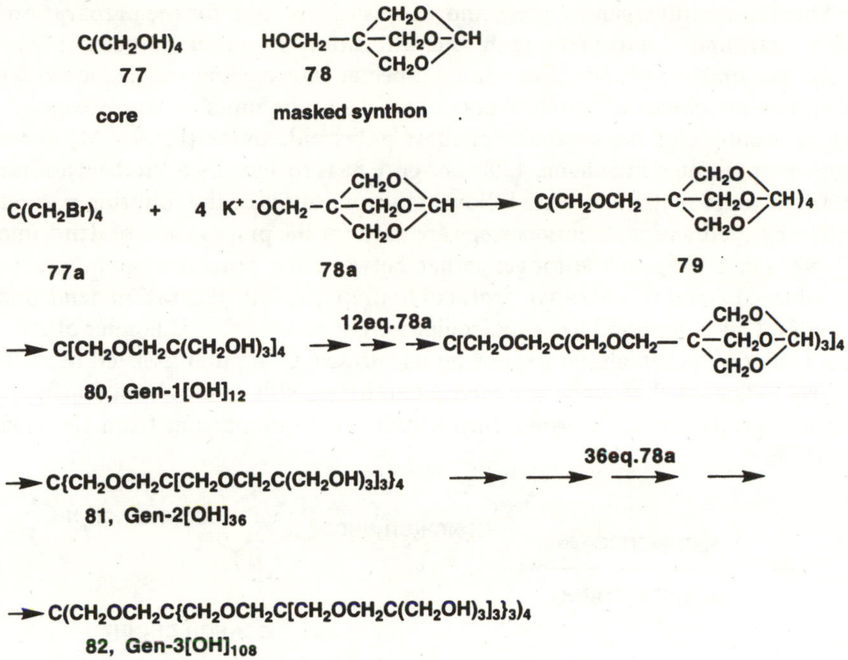

$C(CH_2OH)_4$ 77 core

$HOCH_2-C(CH_2O)_3CH$ 78 masked synthon

$C(CH_2Br)_4$ + 4 K^+ $^-OCH_2-C(CH_2O)_3CH$ ⟶ $C(CH_2OCH_2-C(CH_2O)_3CH)_4$

77a 78a 79

⟶ $C[CH_2OCH_2C(CH_2OH)_3]_4$ ⟶ ⟶ ⟶ (12eq.78a) $C[CH_2OCH_2C(CH_2OCH_2-C(CH_2O)_3CH)_3]_4$

80, Gen-1$[OH]_{12}$

⟶ $C\{CH_2OCH_2C[CH_2OCH_2C(CH_2OH)_3]_3\}_4$ ⟶ ⟶ ⟶ (36eq.78a) ⟶

81, Gen-2$[OH]_{36}$

⟶ $C(CH_2OCH_2C\{CH_2OCH_2C[CH_2OCH_2C(CH_2OH)_3]_3\}_3)_4$

82, Gen-3$[OH]_{108}$

Scheme 4.26

A modified approach to the synthesis of a similar structure can be based on the use of more lengthy tethers between the branching points, like that present in ether **83**.[17c] Reaction of the latter with 4 equivalents of monomesylate **84** proceeded smoothly to give (after removal of the Tr protecting groups) dodecaol **85** (Scheme 4.27). It is to be anticipated that, for this model, similar steric inhibition of the growth will occur at later steps compared with the case shown previously in Scheme 4.26.

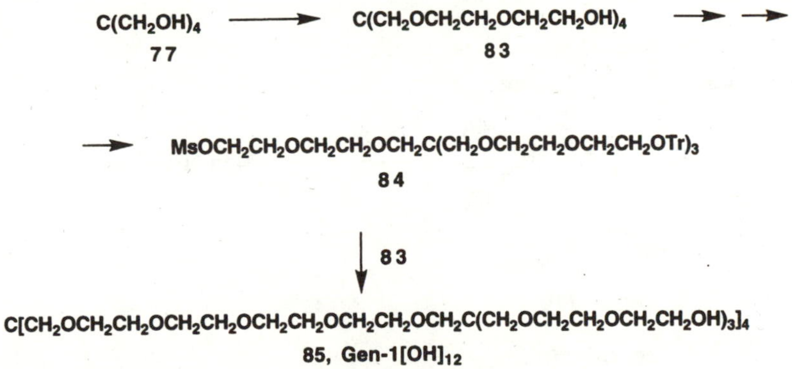

$C(CH_2OH)_4$ ⟶ $C(CH_2OCH_2CH_2OCH_2CH_2OH)_4$ ⟶ ⟶

77 83

⟶ $MsOCH_2CH_2OCH_2CH_2OCH_2C(CH_2OCH_2CH_2OCH_2CH_2OTr)_3$

84

↓ 83

$C[CH_2OCH_2CH_2OCH_2CH_2OCH_2CH_2OCH_2CH_2OCH_2C(CH_2OCH_2CH_2OCH_2CH_2OH)_3]_4$

85, Gen-1$[OH]_{12}$

Scheme 4.27

The stepwise divergent strategy shown was widely used for the preparation of other starburst dendrimers with the utilization of different central cores, monomer units, and tethering chains. Spectacular success was achieved with ammonia or amines as initiator cores and β-alanine units as monomers.[17d] A typical example of the synthesis of these polyamidoamine (PAMAM) dendrimers is represented in Scheme 4.28. The core **86** is formed by a Michael addition of ammonia to methyl acrylate followed by aminolysis of the resulting polyester with ethylenediamine. Identical steps are used for the preparation of dendrimers **87**, **88**, *etc*. Owing to the longer tether between the branching points, it was possible to extend this iterative protocol to prepare ninth generation dendrimers (**Gen-9**). This compound has a molecular weight of 349 883, a diameter of 100 Å, and it contains 1536 amino groups on its surface. Computer assisted modeling clearly indicated that while the **Gen-3** dendrimer still resembles a starfish, its shape is gradually transformed into a ball-like form starting from the **Gen-5** dendrimer.

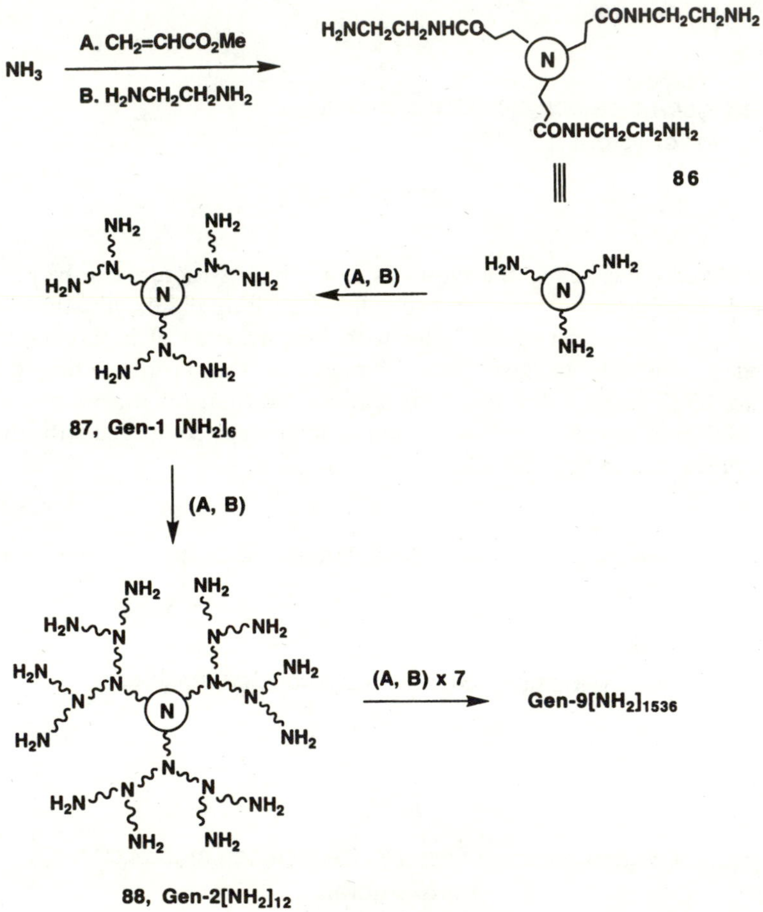

NH_3 —(A. $CH_2{=}CHCO_2Me$; B. $H_2NCH_2CH_2NH_2$)→ **86** [N core with three $CH_2CH_2CONHCH_2CH_2NH_2$ arms] —(A, B)→ **87, Gen-1 $[NH_2]_6$** —(A, B)→ **88, Gen-2$[NH_2]_{12}$** —(A, B) x 7→ **Gen-9$[NH_2]_{1536}$**

Scheme 4.28

Limitations to the starburst growth can be used to advantage in systems specifically designed for multidirectional polymerization. One of the most illustrative examples attesting to the potential of this approach was given in Bochkarev's study of the anionic condensation of $HGe(C_6F_5)_3$ (Scheme 4.29).[17e] The basic reaction used in this scheme involves the deprotonation of precursor **89** followed by reaction of the initially formed anionic species **89a** with the next molecule of **89** (nucleophilic substitution of *para*-fluorine atom) to give monosubstituted adduct **89b** and, ultimately, trisubstituted product **89c** (**Gen-1**). The continuation of this cycle results in formation of the **Gen-2** adduct **89d**. Owing to a tetrahedral configuration of the bonds around Ge atoms and the presence of an increasing number of similarly reactive *p*-F groups at every forming branch, the consecutive three-dimensional growth of the oligomer should lead to the formation of the starburst growing spheroid macromolecule composed of fragments like **90**. The study of molecular models revealed that, after the formation of the **Gen-3** oligomer, the steric screening of the reactive sites increases enormously and further growth is nearly blocked. In fact it was shown that the major component of the mixture formed in the described reaction has a nearly spherical shape (average diameter *ca.* 130–140 Å) and molecular weight of about 174 000 (which roughly corresponds to the composition of the **Gen-3** oligomer). It is especially noteworthy that this highly ordered material is formed as the result of an intrinsically stochastic polycondensation process and that the control of its formation is effected by steric inhibition of further growth at a certain number of branching cycles.

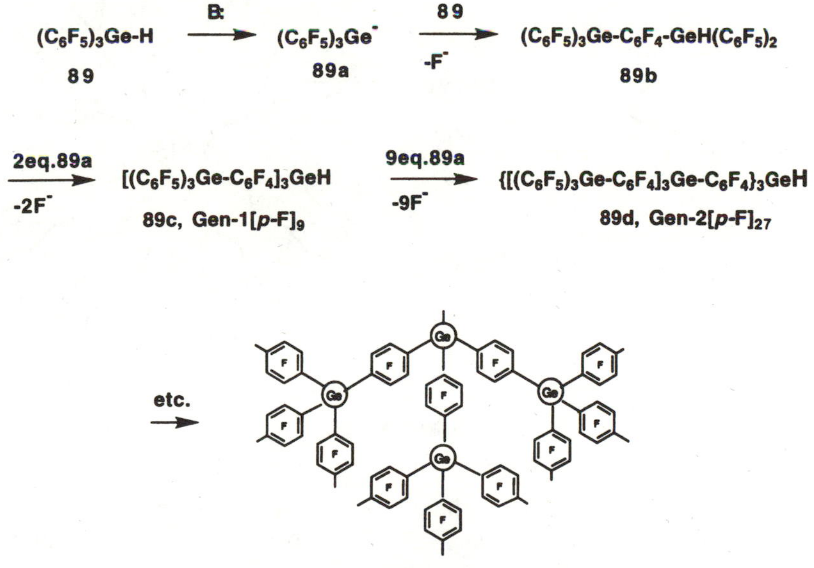

$$(C_6F_5)_3Ge\text{-}H \ (\mathbf{89}) \xrightarrow{B:} (C_6F_5)_3Ge^- \ (\mathbf{89a}) \xrightarrow[-F^-]{89} (C_6F_5)_3Ge\text{-}C_6F_4\text{-}GeH(C_6F_5)_2 \ (\mathbf{89b})$$

$$\xrightarrow[-2F^-]{2eq.89a} [(C_6F_5)_3Ge\text{-}C_6F_4]_3GeH \ (\mathbf{89c}, \text{Gen-1}[p\text{-F}]_9) \xrightarrow[-9F^-]{9eq.89a} \{[(C_6F_5)_3Ge\text{-}C_6F_4]_3Ge\text{-}C_6F_4\}_3GeH \ (\mathbf{89d}, \text{Gen-2}[p\text{-F}]_{27})$$

etc. →

90, Gen-n

Scheme 4.29

An entirely different route to dendrimers less dependent on the restrictions shown above has been suggested based upon the convergent, rather than divergent, topology of the synthetic scheme.[17f] The main drawback of the divergent schemes shown is the multiplication of the number of reactive sites in the molecule with every branching. As a result, the problem of completeness of further transformations becomes more and more acute. The convergent approach outlined in Scheme 4.30 is free from this complication because the number of reactive groups is kept constant along the process of molecular growth. Here the construction starts from the periphery and is directed toward the core. The initial substrate (2 equivalents) containing the fragment of the future surface (**S**) and a reactive functionality (**F_r**) is coupled with monomer (1 equivalent) bearing two coupling sites (**c**) and a protected functional group (**F_p**). After the coupling is performed the latter group is transformed again into **F_r**, and the resulting substrate is ready to participate in the next cycle of operations. The final step for the creation of the spheroid-like structure implies the coupling of four identical branches with a tetrafunctional core. In a way, this sequence of operations mimics the assembling of an artificial Christmas tree from prefabricated branches.

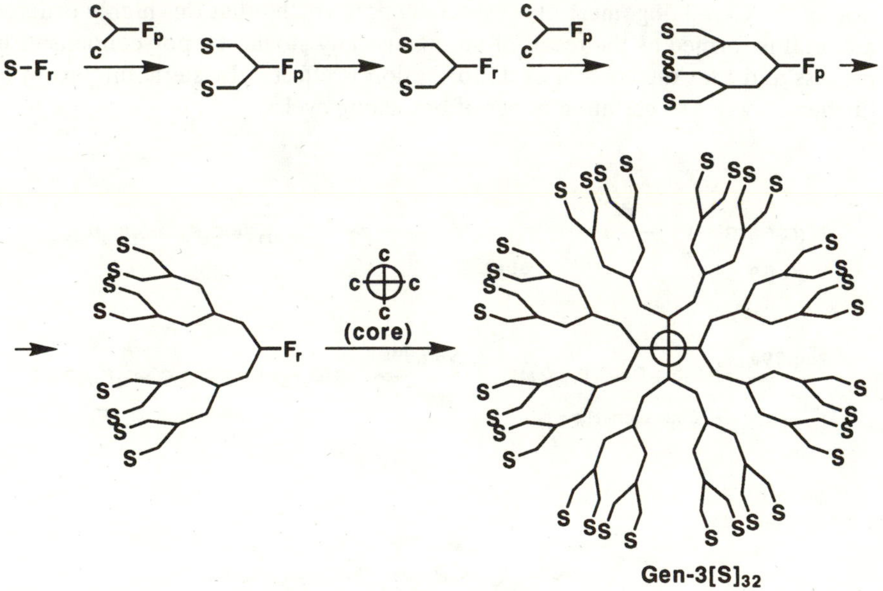

Scheme 4.30

Utilization of this principle for the synthesis of the target tree-like (dendritic) macromolecule is shown in Scheme 4.31. The basic reaction chosen was again very simple, namely the formation of benzyl ethers from phenols and benzylic halides. It was found that the coupling of benzylic bromide **91** with 2 equivalents of 3,5-dihydroxybenzyl alcohol **92** could be carried out without affecting the

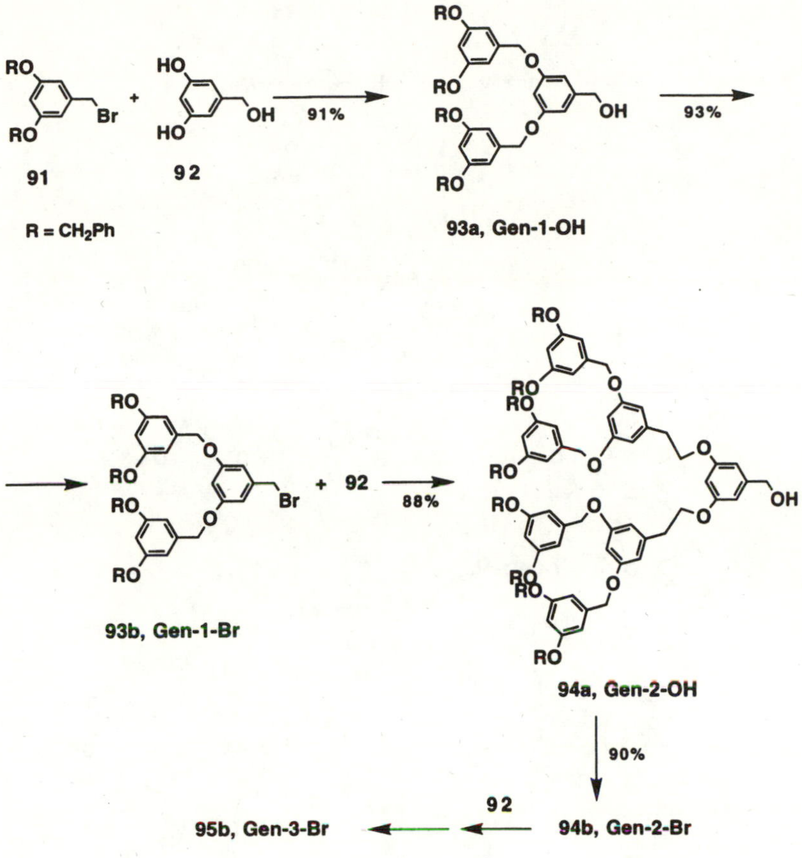

91 + 92 → (91%) 93a, Gen-1-OH → (93%)

R = CH_2Ph

→ 93b, Gen-1-Br + 92 → (88%) 94a, Gen-2-OH

94a, Gen-2-OH → (90%) 94b, Gen-2-Br → (92) → 95b, Gen-3-Br

Scheme 4.31

benzylic hydroxyl in the latter in an almost quantitative yield. The resulting product **93a** was converted into bromide **93b** and two equivalents of the latter were again condensed with the same monomer **92**. The whole sequence was repeated several times, to the level of third generation bromide **95b**. As the central core, the trifunctional compound **96** was chosen and the target dendrites were prepared as exemplified by the coupling of **96** with three equivalents of the bromide **95b** to form **97** (see Scheme 4.31, continued, on the next page).[17f] The fruitfulness of the suggested approach was further ascertained by its successful application in the preparation of more complicated dendrimer structures.[17g]

The most difficult part of this study (and actually of all studies in this field) is the characterization of the homogeneity and structure of the products. Usual mass spectral methods turned out to be applicable for molecular weight determination for **Gen-1** to **Gen-3** dendrimers. The use of a more sophisticated method, the low-angle light scattering technique, permitted the determination of the molecular weight of several key structures up to **Gen-6**. The identity of all compounds was thoroughly checked by a careful study of their NMR spectra,

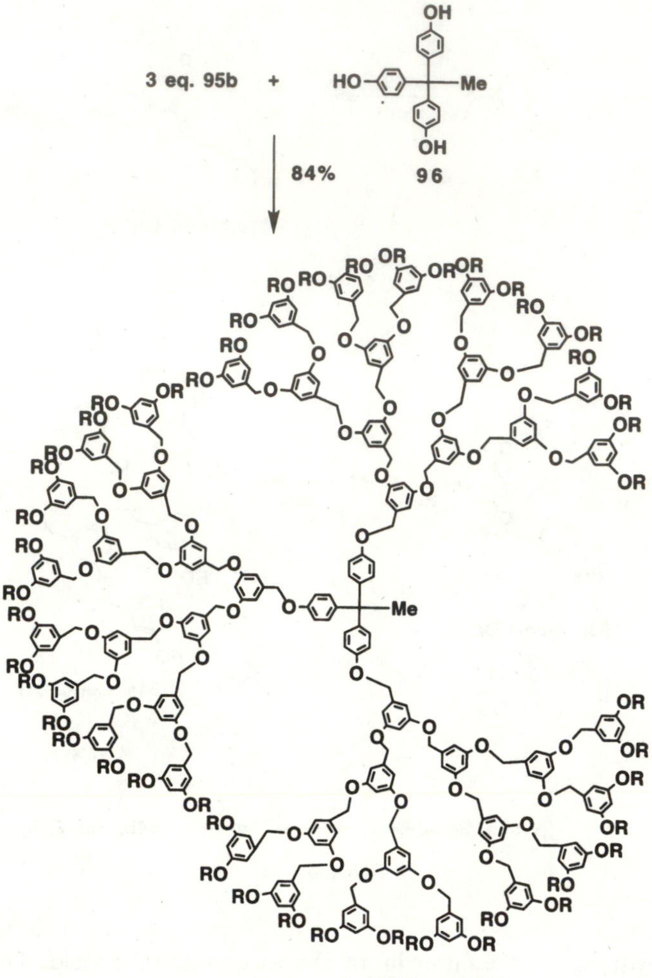

3 eq. 95b + HO–C6H4–C(Me)(C6H4OH)2 (96), 84% → 97, Gen-4[OR]$_{48}$

Scheme 4.31 (*continued*)

but the sensitivity of this method actually precludes its utilization for the evaluation of the homogeneity of the high molecular weight final products.

The area of dendrimer design and synthesis seems to be especially well suited for a nearly unlimited exercise of the chemist's fantasy. One can invent the most unusual shapes of space filling structures and then make an attempt to build the corresponding molecules using the very simple principle of controlled three-dimensional growth. A good example of the promise and potential in this area has been provided recently in a study targeting the preparation of the so-called two-directional arborols with a structure resembling bar-bells, for example **98** (Scheme 4.32).[17h] Compounds like **98** contain two highly hydrophilic moieties separated by a lipophilic chain of variable length. The availability of a regular

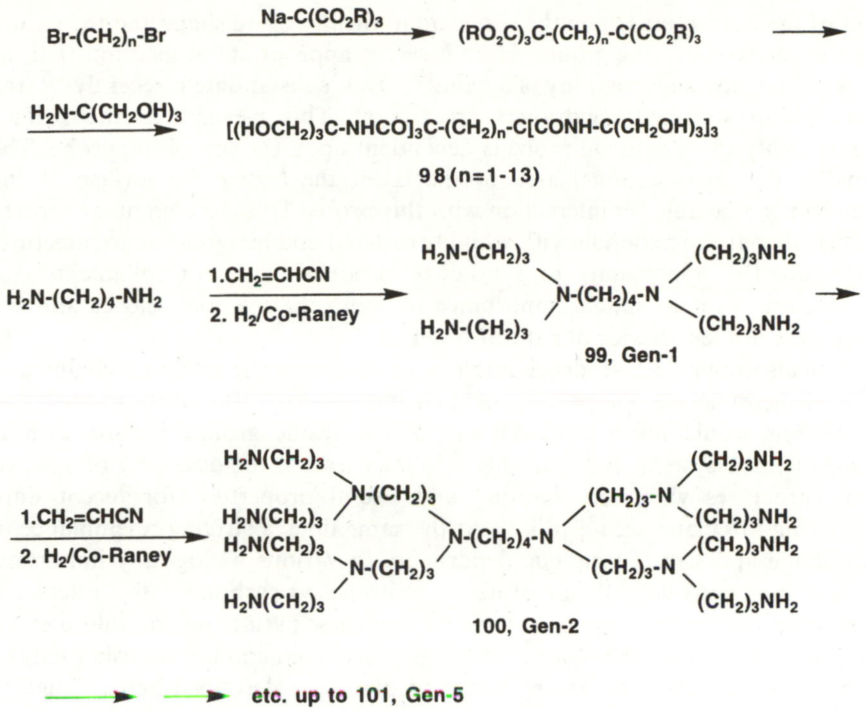

$Br\text{-}(CH_2)_n\text{-}Br \xrightarrow{Na\text{-}C(CO_2R)_3} (RO_2C)_3C\text{-}(CH_2)_n\text{-}C(CO_2R)_3 \longrightarrow$

$\xrightarrow{H_2N\text{-}C(CH_2OH)_3} [(HOCH_2)_3C\text{-}NHCO]_3C\text{-}(CH_2)_n\text{-}C[CONH\text{-}C(CH_2OH)_3]_3$

98 (n=1-13)

$H_2N\text{-}(CH_2)_4\text{-}NH_2$ →(1. $CH_2{=}CHCN$; 2. H_2/Co-Raney)→ $[H_2N\text{-}(CH_2)_3]_2N\text{-}(CH_2)_4\text{-}N[(CH_2)_3NH_2]_2$ →

99, Gen-1

→(1. $CH_2{=}CHCN$; 2. H_2/Co-Raney)→ $\{[H_2N(CH_2)_3]_2N\text{-}(CH_2)_3\}_2N\text{-}(CH_2)_4\text{-}N\{(CH_2)_3\text{-}N[(CH_2)_3NH_2]_2\}_2$

100, Gen-2

→ → **etc. up to 101, Gen-5**

Scheme 4.32

series of these compounds opened an opportunity to study systematically the dependence of such important properties as the interaction with molecules of solvent, aggregation, and gelation upon controllable variation in the molecular shape, number, and nature of terminal groups, tether length, *etc*. Synthesis of two-directional dendrimers **99–101** represents a special interest as these compounds are now manufactured in multikilogram lots and are available in sufficient quantities for broad property studies and prospects of application.[17i]

Now it seems appropriate to make a few comments about the essence of the aforementioned studies. Are there any other reasons besides a scientist's curiosity for justifying such interest in the synthesis of various types of branched structures? To answer this question it is necessary to comment briefly about their peculiar properties.[17a] First of all, such molecules as shown above have a very well defined size and shape, which can be designed *a priori* by the proper choice of the components utilized for the construction of the given compound. Quite often this shape can be directly seen in electron microscope images. It is also very important to recognize that these high molecular weight compounds are individual products, while traditional synthetic polymers are always composed of a set of polymer homologs. The surface of tree-like molecules is covered by the functional groups and their nature, number, and position rigorously defined by the original design. Hence it is possible to evaluate in a very straightforward way the dependence of the chemical and physical properties of these structures upon the variation in their size and shape and the nature of the surface covering groups. Here it seems appropriate to mention that, as was originally suggested by Tomalia[17a] and substantiated recently,[17j] the surfaces of starburst dendrimers are fractal. This means that the surface accessibility to an external probe is dependent upon the size of the probe. The smaller the cross-sectional area of the latter the bigger the surface of the dendrimer available for interaction with this probe. This is a common property of certain natural materials with a highly ordered and hierarchical architecture; therefore the opportunity to synthesize products with controllable fractal characteristics is of special importance for both the material science and the modeling of the behavior of natural systems.

It is also obvious that, depending on the nature of the tethering chains and size of the molecule, the packing of their interior may also be changed. These variations would affect the reactivity of the inside groups toward various reagents. The system studies in this field may lead to the discovery of a set of new structures with very unusual and useful properties (for recent data justifying this claim see ref. 17k,l). At the same time, a striking resemblance in the shape and size of synthetic dendrimers to various biologically important compounds such as globular proteins or branched carbohydrates, offers the opportunity to study and mimic the action of these natural macromolecules. In fact it seems that Mother Nature long ago discovered and ingeniously used the functional benefits provided by the type of design described here. Thus the reserve polysaccharides, such as glycogen, amylopectin and soluble laminaran, which serve as glucose depots for animals, plants, and brown algae, respectively, are assembled as irregular tree-like structures from low molecular weight units. Owing to a nearly globular shape, these natural polymers contain an enormous number of sites (non-reducing ends) available for enzyme attack. Hence enzymatic hydrolysis leading to the release of glucose proceeds at a tremendous rate. This property may be very useful in an emergency, when the organism needs an urgent supply of fresh 'fuel'. The studies of the chemistry and biochemistry of these natural compounds are extremely difficult owing to their complexity and one may expect that the syntheses of simplified and structurally well-defined globular model compounds would greatly facilitate progress in this field.

4.4 COMPOUNDS WITH 'TOPOLOGICAL' BONDS

No one can argue with the statement that the stability of a particular structure depends upon the strength of the chemical bonds holding together the atoms that comprise that particular molecular assembly. The nature of this bonding may vary from fully ionic to purely covalent with little (if any) polarization. They may be very strong or very weak bonds, but *no* chemical bonding between the fragments of a molecule sounds like nonsense. However, the application of purely geometrical considerations to the analysis of molecular constructions[18a–d] led to the paradoxical conclusion that it might be possible to create stable molecules from fragments with no chemical bonds between them whatsoever. Arrangements of some possible types of these fragments are shown in Scheme 4.33, and it is immediately obvious that their existence does not actually necessitate the revision of traditional structural theory.

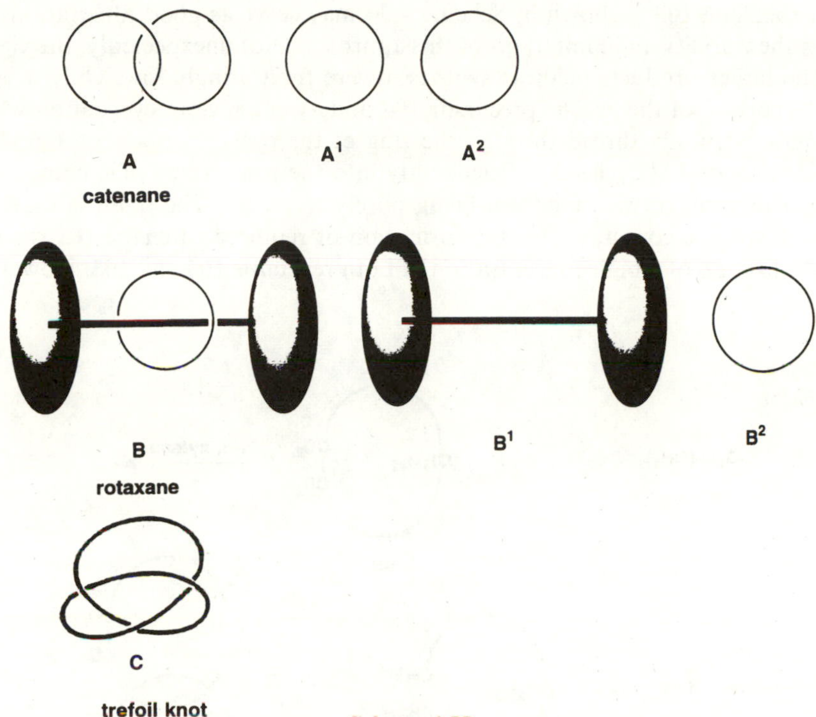

Scheme 4.33

(In part reprinted with permission from *Chem. Rev.*, **1987**, *87*, 399. Copyright (1987) American Chemical Society)

The reasons why these exotic compounds must be as stable as ordinary organic compounds are also obvious. Two cycles, A^1 and A^2 in catenanes (type A) are connected as rings in a mechanical chain. The entire construction can be taken apart only if at least one of the covalent bonds of any ring is broken. The structure of the rotaxanes (type B) can be made stable if the groups, which are placed at the ends of the rod in B^1, are bulky and able to prevent the central ring B^2 from the sliding over the ends. The third arrangement shown, a knot (type C), conventionally belongs to the family of regular organic molecules as it consists of only one fragment. However, its knot-like shape makes this structure very peculiar from both the point of view of its synthesis as well as the properties associated with this unique topology.

So far so good, and appropriate compliments are due to the scientists' imagination which begot these fascinating structures, but how can one envisage the preparation of these compounds? The task seems formidable, especially because the existing powerful arsenal of diverse synthetic methods, perfectly tuned to solve the problem of creating and rupturing covalent bonds in structures of any complexity, does not appear to be useful for assembling molecules of the types shown in Scheme 4.33.

Initial pursuits in this field were based on obvious and rather straightforward statistical considerations. The first successful syntheses of 34,34-catenane **102**[18e] and rotaxane **103**[18f] shown in Scheme 4.34 may serve as good illustrations of both the viability and limitations of this approach. Not unexpectedly, the yields of the target products in these syntheses were frustratingly low. Only a very small portion of the acyclic precursor **104** underwent acyloin cyclization while being accidentally thread through the ring of the cyclic component **105**. The major part of **104** cyclized independently into the respective cyclic compound **106**. This problem was far from being purely technical. The point is that the most favorable conditions for the formation of required catenane **102** require the utilization of a high concentration of both reactants **104** and **105**. However,

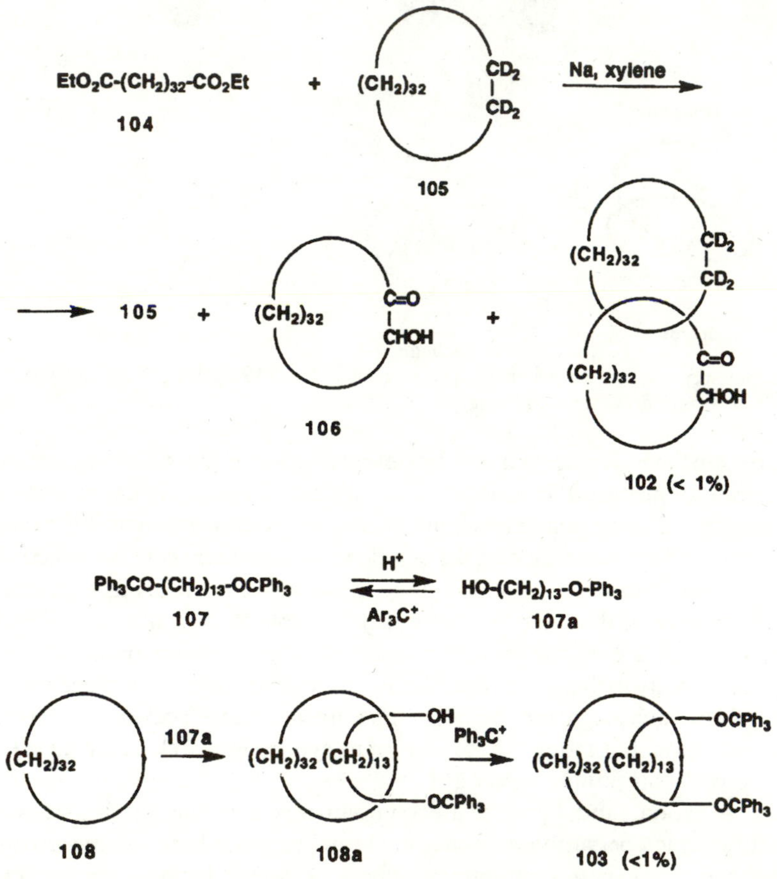

$EtO_2C\text{-}(CH_2)_{32}\text{-}CO_2Et$ (**104**) + $(CH_2)_{32}$ / CD_2–CD_2 (**105**) —Na, xylene→ **105** + $(CH_2)_{32}$ / C=O–CHOH (**106**) + **102** (< 1%)

$Ph_3CO\text{-}(CH_2)_{13}\text{-}OCPh_3$ (**107**) ⇌ (H^+ / Ar_3C^+) $HO\text{-}(CH_2)_{13}\text{-}O\text{-}Ph_3$ (**107a**)

$(CH_2)_{32}$ (**108**) —**107a**→ $(CH_2)_{32}$ $(CH_2)_{13}$ –OH, –$OCPh_3$ (**108a**) —Ph_3C^+→ $(CH_2)_{32}$ $(CH_2)_{13}$ –$OCPh_3$, –$OCPh_3$ (**103** (<1%))

Scheme 4.34

these very conditions are unfavorable for the intramolecular macrocyclization of **104** because of the tendency of **104** to undergo intermolecular oligomerization (*cf.* discussion of this problem in Section 2.18.3). In a similar way, the formation of rotaxane **103** required the intermediate formation of complex **108a** from macrocycle **108** and the monoprotected derivative **107a**, a very unlikely event for the purely stochastic process.

Further developments in this field led to the elaboration of an alternative approach that envisaged the creation of temporary bridges which forcefully held together the parts of the future catenane or rotaxane. Quite a number of catenane and rotaxane systems were synthesized *via* this route.[18b–g] As a rule, though, these syntheses were too cumbersome to be of any preparative importance. It became obvious that the non-trivial character of the target structures necessitates the elaboration of equally non-trivial approaches for their syntheses.

The main problem of catenane or rotaxane synthesis can be formulated as follows: how to hold together the non-reacting partners, like **104** and **105**, or **107a** and **108**, in an orientation to secure the formation of catenane or rotaxane, respectively? Below we will consider several examples illustrative of the modern approach to solving this problem.

Often in this book we refer to *selectivity* as the most burning problem of organic synthesis and mentioned that complexation may be employed as an efficient tool to deal with this problem. This same principle was successfully applied to the elaboration of the directed synthesis of catenane structures.

It is well known that the formation of donor–acceptor complexes is greatly facilitated when one of the components is a macrocyclic ether (see below, Section 4.8). As was shown in the studies of Stoddart's group,[19a] stable complexes are formed between a neutral host (a polyether macrocycle as a donor) and a charged guest (an organic dipyridinium dication as an acceptor, complex **109**, Scheme 4.35). The reverse situation with a charged host (a tetracationic polypyridinium macrocycle as an acceptor) and a neutral guest (a *para*-substituted aromatic diether as a donor) is equally good for the formation of the stable complex **110**. These results prompted the design of a hybrid system, catenane **111**, with the macrocyclic polyether (donor) component of complex **109** combined with the macrocyclic tetrapyridinium (acceptor) component of complex **110**. The idea turned out to be extremely fruitful and was exploited in the most brilliant way. In fact, when bis-pyridinium salt **112** was treated with 1 equivalent of 1,4-bis(bromomethyl)benzene **113** in the presence of a 2.5 molar excess of the macrocyclic ether **114**, the target compound, the tetrapyridinium salt catenane **111**, was formed in 70% yield (Scheme 4.35). The efficiency of this reaction is due to the initial formation of the stable donor–acceptor complex between the reactants **112** and **114**, which reacts with **113** to give intermediate **114a** and then **111**. This complexation sharply reduced the possibility of the reaction of **113** with the non-complexed **112**, which may have led otherwise to the closure of the tetrapyridinium macrocycle, as in **110** without the participation of **114**. This explanation was further validated by a thorough study of the NMR spectra of **111**, which revealed very strong non-bonded interactions

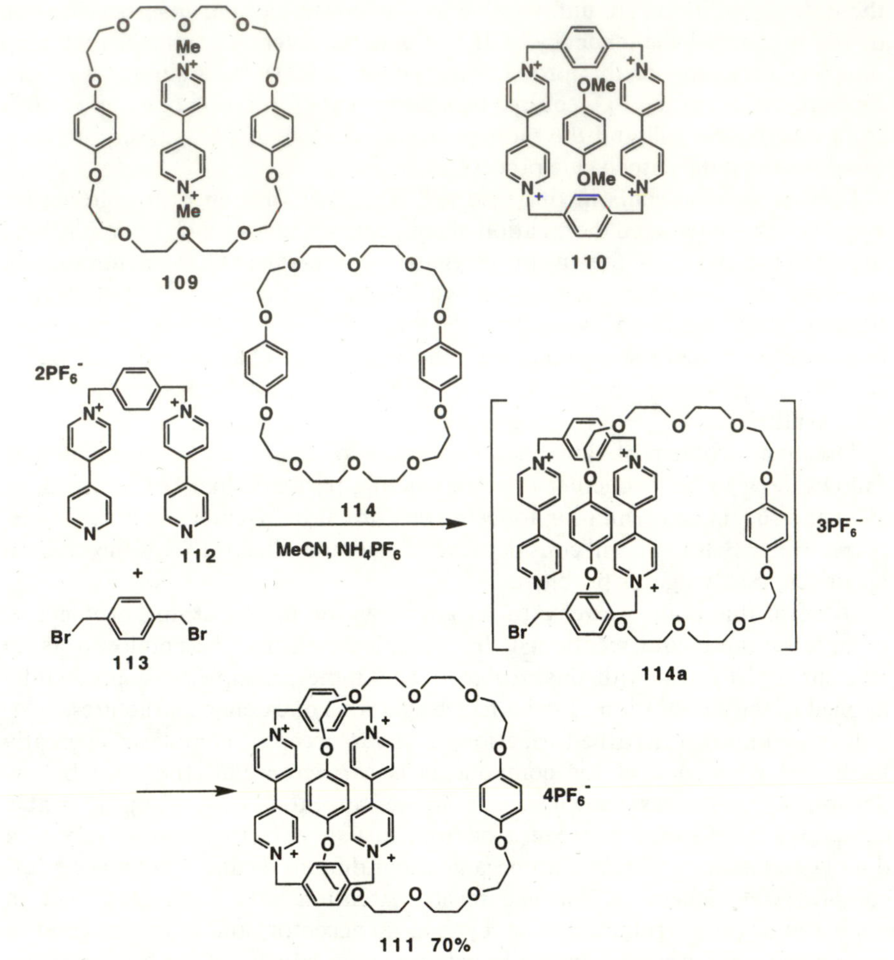

Scheme 4.35

between the two rings of this catenane (the barrier for the rotation of these rings is equal to 12–14 kcal/mol^{-1}).

The ease of spontaneous formation of a structure of that complexity is really striking and the authors compared its efficiency with the well-known but little understood phenomena of self-assembling in complicated molecular systems in living cells. The same principle of self-assembly worked equally well when applied to the synthesis of an analog of **111**, containing 1,5-dinaphtho ether units instead of the hydroquinol units present in **114**.[19b] Stoddart's group has also checked the opportunity to apply the elaborated approach to the synthesis of [3]catenanes, containing two rings on a common macrocycle. Molecular modeling predicted that this goal could be achieved providing additional phenylene units are installed both into the bis-cationic unit, as in **115**, and into the dibromide, as in **116**. This prediction was absolutely correct, and in fact the interaction of equimolar quantities of **115** and **116** in the presence of seven equivalents of macrocycle **114** led to the formation of [3]catenane **117** in 20% yield (Scheme 4.36)[19c]. Among the number of interesting properties of **117**, the most spectacular is probably the one involving the phenomena of translational isomerism. NMR spectral studies of this compound revealed the simultaneous movement of the opposite bipyridyl fragments from one pair of diametrically opposed hydroquinol rings to the other one. 'Molecular trains' is how this system was called and, as is stated in this paper, here 'crashes do not happen'.

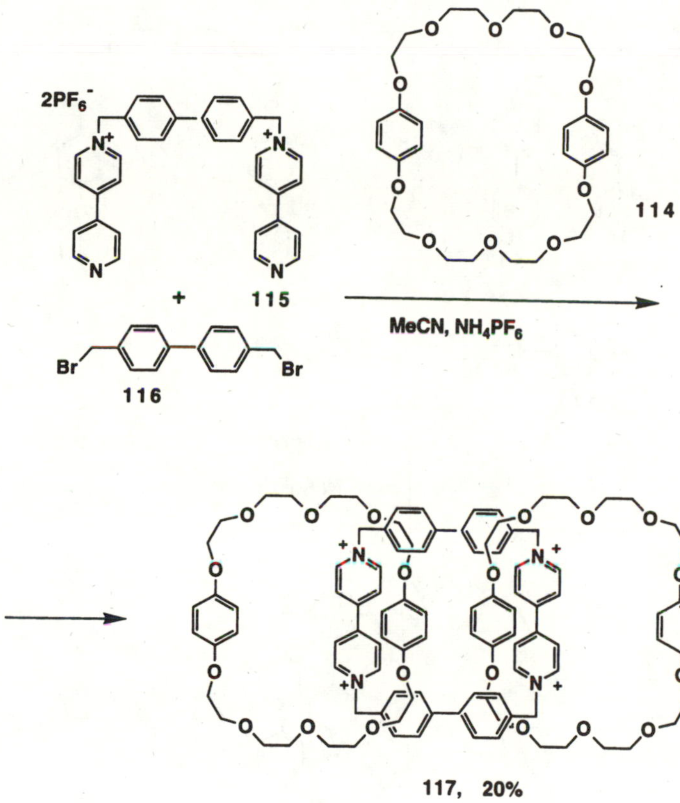

$2PF_6^-$ 115 + 116 114 MeCN, NH_4PF_6 → 117, 20%

Scheme 4.36

The observed ease of formation of [3]catenane **117** is especially remarkable in view of the complete failure to achieve the formation of the parent tetracationic macrocycle from **115** and **116**, in the absence of crown ether **114**. This result prompted the following important conclusion in the cited paper, 'it is easier to construct molecular assemblies than it is to create one of their components by itself. The message is a simple but exciting one — the prospects for self-assembly, as a synthetic paradigm, are extremely encouraging'.[19c]

The efficiency of this self-assembling approach was further demonstrated by its application for the directed synthesis of rotaxane **118** (Scheme 4.37).[19d] The interaction of the same components, **112** and **113**, carried out in the presence of the acyclic polyether **119** (also derived from hydroquinol) gave **118** directly in satisfactory yield. It seems obvious that the feasibility of this coupling is again determined by the initial formation of the complex between **112** and **119**, which is stabilized because of non-covalent binding, as was the case in the above-mentioned synthesis of catenanes. The generality of the scope of assembling other rotaxanes along this route has been amply demonstrated.[19e–g]

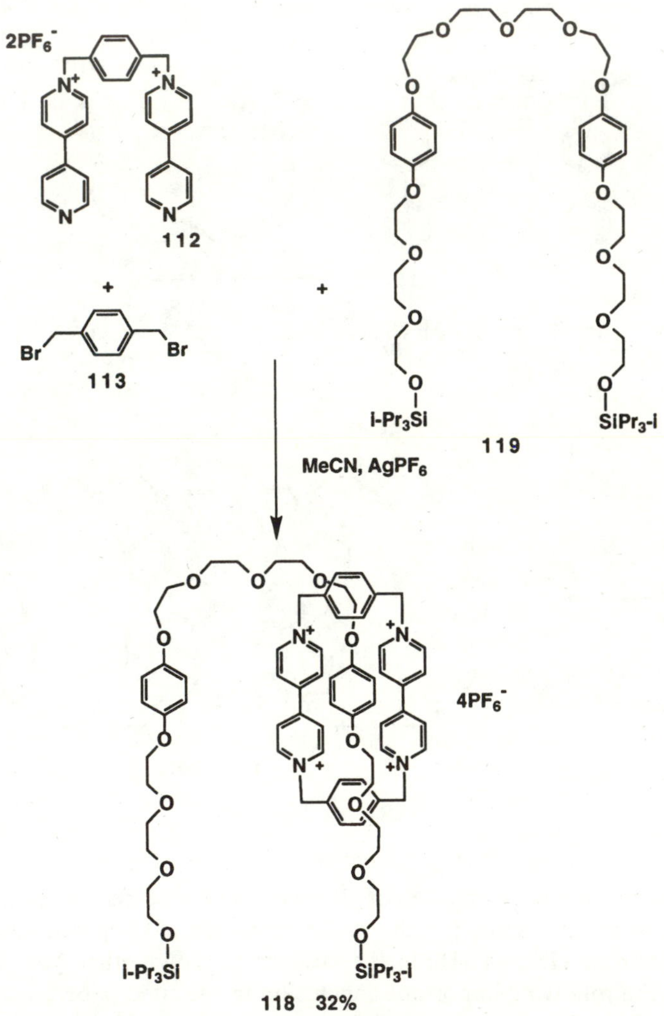

$2PF_6^-$ **112** + **113** + **119** → MeCN, $AgPF_6$ → **118** 32% ($4PF_6^-$)

Scheme 4.37

A noteworthy observation was also made in connection with the structure of **118**. Its central 'string' contains two equivalent donor sites, hydroquinol units, each capable of coordinating the tetracationic 'bead'. Detailed analysis of the NMR spectra of **118** revealed that there exists a substantial barrier for the shift of this 'bead' from one site to another and that the latter 'moves back and forth like a shuttle between two identical stations'. In this particular case this process is degenerate owing to the identity of the 'left' and 'right' ends of the 'string'. However, in the authors' opinion,[19d] 'the opportunity now exists to desymmetrize the molecular shuttle by inserting nonidentical "stations" along the molecular polyether thread in such a manner that these different "stations" can be addressed selectively by chemical, electrochemical, or photochemical means and to provide a mechanism to drive the "bead" to and fro between "stations" along the "thread". Insofar as it becomes possible to control the movement of one molecular component with respect to the others in a [2]rotaxane, the technology for building "molecular machines" will emerge'.[19g] Thus, studies originally initiated as typical 'structure oriented designs' now become remarkably reoriented into 'function oriented designs'.

Quite different and again very efficient 'complexation-based' routes to the synthesis of compounds with a 'topological' bond have been devised by Sauvage's group.[20a] This approach is based upon the well-known ability of transition metals to form three-dimensional complexes with appropriate ligands and for these to serve as templates to secure the spatial pre-arrangement of the reactants required for the formation of this or that particular structure (*cf.* the successful synthesis of cyclooctatetraene from acetylene using the preliminary coordination of Ni, see Section 2.19.4). Taking advantage of this property, several strategic options may be conceived for the construction of the interlocking ring system of catenanes (see Scheme 4.38).

Strategies A and B were successfully used for the synthesis of [2]catenanes.[20a] Here we shall discuss only strategy C, which is especially efficient for the preparation of [3]catenanes (Scheme 4.39).[20b] The crucial part of the route shown in this scheme involved the initial formation of triple complex **120** between macrocyclic polyether **121**, bis-propargyloxy derivative **122**, and a copper(I) salt. This reaction proceeded almost instantaneously at 20 °C and gave **120** in nearly quantitative yield. This is probably one of the most impressive examples of an entropy disfavored reaction being carried out with great ease because of the additional driving force provided in the opportunity to form extremely stable tetradentate coordination between a central copper ion and two bidentate phenanthroline moieties of the reagents.

The next step, dimerization leading to the formation of the [3]catenane system **123a** *via* oxidative coupling of acetylenic moieties, was also rather efficient (although there were no obvious reasons to account for the nearly complete absence of linear oligomeric products or the ease of formation of bis-cationic species *via* coupling of monocations). Finally, the central metal core was removed by oxidation and the target compound **123**, containing interlocked

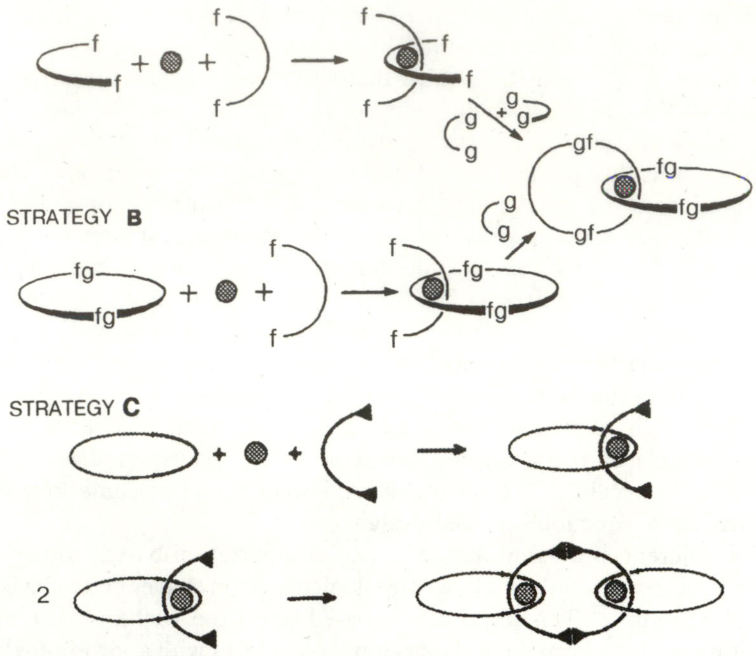

STRATEGY A
STRATEGY B
STRATEGY C
For A and B: f and g designate functional groups capable of forming f-g bond: f-f moiety is capable of serving as a bidentate ligand for the transition metal (circle)
For C: triangles represent identical functional groups capable of forming bonds in cyclodimerization reaction

Scheme 4.38

(A and B reprinted with permission from *Chem. Rev.*, **1987**, *87*, 795; C reprinted with permission from *Chem. Rev.*, **1987**, *87*, 399. Copyright (1987) American Chemical Society)

30–44–30 membered rings, was obtained as the major product in a satisfactory overall yield. It was also found that besides the cyclodimerization leading to **123a** ($N = 2$) cyclooligomerization also takes place. As a result, higher cyclooligomers are formed as by-products. Among the latter, the multi-ring catenanes **123b** ($N = 6$), containing up to six 30-membered rings interlocked to the central ring, were identified (Scheme 4.40).[20c]

While several dozen reports describe the preparation of catenanes with various degrees of efficiency,[18–20] until only recently the knots were still considered as unattainable compounds. In fact, as is stated in a fundamental publication of Frisch and Wasserman,[18a] the probability of knot formation by the cyclization of a C_{60} chain is extremely low (*ca.* 10^{-3}–10^{-2}) and it may be substantially increased only if the chain is composed from several hundred links. As the authors concluded, 'we regard a test for this guess as something for the far future'. However, 10 years later, Sokolov pointed out that an entirely

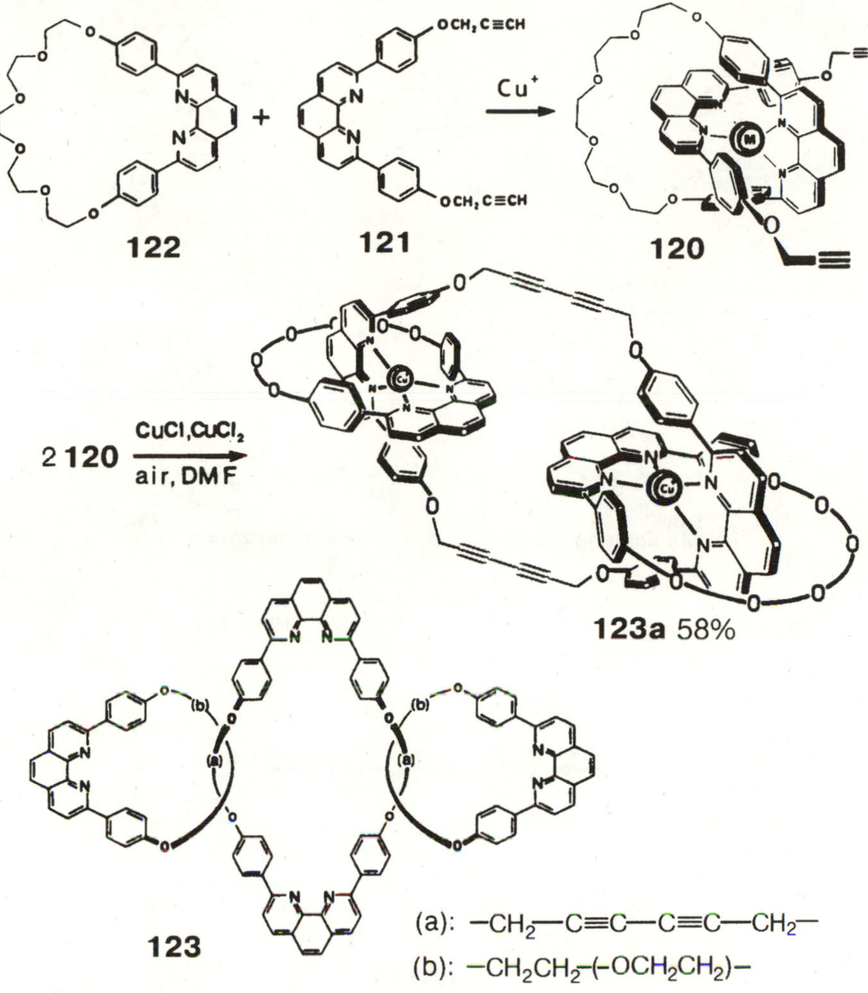

Scheme 4.39
(In part reprinted with permission from *Chem. Rev.*, **1987**, *87*, 399. Copyright (1987) American Chemical Society)

different approach toward the synthesis of knots could be conceived by taking advantage of the ability of transition metals to form three-dimensional complexes with various ligands.[18c] The idea seemed to be promising but its fruitfulness was demonstrated only recently by the first synthesis of the trefoil knot by Dietrich-Bucherer and Sauvage.[20d,e] The strategy of this synthesis is generalized in Scheme 4.41.

This strategy envisaged the formation of a double helix composed of two bis-chelating threads twined by chelation around two metal ions. The choice of the 1,10-phenanthroline unit as the bidentate moiety was dictated by the successful

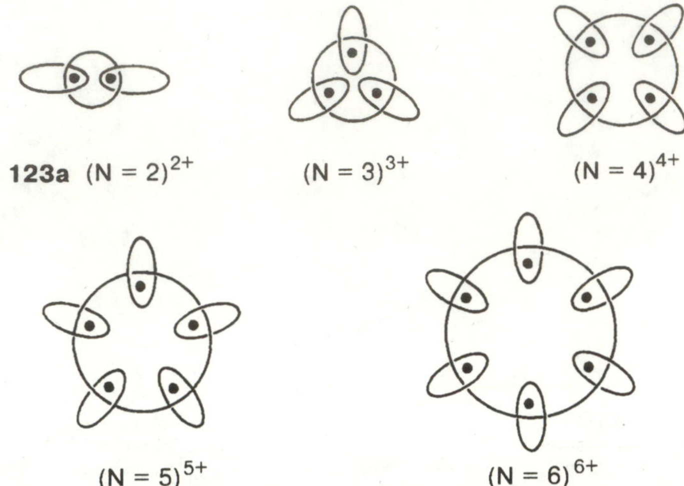

Scheme 4.40
(Reprinted with permission from *J. Am. Chem. Soc.*, **1991**, *113*, 4023. Copyright (1991) American Chemical Society)

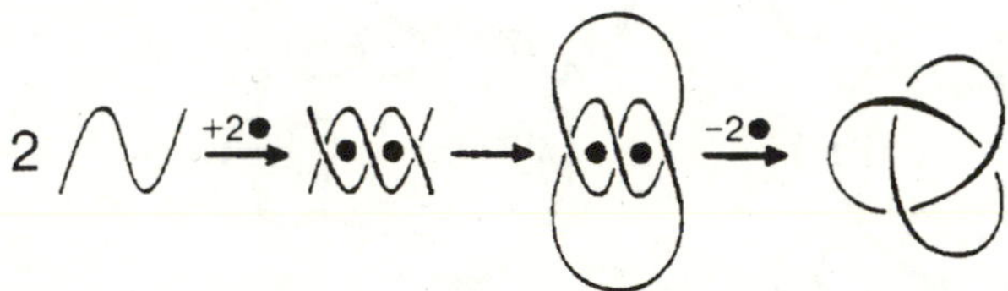

Strategy for making trefoil knot based on the intermediate formation of doubly helixed molecular threads due to the template effect of two transition metal atoms (black circles)

Scheme 4.41
(Reprinted with permission from *Chem. Rev.*, **1987**, *87*, 795. Copyright (1987) American Chemical Society)

previous experience of the same group in the synthesis of catenane systems (see above). Several additional factors were taken into consideration in choosing the structure of the acyclic precursor to minimize side reactions (such as the formation of mononuclear complexes by folding of the bis-chelating thread around one metal ion). These and other more or less obvious considerations ultimately led to the design of compound **124** as a candidate for testing.

Reaction of **124** with copper(I) proceeded readily and gave the desired double-helix complex **125**, but in rather modest yield (*ca.* 15%). Transformation of this bimetallic complex **125** into the knot-like structure required the interconnection of opposite ends of the intertwined threads, as is shown in Scheme 4.41. To achieve this goal, **125** was treated with the diiodo derivative of hexaethylene glycol **126**. This reaction turned out to be non-selective and a

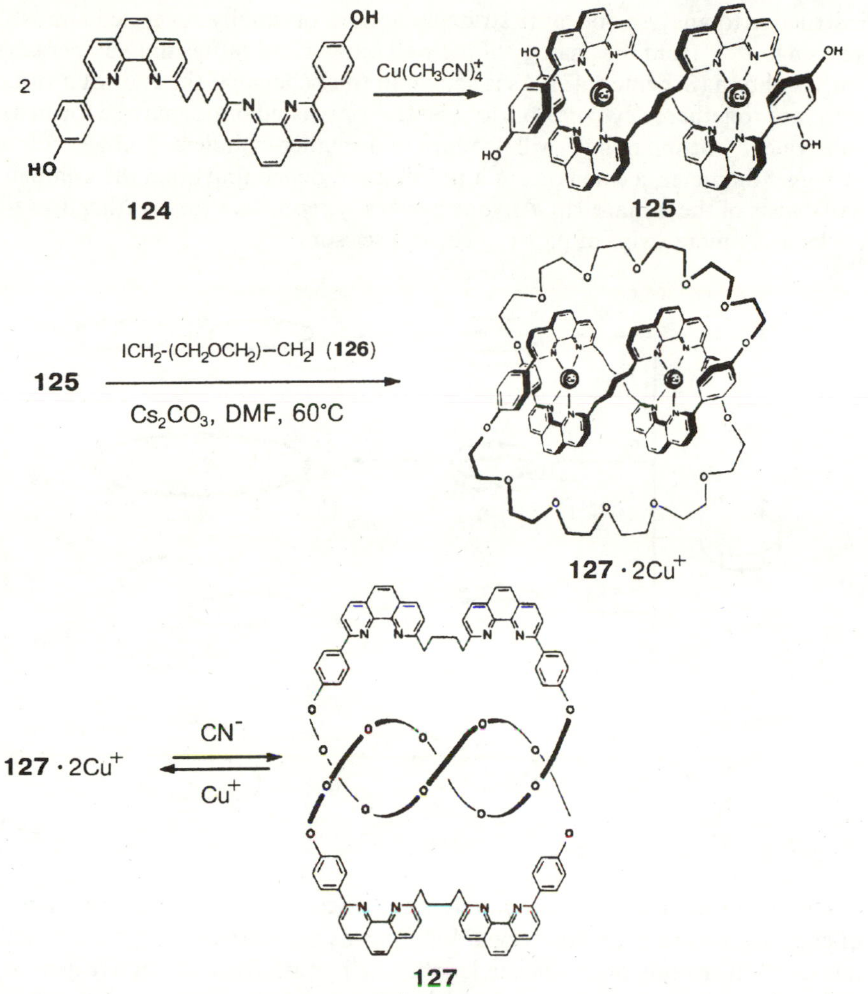

2 **124** $\xrightarrow{Cu(CH_3CN)_4^+}$ **125**

125 $\xrightarrow[Cs_2CO_3,\ DMF,\ 60°C]{ICH_2\text{-}(CH_2OCH_2)\text{-}CH_2I\ (\mathbf{126})}$ **127 · 2Cu$^+$**

127 · 2Cu$^+$ $\underset{Cu^+}{\overset{CN^-}{\rightleftharpoons}}$ **127**

Scheme 4.42

(Reprinted with permission from *Acc. Chem. Res.*, **1990**, *23*, 319. Copyright (1990) American Chemical Society)

number of products were formed. However, a noticeable amount of **127 · 2Cu**$^+$ was isolated and fully characterized as a trefoil-knot structure. Demetallation gave the desired knot **127**. While the yield of the latter was less than spectacular (*ca.* 3%),[20f] its synthesis represents an outstanding achievement both as the solution of a decades-old challenging problem and as the brilliant illustration of the validity of the original synthetic design.[20g]

The non-trivial character and the structural peculiarities of compounds with 'topological' bonds stimulates the elaboration of equally non-trivial approaches to the solution of synthetic problems in this field. In this respect it seems instructive to analyse one more strategic option, originally suggested almost 30 years ago,[18a–c] from the analogy of the well known and rather simple operations with a ribbon (Scheme 4.43). As is shown in this scheme, if the ends of a ribbon are stuck together, a cylinder, a Mobius strip or twisted braid, may be formed. A subsequent 'cutting in half' will produce two smaller cylinders, a single cylinder of bigger diameter, a catenane, or a trefoil knot, depending upon the number of half-twists of the initial strip. Anyone can easily reproduce these rather amazing tricks using mere strips of paper, glue, and scissors.

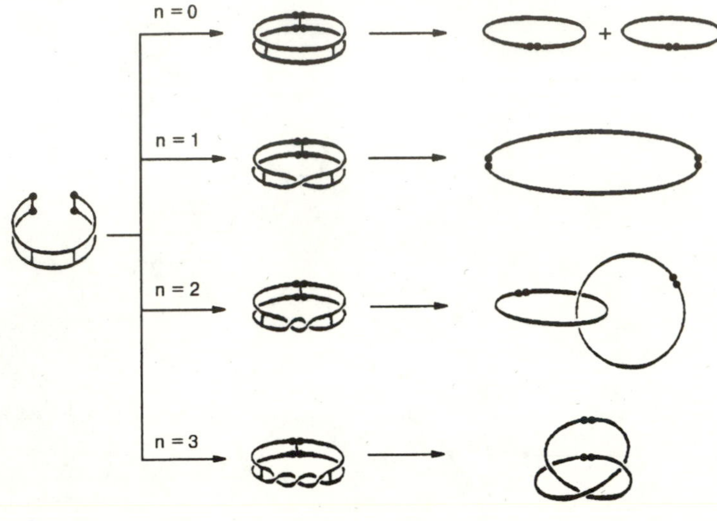

Scheme 4.43

How can one translate these purely mechanical operations into the language of chemical formulas and reactions? First, it was necessary to design a molecular analog of the ribbon of a sufficient length which must: (i) have reactive groups at the ends, to be used for the intramolecular cyclization of this strip, and (ii) consist of two strands connected with temporary bridges which could be destroyed after the ends of the strip are bound together. The structure of the oligomeric polyether **128**, consisting of two strands connected by **C=C** bridges, was chosen by Walba[21a] as a possible candidate satisfying the above requirements (Scheme 4.44).

Inspection of molecular models revealed that an intramolecular cyclization of oligomers **128** may result in the formation of unusual structures if $m > 2$. For $m = 3$, a Mobius strip, II, could be formed (from a half twisted **128**) in addition to the trivial product, cylinder I. Synthetic studies in this field initially produced rather encouraging results. It was shown that a Mobius-shaped structure, **129a**, could be formed in a satisfactory yield from **128a** ($m = 3$) together with the conventional cylinder-shaped product **130a**. Ozonolysis of **129a** led to the rupture of the rungs in this Mobius ladder and resulted in the formation of the macrocylic hexaketone **131**. As was expected, the oxidative cleavage of the

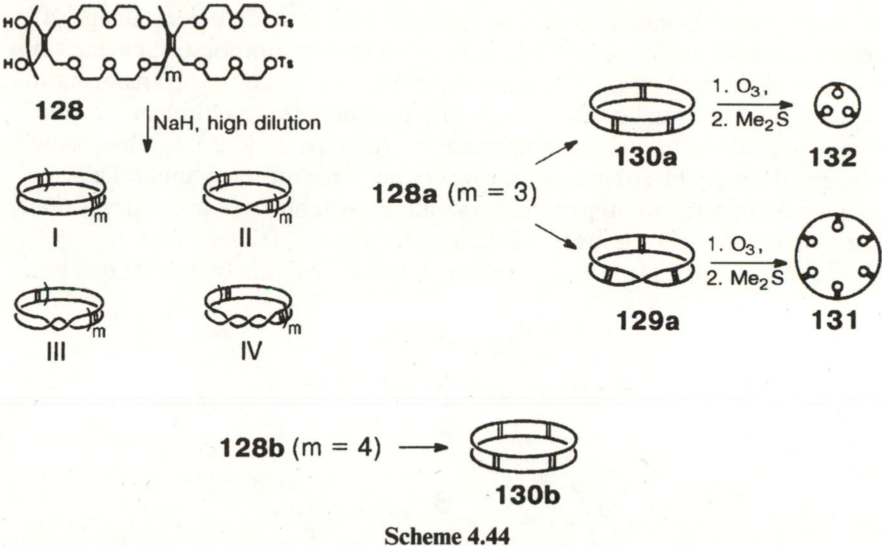

Scheme 4.44

regular cyclization product **130a** produced triketone **132** (Scheme 4.44).[21b] These results provided some grounds to hope that the next oligomer homolog of **128b**, with $m = 4$, would produce not only a Mobius-like structure similar to **129a** but a trefoil knot as well. Rather disappointingly, nothing like that happened. The only identified product of this reaction turned out to be the cylinder **130b**. The absence of reliable tools to control the selectivity of the reaction has actually knocked out an otherwise brilliant project. Nevertheless, the very challenge of the problem and the ingenuity displayed in the attempt to solve it are worthy of deep respect.

The problem of forming topologically connected molecules also has some ramifications in the field of natural macromolecules. It was found long ago that catenanes, and even knots, are formed in the course of DNA replication (see, for the leading references, a review[18d]). While it is still unknown whether this phenomenon has any functional meaning, it occurs very often and a special group of enzymes, the topoisomerases, are produced in the living cell to make necessary repairs to DNA. In fact, the principles of molecular design were applied recently to the design and synthesis of catenated and knotted DNA fragments (required for modeling studies of the properties of natural DNAs of these shapes[21d]).

Experimental data about the properties of Mobius strip-like structures as the 'ladder' **129a** or knots like **127** are rather scarce. However, a theoretical analysis of the peculiarities of these constructions (see refs. 18a–d, 21b, 21c and literature cited therein) led to some conclusions of general importance. Thus it was established that a new phenomenon of topological chirality should be observed for compounds having the shape of trefoil knots or Mobius strips. Normally, chemists deal with chiral objects which can be (in principle) transformed into their mirror image by a continuous deformation. For example, chiral tetrahedron **A** can be converted into its mirror image **B** by deformations of angles *via* a symmetrical planar configuration **C** (Scheme 4.45). No rupture of any bond to the central atom is required in this transformation. While this transformation is energetically forbidden for a chiral sp^3 carbon, it proceeds easily with amino compounds of the type $R^1R^2R^3N$ (which is why additional constraints to prevent the inversion of the nitrogen 'umbrella' should be imposed in order to prepare tertiary amines with nitrogen as a chiral center). For the topological chirality of knots or Mobius ladders (for $n > 3$), no inversion of configuration could happen without a rupture of at least one bond.

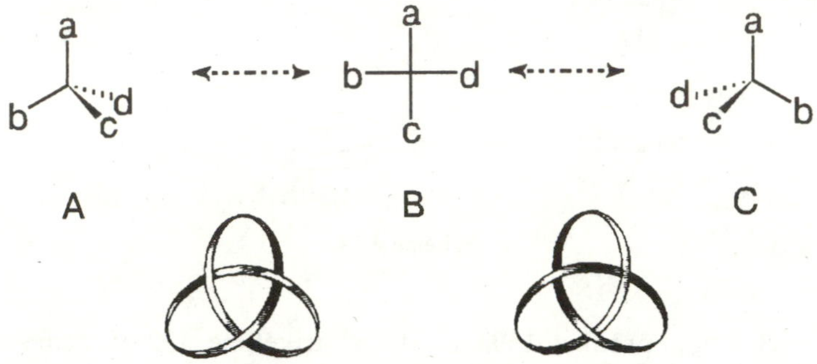

a b c d A B C

Enantiomeric topoisomers of trefoil knot

Scheme 4.45

(Reprinted with permission from *Acc. Chem. Res.*, **1990**, *23*, 319. Copyright (1990) American Chemical Society)

Recent data provided experimental verification of some of these ideas. The analysis of NMR spectra of the knot **127·2Cu**$^+$, recorded in the presence of chiral reagents, clearly demonstrated that this compound is actually a mixture of two enantiomers.[20d] X-ray data revealed that **127·2Cu**$^+$ crystallizes as a mixture of enantiomers.[20e]

The synthesis of the models shown above acquired additional meaning from the absolute novelty of the stereochemistry problems which became available to experimental studies. While it is premature to make definite suggestions about the peculiarities of the chemistry of compounds having 'topological' bonds, the uniqueness of their structure seems to guarantee that a number of non-trivial phenomena, such as unusual chelating properties, regulated catalytic activity, *etc*., will be discovered in this field.[21e]

4.5 'ABNORMAL' STRUCTURES *VS.* CLASSICAL THEORY

A student who starts taking classes in organic chemistry is very likely to get the impression that its theory rests on a set of postulates as unshakable as those of Euclidean geometry. By the end of the first semester one is supposed to learn that four substituents around an sp^3 carbon atom are positioned at the corners of a tetrahedron with an angle of 109.5° between the orbitals, while planar and linear arrangements are associated with sp^2 and sp carbon centers, respectively. These are the cornerstones of organic chemistry. An enormous body of experimental material substantiated their solidness and there are no serious reasons to cast doubts at the general validity of these postulates. However, the human quest for the unknown does not accept any limitations. There is a very special appeal in challenging well-established concepts. It is hardly surprising to see that, in the last few decades, numerous synthetic efforts were centered around exotic structures whose very existence seemed to be questionable, at least from the point of view of classical structural theory. In fact, the design and synthesis of Platonic hydrocarbons, as well as other cage structures, were among these studies. Below we will consider a few examples related to other structural types, chosen rather sporadically, to illustrate the main trends in this field.

4.5.1 Distortions of sp^3 Carbon Configuration. Flattened and Pyramidalized sp^3 Carbon

In the structures considered above, regardless of the great distortions imposed upon the valent angles, the sp^3 carbon atoms of the framework still preserve their favorite tetrahedral configuration. As quickly as this feature is recognized, human nature calls to the forefront whether it is possible to place the sp^3 carbon in a position which forces it to adopt a planar configuration, as the one shown in Scheme 4.45 in a transitional state for racemization[22a])? This seemingly childish curiosity gets in the way of the main thought and prompted the design of molecular constructions especially suitable for probing that question. Theoretical analysis led to the conclusion that such a planar configuration is extremely unstable and could be realized only in a very specific structural context, like that represented by the hypothetical structures **133–135** (Scheme 4.46).[22b] However, no such structures have been synthesized so far. Initially it also seemed possible to achieve the flattening of the central carbon in a somewhat similar but non-aromatic system composed of small ring compounds of the general type **136**, which were named *fenestranes* due to the similarity of their shape to a window pane (fenêtre — window, French).[22b] Semiempirical

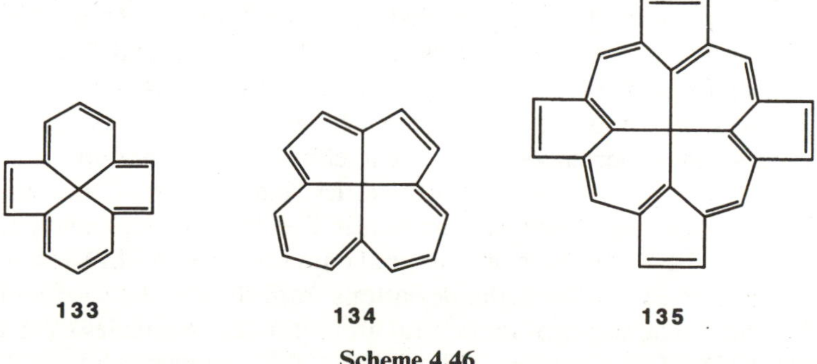

Scheme 4.46

calculations predicted that the framework of [4.4.4.4]fenestrane **137** (Scheme 4.47) might be reasonably stable, provided it is not planar. One of the possible structures for this framework, **137a**, has a quaternary carbon atom with the configuration of a flattened tetrahedron, while its isomer **137b** has the configuration of a pyramid (*vide infra*).

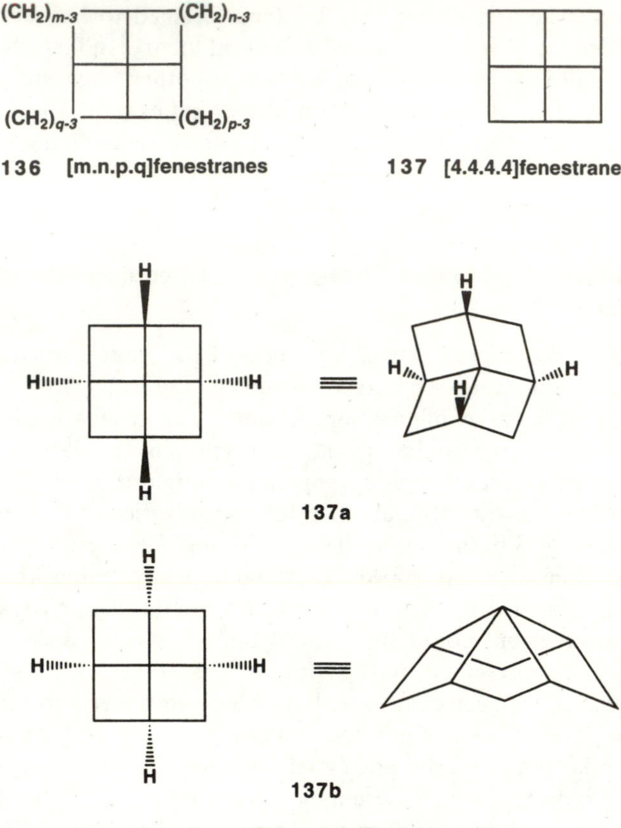

Scheme 4.47

Let us now consider the data provided by synthetic experiments in this field. A wide variety of fenestranes, represented by the general formula **136**, composed of rings of various size, were synthesized.[22b] As for **137**, only its simplified version, composed of three fused four-membered rings (as with **138**, the tricyclics **138a**) succumbed to synthetic efforts. The compound was called 'broken fenestrane', out of line with the strict rules of IUPAC nomenclature (Scheme 4.48). Among tetracyclic compounds, the closest approximation to **137** is the compound **139**. Analysis of structural data for **138a** and **139**, provided by X-ray studies, revealed substantial deviations from the standard values for sp^3 centers in the valent angles of the central atom [for **139** the angles C(1)–C(10)–C(6) and C(3)–C(10)–C(8) are 128.3° and 129.2°, respectively].[22c] While a substantial distortion from tetrahedral configuration has been achieved, there is still a long way to go to the complete planarization of the carbon.[22a,b]

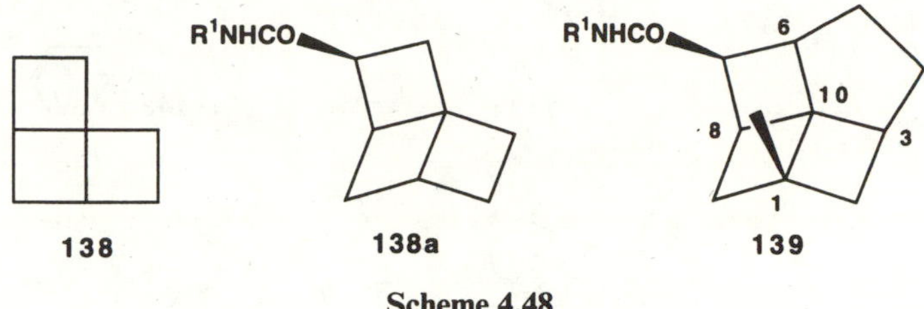

Scheme 4.48

The impetus to prepare a novel type of carbon framework, like that present in the structures **138a** or **139**, prompted intensive studies to elaborate new synthetic protocols. Examples illustrative of the strategy were shown earlier (*e.g.* Sections 2.19.2 and 3.2.6), but the existing data are too scarce to discuss the specifics of the chemical reactivity of fenestranes arising from the planarization effects.

Interesting ramifications emerged from the studies of benzoannulated fenestrane analogs. All-*cis*-tetrabenzo[5.5.5.5]fenestrane **140**, which is accessible in gram amounts in nine steps from 1,3-indanedione **141** and benzylideneacetone **142** *via* intermediates **143a** and **143b**,[22d] can be easily brominated to give tetrabromide **144a** (Scheme 4.49). The latter readily undergoes S_N1-type reactions at the bridgehead positions.[22e] These transformations allowed the preparation of a series of fenestranes, **144b–e**, tetrasubstituted at the bridgehead positions. Friedel–Crafts reaction of **144a** with benzene proceeds with remarkable ease as a four-fold **C–C** coupling to give a highly symmetrical adduct, centrohexaindan **145**, in 50% yield.[22f] The structure of the latter compound is most unusual. The central carbon atom is surrounded by four equivalent quaternary carbons and so the central core represents a regular tetrahedron. Six edges of this tetrahedron are bridged by *ortho*-phenylene moieties. The presence of six benzene rings at the periphery of the highly symmetrical molecule **145** offers numerous opportunities for further functionalization. Rather paradoxically, the studies initially aimed at the preparation of compounds with a flattened sp^3 carbon atom eventually led to the synthesis of compounds with an ideally tetrahedral carbon and new options for the design of novel spheroid-like molecules. Apparently there will always be a drive towards the creation of spheroid-shaped structures, which serve, like a magic magnet, to attract the attention of synthetic chemists!

The reactions shown in Scheme 4.49 imply an ease of formation for carbocationic intermediates with carbenium ion center(s) at the bridgehead position(s). Studies aimed at the generation and spectral investigation of these cationic species under long-life conditions will most certainly be forthcoming as the peculiarity of the fenestrane framework reveals itself in the appearance of new and unusual structural effects.

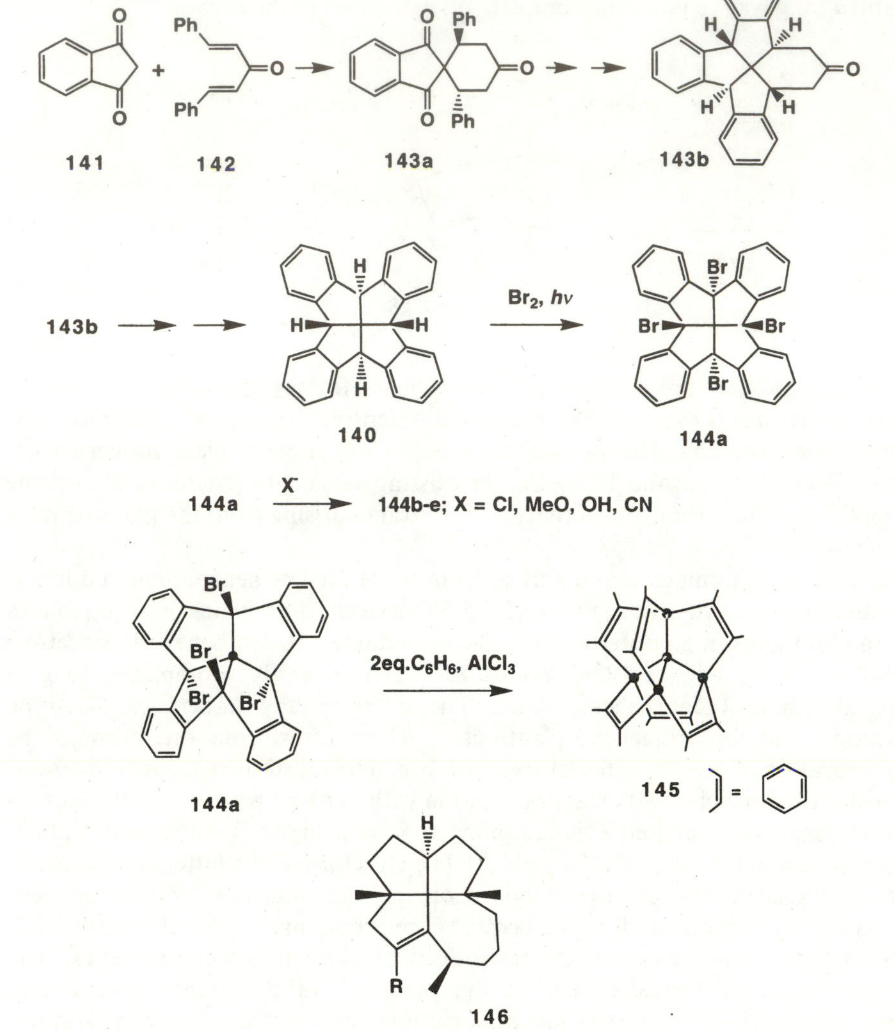

141 142 143a 143b

143b → → 140 Br_2, hν 144a

144a $\xrightarrow{X^-}$ 144b-e; X = Cl, MeO, OH, CN

144a 2eq.C_6H_6, $AlCl_3$ 145

146

Scheme 4.49

It is also relevant to add that while the experimental studies in this area have actually originated from rather abstract considerations advanced by organic chemists, the fenestrane-like design had been employed in Nature for a long while in the creation of the framework of several natural compounds (such as laurenene **146**). What their functions are and how (or if) the functions are related to the specificity of their structure are still questions to be answered. It is also worth noting that synthesis of **146** was achieved rather easily (see refs. in 22b) by employing the methods elaborated in the course of studies aimed at the preparation of the fenestrane framework.

As already mentioned, the framework of the hypothetical [4.4.4.4]fenestrane **137** can exist in two different configurations: **137a** and **137b**. Neither of them may adopt a flattened conformation.[22b] The structure of all known fenestranes corresponds to the shape represented by **137a**. An alternative configuration, **137b**, seems to be rather fantastic as it presumes that the central carbon atom and its four substituents form a tetragonal pyramid instead of the conventional regular or distorted tetrahedron. In fact, **137b** represents the most distorted (actually completely inverted) configuration of the normal tetrahedral sp^3 carbon atom. Thus *a priori* it must have been expected that such a pyramidal configuration should be extremely unstable and consquently the probability of preparing compounds of this type is marginal at best. In fact, though, a whole family of compounds containing inverted carbons were synthesized and this problem proved to be much less painstaking than attempts to create **1** or **137a**. True, it must be admitted that the parent compound, pyramidane **147** (or [3.3.3.3]fenestrane), has never been synthesized and that it is doubtful that this particular structure has any chances for materialization. None the less, other small-ring propellanes, compounds having three non-zero bridges and one zero bridge between bridgehead carbons, *e.g.* **148** (Scheme 4.50),[23a] turned out to be appropriate candidates to explore the opportunities for carbon pyramidalization.

The most interesting of these compounds is [1.1.1]propellane **149**. Initial theoretical analysis of this structure revealed that it must be more stable than the corresponding stretched biradical, but did not draw any conclusions as to whether or not **149** would be capable of existing as an observable species. Later calculations by Wiberg[23b] suggested that, contrary to these initial expectations, this compound should be surprisingly stable, since its strain energy is not much higher than that of the well-known bicyclo[1.1.1]pentane system **149a**. This prediction was confirmed in the successful synthesis of **149** from 1,3-dibromobicyclo[1.1.1]pentane **150**.[23b] An alternative and simpler route starting from the readily available dichloride **151** *via* bicyclobutane derivative **152** was suggested shortly after by Szeimies (Scheme 4.50).[23c] The simplicity and efficiency of the Szeimies' method is truly remarkable (especially in view of the non-triviality of the target molecule). In a similar way, several other representatives of the family were also obtained.[23a,c]

A great deal of spectral data has been accumulated for **149**. Its structure is now firmly established as that of a trigonal bipyramid containing two bridgehead carbon atoms with pyramidal configurations. The uniqueness of **149** and its relatives is obvious. Conventional description in terms of the geometry of sp^3 hybridized carbon atoms, which otherwise can account for a rather substantial degree of distortion from the ideal configuration of a tetrahedron, does not work for these compounds. No such configuration can be realized at all for the bridgehead carbons in **149**, which necessarily have umbrella-like positioned bonds.

Not surprisingly, **149** is a very thermodynamically unstable molecule and it easily undergoes rearrangements upon thermolysis (Scheme 4.50). However, it was rather surprising to learn that the major part of the strain stored in the

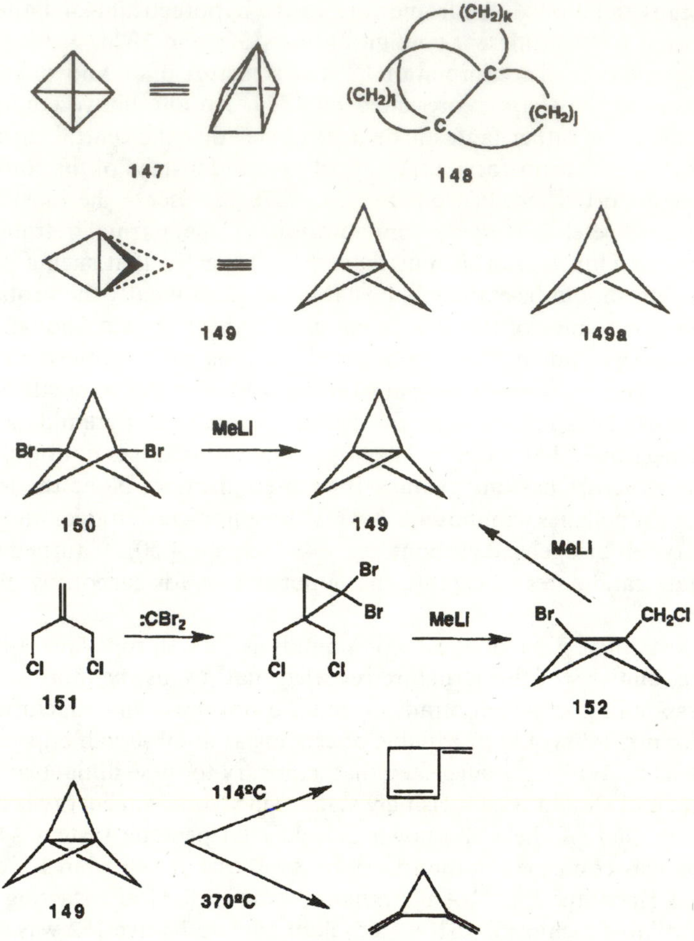

147 148

149 149a

150 MeLi 149

151 :CBr$_2$ MeLi MeLi 152

149 114°C 370°C

Scheme 4.50

framework of **149** was accumulated in the course of creation of the intermediate bicyclo[1.1.0]butane system, like that present in **152**. The energy cost required for the formation of the third ring is not very high (in fact it is almost equal to the amount necessary to form cyclopropane from an acyclic precursor).[23b] These considerations have paved the way for the elaboration of the pathway **151** → **149** shown in Scheme 4.50.

The chemistry of **149** is rather unusual from the point of view of the typical reactivity pattern observed for ordinary small-ring systems. Especially striking is the ease of central bond opening in radical reactions. Thus **149** spontaneously reacts with iodine, thiophenol, and even with carbon tetrachloride to give almost quantitative yields of the respective 1,3-adducts **153a–c** (Scheme 4.51).[23d] A number of other additions, including those leading to the formation of oligomers **153d**, in the presence of radical initiators have also been described. The occurrence of these transformations with [1.1.1]propellane reflects the relative weakness of the elongated central **C–C** bond, which is unusually long (1.596 Å instead of the ordinary 1.54 Å).

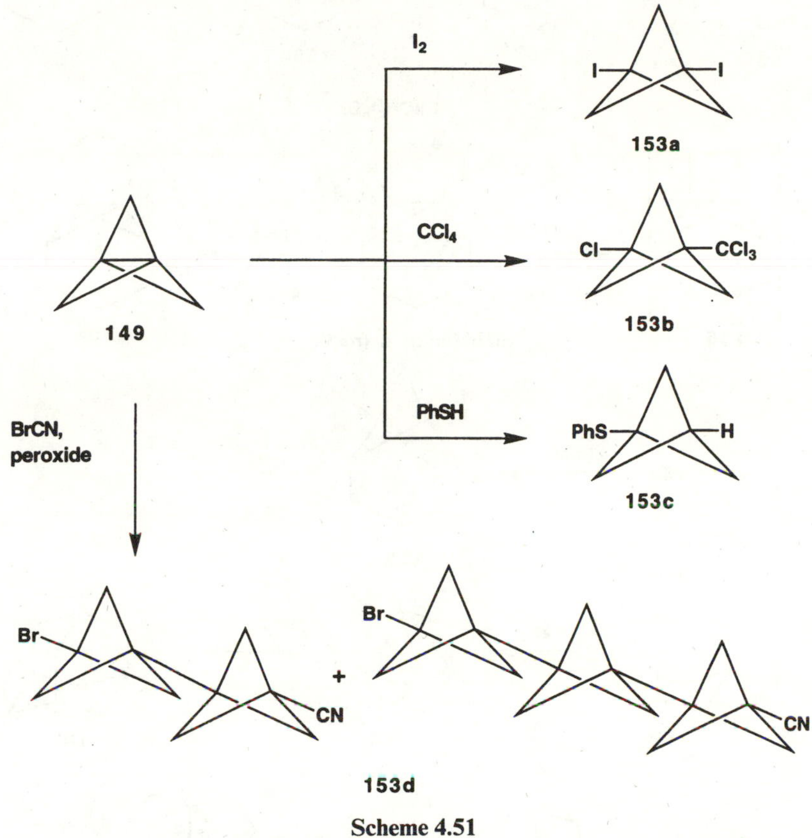

Scheme 4.51

The 'pyramidal' configuration is a common feature for several other small-ring propellanes such as **154** and **155** (Scheme 4.52).[23a] Preparation of bridged [1.1.1]propellanes like **154** was a relatively easy task owing to the surprising ability of the bridged bicyclobutane **156** ('Moore's' hydrocarbon) to undergo metallation to give **157**. Further conversion gave the target compound *via* intermediate **158a**.[23c]

The central **C–C** bond in **154** is also very reactive toward various reagents. Especially remarkable is its ability to react with Grignard reagents to form magnesium derivative **159**. This reaction represents a unique case of cleavage of the **C–C** bond under the action of such agents and once again attests to the peculiarity of the bonding in propellanes. The same derivative **157** was also employed for the synthesis of the bridged [2.1.1]- and [3.1.1]propellanes **155** and **160** *via* intermediates **158b** and **158c**, respectively.[23e] Propellane **155** was too unstable to be isolated but its formation was inferred from the structure of the

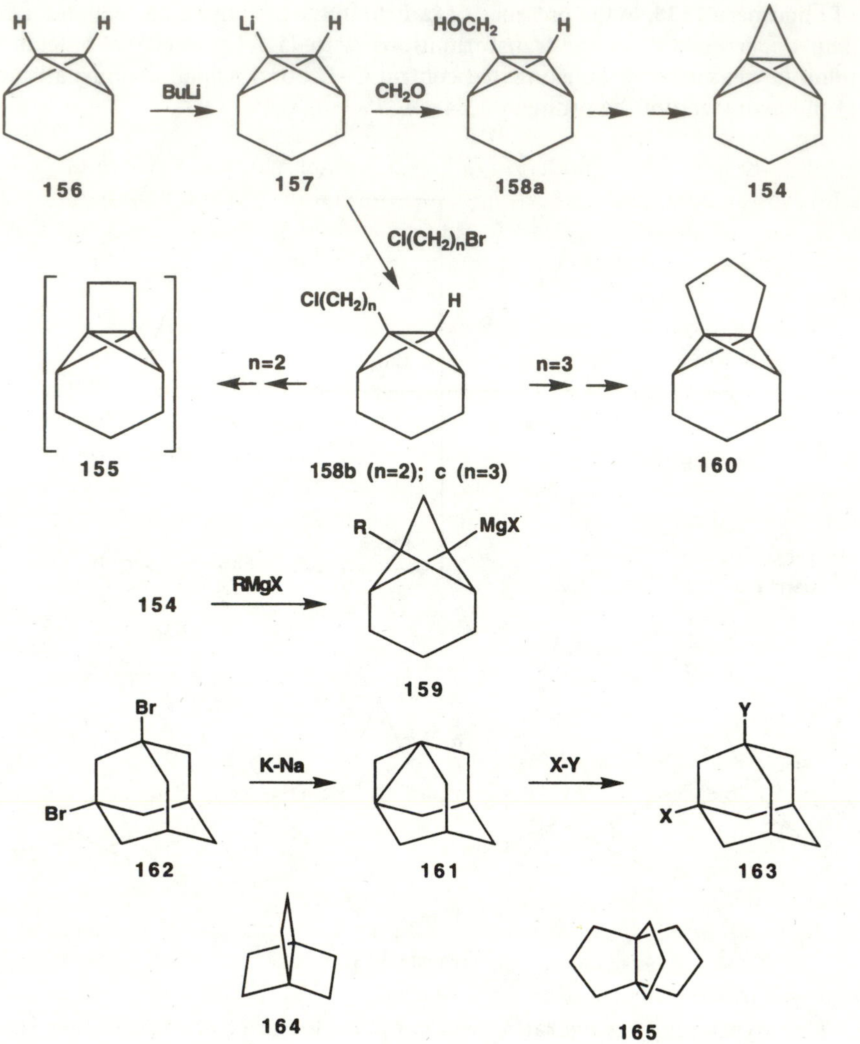

H H Li H HOCH$_2$ H

BuLi CH$_2$O

156 157 158a 154

Cl(CH$_2$)$_n$Br

Cl(CH$_2$)$_n$ H

n=2 n=3

155 158b (n=2); c (n=3) 160

R MgX

154 RMgX

159

Br Br K-Na X-Y Y X

162 161 163

164 165

Scheme 4.52

products of its trapping. Preparation of dehydroadamantane **161** from 1,3-dibromoadamantane **162** serves an additional example of the efficiency of the Wurtz coupling for the synthesis of cyclopropane fragments.[23f] X-ray data for the 5-cyano derivative of **161** revealed that the bridgehead atoms in this structure lie about 0.1 Å above the plane of the three adjacent methylene groups, with the internal 1,3-bond being remarkably long (1.64 Å).[23g] As might have been expected, the rupture of this bond takes place with almost any reagents to restore the stable adamantane system **163**. This transformation occurs easily even with a solution of **157** standing in the presence of oxygen ($t_{1/2}$ 6 h) and gives the respective peroxide and O-bridged oligomers.[23f]

The presence of even one cyclobutane fragment decreases the stability of the propellane system (*cf.* **155**). In fact, unsubstituted [2.2.2]propellane **164** still remains an elusive goal. According to the results of calculations, **164** may not even be an obtainable compound. Its central bond should be very weak and the energy barrier between the covalent and 1,4-diyl structures has been calculated as close to zero.[23a] Introduction of an additional methylene link in all bridges leads to the complete restoration of the normal tetrahedral configuration and [3.3.3]propellanes like **165** do not differ in their properties from the regular cyclopentane derivatives. In fact, quite a number of natural compounds contain this structural fragment.

It seems appropriate to comment here that some peculiarities of the reactivity pattern of three-membered rings were noticed long ago. In fact, the ability of cyclopropane to undergo **C–C** bond scission upon hydrogenation, or the action of such reagents as proton acids or halogens under extremely mild conditions, is a well known phenomenon. This ability necessitated the elaboration of a new concept, the banana-like bonding in this system. The successful synthesis of specifically designed small-ring propellanes offered additional opportunities to study unusual structural effects and the reactivity pattern of the cyclopropanes included in this most 'strange' but still existing framework. So far, no fully consistent explanation has been advanced for the data discussed above. This is still a challenging task for the theoreticians and its solution might lead to a revision or at least a refinement of the very concept of chemical bonding.

4.5.2 Distortion of the Double Bond

It was established long ago that all four substituents on a double bond are coplanar with the double bonded carbon atoms. No doubt they really are in the vast majority of double bond containing compounds, but is it possible to assemble a structure violating this rule and, if the answer is 'Yes', what would happen to the properties of such distorted double bond? Obviously, a distortion of the standard configuration of the alkenic moiety may arise owing either to changes in the valent angles between adjacent substituents (without affecting their coplanarity; in-plane angle distortion) or to a deviation of the planarity of the system (torsional strain and 'pyramidalization') or a combination of both.[24a]

The limiting case of in-plane angle distortion is represented in the structure of cyclopropene **166** (Scheme 4.53). The preparation of this compound was first described in 1897 by Freundler.[24b] It was most unusual result for that time and, not surprisingly, there was a lot of skepticism about its validity. Nonetheless, Freundler's procedure was reproduced in later studies and thus cyclopropene can be properly considered as the oldest member of the 'club of exotic structures'. Owing to its highly strained character, **166** is rather unstable and undergoes polymerization even at $-78\,^{\circ}C$. It is a much more reactive alkene than any other unsaturated compound. Thus its reaction with cyclopentadiene to give the Diels–Alder adduct takes place at room temperature,[24c] while a similar reaction with ethylene requires heating to $150\,^{\circ}C$. Even more revealing is the capacity of cyclopropenes to participate in reactions which are not viable for regular alkenes. Grignard reagents, which are totally inert toward alkenes with non-conjugated double bonds, add easily across the cyclopropene double bond of **166a** (Scheme 4.53).[24d] The resulting cyclopropylmagnesium reagent **167** can be further trapped with a number of electrophiles. This sequence of nucleophile addition/electrophile quenching has been developed as an excellent method for the synthesis of advanced intermediates of the type **168** for the preparation of a series of natural compounds.[24e]

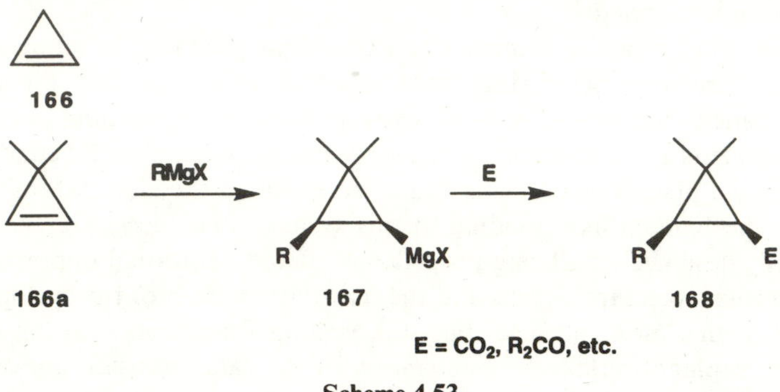

Scheme 4.53

Investigations in the area of pyramidalized alkenes are especially instructive from the point of view of molecular design.[25a] In this area, both theoretical and experimental studies were carried out in parallel. The pyramidalization angle (Φ) is defined as the angle between the extension of the double bond and the plane containing the doubly bonded carbon and the two geminal atoms attached to it (Scheme 4.54). Model calculations on ethylene predicted that when the angles in this planar structure are significantly reduced from the usual 120° owing to the imposed steric constraints, pyramidalized geometry might become favorable.[24a,25a] This prediction has been tested for several types of framework containing a small-ring cycloalkene moiety. According to calculations, for bicyclo[*n*.1.0]alkenes, a non-planar geometry is more stable than the planar one. The planar configuration might actually represent a transition state connecting two pyramidalized configurations. So far, the parent hydrocarbon **169** (n = 1) has not been obtained. Bridged derivatives containing this fragment, such as **170**, were prepared as discreet, albeit transient, species. Treatment of bromide **171** with alkyllithium reagents proceeded *via* intermediate formation of **170**. Ready addition of excess reagent at the double bond of **170** generated *in situ* gave a lithiated product to be further quenched with an electrophile to yield the final adduct, *e.g.* **172**. Alternatively, **170** may be generated under the action of non-nucleophilic bases in the presence of a 1,3-diene. In this case the respective Diels–Alder products (*e.g.* **173**) are obtained in good yields.[25b]

Pyramidalization should also be expected for compounds containing the highly strained bicyclo[2.2.0]hex-1(4)-ene moiety. The parent compound **174** of this series was prepared by electrochemical reduction of **175**. Its reactivity

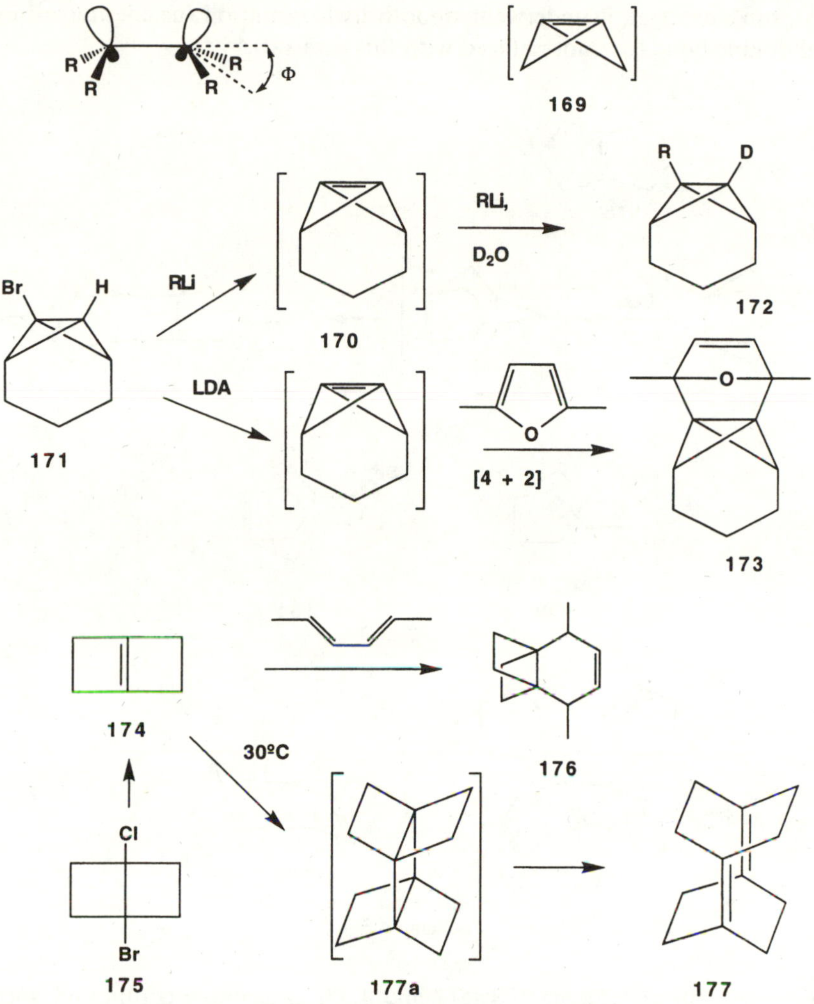

Scheme 4.54

profile parallels that expected for pyramidalized alkenes. It is readily oxidized in the presence of air, gives 1,2-*cis* adducts with CCl_4, and forms Diels–Alder adducts like **176**.[25c] All of these reactions take place at low temperature. Let us also note that cubene **11**, discussed earlier, belongs to the same category of pyramidalized alkenes containing the bicyclo[2.2.0]hex-1(4)-ene fragment (Φ approximately 90°). Alkene **174** also exhibits a unique propensity to undergo [2 + 2] cycloaddition upon standing in dilute solutions at room temperature, to form dimer **177** *via* a presumable intermediate **177a**.[25d] X-ray analysis indicated that **177** contains pyramidalized double bonds with Φ approximately 27°. Not surprisingly, this alkene was also very reactive toward oxygen and formed Diels–Alder adducts readily. It was also found that, in the presence of

Wilkinson's catalyst, it underwent smooth hydrogenation (usually tetrasubstituted double bonds are not reduced with this catalyst).[25d,e]

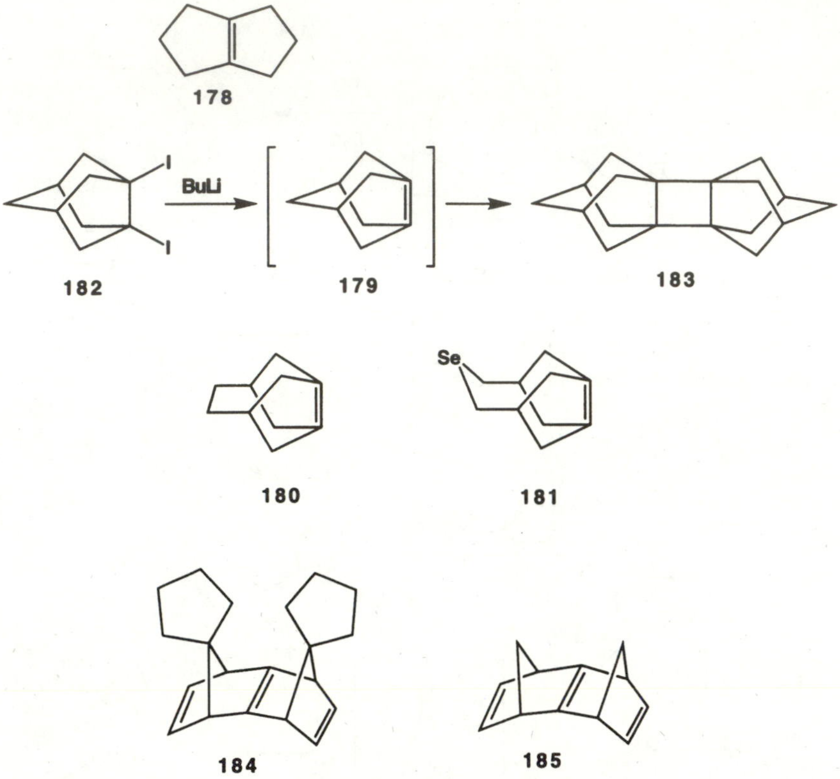

178

BuLi

182 179 183

Se

180 181

184 185

Scheme 4.55

Bicyclo[3.3.0]oct-1(5)-ene **178** (Scheme 4.55) is a stable compound with a flattened alkene fragment and exhibits a regular pattern of reactivity. Computational studies revealed, however, that installation of a short 3,7-bridge should lead to noticeable pyramidalization of the double bond.[25a] Compounds like **179–181** were synthesized to check this prediction. Tricyclic hydrocarbon **179**, with the smallest possible bridge, was generated as a transient species from diiodide **182**. The formation of **179** is implicated by the isolation of its cyclodimer **183** (or respective Diels–Alder adduct if the reaction is carried out in the presence of a 1,3-diene).[25f] The next member of this series, **180**, is more stable. In fact, the formation of **180** was ascertained not only from the structure of the final products (as was done for **179**), but also by its matrix isolation and analysis of spectral data.[25g] The selenium derivative **181** was found to be stable at ambient temperature in the absence of oxygen. X-ray data confirmed a noticeable pyramidalization of the double bond in **181** but the distortion was different: $\Phi_1 = 20.3°$ and $\Phi_2 = 12.3°$ for the two doubly bonded carbons.[25h]

Deplanarization of the double bond may also occur in the classical norbornene (bicyclo[2.2.1]heptene) system. A noticeable distortion of the double bond planarity (up to 7.5°) is observed in some of the norbornene derivatives.[25i] This feature is probably responsible for the increased reactivity of the double bond in these compounds and the preponderance of attack from the *exo* face. This deviation, negligible at first glance, was not overlooked and a series of studies followed to design and synthesize specific structures in which geometrical restraints force the deplanarization of the double bond to a larger extent. So far the maximum distortion of the double bond in this system has been found in compound **184** (Scheme 4.55). Repulsion of the two voluminous substituents at the bridged carbons causes significant distortion of the central double bond of this compound. According to X-ray analysis,[25j] the pyramidalization angle Φ is 32.4°! One might have expected that various additions across this double bond would proceed with amazing ease, but in fact the majority of standard reagents are rather unreactive toward this substrate. The reason for this apparent inertness seems to be trivial: the substituents, introduced to impose the unnatural folding of the framework, blocked, at the same time, the approach of reagents to this reactive site. Nevertheless, it was possible to gain some knowledge about the reactivity pattern of the distorted double bond from the earlier data for a similar model **185** with a lesser degree of non-planarity.[25k] The central double bond in **185** revealed enhanced reactivity toward a number of reagents. If a chloroform solution of **185** is exposed to air, complete conversion into the monoepoxide occurs within several minutes. In fact, this sensitivity made the handling of this compound extremely difficult. The authors were unable to determine the exact geometry of **185** by X-ray studies and were forced to elaborate the similar but more complex model **184**.

There are quite a number of other structures specially designed to induce the pyramidalization of the double bond. Studies in this field are being pursued very actively,[25a] It is relevant to note that dodecahedradiene **35** and dodecahedrene **36**, mentioned earlier, both belong to the class of non-planar alkenes (Φ approximately 45°).[6c] These compounds are extremely reactive toward oxygen and are converted easily into the respective epoxides. Owing to the severe distortion in the dodecahedrane frameworks, the double bond exhibits unsually high reactivity as a dienophile. For example, [4 + 2] cycloaddition of furan with dodecahedrene derivatives proceeds with amazing (for an otherwise non-activated double bond!) ease at 20 °C and gives the respective adducts in high yield.[6c]

4.5.3 Non-planar and Still Aromatic?

'Benzene is flat. Every undergraduate student learns early in his career that the extraordinary stability of benzene is associated with the cyclic overlap of six orbitals. Naturally, chemists have addressed themselves to the question of how much bending a benzene ring can withstand without giving up its aromatic character'. That is how the paper which describes the first synthesis of [6]paracyclophane **186** begins.[26a] According to calculations, introducing the short hexamethylene bridge between *para* positions should lead to noticeable out-of-plane deformation of the benzene ring. X-ray data obtained later for 8-carboxy[6]paracyclophane **186a** revealed that the benzene ring in **186a** has a boat conformation with the deviation of the ring segments containing *para*-substituted carbon atoms from the base plane of this boat being as high as 20.5° (*cf.* calculated value 22.4° in ref. 26a)[26b] The question arises: is this compound aromatic or not? The answer depends upon which criterion of aromaticity one would like to apply. According to spectroscopic data, **186** is still aromatic, since no dramatic perturbations were disclosed when compared to the NMR, UV, and IR data for common aromatic compounds. At the same time, its reactivity pattern has little in common with benzene derivatives and is reminiscent of the properties of 1,3-dienes. For example, bromination of **186** gave, quantitatively, 1,4-dibromide **187**. Active dienophiles react with **186** to form the respective Diels–Alder adducts.[26b] Upon irradiation, an equilibrium mixture of **186** and the Dewar isomer **188** is formed in a ratio of 1:3.

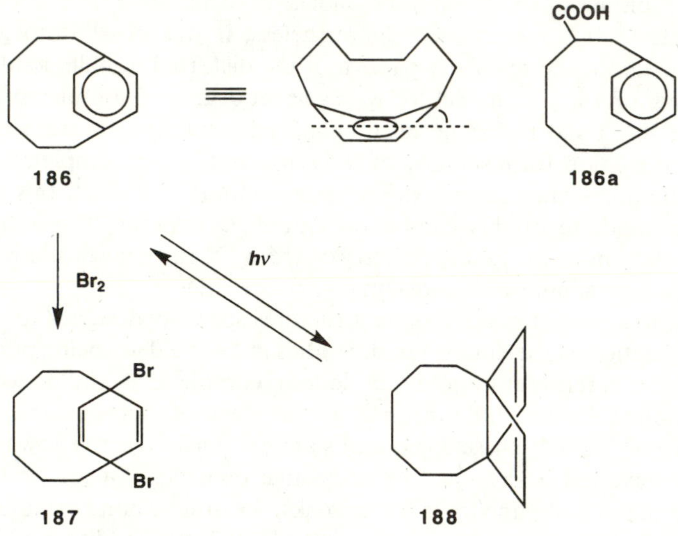

Scheme 4.56

An even greater deviation from planarity was disclosed in the series of metacyclophane derivatives. For the limiting case of [5]metacyclophanes such as **189a** and **189b** (Scheme 4.57),[26c] it has been established that in the bridging region the bending angle of the aromatic ring is increased to 26.8°! Yet this distortion apparently did not seriously affect the aromaticity of this compound (if evaluated from physicochemical data). X-ray data for **189a** clearly showed that the lengths of all aromatic carbon–carbon bonds are identical and typical for aromatic compounds with delocalized bonds.[26c] Nothing unusual was found in other spectral data for **189a** and **189b**. Their reactivity pattern, though, stands in striking contrast to that of regular aromatic derivatives. Compounds like **189** are able to undergo a 'number of unusual rearrangements, addition and substitution reactions, which find no counterpart in ordinary aromatic chemistry'.[26d] Some of the most striking examples are shown in Scheme 4.57.

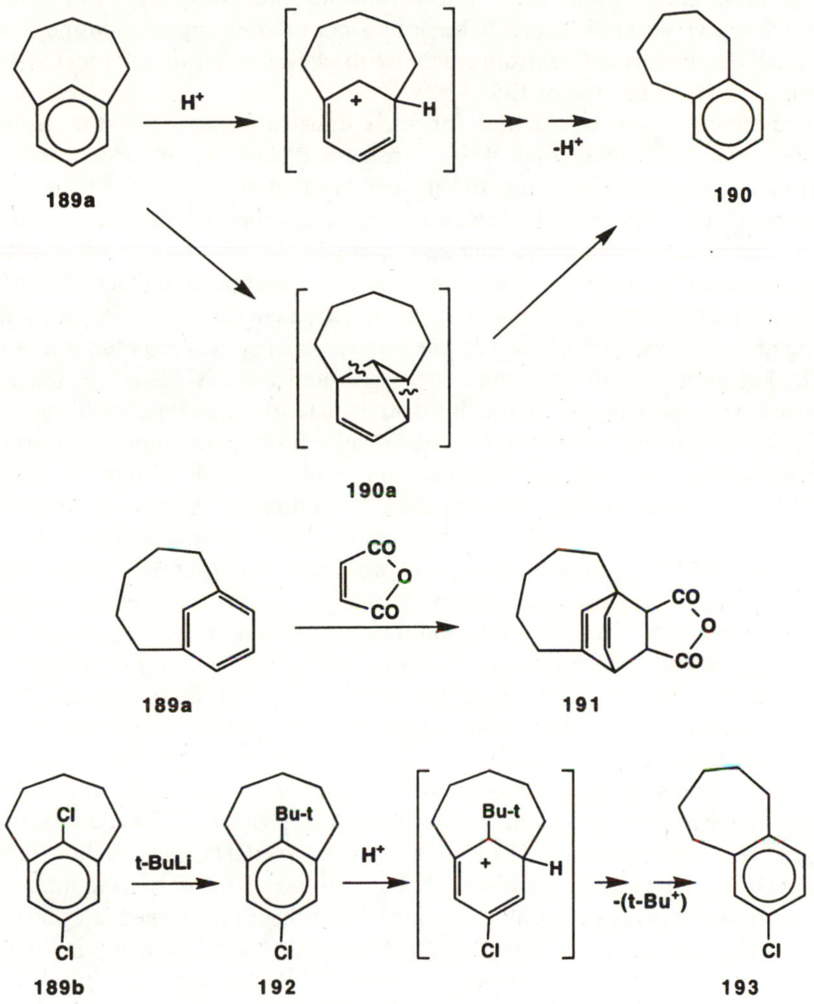

Scheme 4.57

Let us comment briefly on these reactions. The rearrangement of **189a** to **190** proceeds extremely easily under the action of catalytic amounts of CF_3COOH at 20 °C, while most usually alkylated aromatic compounds are able to undergo such rearrangements under much harsher conditions. Upon irradiation of **189a** a formally similar transformation occurs, but special experiments have shown that the latter reaction proceeds *via* formation of a tricyclic intermediate of the benzvalene type **190a**. Diels–Alder reaction with conventional benzene derivatives never occurs unless rather forcing conditions are used. Compound **189a** reacts with maleic anhydride at 25 °C within 15 min to produce adduct **191** in a nearly quantitative yield. Another surprising reaction takes place upon the treatment of dichloride **189b** with *tert*-BuLi at −70 °C. A formal nucleophilic substitution occurs under these mild conditions and leads to the formation of the *tert*-butyl derivative **192**. The latter compound, upon treatment with CF_3COOH, undergoes rearrangement with the elimination of the *tert*-butyl group and the formation of **193**.

The simplest way to account for such unusual behavior in the aromatic system was to suggest that it is transformed into a non-aromatic 1,3,5-cyclohexatriene structure due to the enforced non-coplanarity of the bent benzene ring in **186** or **189**. However, this suggestion contradicts the spectral data, mentioned above, indicating an unperturbed aromatic character of these compounds. The most plausible explanation of this abnormal reactivity pattern assumes that the ultimate reason lies in the tremendous steric strain of these bridged structures. In other words, the internal energy accumulated in **189** is so high that only a small additional energy of activation is needed to reach the transition complex necessary for this or that reaction. The relief of the strain in the initial stages of the additions to the π-bonds more than compensates for the energy needed to destroy an otherwise very stable aromatic system.

The discrepancy between physicochemical and reactivity criteria for aromaticity was known long ago. Yet the data procured by the studies of compounds like **186** or **189** shed a new light on this problem. The observed deviation from planarity did not cause any substantial changes in the spectral parameters of these compounds. Thus they still should be considered aromatic from the point of view of physical chemists. At the same time, this distortion of geometry so dramatically affected their reactivity pattern that, in accordance with this criterion, cyclophanes **186** or **189** should best be referred to as non-aromatic compounds, derivatives of 1,3,5-cyclohexatriene.[26e]

When discussing the properties of the distorted aromatic nucleus it is impossible not to mention the unique system of $[2_6]$(1,2,3,4,5,6)cyclophane or superphane **194** (Scheme 4.58), which contains six CH_2–CH_2 carbon bridges connecting two benzene residues.[27a] This structure is of special interest to chemists as it represents the ultimate member of the multibridged [2]cyclophane series. Its synthesis was achieved *via* a short and elegant pathway involving a sequence of high-yielding thermal reactions.[27b] Structural data revealed that both benzene rings in **194** are planar, normal hexagons. It was also disclosed that the sp^2–sp^3 carbon–carbon bonds deviate from coplanarity with the benzene ring by 20.3°. The π-orbitals of these rings are not perpendicular to the plane but deflected by approximately 10°. Here we are dealing with a distortion that does not affect the geometry of the ring composed of six sp^2 carbon atoms but, rather, the configuration of the π-electron orbitals. Surprisingly, this peculiar geometry does not create any problem for orbital overlap. According to spectral data, **194** is a typical aromatic structure. Superphane is a very strained but remarkably stable compound. Unlike other $[2_n]$cyclophanes, **194** is unreactive toward the usual dienophiles and to Birch reduction. Yet a

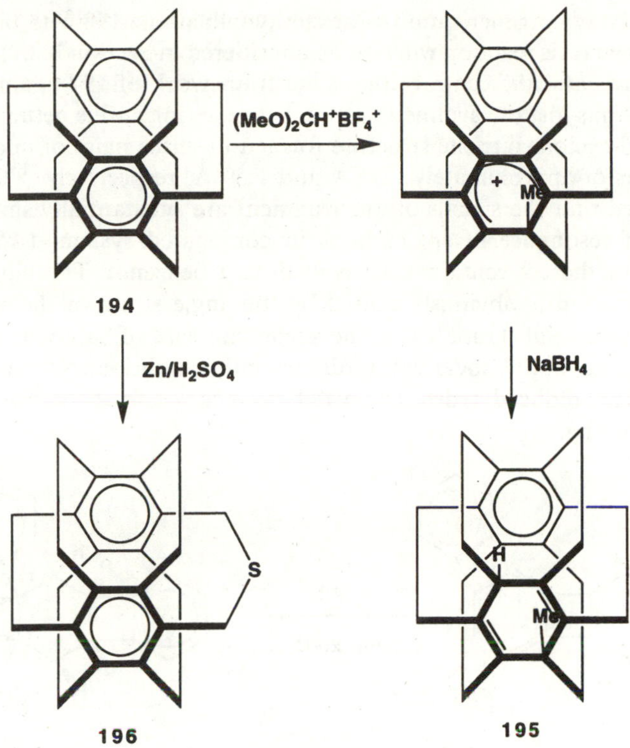

Scheme 4.58

number of other chemical properties of **194** are rather unusual. This aromatic compound can be easily transformed into the reduced derivative **195** by treatment with the very mild electrophile $(MeO)_2CH^+BF_4^-$ and hydride reduction (Scheme 4.58). No such reaction is known for ordinary aromatic compounds. Several other non-trivial additions have also been reported for **194**.[27b] The most interesting finding relates to the reactivity of the benzylic **C–C** bonds of the bridge. An attempt to reduce **194** by zinc in concentrated sulfuric acid resulted in the unexpected insertion of a sulfur atom between the two carbon atoms of the bridge to form compound **196**. This reaction is absolutely unprecedented and its occurrence suggests the operation of unusual effects of the distorted aromatic rings on the reactivity of the adjacent **C–C** bonds. So far there is no consistent interpretation of all these reactions other than the traditional references to the unusually strained character of this structure as the feature responsible for the peculiarities of its reactivity.

All previous examples referred to the anomalies due to out-of-plane distortions of the aromatic rings. Properties of aromatic compounds could also be affected by in-plane angle distortions caused by a fusion of a small ring and an aromatic residue. Among various compounds which have been designed and synthesized for investigations into these effects, the most interesting data were obtained for tris(benzocyclobutadieno)benzene **197**, easily prepared by the cobalt-catalysed cotrimerization of hexaethynylbenzene **198** with bis(trimethylsilyl)acetylene (this reaction will also be considered in Section 3.2.7).[27c] According to X-ray data, the central ring in the framework of **197** is almost planar. However, it has lost the symmetry of the benzene ring and is better represented by the 1,3,5-cyclohexatriene structure formed by three pairs of unequal bonds with lengths of approximately 1.33 Å and 1.50 Å, respectively. ^{13}C NMR has also revealed that the signals of this fragment are substantially shifted toward the area of resonance of sp^2 carbons in conjugated systems ($\Delta\delta$ -18 ppm compared to the conventional hexasubstituted benzene). The high degree of bond localization is obviously caused by the angle strain of the system. This representation is substantiated by the surprising ease of catalytic reduction of the central ring of **197** under extraordinary mild conditions.[27d] It is also worth noting that the reduced hydrocarbon **199** has a cup-shaped structure with a flat cyclohexane ring.

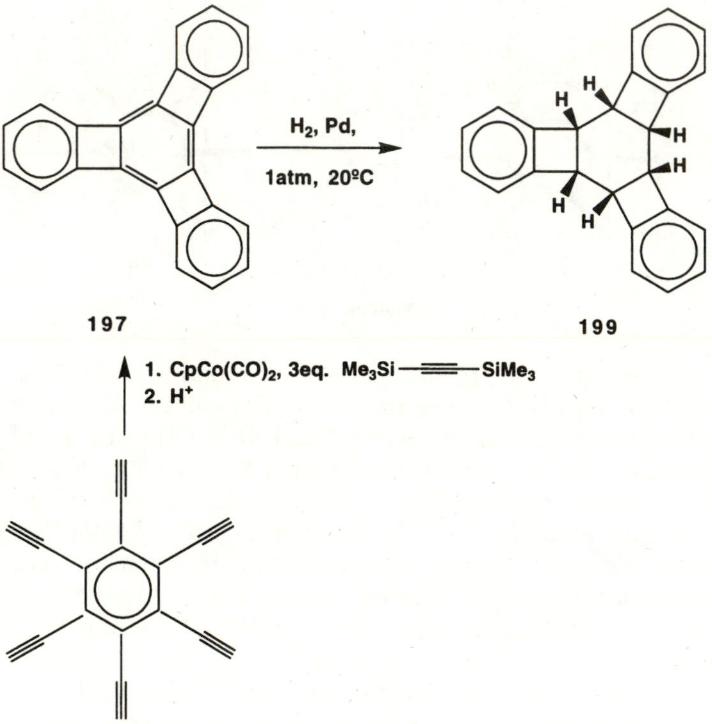

H_2, Pd, 1atm, 20ºC

197 **199**

1. $CpCo(CO)_2$, 3eq. Me_3Si—≡—$SiMe_3$
2. H^+

198

Scheme 4.59

4.5.4 How to Increase the Reactivity of the Regular C–H Bond in Saturated Hydrocarbons

The low reactivity of **C–H** bonds in normal alkanes and cycloalkanes is actually one of the reasons why the chemistry of these compounds is not especially exciting. Aside from the discussion of oxidation at elevated temperatures and certain radical substitution reactions, very little can be said about the transformations of these compounds without referring to heterogeneous catalysis. Earlier in this section we came across several cases when the reactivity of the standard **C–H** bond turns out to be rather peculiar owing to the specific distortions of bond angles and/or lengths in certain exotic frameworks (*vide supra* for examples from dodecahedrane, cubane, and bicyclobutane chemistry). It is also relevent to mention that under the forcing conditions of superacid media, even acyclic alkanes are able to undergo various reactions with strong electrophiles, especially in cases when stable *tert*-carbenium ions can be formed.[28] For adamantane, similar reactions may occur even under ordinary conditions. Thus the stable 1-adamantyl cation intermediate **15** can be easily generated upon interaction of adamantane with concentrated sulfuric acid. The ease of this reaction is considered to be a consequence of the rigid geometry of adamantane, which ensures additional stabilization of the cationic center in **15** (*vide supra*).

An entirely different and rather unusual opportunity for the activation of regular **C–H** fragments was revealed in the course of investigations of certain bicyclic as well as medium-ring-size hydrocarbons. For example, saturated bicyclic hydrocarbon **200** reacts with trifluoroacetic acid, or even acetic acid, at 40 °C with the evolution of hydrogen and the formation of the extremely stable carbocation **201**. With the use of trifluoromethanesulfonic acid, a stronger acid, this process takes place almost like a titration even at 0 °C! (Scheme 4.60).[29a]

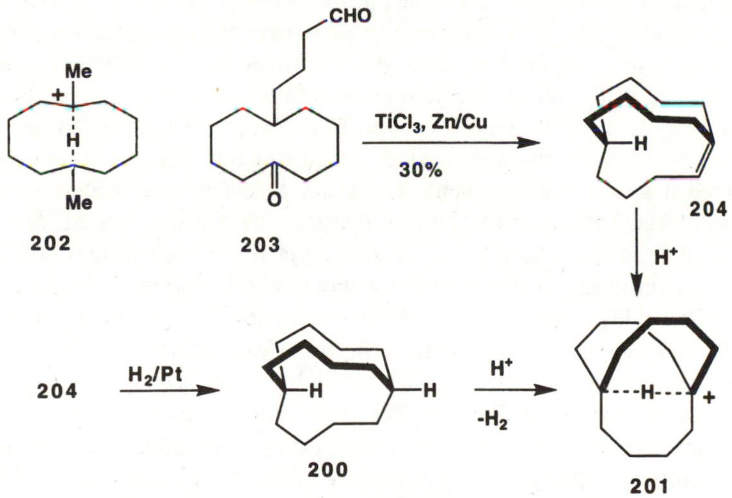

Scheme 4.60

Initially, this result may seem to be very strange and difficult to account for, because there are no noticeable length and/or angle distortions in the structure of **200**. Is this another example showing the role of serendipity in science? Not at all! Perhaps it is more appropriate to consider this case as a good illustration of the power and fruitfulness of modern molecular design. In the early 1960s it was observed that 1,4- or 1,5-hydride shifts may occur easily in medium-size rings (transannular effects). Later it was found[9d,29b] that, in the cyclodecane series, under superacid conditions at $-70\,°C$, it was possible to observe the formation of a stable cationic species, the so-called μ-hydrido-bridged cation, for example **202**, containing a transannular **C–H–C** three-center, two-electron bond. This assignment was based upon the observation of a single very high-field signal in the ^1H NMR spectrum ($\delta = -3.9$ ppm). Analysis of the data led to the conclusion that the transannular interactions leading ultimately to the formation of the μ-hydrido-bridged cations should be enhanced if the cyclodecane system contains an additional bridge that forces one of the bridgehead hydrogens into a position inside the ring system. How can one prepare the required bicyclo[4.4.4]tetradecane system containing *in,out* bridgehead hydrogens? Here, McMurry's reaction of low-valent titanium-induced coupling of carbonyl compounds turns out to be the method of choice (for pertinent data on the McMurry reaction, see Section 3.2.4).[29c] The application of this method to ketoaldehyde **203** led to a smooth intramolecular coupling to form *in*-bicyclo[4.4.4]tetradec-1-ene **204**. As was to be expected, the protonation of the double bond in **204** proceeded with nucleophilic participation of the inside hydrogen to give cation **201**, a stable species under ordinary conditions.[29d] NMR parameters of **201** were very close to those reported earlier for **202** (for example, ^1H NMR revealed the presence of a heavily shielded single proton signal at $\delta = -3.46$ ppm).

With this information at hand, the authors were in a position to try to achieve the generation of **201** *via* a less trivial route from the saturated precursor **200**, which was easily prepared by catalytic hydrogenation of **204**. The final and successful step of this study, the demonstration of the ease of protonolysis of the **C–H** bond in the saturated hydrocarbon **200**,[29a] represents a rare example of the ultimate success which was well planned in advance. The authors also performed a series of experiments necessary to clarify the mechanism of this process. In the summary to this publication the authors stated: 'This work provides a clearcut example of stoichiometric hydrogen evolution in an alkane protonolysis reaction and provides good evidence that the $\mathbf{RH \rightarrow RH_2^+ \rightarrow R^+ + H_2}$ pathway is available for cation formation in simple alkanes'.[29a] It should be emphasized that the framework of **200** seems to be rather ordinary, especially in comparison with such 'monsters' like [1.1.1]propellane or tetrahedrane. The framework of **200** does not indicate any unusual steric constraints and, from this point of view, nothing peculiar could be expected for its chemistry. The surprising reactivity of the *out* bridgehead **C–H** bond in **200** to proton attack is determined entirely by the ability of the *in* bridgehead **C–H** to stabilize the incipient carbocationic ion center due to the enforced proximity of these centers. Even the **C–H** bond, if placed properly, may play the role of a powerful electron-donating group!

4.6 CONCLUDING REMARKS

It would not be very difficult to continue with a list of impressive results published in the area of structure-oriented design. We believe, however, that at least some of the general tendencies and ramifications can be seen in the examples given above.

One cannot help being amazed by the imagination and skill of chemists able to design and prepare an enormous variety of molecules of unusual shapes. In addition to the well-known forms like threads (linear polymers), nets (cross-linked polymers), rings (cyclic molecules), triangles (cyclopropanes), and rectangles (cyclobutadiene), a set of new structures, polyhedranes (cages), chains (catenanes), hollow spheres, tree-like structures, *etc*., have made their appearance recently. Carbon and carbon-containing fragments served as building blocks for the creation of wonderful constructions appealing both from the aesthetic and scientific points of view. In a way these are materials that are as pliant as clay, enabling the creatively minded masters to exercise their fantasy and command in experimentation in the pursuits of the most daring ideas. It may even seem that whatever is possible, has already been created. In fact, there are absolutely no reasons to suspect that more eccentric molecules cannot be designed as the next goals to challenge the skills of organic chemists. This ever-expanding area of organic chemistry may serve as the best illustration of Berthelot's statement (see Section 1.5) about the creative ability of this science.

The obvious result of this expansion is the development of the theory of organic chemistry. The more we learn about the unusual properties of compounds having unprecedented molecular architecture, or those compounds simply containing normal structural elements in an unusual surrounding, the better we realize how flexible and alive structural theory is. Each step in the studies aimed at expanding the set of new structures leads to a better understanding of this theory. The classical set of its strict definitions and rules is undergoing permanent modification. It seems as if there is not one single concept which has not been challenged and found to be viable only within a very definite, albeit rather broad, limit of applicability.

With an in-depth understanding of the flexibility of the properties of carbon compounds it becomes more and more difficult to teach this discipline in a traditional way. It is our belief that the basics of this science could (and should) be presented in regular textbooks in conjunction with its ever emerging problems and paradoxes. Special emphasis should be given to the nearly unlimited opportunities to create various molecular assemblies from simple building blocks and to the striking opportunities to affect the reactivity patterns of the common functionalities by varying their structural context. To create such an exciting book, illustrating the true nature of organic chemistry as both a mature and young science, rapidly developing and full of promise, would be a problem of equivalent challenge to the design of novel structures.

PART II FUNCTION ORIENTED DESIGN

This section will introduce studies aimed at the creation of molecules with a well-defined set of properties which could secure practical usefulness of the designed compounds. If these properties could be unambigously correlated with their structure, then it would not be especially difficult for chemists to fulfill almost any order for the 'customers'. Unfortunately, as we mentioned in the first chapter (Section 1.3), in most cases it is impossible to predict exactly that a given structure, already existing or newly designed, will be able to do a particular job. Mere analogy may be a more or less reliable tool for the design of compounds with a required set of simple properties (like solubility, boiling point, light adsorption parameters, *etc.*), provided the search is limited to a series of closely related structures. This approach, however, may lead to erroneous conclusions if more complex properties are considered. A plethora of examples can be cited to illustrate this point, but only a few will be given here.

Ethyl alcohol is among the few organic compounds which were known with a certainty many centuries ago. However, if, owing to some mysterious reasons, ethyl alcohol were unknown until now, no amount of data about the properties of other representatives of the aliphatic alcohol series would enable us to suspect its highly specific, both detrimental and beneficial, effects (depending on the dose!) on the life of humans. It happens quite often that even well-known compounds may stay overlooked for decades right up to the moment when, owing to some accidental finding, they turn out to be extremely important.[30] Neither the capacity of THF to serve as a stabilizing solvent for vinyl Grignards, nor the anesthetic properties of cyclopropane, nor the ability of cholesterol benzoate to form liquid crystals, nor the unique set of the physical and chemical properties of polyfluoroethylene (PTFE) could have been predicted in advance from a mere consideration of the structure of these molecules. Therefore it remains a truly formidable task to elaborate the general principles of molecular design for the purpose of creating a new structure with a predetermined set of properties. Nevertheless, in many important areas a rational approach, based upon the ideology of molecular design, has already demonstrated its viability.

Logistically, function oriented design is a much more complicated task than the mere creation of molecular assemblies of structure oriented design. It is a multistep procedure that may involve several discrete stages. First, it is necessary to translate the 'orders of the customers' into the language of molecular structures. The required property of the target material must be somehow correlated with the structure of organic compounds. Quite often, the goals of these studies are formulated by non-chemists, and it is therefore of utmost importance that representatives of both sides be able to achieve a complete, mutual understanding of the ends and means of the entire project. Next, using a set of structural criteria, a certain specific molecular construction is generated and its structural parameters are optimized. The third phase involves the synthesis of target structures and the investigation of their actual properties. This is followed by variations of the structural parameters to fine tune the performance of the molecule to its required role. It is obvious that such a strategy is potentially much more fruitful than the traditional 'trial and error' method for the target-focused search. However, if the trial-and-error method merely encompasses an enormous number of routine trials, *i.e.* the preparation and screening of hundreds and thousands of compounds and a sort of special luck in finding the most efficient candidate, the molecular design approach requires application of both imagination and deep comprehension of the essence of the problem as well as the ability to exploit a vast multi-faceted knowledge. It is also to be admitted that even now, more often than not, the initial impact for function oriented design is provided by some serendipitous or at least unexpected discovery.[30] Nevertheless, methods of modern molecular design serve as powerful tools to facilitate the elaboration of the accidental findings into objects of carefully planned and target oriented investigations.

By definition, the final purpose of function oriented design is the creation of a molecule with useful properties. This usefulness may refer to a very diverse area of application and therefore it was virtually impossible to suggest a reasonable set of classification guidelines to present the material in this section. Any attempt to give a more or less complete coverage of even the main trends of studies in this area was actually precluded by the enormous amount of relevant material. Thus we were forced to restrict ourselves to the analysis of a rather limited number of examples, chosen mainly to illustrate some of the trends of function oriented design with the criteria of instructiveness and simplicity (and our own personal preferences, for sure!).

4.7 Design of Tools for Organic Synthesis

In Chapter 2 we have already considered some aspects of this problem. In fact, the identification of a synthon as a fragment of a target molecule possessing a certain functionality and polarity dictates that a set of requirements for the choice of the corresponding reagents is formulated. The elaboration of a specific reagent capable of satisfying these requirements represents the typical problem of function oriented molecular design. For example, the need to employ carbonyl anion synthons in planning a synthesis led to the development of a manifold of reagents specifically designed to fulfill the required synthetic task. The methodology of the synthon approach, and the corresponding diversity of reagents resulting from the directed studies in this area, was thoroughly discussed earlier and will not be repeated here. Below we consider a few additional examples to illustrate other approaches to the design of reagents with well-defined properties.

Strong bases are widely exploited in organic synthesis. Very often it is also necessary, though, to use bases which are non-nucleophilic. These bases must be able to abstract a proton but, at the same time, be almost totally inert toward other electrophiles. The need for such bases had actually stimulated the development of such reagents like lithium diisopropylamide and related compounds, but in certain cases it might be also necessary to to have bases which are incapable of abstracting a proton from **C–H** acids. This requirement can be illustrated in the following example. Di-*tert*-butyl ether **205** (Scheme

4.61) is a fairly strained compound. In contrast to most ethers, it is very unstable and can undergo hydrolysis even under the action of such a weak reagent as carbonic acid. Synthesis of **205** requires the alkylation of a sterically hindered hydroxyl of *tert*-butyl alcohol with a sterically demanding *tert*-butyl electrophile. Thus it is necessary to use a very active electrophile like the *tert*-butyl cation (for example, *tert*-$Bu^+SbF_6^-$) and to carry out the reaction in the presence of base to quench the strong acid released. This base (i) must not possess nucleophilic properties (otherwise the electrophile would attack this base instead of the sterically hindered target) and (ii) should not have a very high kinetic basicity (a kinetically strong base may cause the elimination of a β-proton from the carbenium ion such as the electrophile, *tert*-$Bu^+SbF_6^-$). These contradictory conditions are actually mandatory requirements for the success of this otherwise very simple transformation.

In the language of molecular structures, the above requirements imply the design of compounds containing a basic center capable of reacting with a protic acid while at the same time surrounded by sterically demanding substituents to preclude its interaction with other electrophilic species. One of the most spectacular solutions of this problem utilizes 1,8-bis(dimethylamino)naphthalene **206** as the base.[31a] This compound contains a pair of basic tertiary amino groups. Owing to steric constraints, however, the tertiary amino groups are forced to adopt a conformation with the methyl substituents turned outwards and the lone electron pairs (responsible for both the basicity and nucleophilicity) turned into the cavity inside this fragment (Scheme 4.61). It is still possible for a proton to enter this cavity and become attached *via* both covalent and hydrogen bonding, as shown in **206a** (**206** turns out to be a rather strong base in a thermodynamic sense). On the other hand, the size of this cavity is so small that no electrophile larger than a proton can approach the basic center. It cannot be alkylated by even the powerful and rather small reagent methyl

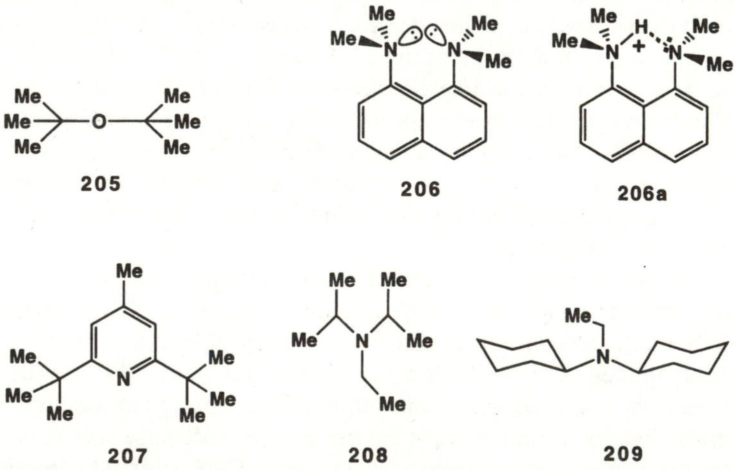

205 **206** **206a**

207 **208** **209**

Scheme 4.61

fluorosulfonate. As a result, **206** turns out to be a truly non-nucleophilic base and virtually devoid of the properties of a kinetic base. The suggested trade name 'Proton Sponge' is thus very appropriate for **206**.

In spite of the elegant solution to the problem as represented in the structure of **206**, it is rather far from being optimal. In fact, the presence of the naphthalene system in this base made the latter rather vulnerable to electrophilic attack directed at aromatic rings. Owing to this complication, **206** is used in syntheses less often than other sterically hindered bases[31b] like 4-methyl-2,6-di-*tert*-butylpyridine **207**, ethyldiisopropylamine, **208** (Hunig's base) or ethyldicyclohexylamine **209**. All of these reagents are now manufactured commercially and widely used in cases where it is essential to carry out a reaction with strong electrophiles under strictly non-acidic conditions with the removal of proton acids as they are formed. In fact, the preparation of **205** was succesfully carried out in the presence of **208**.[31c]

Another example, given in Scheme 4.62, refers to an entirely different type of reaction but actually to the same problem: how can one perform a reaction with a strong electrophile under essentially neutral conditions? The acylation of cyclohexene with $CH_3CO^+SbCl_6^-$ in the absence of added base gave only 22% of the desired product **210a**. The remainder was a mixture of unidentified products and the α,β-conjugated isomer **210b** (9%).[31d] These complications arose because of the formation of a very strong acid, $HSbCl_6$ (as the result of initial attack of the acetyl cation across the double bond with the elimination of a proton), which can induce a number of side reactions with acid-sensitive compounds like **210a** and **210b**. It was imperative to remove this acid. Hindered amines, such as **208** or **209**, are perfectly suited to do this job. The same reaction performed in the presence of these bases proceeded cleanly and gave the target material **210a** in 80% yield.[31d]

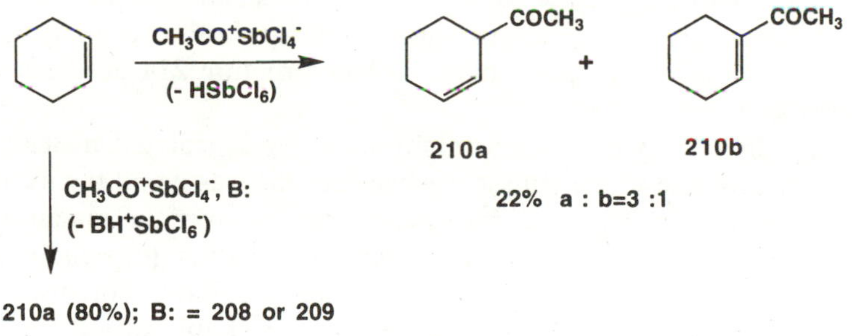

$CH_3CO^+SbCl_4^-$ / (- $HSbCl_6$)

$COCH_3$ + $COCH_3$

210a 210b

22% a : b=3 :1

$CH_3CO^+SbCl_4^-$, B: / (- $BH^+SbCl_6^-$)

210a (80%); B: = 208 or 209

Scheme 4.62

The vast utilization of hindered secondary amines like diisopropylamine or dicyclohexylamine in carbanion chemistry is also based on the difference in their behavior toward protons and other electrophiles. Thus quite a number of the methods depend upon the use of the lithium or magnesium salts of these amines for the generation of carbanionic species. These salts are very strong kinetic bases and therefore are able to abstract a proton from a variety of **C–H** acids. At the same time, the parent secondary amines formed at this step are rather inert (owing to the presence of bulky substituents) toward the attack of electrophiles (other than protons) and hence do not interfere with the quenching of the carbanions generated by various electrophiles. Numerous examples of these reactions have been given in previous sections.

4.8 CROWN ETHERS. FROM SERENDIPITY TO DESIGN

In 1967, a short and rather unpretentious communication by Charles Pedersen[32a] announced the appearance of a new class of chemicals, macrocyclic ethers, as entirely new vehicles for the formation of complexes between organic compounds and inorganic cations. The structures of crown ethers were not specially designed for this purpose. The original intention was a much more modest and very practical one. For years, Pedersen was working for the Du Pont company on the problem of retarding the autoxidation of petroleum products and rubber. Since the autoxidation is catalysed by traces of transition metal salts, the removal of these impurities was essential. Traditional methods based upon the chelation of heavy metals by organic polydentate complexons were especially suitable for this purpose and studies were initially centered around the elaboration of more efficient and practical analogs of the then-known reagents. As a target, the polyether **211** (Scheme 4.63) was chosen because it possessed five ether oxygens. It was expected that these oxygens should provide an especially good complexation set for the vanadyl group, VO^{3+}. The synthesis of **211** was planned *via* trivial transformations involving the alkylation of a monoprotected derivative of catechol **212a** with dichloro ether **213** followed by the deprotection of the presumably formed intermediate **211a**. Rather disappointingly, an intractable gum was formed as a result of this experiment. Instead of product **211**, a very small amount of crystalline material (in a yield less than 1%) was obtained. Its properties had very little in common with those expected for the target compound. Spectral and analytical data showed unambiguously that this material had structure **214** of a macrocyclic polyether (Scheme 4.63).

Typically, in cases where some minor impurity is isolated instead of the expected product, the whole stuff is washed into the waste and the experiment repeated with purified reactants and greater control over the reaction conditions. If Pedersen had followed this pattern of behavior (especially because product **214** was unable to complex the VO^{3+} cation and therefore useless from the viewpoint of the initial practical request!), he probably would never again have had the chance to go to Stockholm to receive the Nobel Prize subsequently awarded to him (together with Donald Cram and Jean-Marie Lehn) in 1987. Fortunately, peculiarities displayed by **214** did not escape Pedersen's attention. In fact, while ether **214** was only slightly soluble in methanol, its solubility increased dramatically in the presence of sodium hydroxide. Additional tests showed that this effect was not base dependent and could be observed with many sodium salts, as well as with many other salts of inorganic cations.[32b,c] Even more intriguing was the observation that inorganic salts which were

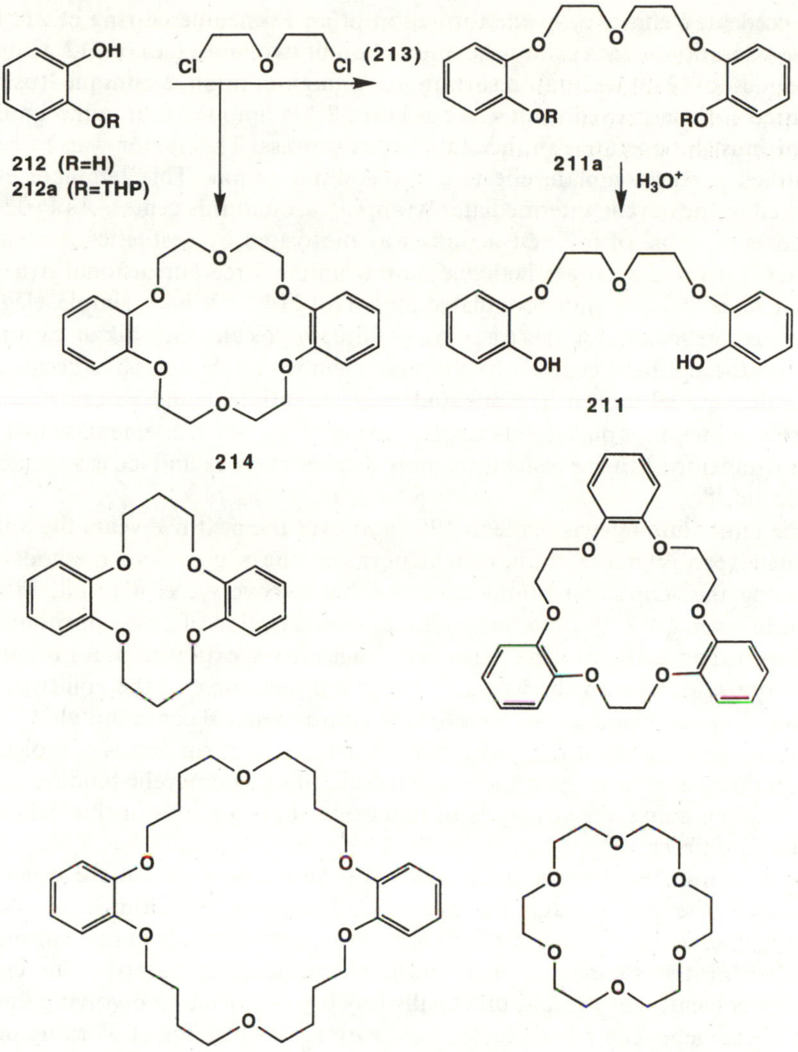

212 (R=H)
212a (R=THP)

(213)

211a

H_3O^+

211

214

Scheme 4.63

virtually insoluble in apolar organic solvents became noticeably soluble in the presence of **214**. These data prompted Pedersen to advance a truly insightful hypothesis about the origin of the observed phenomena. He suggested that the presence of a macrocyclic polyether system with a hole in the center (represented in the structure of **214**) enables **214** to capture an inorganic cation of an appropriate size and hold it by strong ion–dipole interactions between the positive center and the lone electron pairs of the six oxygen atoms. The mystery of the formation of **214** was then easily resolved. It is due to the presence of unprotected catechol **212** as an impurity in the sample of **212a**. When pure **212** was used instead, the yield of **214** increased to a respectable 45%. The unprecedented efficiency of the formation of an 18-membered ring of **214** in a single operation, as a result of the interaction of two molecules of **212** with two molecules of **213** without resorting to a high-dilution technique (usually required for macrocyclizations, see Section 2.17), implied that some unusual factor must be operative in facilitating this process. This factor was correctly identified as the template effect of the sodium cation. This template effect resulted in the acyclic intermediates wrapping around this center. As Pedersen stated later: 'One of my first actions was motivated by aesthetics more than science. I derived great aesthetic pleasure from the three-dimensional structure as portrayed in a computer-simulated model (of **214**)... What a simple, elegant, and effective means for the trapping of hitherto recalcitrant alkali cations! I applied the epithet "crown" to the first member of this class... because its molecular model looked like one and, with it, cations could be crowned and uncrowned without physical damage to either[32b].... My excitement, which had been rising during this investigation, now reached its peak and ideas swarmed in my brain'.[32c]

The initial finding was made in 1962 and over the next five years the author devoted very intensive studies at elaborating the preparative methods and outlining the scope and ramifications of his discovery. As a result, his full publication in 1967[32d] contained an in-depth description of a new phenomenon and opened an entry into an enormously huge area of exploration for hundreds of laboratories all over the world:[32e–g] a striking example of the ability of one outstanding chemist to create jobs for the entire chemical community!

Below we will concentrate our attention only at the main trends of molecular design that were based upon the original concept of macrocyclic binding. From the very beginning, the strategies of numerous investigations in this field were distinctly different.[32e]

Quite a number of structurally diverse crown ethers and, more generally, *coronands* (the generic name for macrocycles containing various donors sites) were synthesized as a result of efforts aimed at the creation of complexing agents suitable for the solubilization of inorganic reagents or catalysts in regular organic solvents. As a result, practically any organic reaction involving the use of these reagents can be conducted now with all the benefits of working under homogeneous conditions in the presence of crown ethers (see, for representative examples, refs. 32e and 32f). Strange as it may seem, the first prepared crown ether **214** and several of its simple analogs (like those shown in Scheme 4.63) turned out to be among the most acceptable in terms of efficiency and cost for very many practical applications in synthetic organic chemistry.

At the same time, the concept of multidentate binding within a macrocyclic framework was intensively explored for the creation of complexing agents able to capture cations of various sizes. Special attention was given to the efficiency and selectivity of complexation. It is worth noting that a very simple and reliable experimental procedure was developed for the evaluation of the complexing capacity of the various ligands. It runs as follows: a certain amount of the macrocyclic complexation agent is added to a two-phase system of an organic solvent (like methylene chloride) and an aqueous solution of metal salt having a highly colored organic counterion (most usually picrate). The appearance of the color in the organic phase can be monitored colorimetrically. The intensity of the color serves as a rough estimate of the transfer of the metal cation to the organic solvent and, hence, the efficiency of complex formation. This simple protocol greatly facilitated investigations aimed at the primary evaluation of the complexing capacities of numerous compounds.

The molecular design in this area was initially based upon the simplified assumption that the size of the central hole is actually the main factor controlling the preference in binding cations of a given ion radius. It was recognized that, for a flexible macrocyclic system, the real dimensions of the internal hollow may vary owing to the conformational mobility of the ring. Thus, it was not surprising that no simple relationship was found between the size of the ring or the number of complexing sites and the selectivity of inorganic ion complexation. The variability of the factors involved can be illustrated by the difference in the selectivity patterns of **214** and its closely related analogs **215** and **216** (Scheme 4.64).[32g] Additional examples are included in this scheme to illustrate typical approaches to the design of coronands possessing specific properties. Thus **217**, consisting of two crown moieties linked with a flexible tether, is capable of providing an efficient 'sandwiching' of a metal ion between the two polyether rings. Some of the compounds from this series turned to be especially good ligands for selective complexation with Na^+.[32h] The problem of the selective extraction of uranium is of obvious practical importance. Analysis of X-ray data of simple inorganic UO_2^{2+} complexes revealed that they adopt either a pseudoplanar pentacoordinate or hexacoordinate structure. Therefore, a specific ligand for this cation (uranophile) must have five or six ligand groups in a nearly planar arrangement. Crown ether **218**, which was designed to fulfil these requirements, gratifyingly exhibited a rather high preference toward UO_2^{2+}. Interestingly, a nearly ideal architecture for the design of these uranophiles was found in the rigid structure of the easily available macrocycles (calixarenes, *vide infra*) like **219**.[32i] These macrocycles revealed outstanding selectivity toward complexation with UO_2^{2+}. The ability to bind UO_2^{2+} was 10^{12}–10^{17} times higher than that of binding cations like Ni^{2+}, Zn^{2+}, and Cu^{2+}! In this case, selectivity is due, not to the appropriateness of the cavity size, but to the arrangement of the required number of coordinating centers on a moderately rigid backbone that still possesses some conformational flexibility which secures an opportunity of an induced fit.

Shortly after Pedersen's discovery, Jean-Marie Lehn, at the Institute de Chimie in Strasbourg, initiated studies aimed at constructing a totally novel type of macrocyclic ligands. It was anticipated that both the overall efficiency and selectivity of the binding should be significantly enhanced if a three-dimensional spatial arangement of the binding sites could be secured. This reasoning led to the design and synthesis of several sets of ligands having more or less rigid bi- or polycyclic frameworks, the so-called *cryptands* (from the Greek word *cripta* — hidden).[33a] Their properties turned out to be truly remarkable. As was to be expected, cryptands revealed excellent complexing properties for various compounds, ions, and covalent organic compounds. In

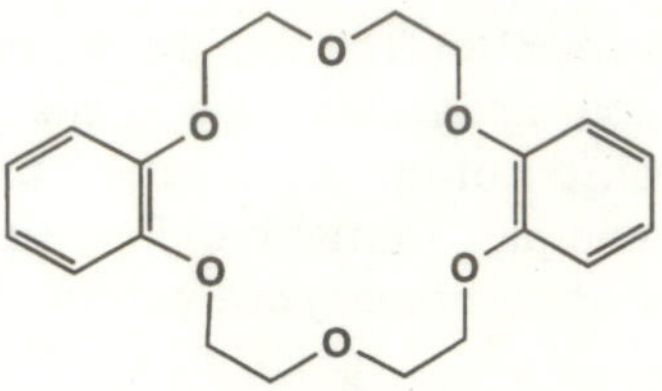

214 $K^+ > Rb^+ > Cs^+ > Na^+ > Li^+$

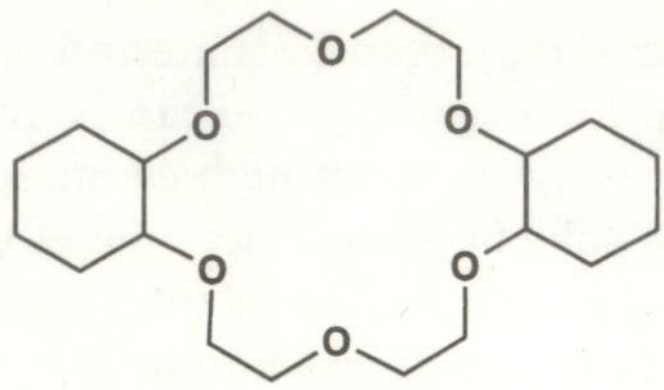

215 Sr^{2+}

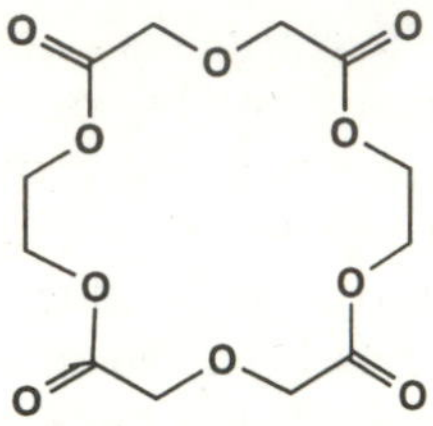

216 Li^+, Co^{2+}

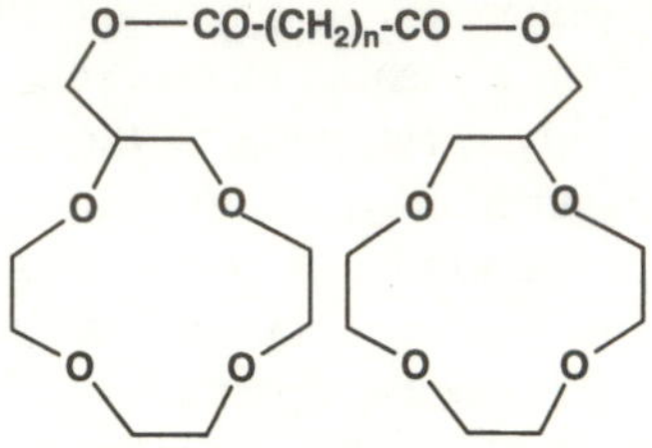

217

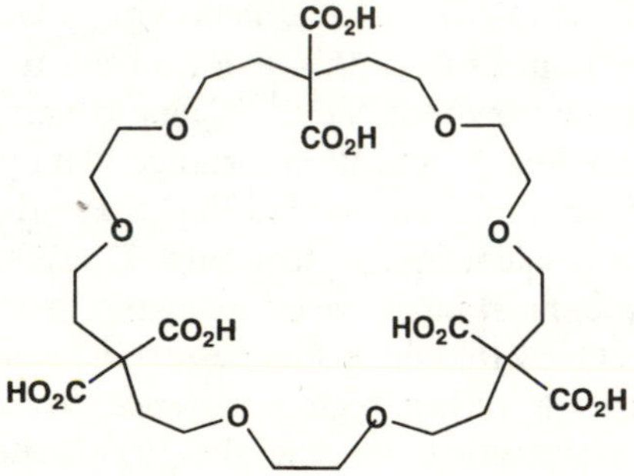

218

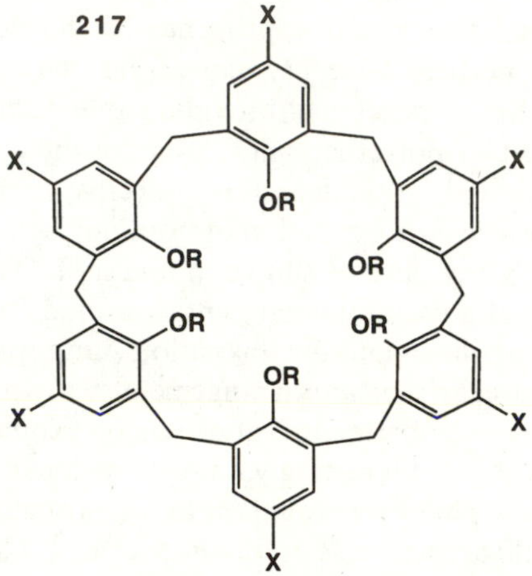

219 R = CH_2COOH; X = SO_3Na

K_{uranyl}/K_M = 80-110 for 218; 10^{12}-10^{17} for 219

M = Ni^{2+}, Zn^{2+}, Cu^{2+}

Scheme 4.64

addition, their structure offered welcome opportunities to affect the selectivity of binding by alterations in the structural parameters of the basic framework. Thus, a variation of the size of the internal cavity of a given type of cryptand can be achieved by changes in the tether length connecting binding sites. These changes may affect the specificity of cation binding in an efficient and predictable way. For example, cryptand **220a** (Scheme 4.65) is a very specific ligand for Li^+. Its analogs **220b** and **220c**, which contain an additional one or two OCH_2CH_2 unit(s) in the bridging chain(s), reveal a similar high selectivity for Na^+ and K^+ correspondingly.[33b,c] As was to be expected, the stability of these complexes was greatly increased in comparison to their monocyclic counterparts (cryptate effect). Thus **220c**$\cdot K^+$ was more stable by factor 10^5 compared to **214**$\cdot K^+$.

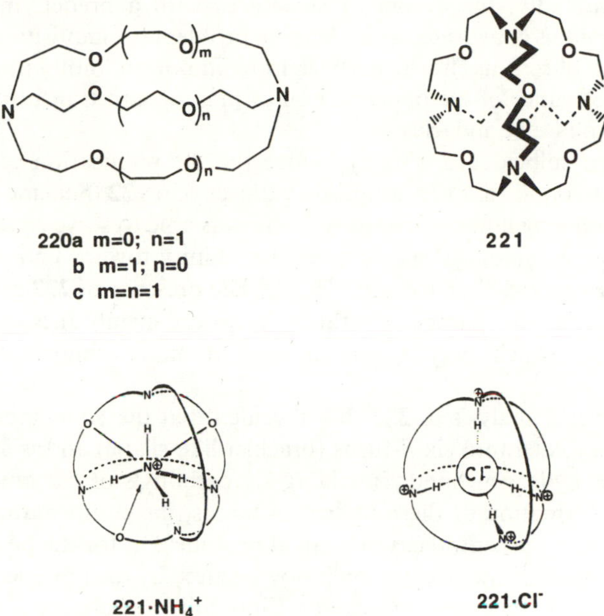

Scheme 4.65

Additional opportunities were realized with the synthesis of cryptands possessing entirely different arrangements of binding sites. In all previously mentioned cases the specificity of the binding is determined primarily by the size of the guest. Cross-bridged cryptand **221** was specifically designed as a nearly spherical molecule with a tetrahedral arrangement of nitrogen binding sites that would recognize not only the size but also the shape of the encapsulated species.[33d] It was rewarding to find that **221** proved to be an excellent ligand for NH_4^+ (tetrahedral recognition as shown in **221**$\cdot$**NH_4^+**) and a very poor ligand for K^+, which has about the same ion radius but a spherical shape. Moreover, the same cryptand in its tetraprotonated form becomes a geometrically optimal receptor for spherical anions, such as halide ions of a size compatible with the size of molecular cavity.[33e] Thus it was found that **221**$\cdot$**NH_4^{4+}** is able to hold inside Cl^- and Br^- (**221**$\cdot$**Cl^-**) but not I^-. Moreover, it also exhibits a very high Cl^-/Br^- selectivity (>1000), which reflects the better fit of the former ion to the size of the internal hollow space. The versatility of this approach to the design of various structures with the shape and arrangement of the binding site tailored for a specific cationic or anionic species is well documented in numerous publications.[33e,f]

Now it seems appropriate to comment, at least very briefly, on one of the major reasons underlying the increased interest toward the various aspects of complexing phenomena. Why it is considered to be important to pursue studies in this area with such zeal and to produce, in ever-increasing numbers, ingeniously constructed molecular systems capable of forming various complexes? Undoubtedly, the design of structures with a predetermined set of complexing abilities constitutes a challenging problem for inquisitive minds and *per se* is worthy of serious efforts. An even more important motivation lies in the biochemical relevance of the phenomenon of binding and transport of various species across phase boundaries.

By a strange coincidence, shortly before the the publication of Pedersen's paper[32a] it was found that the antibiotic valinomycin **222** (Scheme 4.66) had a powerful complexing affinity toward K^+ and was able to serve as the carrier of this cation across biological membranes. Its affinity toward complexing with K^+ is 10^4 times greater than toward Na^+.[33f] The presence of **222** greatly affects the performance of the 'pumps', operating in the cell membranes to control the K^+/Na^+ ratio, which play a crucial role in many important biological functions.

X-ray structural analysis of **222·K**$^+$ revealed that the 36-membered ring of valinomycin is folded into six β-turns (bracelet-like shape) and is stabilized by intramolecular hydrogen bonds involving six carbonyls of the amido groups. Owing to the formation of these hydrogen bonds, the conformation of **222** is almost frozen and the central cavity is an almost ideal fit for the potassium ion. Efficient binding of the potassium ion is now secured by the presence of six ester carbonyls oriented inwards in this cavity. Lipophilic side chains of the residues are directed outwards both to ensure the shielding of the hydrogen bonds from the interaction with the solvent and to ensure the solubility of the complex **222·K**$^+$ in lypophilic media. Another type of natural ionophore (*i.e.* ion-transporting molecule) is represented by the structure of nonactin **223**.

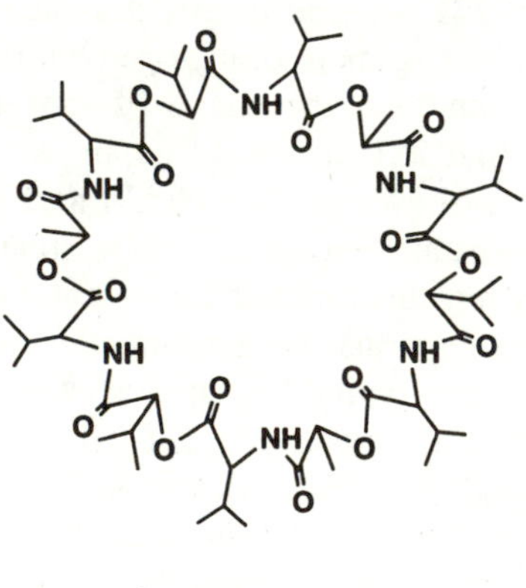

222

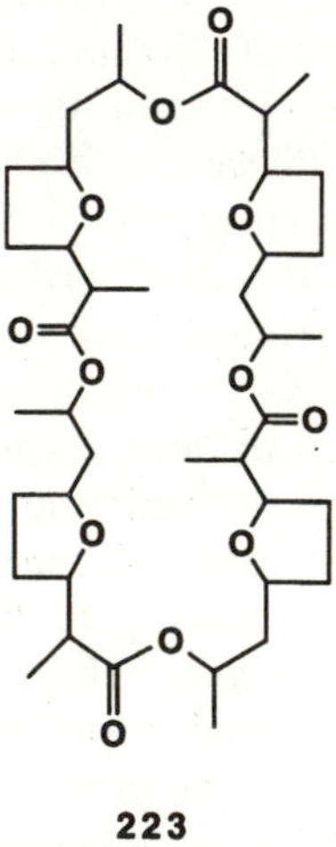

223

Scheme 4.66

Both in their general pattern of structure and mode of cation binding, **222** and **223**, as well as many other naturally occurring ionophores, are similar to crown ethers. In fact, the discovery of the crown ethers provided scientists with long-sought after artificial models that enabled them to mimic the selective binding of cations and their efficient transport across phase borders. It was not surprising to see that Pedersen's discovery was immediately recognized as a breakthrough in the understanding of the biological phenomenon of ion transport. In a matter of a few months, numerous investigations were targeted toward the design of artificial multidentate compounds as models for elucidating the mechanism of action as well as the structure–activity relationships for natural ionophores. The ultimate goal of this research was to create artificial mimics with a promise of medicinal applications.

Overshadowing even these immediate ramifications was the opportunity to accumulate knowledge essential for the understanding of one of the most challenging biochemical problems: the problem of molecular recognition. All the major biochemical events that occur in living organisms are somehow related to the initial event of recognition. The recognition of various endo- or exogeneous compounds by certain receptors in the cell wall or in the organelles in its interior is followed by a specific response triggered by the incoming chemical signals. Examples attesting to the universal significance of molecular recognition can be found at all levels of biological organization. Some examples, randomly chosen for illustration, include: specific enzyme–substrate or antigen–antibody interactions, the formation of complementary DNA and RNA chains, olfactory and taste receptors, and hormone and pheromone interactions with the respective receptors, as well as the more general phenomenon of chemical communication. A spectacular example that shows the tremendous importance of the accuracy of recognition can be found in the area of chemical communication between insects. For example, the antennae receptors of a given insect are able to identify molecules of a specific pheromone unmistakenly among thousands of alien molecules present in the immediate environment. As the result of an interaction between this receptor and just a few molecules of pheromone, a dramatic change in the behavior of the entire organism can be triggered. This signal–response event corresponds to an amplification of the initial signal by an incredible magnitude (up to 10^{24} times!).

Early into their endeavors in the field of crown ethers, the explorers in this field realized that the creation of artificial systems able to mimic the phenomena of biological binding and recognition may lead to far-fetched consequences. As Lehn stated, referring to the unique property of the tetrahedral recognition by **221** toward NH_4^+, 'it represents a state-of-art illustration of the *molecular engineering* involved in achieving the goal of *abiotic receptor chemistry*: the design of synthetic receptor molecules, by correct manipulation of geometric (receptor structure) and energetic (binding sites, intermolecular interactions) features so as to achieve high receptor–substrate complementarity'.[33d]

We have no opportunity to list here even the major routes of molecular design studies in this immense area. For us as chemists the most interesting is the application to the problem of mimicking enzyme action. Below we shall concentrate our attention on these studies. It is appropriate, first, to formulate the chemical aspects of this problem.

4.9 ENZYME MIMICKING

4.9.1 Outline of the Problem

Modern organic chemistry may proudly claim that it is able to synthesize compounds of enormous complexity unknown in Nature, and to master a most diversified set of methods enabling one to carry out almost any kind of chemical transformation. This claim is well substantiated by the number of outstanding achievements in organic synthesis during the last few decades. However, this claim quickly loses its impressiveness when compared to the performance of the chemical machinery of the most primitive living cells. Thousands of compounds, both simple and extremely complex, are produced by enzymes at any moment in any living cell under ambient conditions. These products are manufactured in water, within a very narrow range of pH, without referring to high temperature or pressure, and without the help of all those diversified reagents carefully designed by chemists for any single chemical operation and supplied in readily available quantities by chemical companies. Every cell is permanently involved in carrying out total multistep syntheses of the plethora of diverse organic compounds required to support its life cycle. These compounds are produced in a matter of minutes, in quantitative yield, with complete regio- and stereoselectivity! What this means is that the most cumbersome problems of strategy and tactics of modern organic synthesis have already been solved by the 'chemical plants' operating in any living system. The apparent ease and perfection of this performance cannot help but make chemists experience feelings of both surprise and admiration mixed with a sort of inferiority complex.

Experimental studies of the pathways of biosynthesis provided the explorers with information about the chemistry of these processes. This knowledge provided a firm ground for an entry into the area of biogenetically patterned syntheses of various natural compounds, which employed the strategic principles designed by Nature (see, for example, the synthesis of morphine, described in Section 3.2.1). Still, very little is known about the mode of enzyme action as catalysts, despite numerous experimental data referring to the mechanism of basic biochemical transformations. Quite a number of more or less viable hypotheses were suggested to account for enzymes' impressive capacity to perform as highly effficient and selective catalysts. These aspects have been the subject of numerous investigations on specifically designed chemical models (*vide infra*). Yet, there are other, and by far much more important, facets of enzyme functioning, namely: (i) the ability of enzymes to recognize the proper substrate at the proper moment among thousands of organic compounds already present in the cell and (ii) the problem of regulating enzyme activity. The operation of hundreds of enzymes in any living system must be strictly regulated so that, at any given moment and site, only a well-defined set of compounds is produced in the amounts required by the current needs of the organism. Moreover, during the process of growth, differentiation, and multiplication, the demands for the chemicals produced in a cell may vary dramatically. Therefore the respective regulatory mechanisms must be both very rigorous at the given moment of the life cycle and sensitive to changes in the chemical environment. Mistakes in regulation may have lethal consequences and therefore the correct functioning of the command systems is of vital importance at all levels, at all times, within a cell, tissue, organ, and organism as a whole.

Thus, for modern chemistry there is no more challenging problem than the creation of an *artificial molecular system capable of functioning with an enzyme-like efficiency, selectivity, and amenability to control.* It is not an exaggeration to claim that the creation of artificial catalysts of that type will be a truly revolutionary breakthrough in chemistry and will lead to in-depth changes in laboratory and industrial syntheses, affecting the whole of civilization.

Enzymes are very complex organic molecules consisting of a huge polypeptide backbone ornamented with a wide array of reactive chemical groups with enormously complicated and highly variable conformations of the whole molecular construction. There are actually no compelling reasons for chemists to synthesize some monstrous artificial molecules of this type to be similar in its properties to its natural counterpart. Aside from being a truly formidable task, such a synthesis is rather unnecessary, unless we are going to create artificial life. The most common (and realistic!) strategy of the studies in this field envisions the design of comparatively small and diversified structures that serve as models to simulate separate aspects of enzyme activity. Investigation of these artificial models provides an opportunity to reproduce this or that peculiarity of enzyme action on well-known molecular structures and thus to pave the way for the understanding of the basic chemisty of enzyme catalysis. The ultimate goal of these investigations consists of the elaboration of purely synthetic systems comparable in their enzyme-like efficiency to that of the natural prototypes.

Design in this area is following several fairly different pathways. Below we shall discuss just a few approaches closely related to the topic of this chapter.

4.9.2 Selectivity and Regulation of Binding

In Section 4.8 we discussed new options for the selective binding of cations and anions that arose because of the discovery of crown ethers. Now we shall continue the discussion of this topic with more emphasis given to its relevance to enzyme mimicking.

In this respect it is especially instructive to consider first the results of studies aimed at the application of multiligand binding to design artificial receptors for species larger than inorganic cations and anions. An instructive example is represented by hexaaza macrocycles of the general formula **224**, composed of two triamino binding sites connected by polymethylene chains. This structure was specially designed for the recognition of linearly shaped doubly charged species (Scheme 4.67).[33g] These serve as efficient ligands for dicarboxylate bis-anions $^{-}O_2C(CH_2)_mCO_2^{-}$ since the latter can be nicely incorporated between the ligand sites, as is schematically represented in **224a**. Even more important, however, is that the selectivity of this binding reveals a good correlation between the length of the $(CH_2)_n$ bridges in macrocycle **224** and that of the $(CH_2)_m$ chain of the substrates. In a similar way, receptor **225** (in which two macrocycles with donor sites are separated by bridges composed of aromatic residues of variable size) can effectively recognize and bind specifically bis-ammonium salts $\mathbf{NH_3{}^+}$–$\mathbf{R{-}NH_3{}^+}$, depending upon the length of the tether (R) connecting the ammonium centers.[33h]

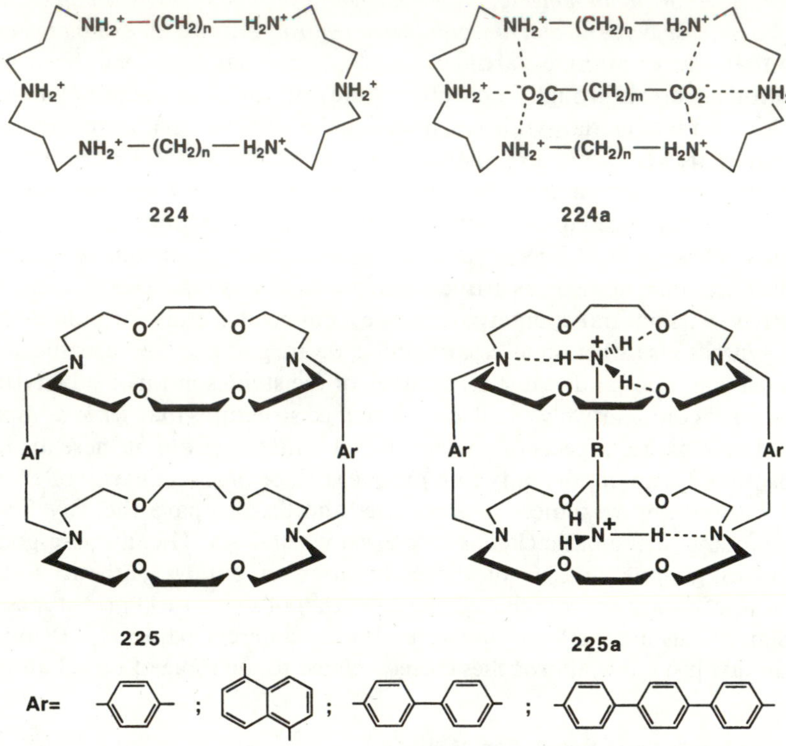

224

224a

225

225a

Ar=

Scheme 4.67

In fact, systems like **224** and **225** represent simple models of artificial receptors with a variable pattern of substrate recognition. It is also noteworthy that the observed properties of these ligands did not come as a surprise but, rather, as the anticipated result of a thoughtful choice in structural parameters for the designed structures.

The problem of binding purely covalent substrates is expectedly difficult because of the lack of centers capable of providing a strong electrostatic attraction. For these compounds it is necessary to elaborate receptors able to bind substrates effectively *via* van der Waals interactions, which are much weaker than the Coulomb forces operating in the previously mentioned complexes. A set of models designed for this purpose is shown in Scheme 4.68.[34a] Cyclophanes of the general formula **226**[34b] have been synthesized as hosts for aromatic hydrocarbons. The presence of a hydrophobic cavity of sufficient size endows **226** with the capacity of complexing hydrophobic aromatic molecules. Quaternary ammonium residues located on the periphery ensure the water solubility of the models. In the series **226a–d**, variations in the chain length regulate the size of the cavity while additional changes to its shape and hydrophobicity are possible by introducing methyl groups, as illustrated in the series **226e–h**. These compounds are able to form rather strong complexes with aromatic compounds in water solution. It was shown that the stability of these complexes and the selectivity of the binding with various guests are very sensitive to the above-mentioned variations of the structural parameters of the host. It was even possible to study the transfer of aromatic compounds between two lypophilic phases across an aqueous phase using additionally modified derivatives of a similar structural type.

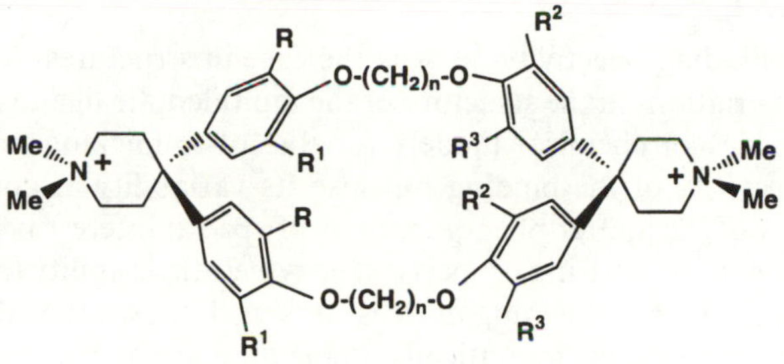

226a-h

	n	R	R^1	R^2	R^3
a	2	H	H	H	H
b	3	H	H	H	H
c	4	H	H	H	H
d	5	H	H	H	H
e	4	Me	H	H	H
f	4	Me	Me	H	H
g	4	Me	Me	Me	H
h	4	Me	Me	Me	Me

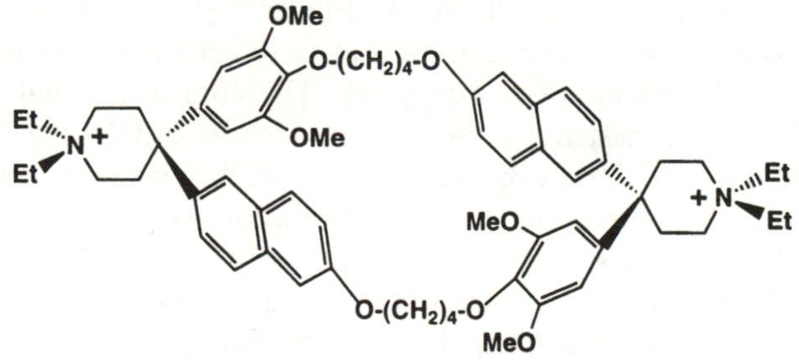

227

Scheme 4.68

In order to accommodate substrates larger than simple aromatic compounds, analog **227** was synthesized; this contains naphthalene rings instead of the benzene nuclei.[34c] As a result of this simple modification, ligand **227** was able to form complexes with molecules as large as steroids. Its ability to bind substrates of lesser size, however, was sharply reduced. Depending upon the variations in the structure of the steroid derivatives, the binding energy with **227** can vary within almost two orders of magnitude. Therefore complexation with **227** may serve as a tool for the preferential binding and transport of certain steroids from their mixture.

These examples demonstrate that molecular design, based on a rather straightforward analysis of the size and shape of the substrate and receptor, nevertheless might be an efficient strategy for achieving a high selectivity in recognition, binding, and transport of various compounds. With this knowledge, one can thus create workable models to study these aspects of enzyme action.

Control over binding selectivity, as was the case in structures like **224–227**, was exerted *via* variations in the structure of the multidentate ligands. However, is it possible to design chemical models capable of mimicking not only the enzyme-like selectivity of the binding but also its variability as controlled by external conditions? The latter phenomenon is of special interest because of its obvious relevance to a surprising property of enzymes: their ability to operate as variable or even as on–off switching devices. It is well known that the catalytic activity of enzymes may vary dramatically, depending upon changes in external conditions like pH, the addition of metal ions, low-molecular weight regulators, *etc*. A reasonable assumption implies that this sensitivity is due to the ability of enzyme molecules to undergo reversible and condition-dependent conformational changes. There are also extensive data showing that the conformation of the active center of the enzyme, responsible for its catalytic activity, may be changed by variations at sites far remoted from this center (allosteric effects). The dependence upon allosteric effects is of special importance as it represents one of the basic regulatory mechanisms operating in the living system, controlling the functioning of the enzyme system with the help of chemical signals produced endogeneously in the cell or transported in from its exterior.

While the design of 'controllable' molecules is still in its infancy, it seems instructive to consider some examples that have been designed with the aim to mimic this unique peculiarity of enzymes. The conformational lability of the polyether ligand suggested its use as a model to enable the insertion of a switchable group into a basic molecular framework. The approach that is based upon the photoregulated change in the shape of the complexation cavity by geometrical isomerism is shown in Schemes 4.69 and 4.70. The cyclic polyether **228a** can easily undergo light-induced isomerization into the *cis* isomer **228b**. As a result of this transformation, the binding ability for alkali metal ions is noticeably increased (approximately 3–4 fold).[35a]

Similar results were observed for bis-crown ether **229** containing an azo linkage between the binding sites.[35b] In this case the photochemical conversion **229a** → **229b** produces strikingly different effects on the relative ability to complex Na^+ and K^+. The binding ability for Na^+ drops almost sixfold, while the affinity for the K^+ cation shows a 42-fold increase. In the dark, **229b**

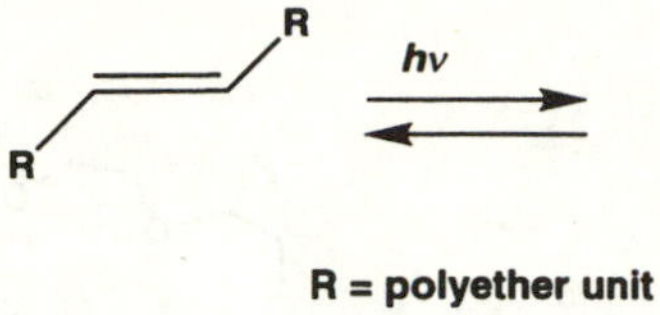

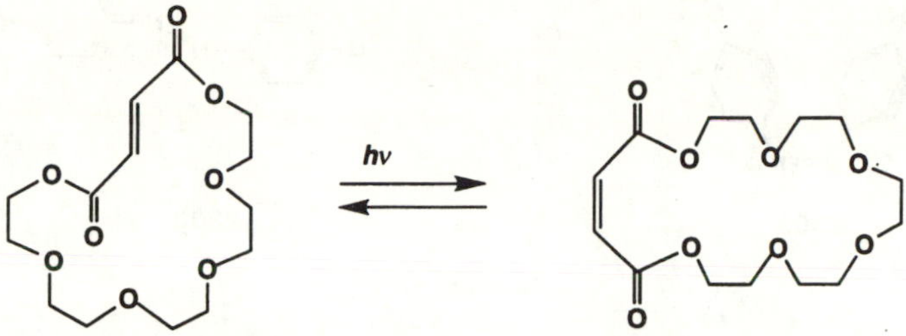

R = polyether unit

228a 228b

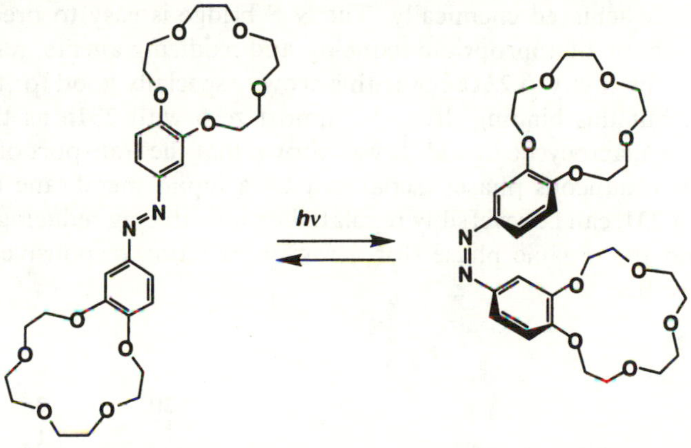

229a 229b

Scheme 4.69

undergoes the reverse isomerization into **229a** and the original pattern of binding is restored. While the suggested explanation of the difference in the binding effects between **229a** and **229b** is not fully consistent, the revealed behavior of this system clearly attests to the potential of creating various types of photoresponsive molecular switches based upon a similar approach.

If an azo group is included into the macrocyclic ring, the changes in the binding capacity brought by *cis–trans* isomerization are even more spectacular.[35c] Thus **230a** reveals a pattern of binding similar to that of **214** (high preference for K^+), while *trans* isomer **230b** is not able to form complexes with K^+ whatsoever. (The presence of the *trans* double bond in the ring prevents the folding into the shape necessary for complexing K^+.) This is a very rare case of an 'all-or-nothing' switch in a purely synthetic model (Scheme 4.70).

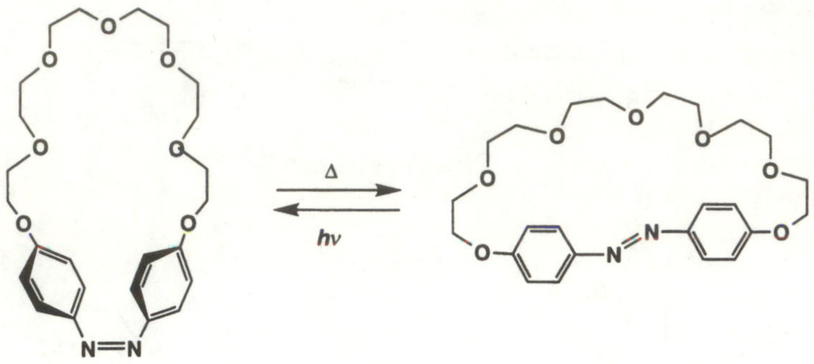

230a

230b

Scheme 4.70

The idea of utilizing redox processes to effect reversible cation binding was investigated in the model **231** (Scheme 4.71). Here the transformation of **231a**, having a closed polyether ring, into the open form **231b** and the reverse process can be easily achieved chemically. The S–S bridge is easy to break and form under the action of appropriate reducing and oxidizing agents, respectively.[35d] The size of the cavity in **231a** made this crown especially good for binding Rb^+ and Cs^+, but this binding drops to almost zero with **231b** as the reduction destroys the macrocyclic ligand. It was shown that the transport of the Cs^+ ion between two aqueous phases, separated by a liquid membrane (chloroform) containing **231**, can be reversibly regulated by introducing reducing or oxidizing agents into the organic phase (for a review of redox responsive macrocyclic ligands, see Beer[35e]).

231a

231b

Scheme 4.71

Additional and very promising opportunities to exert control over the selectivity of binding can be achieved by the design of hybrid molecules containing two different types of coordination sites. Compound **232** contains two types of bonding sites: a polyether chain capable of serving as a ligand for metal cations, and aromatic rings capable of forming charge-transfer complexes with strong acceptors (such as tetracyanoquinodimethane **233**) (Scheme 4.72).[35f] In acetonitrile solution, the revesible formation of the sandwich-like 1:1 complex **234** between **232** and **233** is observed. The addition of 1 equivalent of a potassium salt leads to a sharp increase in the stability of **234** (the association constant exhibits a 35-fold increase). The reverse effect is observed if 2 equivalents of the same salt are added. The interpretation shown in this scheme is based upon a reasonable assumption that the binding of the metal to the polyether chain serves as an additional driving force to change the initial extended conformation of **232** into the folded **232a**, which is better adjusted to form the sandwich complex with **233**. As a result, the stabilized binary complex **234a** is formed. The presence of a second metal ion results in the formation of

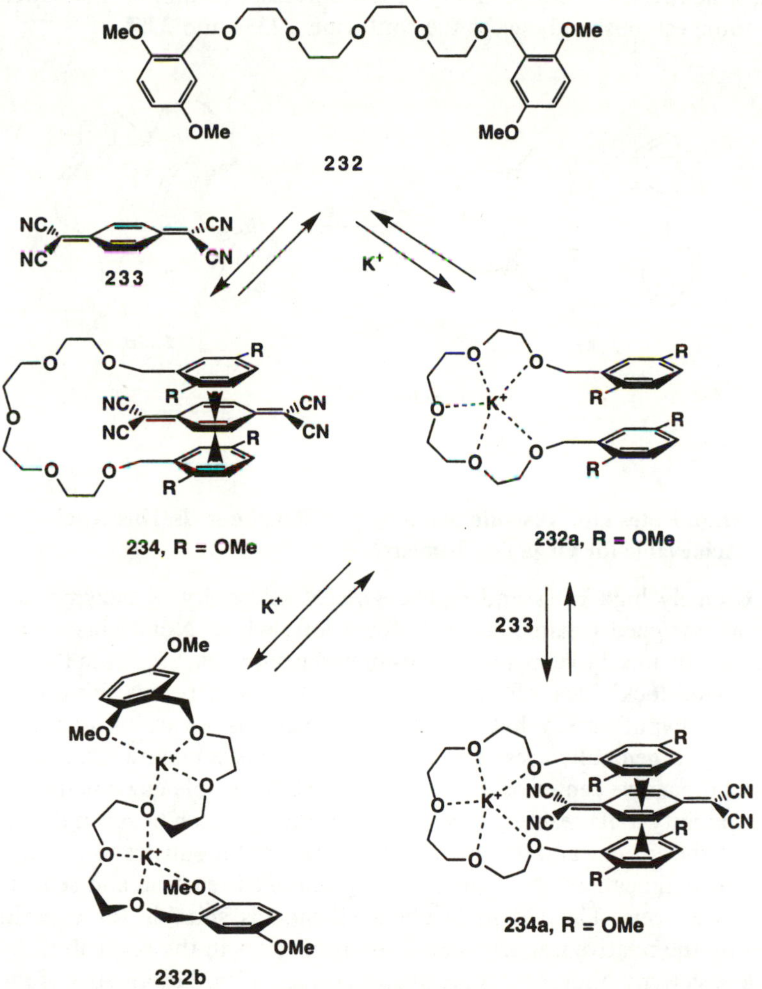

MeO O O O O O OMe

OMe MeO

232

NC NC CN CN

233

K⁺

234, R = OMe

232a, R = OMe

K⁺

233

232b

234a, R = OMe

Scheme 4.72

the bicyclic chelate **232b** with the disposition of aromatic rings unfavorable to sandwiching **233**. The authors[35f] reasonably consider the observed effects as model examples of allosteric regulations of enzyme activity.

A second example also illustrates the effects produced by complex formation at one binding site on the complexation ability of another site. Compound **235** was designed as a ditopic ligand capable of forming complexes of two types: (1) with alkali metal ions because of the presence of the polyether ring fragment and (2) with transition metals at the bipyridyl unit. The latter fragment is used as the 'gadget' to control the complexing ability of the former group. It was shown that, in the absence of heavy metals, **235** is a slightly more selective in complexing K^+ in comparison to Na^+. In the presence of tungsten derivatives, this selectivity is reversed (Scheme 4.73). This difference is ascribed to the effects brought by the binding of two pyridyl units as chelating ligands to the transition metal, which restricts the flexibility of the polyether chain and thus affects the equilibrium established between the conformers **235a** and **235b**.[35g]

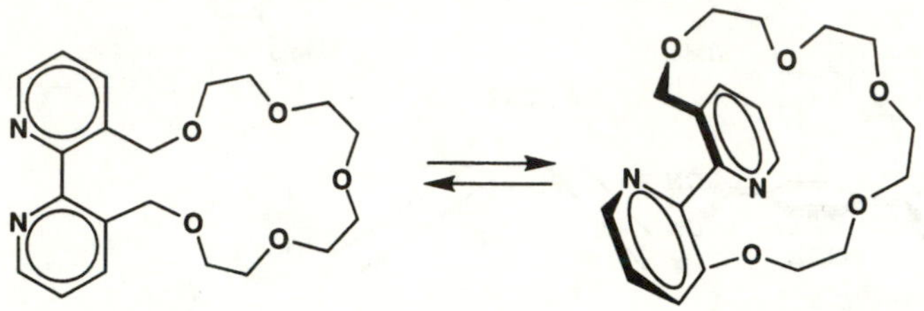

235a **235b**

Scheme 4.73

4.9.3 High Rates and Absolute Selectivity of Reactions. Is This Goal Achievable for Organic Chemists?

The extremely high rates and high degree of selectivity of enzyme-catalysed reactions intrigued organic chemists for a long while. Numerous suggestions have been advanced to account for this phenomenon, starting from the century-old 'key-and-lock' idea of Emil Fisher to the more recent 'induced fit' of Koshland's hypotheses. Whatever different the details of various interpretations might be, it is generally presumed that a substrate is somehow fixed within the cavity of the active center of the conformationally flexible enzyme molecule, in close proximity to its reacting groups. The resulting match between the reacting centers of the enzyme and the reacting conformer of the substrates is supposedly one of the main causes of the observed high rates of reaction and selectivity of enzyme reactions. The design of chemical models suitable for experimental studies on the relative importance of various factors in the control of the rates and the selectivity of organic reactions has been, and still is, an area of the most intensive interest.

Enzyme-catalysed chemical reactions can be up to 10^{12} times faster than their counterparts in organic chemistry. These amazing rate accelerations represent one of the most controversial issues of enzyme catalysis. No truly consistent explanation for this phenomenon has yet been advanced.[36a–c] Attempts to mimic this aspect of enzyme catalysis at specially designed artificial systems are numerous, but generally the interpretation of these results is not free from ambiguities. Nevertheless, it is appropriate to consider at least briefly some approaches illustrative of the general trends of these studies.

First of all, it is necessary to emphasize that accelerations up to 10^8 times can be observed quite often for intramolecular reactions when compared to their intermolecular counterpart. The binding of a substrate to an enzyme to form an enzyme–substrate complex is the initial step of every enzymatic process. Therefore, there should exist a close analogy between the enzyme–substrate complexation and the effects caused by the intramolecularity of chemical reactions. It has to be admitted, however, that there is no generally accepted interpretation of the effects of intramolecularity on reaction rates (see, for discussion, refs. 36b–d). Yet it is instructive to have a look at one of the simplest examples showing how much can be gained in mimicking the high reactivity of enzymes in the design of models for the respective intramolecular reactions.

Acid-catalysed hydrolysis of unactivated amides usually requires prolonged heating in the presence of concentrated mineral acid. At the same time, chymotrypsin is able to cleave amides readily at neutral pH and room temperature. In the search for a chemical model to simulate the latter reaction, Menger investigated the intramolecular hydrolysis of the carboxamide function in compound **236**.[36e] This choice was dictated by the results of molecular mechanics calculations, which predicted that in both preferable conformations, **236a** or **236b**, the carboxyl groups are poised within van der Waals contact distance of the amide carbonyl group. In the case of **236b**, the close proximity of the hydroxyl oxygen to the reacting site secures the opportunity for a synchronous proton transfer and nucleophilic attack at the amidic function. Hence it was suggested that the cleavage of the amide function should proceed very easily (Scheme 4.74).

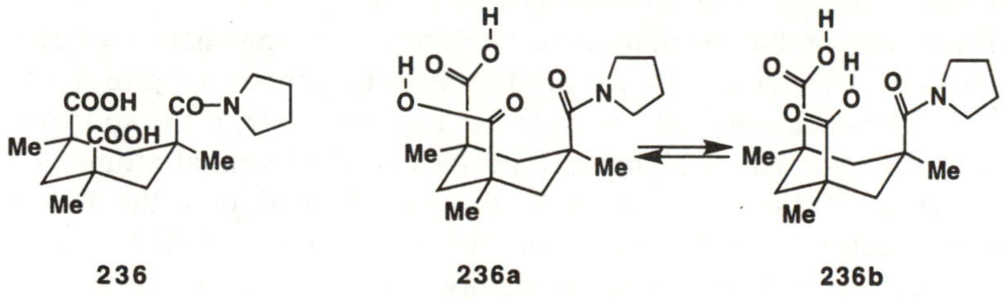

COOH CO–N, COOH, Me, Me, Me

236 **236a** **236b**

Scheme 4.74

These predictions were substantiated experimentally in a spectacular way. In fact, it was demonstrated that the intramolecular cleavage of the amidic bond in

236 occurs at ambient temperature, neutral pH, and without any additional catalysts, at a chymotrypsin-like rate! The acceleration of the reaction, compared to its intermolecular counterpart, was estimated to be as high as 10^{12}. These and similar results led the author to suggest that if a proteolytic enzyme can bring 'its aspartate carbonyls adjacent to an amide substrate with a geometry portrayed in **236a** or **236b**, very little additional catalytic power would be necessary'. As is also stated in Menger's review,[36b] 'it is tempting to conclude that one need never explain enzyme catalysis with some sort of esoteric mechanism'. Probably it is also relevant to mention that this very review begins with a reference to 'an anonymous, yet prominent biochemist', who would claim that 'it does not matter how elegant an enzymatic model you organic chemists construct, no biochemist will ever pay much attention to it'.

Impressively high rates of enzyme-catalysed reactions cannot overshadow an even more impressive peculiarity of these processes, namely their unique chemo-, regio-, and stereoselectivity pattern. In fact, enzymes are able to carry out reactions involving a specific site of the molecule without any side reactions competing at the alternative and almost identical centers, and with total control over the absolute streochemistry of the product. In this respect, modern organic chemistry, all its powerful arsenal notwithstanding, is still unable to compete with Nature.

One of the most challenging problems refers to the chemical modeling of the regioselectivity pattern of enzyme action. Thus, the enzyme unsaturase converts stearic acid **237** into oleic acid **238** and no carbon atoms, except the C-9 and C-10 pair of the long aliphatic chain, are affected in this reaction (Scheme 4.75). From a chemical point of view, this result implies that the conformationally flexible chain of substrate **237** is locked inside the cavity of the enzyme's active center in such a way that only the central $–CH_2–CH_2–$ moiety is exposed to the assault of the oxidizing moiety present in the active site of this enzyme. So far, no successful attempts to simulate this absolute CH_2 group selectivity at the undifferentiated centers of long-chain aliphatic compounds have been described for the chemical models. Yet there is experimental evidence showing that factors affecting the shape of long-chain molecules may significantly alter the selectivity of the non-enzymatic reactions at the alternative positions in these systems.

It is well known that the homolytic chlorination of long-chain aliphatic acids in solution occurs in almost a non-discriminating mode to give a complex mixture of monochlorinated products. At the same time, if the substrates are applied to the surface of alumina, an attack at the terminal atoms is highly preponderant.[37a] In a related study it was disclosed that the mixture of monochlorinated acids formed upon the chlorination of **237** under these heterogeneous conditions consisted mainly of 18-chloro- and 17-chlorostearic acids (**239a** and **239b** correspondingly, ratio **a**:**b** = 0.8:1, total content >90%).[37b] The observed effect was interpreted as a result of the formation of a densely packed monolayer of the adsorbed acid in which the polar carboxylic groups are attached to the surface and the lypophilic terminal groups are stretched outside and therefore more susceptible than the internal CH_2 links to

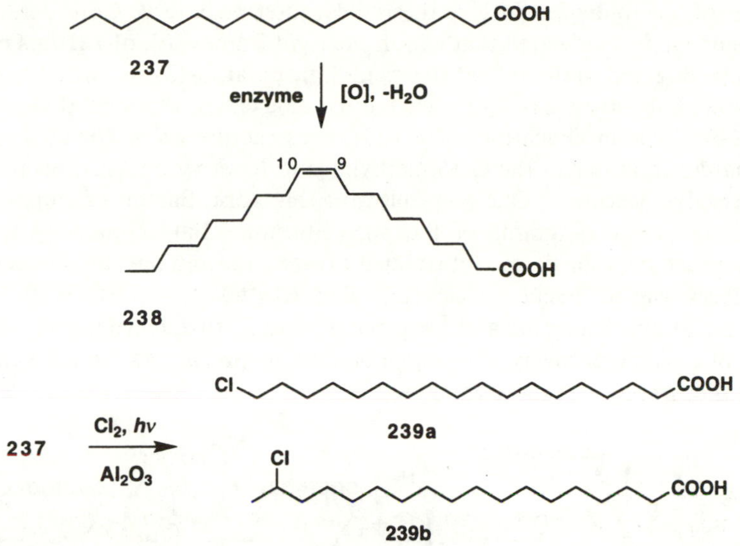

COOH

237

enzyme [O], $-H_2O$

10 9

COOH

238

Cl COOH

239a

237 Cl_2, $h\nu$ / Al_2O_3

Cl

COOH

239b

Scheme 4.75

the attack of chemical reagents. In a way, this example serves as an illustration of a purely mechanistic approach to the solution of the selectivity problem. Obviously it is of limited applicability.

Organic molecules with a rigid framework are much more suitable for the design of workable models capable of simulating the enzyme-like selectivity of chemical functionalization at unactivated centers. Intramolecular reactions seem to be especially useful in this context. In fact, the rigidity of the skeleton offers an opportunity of purposefully placing a reacting function in close proximity to the desired site of functionalization, thus making preferable the attack at this center.

The fruitfulness of this approach was first demonstrated in the 1960s in Barton's studies of the synthesis of the highly active natural hormone aldosterone **240.**[37c–e] Aldosterone was available in only milligram quantities from natural sources and, for this reason, the elaboration of a chemical synthesis of **240** was an extremely important task. The most distinctive feature of the structure of **240** is the presence of a functionalized substituent at the C-13 position (as opposed to the typical methyl group found in the structure of the majority of sterols). At that time, numerous partial and total syntheses of various steroid compounds unsubstituted at C-18 had been described (*e.g.* Scheme 3.1). Barton suggested[37c,d] that a practical pathway applicable for the preparation of aldosterone from other sterols could be elaborated *via* a site-selective oxidation of the C-18 methyl group. The essence of Barton's approach is illustrated by the transformation of cortisone acetate **241** into aldosterone **240**, as is shown in Scheme 4.76. The mechanism of the novel photochemical reaction specifically designed to achieve this result is also presented in this scheme. The key steps include the initial formation of radical **A**, followed by the transfer of a δ-hydrogen to give the radical intermediate **B**. Consideration of molecular models indicated that within the rigid framework of **241** this process should be directed preferably at the methyl group at C-13. As a matter of fact, oxidation of substrate **241** also occurred to some extent at the methyl group at C-10. Additional modification of the substrate structure was needed to bring the reacting group closer to the C-13 methyl group to carry out this reaction as a truly selective process.[37e] One year following this work, Barton's group was able to prepare about 60 grams of the pure hormone **240**, a quantity that far surpassed the combined amount isolated previously from natural sources. This final success was the result of a carefully planned study that started with a clear-cut formulation of the goal and the general strategy to achieve it, the judicious design of a viable pathway, the choice of suitable precursors, and the elaboration of the appropriate reaction.

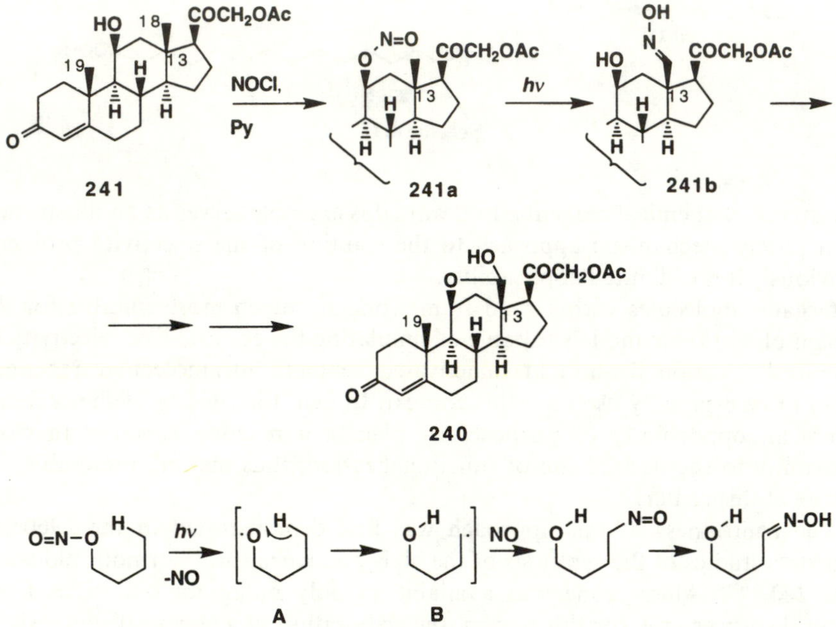

HO 18 $COCH_2OAc$ 19 H 13 H H O **241** NOCl, Py N=O O $COCH_2OAc$ H 13 H H **241a** *hν* OH N HO $COCH_2OAc$ H 13 H H **241b**

HO O $COCH_2OAc$ 19 H 13 H H O **240**

O=N–O H *hν* -NO ·O H **A** O H **B** NO O H N=O O H N-OH

Scheme 4.76

In the course of sterol biosynthesis, enzymes can effect the functionalization of the main hydrocarbon backbone at almost any center to produce selectively diverse functionalized derivatives. Is it possible to reproduce such flexibility and selectivity on purely chemical models? Barton's synthesis of **240** is a case in which the functionalization is achieved at the saturated center positioned in *close proximity to the reacting center*. The goal of Breslow's group was much more challenging, namely to develop a general route for *controlled remote functionalization*. The idea of this approach seems to be surprisingly simple.[37f] It was suggested that a single function present in cholestanol **242** could be used as a site for attaching a temporary tether (a spacer) containing a functionality capable of reacting with a non-activated C–H bond. By this anchoring, the intermolecular reaction was transformed into a intramolecular process and it was conceived that the regioselectivity of the latter process could be controlled by variations in the length of the spacer. The design of the appropriate system required extensive utilization of molecular modeling to determine the optimal nature and size of the tether. Two chemical reactions were employed for hydrogen abstraction, namely benzophenone-mediated photochemical oxidation and homolytic halogenation. Examples of the former reaction are shown in Scheme 4.77.[37f]

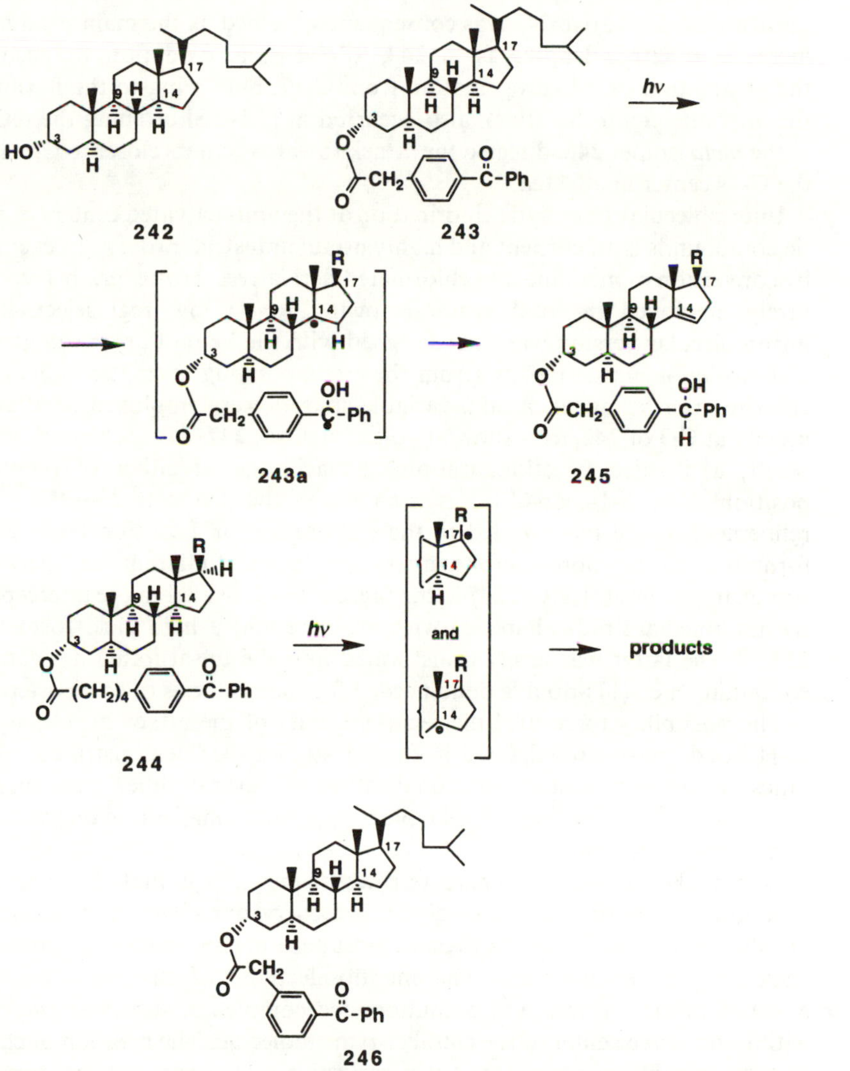

242 **243** $h\nu$ **243a** **245** **244** $h\nu$ and products **246**

Scheme 4.77

Benzophenone-mediated photooxidation is a well-known reaction that involves the initial formation of a radical species from the photoexcitation of a carbonyl group, the subsequent abstraction of a hydrogen atom from any available source and, finally, the quenching of the radical intermediates by hydrogen transfer or elimination processes. In the ester **243**, the benzophenone moiety is fixed to C-3 with the help of a three-atom spacer, while compound **244** has the same attachment but with six atoms in the tether. This difference was sufficient to alter dramatically the selectivity of the intramolecular photoinduced oxidation. In fact, the reacting species produced from the carbonyl function of **243** was able to reach C-14 and thus effect an oxidation at this site with high selectivity by way of the presumed biradical intermediate **243a**. An unsaturated derivative **245** was consequently formed as the main product. The increase in tether length, as in **244**, allowed the oxidation to occur with the abstraction of a hydrogen atom from C-17, but owing to the flexibility of the aliphatic chain the attack also occurred at C-14. Shortening the tether, as in the *meta* isomer **246**, directed the attack towards centers closer to C-3, leaving the C-14 center unaffected.

Intermolecular homolytic chlorination at the non-activated centers of aliphatic compounds is an efficient and highly useful industrial process, as exemplified by conversion of methane into chlorinated derivatives. However, this reaction is rarely employed in total synthesis owing to its low regioselectivity. An intramolecular version was thus designed with the help of an *in situ* generated dichloroiodobenzene moiety (from the corresponding aromatic iodide) as the chlorine transfer agent. Again, variable spacers were employed to attach this moiety at C-3 of **242**, as is shown in the structures **247–249** (Scheme 4.78). As a result, a directed functionalization was achieved at either of the tertiary positions C-9, C-14, or C-17, as is shown in the products **250–252**.[37f] Later refinement of the reaction led to the elaboration of a version leading to the formation of the more useful mercapto or bromo functionalized derivatives. For example, photolysis of **247** in the presence of CBr_4 led to the interception of the intermediate radical species with bromine and a high yield formation of **253**.[37g] The latter was easily transformed into the unsaturated derivative **254** containing the 9(11)-double bond needed for the synthesis of corticosteroids.

The possibility to control the regioselectivity of the attack at non-activated C–H bonds was also validated in further studies that demonstrated opportunities to achieve remote functionalization at several other sites in steroid derivatives,[37h–j] including the valuable case of functionalization on the aliphatic side chain of cholesterol.[37i]

One of the outstanding characteristics of enzymes is that they are able to function as enantioselective catalysts and carry out chemical reactions with absolute stereospecificity. As a result, most natural compounds are produced as optically pure enantiomers. The enantioselectivity of enzymatic catalysis is ascribed to the formation of a multibonded complex of substrate and reagent within the active center of the chiral enzyme molecule. The creation of chemical systems capable of serving as enantioselective catalysts represents one of the central and most challenging problems of modern chemistry. In the last decades,

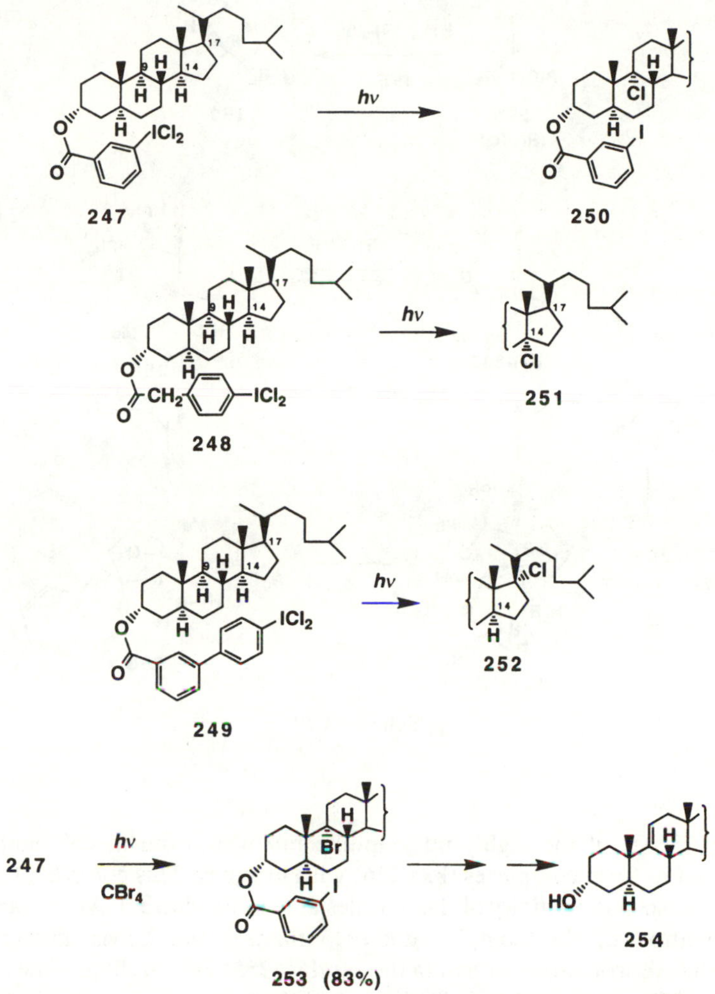

247 → (hν) → 250

248 → (hν) → 251

249 → (hν) → 252

247 → (hν, CBr_4) → 253 (83%) → → 254

Scheme 4.78

quite a number of such catalysts have been elaborated, many of which are now widely used in organic synthesis.[37k] The developments in this area were mainly due to empirical studies guided by general considerations of the prerequisites required for enantioselectivity of catalyst action. In only a few cases was success achieved as a result of a thoughtful molecular design of promising candidates.[37l] Probably the most explicit illustration of this approach can be found in the studies of Corey's group which ultimately led to the elaboration of a series of extremely efficient catalysts. The chiral oxaazaborolidine **255**, used for the enantioselective reduction of the carbonyl group with boranes, serves as a good example (Scheme 4.79).[37m]

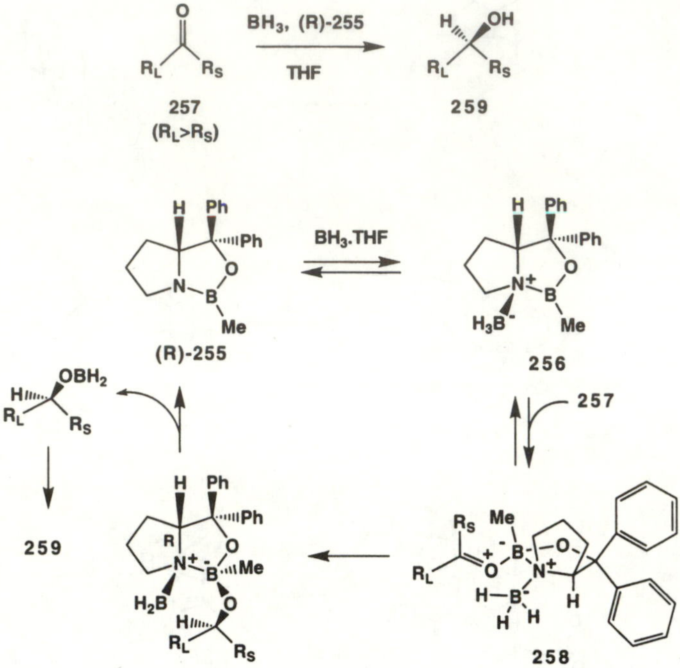

Scheme 4.79

The presence of the highly nucleophilic nitrogen atom in **255** enables this compound to form complexes (like **256**) with boranes. This complexation leads to an increase in the activity of the boranes as hydride donors. At the same time, the formation of this complex greatly enhances the Lewis acidity of the electrophilic boron atom present in the catalyst **255**. The binding of the carbonyl substrate **257** at this center is facilitated and an intermediate triple complex, substrate–catalyst–reagent, is formed. Molecular model considerations revealed that owing to the rigidity of the framework and steric effects of the substituents in catalyst **255** and intermediate complex **256**, the latter should exhibit a high preference toward binding to the least hindered enantiotopic side of the carbonyl group of substrate **257**. Hence, when the borane reagent and **257** are brought together in a three-dimensional chiral environment, as shown in structure **258**, an intramolecular hydride transfer should occur in a highly stereospecific mode. In line with these expectations, the reduction of the carbonyl group of various ketones results in a highly preferable formation of one enantiomer of the corresponding alcohols **259** with ee $\geq$ 83–97%. The sequence shown in Scheme 4.79 resembles the operation of an assembly line. The term 'molecular robot', suggested by Corey[37n] for catalysts like **255**, seems to be quite appropriate. It should not be overlooked that the principles employed in the design of this catalyst were based upon assumptions previously formulated to account for the stereospecificity of a similar biochemical reduction of carbonyl groups by reductase.[36a] One additional advantage of the artificial chiral catalysts is also worth of mentioning. Enzymes are specifically tailored to catalyse the conversions of the specific substrate which fits the shape and size of the pocket of their active sites. Quite often the scope of their preparative application is thus very limited. On the other hand, chiral catalysts of the type described above are not substrate specific and thus can be widely used.

Similar approaches to that outlined in Scheme 4.79 have been successfully developed to produce a set of highly efficient chiral catalysts for a number of **C–C** bond-forming reactions, like, for example, the ethynylation of aldehydes,[37o] the Diels–Alder reaction, and the aldol reaction,[37p] as well as for vicinal hydroxylation of certain alkenes.[37q]

In the previous sections we gave a number of selected examples referring to enzyme-mimicking systems to illustrate the fruitfulness of molecular design for both acquiring information about the specifics of enzyme action and for the elaboration of biomimetical approaches applicable for preparative utilizations. At present, it is well established that the peculiar features associated with enzyme-catalysed processes, such as rate enhancement, selective binding and recognition, allosteric effects, and the control of selectivity, can be imitated with carefully designed artificial systems. At the same time, it is no less obvious that an in-depth understanding of the intricate mechanism of enzymatic catalysis cannot be gained by the study of such simplified models as those considered above. Actually, this problem requires the multifaceted design of much more sophisticated models using the concept and methodology of a newly emerging science, supramolecular chemistry (see below and relevant discussions in refs. 33a, 35g, and 36a).

4.10 LIGANDS WITH A PREDETERMINED SELECTIVITY. DESIGN AND CREATION OF MOLECULAR VESSELS

'The Design of Molecular Hosts, Guests, and Their Complexes' was the title of the Nobel Lecture of Donald Cram.[38a] Ten years earlier, in a review entitled 'Design of Complexes between Synthetic Hosts and Organic Guests', he compiled a list of useful guidelines (18 in total!) for the rational design of hosts.[38b] The most crucial was the general concept of creating a ligand system containing a rigid three-dimensional framework with a set of complexation sites already pre-organized to match the binding sites of the guest. The choice of possible candidates and the optimization of the structural parameters required the extensive use of molecular modeling. As was stated in the aforementioned lecture, 'From the beginning, we used Corey–Pauling–Koltun (CPK) molecular models, which served as a compass on an otherwise uncharted sea full of synthesizable target complexes. We spent hundreds of hours building CPK models of potential complexes, and grading them for desirability as research targets'.

Among the numerous types of hosts designed by Cram's group, by far the most spectacular and promising are *spherands*.[38a,c] The basic framework of these ligands is composed of six *meta*-bridged aromatic moieties bearing six oxygen centers, as represented in the structure of **260** (Scheme 4.80). Not surprisingly, this arrangement turned out to be very suitable for the formation of complexes with inorganic cations of the proper size. However, it was really remarkable that, according to X-ray studies, the conformation of free spherand **260** and its lithium or sodium complex **260**$\cdot$**M**$^+$ are practically identical,[38c] while in all previously reported cases substantial conformational changes caused by complexation were observed. As is stated in the first publication in this area,[38c] 'Thus the full burden of binding-site collection and organization in this spherand is transferred from the complexation process to the synthesis of the ligand system, whose conformation is enforced by its rigid support'. In fact, that was exactly the reason why this particular molecular framework had been chosen as an initial target.[38d]

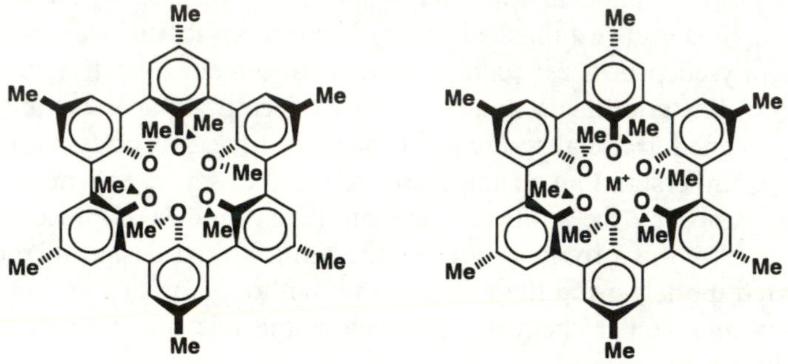

260

260·M$^+$
M = Li, Na

Scheme 4.80

The size of the cavity and rigidity of its shape made **260** a very good host for complexing small cations like Li^+ and Na^+, but it completely rejected other cations as guests. The magnitude of the selectivity for closely related cations was truly unprecedented. In fact, **260** is able to bind Na^+ 10^{10} times better than K^+. This property enabled the authors to use the framework present in **260** for the design of chromogenic ligand **261** (Scheme 4.81) as a specific indicator for sodium and lithium ions.[38e] Spherand **261** contains an additional azo substituent as a chromophore in the *para* position to the free hydroxyl group. Its solutions have a faint yellow color. Ionization of the phenolic hydroxyl causes an immediate change of color to green and deep blue. Free **261** has a pK_a of 13, but this value drops to 5.9 if complexed with Li^+ and to 6.9 when complexed with Na^+. In other words, the complexation with these metals acidifies the system by a factor of about 6–7 powers of ten. Addition of K^+, Mg^{2+}, or Ca^{2+} does not substantially affect the pK_a value of **261**. Therefore a noticeable color

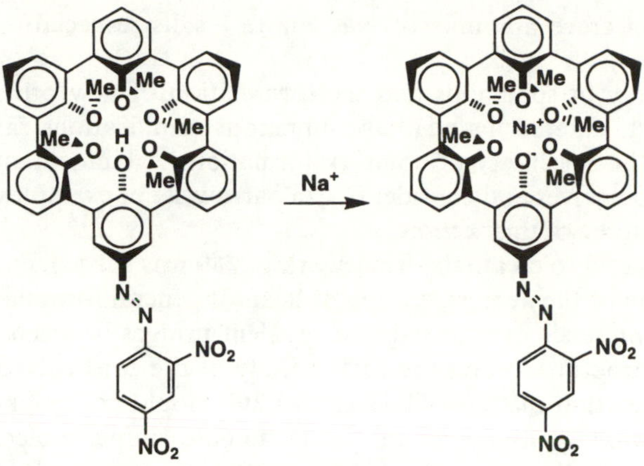

261 **261·Na⁺**

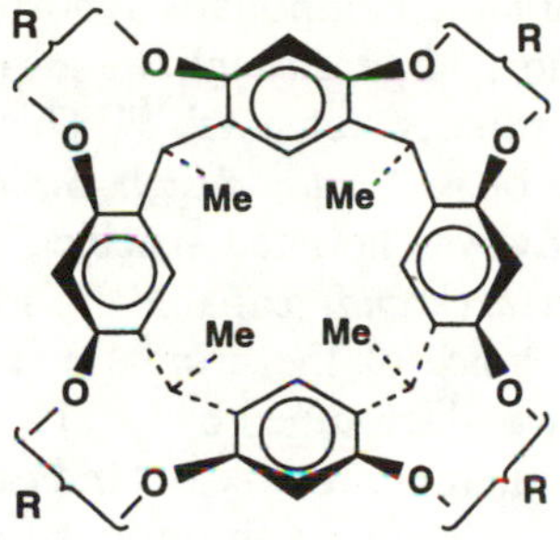

262

Scheme 4.81

changes occurs readily if **261** comes into contact even with trace amounts of Na^+ or Li^+, which are capable of forming complexes, whereas no such effects are observed with K^+ salts. The sensitivity of these color reactions is truly amazing. Li^+ and Na^+ ions can be detected at concentrations as low as 10^{-8} mol L^{-1} in the presence of other cations. It means that the presence of even 5×10^{-4} mg of sodium salt per 1 L of solvent (or 0.5 ppb) renders this particular batch of solvent too 'dirty' to use with **261**. This sensitivity creates a very unusual problem in handling. Regular organic solvents, including $CHCl_3$ and CH_2Cl_2, stored in glass containers may contain enough Na^+ to cause the color change when trace amounts of **261** are added. Thus special precautions must be taken to avoid the possible contamination of both solvents and reagents. All operations with solutions of **261** should be carried out with the use of specially pretreated quartz, polypropylene, or PTFE containers and pipettes. This unique sensitivity may be especially useful for problems requiring the analysis of trace amounts of Na^+ or Li^+ salts, especially in biological systems.

The discovery of spherands and the formulation of new principles for the design of multicentered ligands have numerous ramifications far beyond the problems of the selectivity of complex formation.[38a] Unfortunately, here we have to limit ourselves and consider only a narrow set of examples most closely related to the topic of this section.

The design used to create the framework of **260** was subsequently applied to the elaboration of the next generation of hosts of general formula **262** (Scheme 4.81).[38f] The presence of four pairs of covalent bridges between the aromatic rings in this structure secures the total rigidity of the construction. Molecular models revealed that the bowl-like shape of **262** should possess a cavity (hence the name *cavitands*) large enough to accomodate simple molecules. As was shown experimentally, cavitands of the type **262** form very stable stoichiometric complexes with uncharged small molecules, like acetone or chloroform.[38f]

Even most spectacular results were obtained when a pair of bowls like **262** were stitched together to form a spheroid-like host. The paper describing the synthesis of the first such molecule **263** (Scheme 4.82) begins with the words: 'Absent among the over six million organic compounds reported are closed surface hosts with an enforced interior, large enough to imprison behind covalent bars guests of the size of ordinary solvents'.[38g] This outstanding achievement may serve as an example of *molecular design par excellence*. In fact, from the very beginning, the study was targeted specifically toward the synthesis of a closed-shell structure (*carcerand*) capable of including host molecules in its hollow interior. The choice of the aromatic ring as a basic construction element for this structure was determined by two factors: previous experience in the design of various spherands and cavitands and the high reactivity of the ring, which guaranteed a flexibility in options for the various chemical manipulations required for the closure of two cavitand shells.

The synthesis of cavitands employed to assemble the carcerand structures begins with the acid-catalysed condensation of resorcinol **264** with acetaldehyde to give octaol **265**. Basically this reaction is very simple. It was discovered almost 50 years ago in the course of studies of industrially important processes of phenol–formaldehyde condensations. Since then, these cyclooligomers (calixarenes) have been extensively studied. However, it is noteworthy that the interaction of **264** with acetaldehyde gave, as the main product, cyclotetramer **265** instead of the usual mixture of oligomers and polymers formed in similar couplings with other phenols. Obviously the folding required for the formation of this structure is favored owing to strong intramolecular hydrogen bonds, as shown in the structure of **265**. Subsequent transformations involved the formation of four methylene bridges between the adjacent phenolic hydroxyls to lead to the formation of the nearly hemisphere-like shape of cavitand **266a**. It was then necessary to introduce functions into the benzene rings which serve as the 'gadgets' to connect the two hemisphere halves. A series of trivial functional group transformations finally furnished the required components **266e** and **266f**. The crucial step, the coupling of the two halves, was conducted at

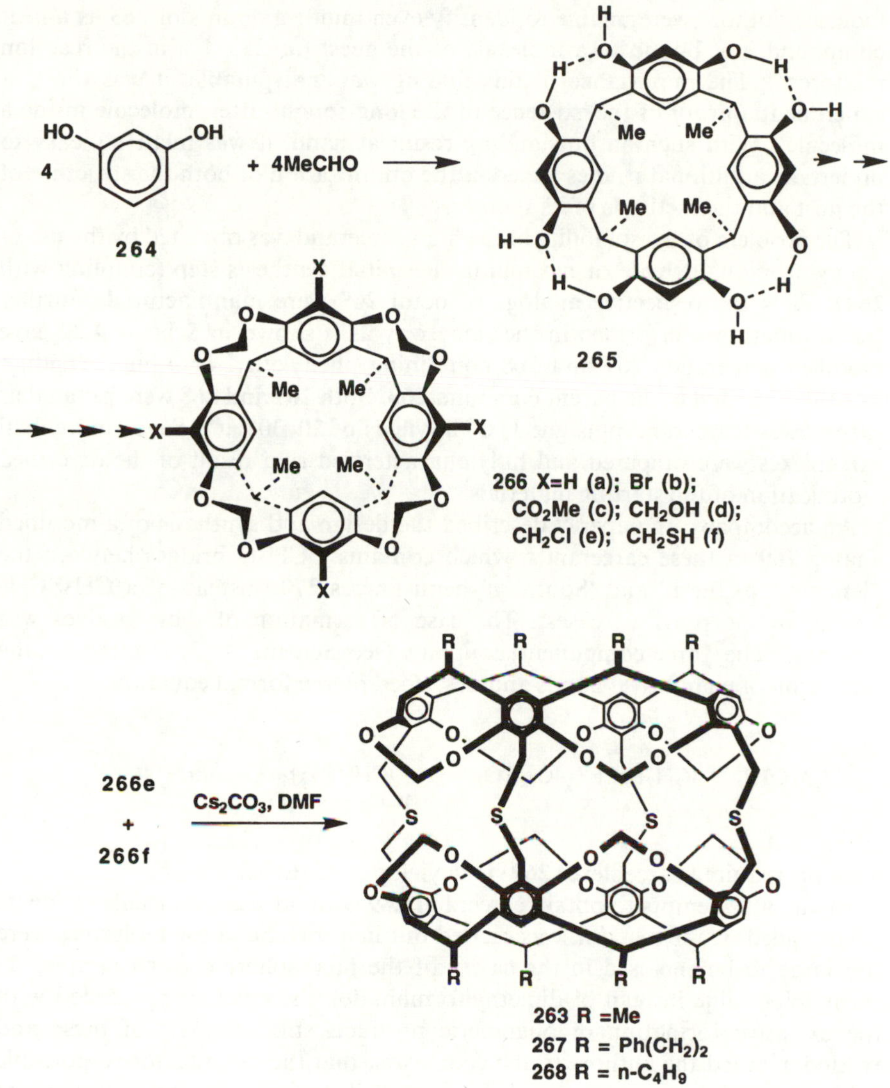

Scheme 4.82

moderately high dilution conditions in the presence of Cs^+ salts in appropriate solvents. In this particular case, most of the material underwent oligomerization. Nevertheless, the required closed-shell structure **263** was also produced as a mixture of related products (*vide infra*) in a satisfactory yield (approximately 29%). This mixture was virtually insoluble and hence its separation into individual components was hardly achievable. By far more important than yield and purity was the following major result: the formation of the closed shell was accompanied by the incorporation of a guest molecule of the solvent [$(CH_3)_2NCHO$ or $(CH_2)_4O$] and/or Cs^+ into the hollow cavity of the host! In fact, the authors were unable to identify even minor amounts of **263** as a neat compound, not bound to a molecule of the guest (as **263 · G**), in the reaction mixture.[38g] The importance of this finding was indisputable: it was the first experimental proof of the existence of the long-sought-after 'molecule inside a molecule'. With such an outstanding result at hand, it was relatively easy to undertake additional studies aimed at the optimization of both the structure of the host and the methods of its synthesis.

The problem of the solubility of the final carcerand was obviated by the use of phenylpropionaldehyde or hexanal in the initial synthesis step (coupling with **264**). Thus the respective analogs of octol **265** were manufactured. Further transformations performed in the same way as is shown in Scheme 4.82 gave modified carcerands **267** and **268** containing phenylethyl or *n*-butyl residues instead of methyl in the parent comound **263**. Both **267** and **268** were isolated as *carceplexes* (carcerand plus guest, **G**) in yields of 20–30%.[38h] Seven individual carceplexes were prepared and fully characterized as a result of the described modification of the starting materials.

An accompanying paper[38i] described the design and synthesis of a modified analog **269** of these carcerands, which contains OCH_2O bridges between the identical 'Northern' and 'Southern' hemispheres **270** (instead of a CH_2SCH_2 link as in the previous cases). The ease of formation of these bridges was amazing. The three-component coupling (see Scheme 4.83) leading to the formation of eight novel bonds and described by the formal equation:

$$2(ArOH)_4 + 4CH_2BrCl + 4Cs_2CO_3 \rightarrow (ArOCH_2OAr)_4 + 4CsBr + 4CsCl + 4CO_2 + 4H_2O$$

gave the required carceplexes **269 · G** in yields of 49–61%.

Again, all attempts to obtain carcerand **269** with no guest molecule inside its cavity failed. If the reaction was carried out in a solvent whose molecules were too large to be enclosed in the cavity of the final spheroid (for example, *N*-formylpiperidine instead of dimethylformamide), the reaction proceeded with the exclusive formation of oligomeric products. The analysis of these and related data led the authors to the conclusion that the main factor responsible for the ease of this entropically prohibitive shell closure is the template effect of the solvent in solvent–solute interaction with two molecules of the reacting hemispheres. In a continuation of these studies, Sherman investigated the dependence of the efficiency of **269 · G** formation in a solution of *N*-methylpyrrolidin-2-on (NMP) on variations in the nature of the guest molecules.[38j] More than 20 small molecules were checked as templates in comparison to NMP as the 'poorest' guest template. It was found that for the best template, pyrazine, the formation of the respective **269 · G** can be carried out in 75% yield even with 1 equivalent of this guest for two molecules of tetrol **270**. As was deduced from a series of competitve reactions, the entrapment selectivity, and hence efficiency of carceplex formation, can vary over a range of 10^6. These effects are not dependent upon polarity, basicity, or other conventional parameters routinely

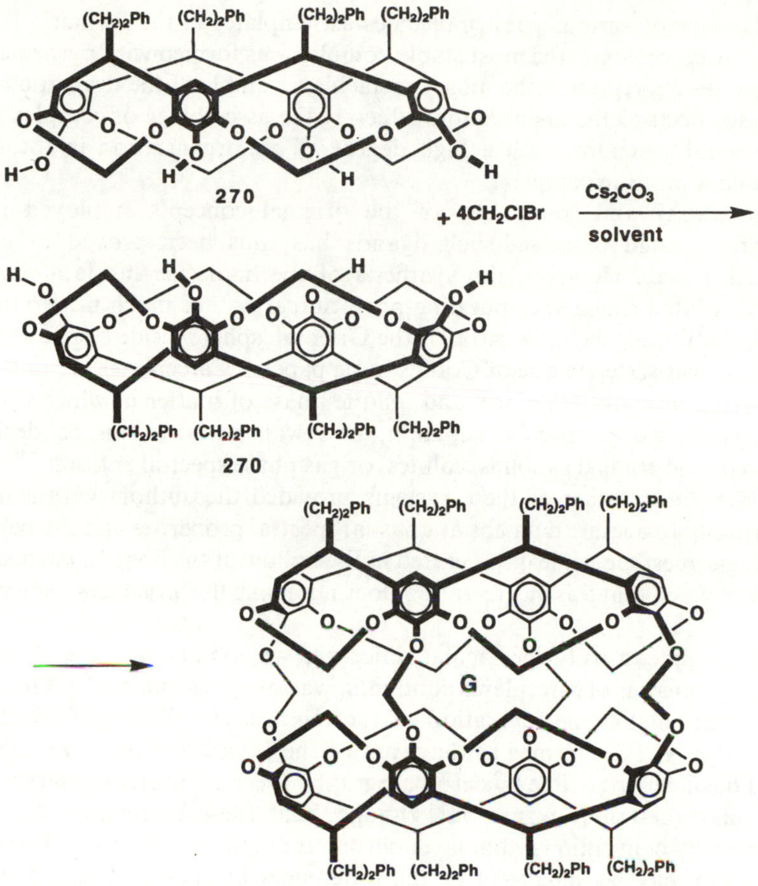

270

+ $4CH_2ClBr$ $\xrightarrow[\text{solvent}]{Cs_2CO_3}$

270

G

269·G; G = solvent (Me_2SO; Me_2NCOMe; Me_2NCHO)

Scheme 4.83

used for the description of solute–solvent interaction. The size, shape, and symmetry of the guests seem to be the main factors. The observed trends prompted the study of the interaction between **270a** (which differs from **270** by the presence of an ethyl group instead of the benzyl substituent) and various guest molecules, with the help of NMR spectroscopy.[38k] As a result, it was shown that a 2:1 complex is formed between tetrol **270a** and a guest molecule in $CDCl_3$ solution in the presence of base. This is one of the very few cases when the occurrence of the phenomenon of self-assembly and molecular encapsulation is substantiated by direct experimental evidence. The ease of formation and stability of these complexes are ascribed to the combined effects of strong hydrogen bonds between the bowls and van der Waals and electrostatic interactions between the guest and the walls of the cavity formed by the two bowls. A remarkably good correlation was also established between the stability of the complexes formed with various guests and previously reported data[38j] on the efficiency of various guest molecules as templates in the formation of the respective carceplexes (the most stable complex was formed with pyrazine).

These results represent the most spectacular example of the dramatic role of template effects as the main factors affecting the assemblage of complex three-dimensional structures with a high degree of pre-organization in the multi-component reaction complex.

The validity and fruitfulness of the original concepts employed in the molecular design of closed-shell ligands has thus been proved in a very convincing way. However, the synthesis of the 'molecule inside a molecule system', while being a very puzzling problem *per se*, has much more meaning than just imitating a child's rattle or the Oriental 'sphere inside a sphere' pieces of art. As was stated in one of Cram's basic papers, 'Carceplexes are molecular cells whose interiors are a new and unique phase of matter in which different amounts of space, space occupation, and wall surfaces can be designed, prepared, and studied as solids, solutes, or gas phase spectral entities'.[38i]

In fact, the creation of these systems provided the authors with a unique opportunity to acquire data about unusual spectral properties and the behavior of a single guest molecule incarcerated in the hollow of the host. In particular, it was discovered that this incarceration does not make the 'prisoners' *incommunicado*.

For example, a truly spectacular difference was observed in the chromatographic properties of carceplexes containing various guest molecules. One might have expected that the separation of the mixture **269·$(CH_3)_2NCOCH_3$** and **269·$(CH_3)_2NCHO$** (formed as the result of shell closure in the mixed solvent) would be hardly posssible by chromatography (because of the similarity of the shells and guests). It turnes out, though, that these carceplexes are easily separated by conventional thin layer chromatography on silica gel. These data were interpreted as indicative of the differences in guest-attenuated dipole–dipole interactions between the shell and silica gel surface.[38i] Numerous data were also collected on the spectral properties of the guest molecules. These data revealed many peculiarities in comparison with the respective data taken in solution or in the gas phase.

Now, what about the chemistry of the isolated molecules inside these artificially created molecular vacuum chambers? This might be the most exciting aspect of the whole story, the aspect likely to produce truly dramatic effects on the understanding of such major phenomena as solvent effects and the intrinsic reactivity pattern of isolated molecules.

'The Taming of Cyclobutadiene' was the title of a brief paper published in the August 1991 issue of *Angewandte Chemie*.[38l] As we already mentioned above, the closed-shell systems were formed as carceplexes and there was no way to remove the incarcerated guests from carceplexes like **269·G** to obtain free carcerands. The situation can be changed, however, if a coupling like that shown in Scheme 4.83 was carried out with the analog of cavitand **270** having only three hydroxyl groups. In this case, only three O–C–O bridges can be formed. As a result, the product **271** (Scheme 4.84) has a portal that allows guests of the proper size to enter the interior space. In the cited study, α-pyrone

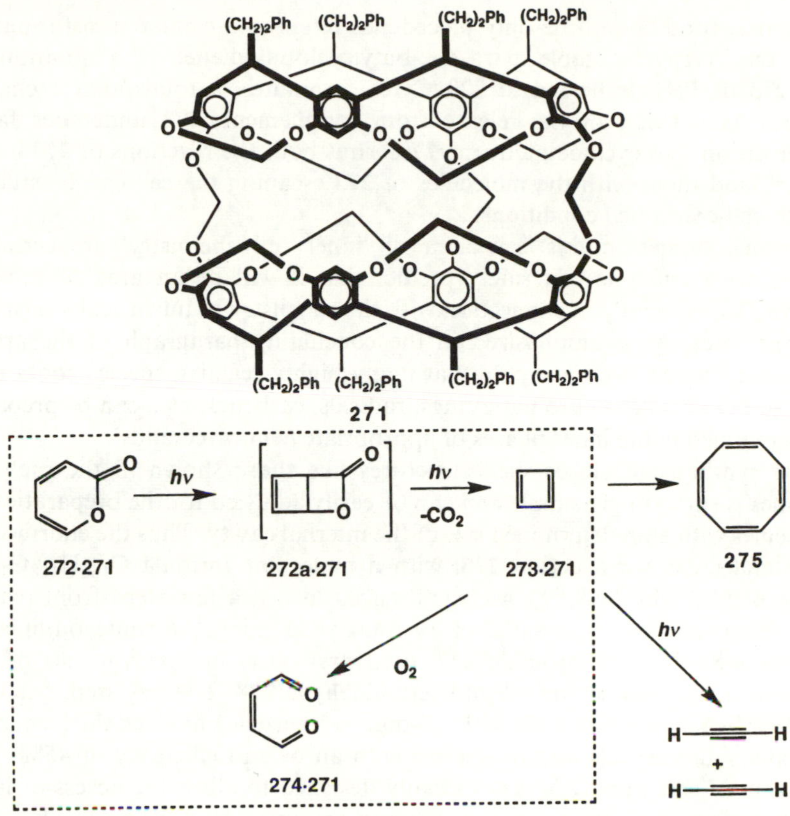

Scheme 4.84

272 was chosen as the guest and its 1:1 carceplex with **271** was prepared simply by heating the components.

When the solution of **271 · 272** was irradiated at 25 °C for 30 minutes, NMR analysis showed a rapid disappearance of the signals of **272** with an accumulation of the signals of cyclobutadiene **273**. Nothing is especially surprising about this reaction in its own right, as it represents a long-known route for the generation of **273** from **272** *via* intermediate **272a**. However, **273** is rather unstable, both thermodynamically and kinetically. Therefore, it was formed in this reaction under all previously used conditions as a fleeting intermediate which underwent immediate dimerization or was intercepted by some quencher. Only when the photolysis of **272** was carried out at 8 K in argon matrix was the formation of free **273** registered by spectral data (see references cited by Cram *et al.*).[381] In striking contrast to the previous data, **273** turned out to be a kinetically stable compound when it was generated and confined in the inner phase of **271**. A greater irradiation time converted **273** into acetylene (a similar reaction was also observed in an argon matrix at 8 K). If oxygen is bubbled through a solution of **271 · 273**, the imprisoned **273** is oxidized and a complex of malealdehyde **274** with **271** is formed. No such reaction was described previously for **273** and the only precedence refers to a similar transformation with the relatively stable tetra-*tert*-butylcyclobutadiene. If a solution of **271·273** in THF is heated at 220 °C for 5 minutes, a total guest exchange occurs. As a result of the release from confinement, **273** undergoes facile dimerization into cyclooctatetraene **275**. Thus both the reactions of **273** inside the cell and those with the molecules of **273** escaping the cell can be studied under well-controlled conditions.

Further studies in this new area of 'inner cell chemistry' are certainly forthcoming and one can safely predict that it will be an area of exciting discoveries, especially in connection with the chemistry of intrinsically unstable intermediates. As is emphasized in the concluding paragraph of the article discussed above: 'We anticipate that many highly reactive species containing bent acetylene, allenic, aromatic rings, radicals, carbenes, *etc.*, can be prepared and examined in the inner phases of appropriate hemicarcerands'.

The synthesis of closed-shell structures like those shown in the previous schemes is surprisingly simple and can be easily adjusted for the preparation of molecules with a predetermined size of the internal cavity. Thus the enormously huge molecular construction **276**, with a molecular formula $C_{160}H_{128}O_{16}N_8$ (molecular weight 2418.69), was synthesized in just a few steps from readily available precursors in a satisfactory total yield using the route outlined in Scheme 4.85.[38m] Tetrabromide **277**, used previously in the synthesis of **267** (Scheme 4.82), was converted into tetraaldehyde **278**. The key step, stitching together two molecules of **278** with 1,3-diaminobenzene, involved the formation of eight double bonds and proceeded with an overall efficiency of 45%. The structure of carcerand **276** was specially designed to allow the access of large molecules into the interior space of the host through four huge portals between its 'Northern' and 'Southern' hemispheres. It was thought that sufficiently voluminous guests would be further held inside by multisite interactions within the shell. In fact, more than a dozen guest molecules (some of them shown in Scheme 4.85) formed reasonably stable complexes with **276** (half-lives varied from 3.2 hours at 25 °C for hexachlorobutadiene to up to 19.6 hours at 112 °C for ferrocene).

For results like those presented in Schemes 4.82–4.85, the routine, matter-of-fact style of regular publications seems almost inappropriate. Here we are dealing with an outstanding discovery. The previously unimaginable creation of a molecular vessel to perform chemical reactions between isolated molecules was finally achieved and thus brought unprecedented opportunities for the future development of organic chemistry.[38n]

As we have seen, studies initially conceived as in-depth investigations of options to create specific ligands turned out to be far more important than anyone could have dared to predict. Besides the numerous practical applications that stemmed from the initial discovery of crown ethers, an absolutely novel area of organic chemistry was created. The basic concepts initially formulated from a thorough analysis of complexation phenomena opened the entry for experimental studies aimed well beyond the borders of traditional molecular chemistry.

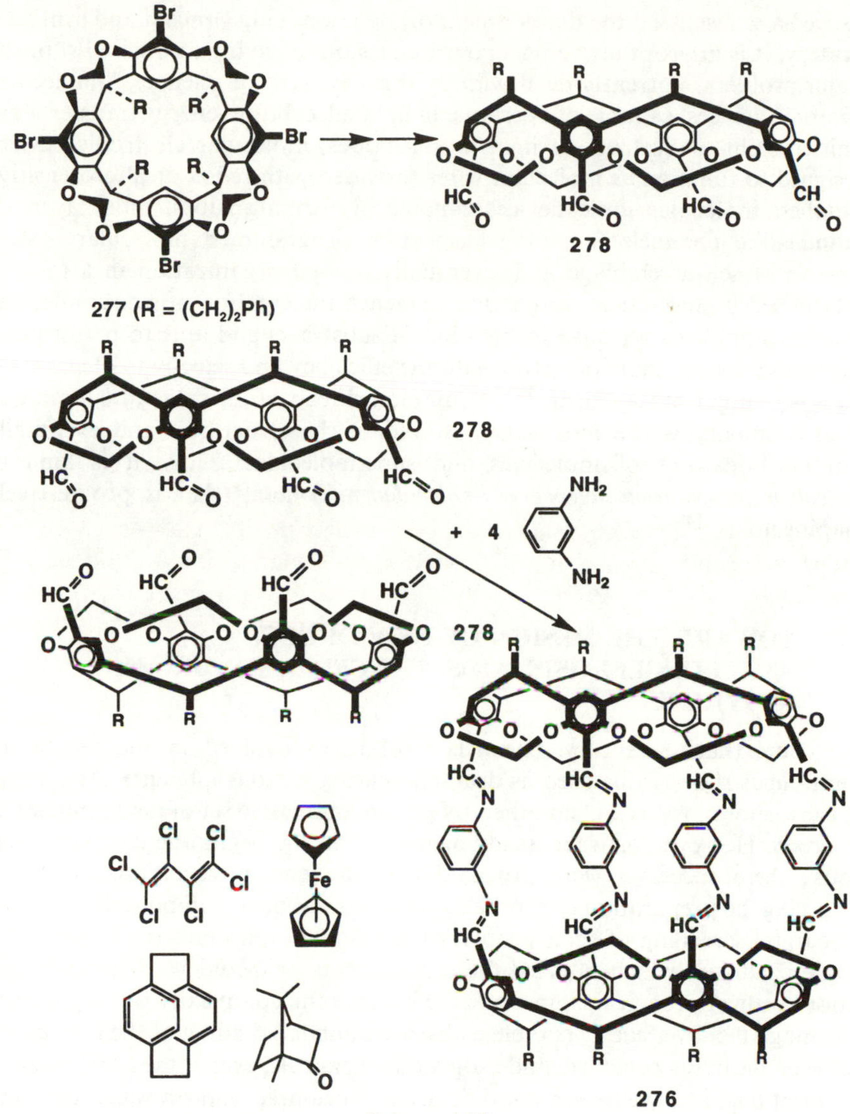

277 (R = $(CH_2)_2Ph$)

278

276

Scheme 4.85

As defined by Lehn,[38o] 'Supramolecular chemistry, the chemistry beyond the molecule, is the designed chemistry of the intermolecular bond, just as molecular chemistry is that of the covalent bond. It is a highly interdisciplinary field of science covering the chemical, physical, and biological features of chemical species held together and organized by means of intermolecular (noncovalent) binding interactions'.

However, supramolecular chemistry, its exciting promise and thrilling interest notwithstanding, lies beyond the limits of our book and readers are addressed to more specialized publications dealing with this subject.[32g,33a,38a,o]

As we have discussed the development of the underlying concept and synthetic strategy, it is appropriate to finish our discussion by giving a concise list of the major problems currently dealt with by this new area of science:[38o] molecular photonic devices, capable of operating in light adsorbtion–energy transfer–light emitter, light-to-electron, or light-to-ion modes; molecular electronic devices designed to function as molecular wires provided with redox or photosensitive switches; molecular ionic devices capable of forming tubes, monolayer, or chundle-like channels for ion transport; a programmed molecular system capable of self-assemblage and, eventually, of self-organization in a fashion determined by molecular recognition elements; the creation of supramolecular systems capable of selective recognition of substrates and able to perform the required chemical transformation with the efficiency and selectivity of enzymes. As was pointed out by Lehn,[38o] 'a common thread of all areas of supramolecular chemistry is the *information*, stored in the structural (and eventually temporal) features of molecules and supermolecules. Thus, it is kind of *molecular information science* or *molecular informatics* that is progressively shaping up'.

4.11 TOWARD THE DESIGN OF NEW DRUGS. ATHEROSCLEROSIS, AIDS, CANCER, AND ORGANIC SYNTHESIS

For more than a century, chemists have been involved in the search for compounds that can be used as drugs for curing various ailments. As a result of these efforts, today's chemotherapy has an impressive set of achievements to its credit. However, as was already mentioned in the beginning chapter of our book, these successes were procured from an enormous amount of labor involving the preparation of thousands and thousands of compounds, followed by careful screening of their properties and activity parameters to find just a single candidate which satisfied the requirements of medicine. This time- and labor-costly approach was unavoidable because the complexity of the problem was magnified by a nearly complete absence of understanding of the interaction between the living organism and exogenous chemicals, even if the latter referred to traditional and well-established drugs. For example, aspirin was introduced to the medical practice in the 1870s and has been widely used since then as an efficient analgesic and fever-relieving drug. More than forty million pounds of this compound are being manufactured yearly in the United States. Yet numerous studies of the mechanism of aspirin action have failed to produce a consistent explanation of the multifaceted pattern of aspirin's effects on the functioning of our organism.

These problems, as well as others of no less importance, like short- and long-term side effects, transport to the ailing tissue, prolongation of the action, compatibility of various drugs, allergenic effects, *etc*., have been under active scrutiny during the whole era of chemotherapy. As a result, a tremendous amount of factual information has been accumulated and preliminary assessments of structure–activity relationships within a series of related compounds greatly facilitated.

The successes achieved in the course of the last two decades, owing to the combined efforts of molecular biology, medicinal chemistry, and chemistry, brought truly dramatic changes to this area. As a result, it is possible to describe the major biochemical events occurring in the cells, tissues, or organs in terms of molecular biology and to identify the systems most affected in the deceased organism. The understanding of the underlying causes of the malfunctioning of biochemical systems paved the way for the elaboration of a more rational approach to the search for new drugs. The basic principle of this approach involves the identification of the targets to be affected by the potential drug, followed by the design of a structure capable of efficient interaction with this target. In other words, the general problem of elaborating the appropriate drug may become formulated in more definite terms, like the creation of inhibitors for a particular enzyme system, or agents capable of affecting DNA synthesis, replication, or gene expression, or factors affecting the hormonal system, or whatever else is specifically needed for the restoration of a normal functioning of the affected biochemical system. The orders of the medicinal 'customer' can be translated, at least to some approximation, into the language of chemical structures.

These new opportunities open the route for the application of the principles of molecular design in an area where traditionally a huge expenditure of time and effort, combined with a touch of mere luck, were the necessary prerequisites for final success. Sure enough, even a modern science with its powerful instrumentation and enormous capacity for information acquisition and processing is still unable to make absolutely exact predictions about the structure of a compound optimal for a drug with a specific pattern of action. However, at the same time, its methods ensure an opportunity to focus the investigation on a rather narrow set of target structures and to avoid the preparation of hundreds of analogs for blind testing.

Actually, any truly serious study in the field of physiologically active compounds currently involves elements of molecular design. It was a tough task for us to choose appropriate examples to be discussed in this section. Finally, we decided to limit ourselves to just a few cases which seemed to be representative of the general trends of molecuar design in this area.

The development of arteriosclerosis, the main factor causing coronary heart disease, depends dramatically upon plasma levels of cholesterol associated with low-density lipoprotein. Hence a plausible route toward the creation of drugs to treat hypercholesterolemia may involve searches for compounds capable of affecting cholesterol biosynthesis. The lengthy sequence leading to the formation of cholesterol from acetyl coenzyme A was elucidated in the early 1960s. As one of the key steps of this sequence, the reduction of 3-hydroxy-3-methylglutaryl coenzyme A (HMG-CoA) **279** to mevalonate **280** was identified (Scheme 4.86). An intensive search for compounds capable of blocking this step led to the discovery of the fungal metabolites compactine **281a** and mevinoline **281b**, efficient inhibitors for HMG-CoA reductase. Compound **281b** was introduced

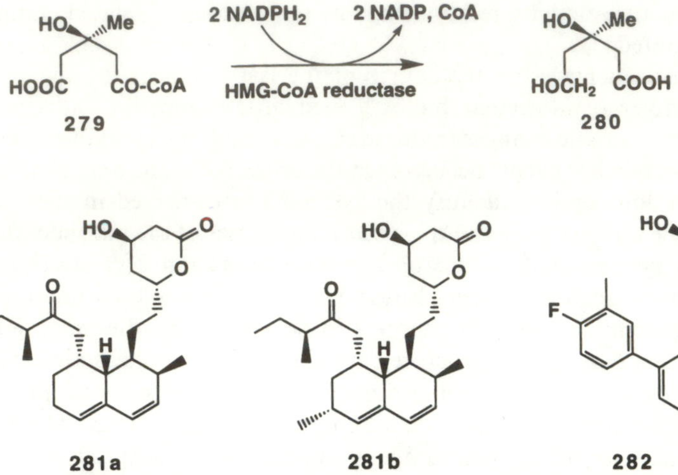

HO Me
HOOC CO-CoA
279

2 NADPH$_2$ → 2 NADP, CoA
HMG-CoA reductase

HO Me
HOCH$_2$ COOH
280

281a **281b** **282**

Scheme 4.86

into clinical application as an efficient drug, capable of lowering the cholesterol level in blood plasma.[39a]

A detailed investigation of the mechanism of action of these inhibitors led to the suggestion that their biological activity depends mainly upon the presence of a β-hydroxylactone moiety, obviously owing to the structural similarity of this fragment to that of the enzyme substrate **279**. This lead was used for the design and synthesis of a set of artificial and more available analogs. Some of them, for example **282**, exhibited remarkably potent activity as hypocholesteremic agents.[39b]

It was long recognized that blood platelets can adhere easily to almost any material, but they do not stick to the lining of healthy blood vessels. At the same time, it was also known that if the endothelial cells that line the blood vessels are damaged or diseased, platelet aggregation occurs readily. The ultimate effect may be beneficial, if it occurs to arrest a hemorrhage, or deleterious, if it leads to the formation of a thrombotic plug (and thus may provoke a heart attack or thrombotic stroke). A consistent explanation of these phenomena emerged only in the late 1970s with the identification of some new prostanoids, the direct participants of the described events.[39c] The first compound, thromboxane A_2 (TxA_2) **283** (Scheme 4.87), produced by blood platelets, was shown to cause the constriction of the blood vessels and platelet aggregation. The second compound, prostacyclin **284**, manufactured by the inner lining of the blood vessels, exhibited the properties of a vasodilator and inhibitor of platelet aggregation. Both compounds are produced in minute quantities. They are extremely potent and very unstable in solution *in vitro* (half-life for **283** and **284** at 37 °C is 37 seconds and a few minutes, respectively). Obviously these two compounds act as opposing agents in the regulatory mechanism of platelet aggregation. The identification of **283** and **284** and the elucidation of their role in the functioning of the cardiovascular system stimulated intensive studies aimed at the creation of stable analogs as potentially useful drugs to treat such serious conditions as stroke, angina, *etc*. Several approaches have been used in the design of synthetic targets.

The preparation of compounds structurally similar to **283** and **284** in order to arrive at their stable mimetics or antagonists seemed to be the most obvious and realistic goal. As a result of these pursuits, a number of prostanoid analogs were synthesized. The bicyclic compound **285** was especially efficient as a stable mimetic of **283**.[39d] The hybrid structure **286** containing both α- and β-prostanoid side chains attached to a terpenoid backbone[39e] exhibited significant activity as an antagonist to **283** (Scheme 4.87).

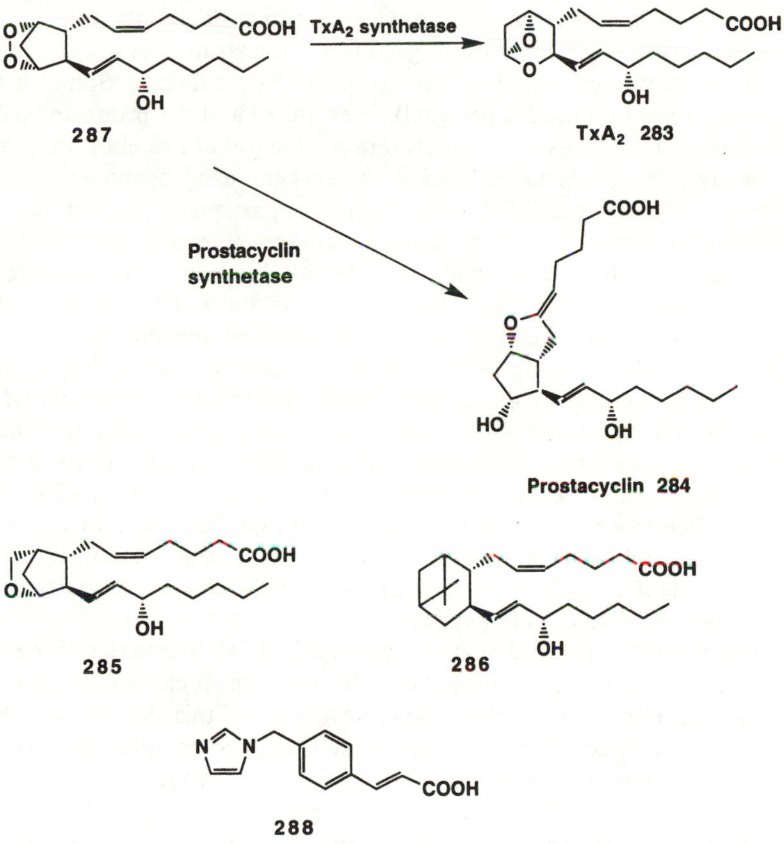

COOH TxA_2 synthetase COOH

OH OH

287 TxA_2 **283**

COOH

Prostacyclin synthetase

HO OH

Prostacyclin **284**

COOH COOH

OH OH

285 **286**

COOH

288

Scheme 4.87

A more promising and, at the same time, more sophisticated approach envisioned the elaboration of inhibitors of TxA_2 synthetase, which converts the peroxide precursor **287** into **283**. Here structure–activity correlations were rather complicated. In the absence of solid evidence on the structure of enzyme active centers, the searches were based primarily upon analogy considerations. TxA_2 synthetase was shown to belong to the family of cytochrome P_{450}

enzymes.[39c] A number of compounds containing the *N*-substituted imidazole fragment were identified earlier as active inhibitors of this system as they exhibited a noticeable suppressing action on TxA_2 synthetase. However, the cytochrome P_{450} enzymes, that are targeted here, are also involved in many other important biochemical transformations. Hence, selectivity was the main problem in the design of candidates for the TxA_2 synthetase inhibitor. It was reasoned that beneficial changes toward the required selectivity pattern may be secured *via* the introduction of a carboxyl group in the structure of the cytochrome P_{450} enzyme inhibitor, because: (i) it mimicks the presence of the acid moiety in **287** and thus increases the affinity of an inhibitor to TxA_2 synthetase, and (ii) the presence of the carboxyl group dramatically alters the lypophilicity of the molecule and thus may reduce its affinity toward other enzymes of this family. This reasoning led to the preparation of a set of efficient inhibitors with well-expressed selectivity for TxA_2 synthetase. Some of these inhibitors, like compound **288** (ozagrel),[39f] exhibited a rather promising pattern of activity in clinical trials as a drug preventing diseases like cerebral vasospasm.

Nowadays, the problem of finding an efficient drug against the human immunodeficiency virus (HIV) has acquired a paramount importance. Not surprisingly, enormous efforts are being spent in the attempts to find an ultimate solution to this problem of enormous complexity. First it was necessary to gain detailed information about the structure of the virus components and the biochemistry of their interaction with host cells. The accumulated data were used further to identify the most vulnerable component of HIV as a target for attack by a pre-designed chemical agent. A specific protease of this virus (HIVP) was established to be one such target and an intensive search for efficient inhibitors of this enzyme has started. As is usually the case, fairly different approaches are employed to design an optimal structure for specific HIVP inhibitors. Below we are going to discuss only one, but one that is a rather instructive example of these studies. We chose this example because the non-triviality of the approach led to the discovery of exotic enzyme inhibitors for HIVP among derivatives of fullerenes.

Structural studies revealed that an active site of HIVP has the shape of an open-ended cylinder with the inner lining formed almost exclusively by hydrophobic amino acids. The radius of the hollow space of this cylinder was shown to be appoximately equal to the radius of the C_{60} fullerene molecule. Extensive computer modeling studies by Kennyon's group[39g] revealed that C_{60} fits nicely into the active site of HIVP and is able to bind tightly to its surface through hydrophobic interactions. Thanks to effective van der Waals contact between C_{60} and the active site surface, a major portion of the latter (approximately 298 Å^2) is removed from solvent exposure and, thereby, virtually blocked. The disclosed steric and chemical complementarity made further searches of HIVP inhibitors among C_{60} derivatives a well-warranted endeavor.

First to be checked was a set of methanofullerene derivatives **289a–c**, readily available *via* addition of substituted diarylcarbenes (generated *in situ*) to one of the double bonds of fullerene **59** followed by a conventional transformation of the carboxamido groups of the side chains (Scheme 4.88).[39h] Bis-succinoyl-

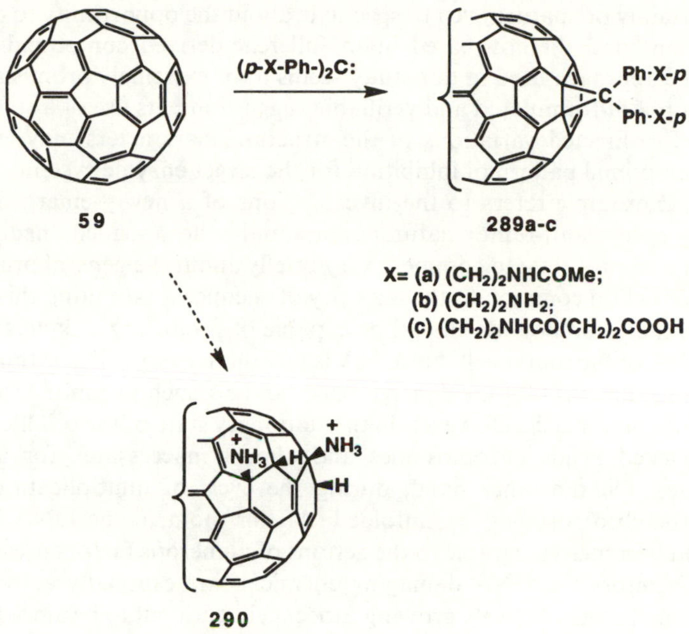

$(p\text{-X-Ph-})_2C$:

59

289a-c

Ph·X-*p*

X= (a) $(CH_2)_2NHCOMe$;
(b) $(CH_2)_2NH_2$;
(c) $(CH_2)_2NHCO(CH_2)_2COOH$

$\overset{+}{N}H_3$

290

Scheme 4.88

amido derivative **289c** was found to be soluble in water at pH $\geqslant 7$ and seemed to be an especially promising candidate for the evaluation of its activity. Computer modeling showed that in the complex of HIVP with **289c** the fullerene core is positioned in the center of the active site, with the hydrophilic side chains extending outside into the solution.[39g] Experiments with **289c** have revealed that the latter shows significant activity as a competitive inhibitor of HIVP. In addition, it was also found that **289c** exhibited an inhibiting action in the HIV-1 infected human peripheral blood mononuclear cells with no cytotoxicity toward uninfected cells.

These results substantiated the validity of the initial binding concept and encouraged the authors to seek an opportunity to improve the inhibiting effects. A more detailed analysis of the model suggested that the binding of fullerene with HIVP could be greatly enhanced if van der Waals contacts are complemented by salt bridges with two catalytic aspartic acid residues also present in the active site of the enzyme. Hence the molecule of the fullerene inhibitor should be furnished with appropriately positioned basic functions capable of interacting with carboxyl groups. Molecular modeling performed for 1,4-diamine **290** demonstrated that, in the complex of this compound with HIVP, amino groups are in close proximity to the aspartate carboxyl groups and are able to form the desired salt bridges. Hence the overall binding should be substantially increased. It is expected that **290** should be a better HIVP inhibitor than the initial test compound.

It is certainly premature even to speculate about the opportunity to develop a practical antiviral therapy based upon fullerene-derived compounds. Nevertheless, the approach used in this study seems to be extremely promising as it is based upon well-formulated and verifiable suggestions. Its use of a step-by-step protocol for directed variations of the structural parameters may eventually secure the optimal pattern of inhibition for the target enzyme system.

The next example refers to the investigations of a newly emerged class of extremely potent anti-tumor natural compounds, the so-called enediyne antibiotics. It is appropriate to comment very briefly about the general principles of anti-tumor action common to the majority of chemicals exhibiting this activity. Basically, these compounds should be capable of producing a damaging effect on the DNA of the tumor cell, but DNA is present in every cell and thus it is not immediately obvious that the damage done can be somehow limited to only the DNA of the tumor cells. However, in the stationary state of the cell life, DNA is closely packed inside chromosomes and almost inaccessable for any alien interactions. On the other hand, during the cycle of multiplication (at the mitosis) the chromosomes are unfolded. At this moment the DNA is almost naked and becomes vulnerable to the actions of numerous factors present in the media. Therefore the DNA-damaging chemicals are especially active against proliferating tissues. Actively growing cancer cells turn out to be among the first targets attacked.

The enediyne's story[40a,b] began in 1965 with the isolation of a novel antitumor antibiotic, neocarcinostatin, from the culture filtrates of *Streptomyces carzinostaticus* var. F-41. This antibiotic was initially identified as a simple protein (molecular weight *ca.* 11 000),[40c] whose primary structure was established shortly afterwards.[40d] Owing to its potency in combination with relatively low toxicity, neocarcinostatin soon found an application in the treatment of pancreatic cancer, gastric cancer, and leukemia in humans. It was established that DNA is a primary target for neocarcinostatin attack, which induces DNA strand scission both *in vivo* and *in vitro*. The presence of thiol-containing cofactors (like thioglycolate) is essential for its activity. Most surprisingly, the effects caused by this antibiotic closely paralleled those of ionizing radiation, the classical DNA-damaging, radical-generating agent.[40a,e] This was the most unexpected and unaccounted for mode of activity for the antibiotic, which presumably possessed the structure of a more or less usual polypeptide! However, in 1979 it was discovered that no mystery is associated with this protein, since after more careful purification the activity of the latter dropped to zero. All the above-mentioned effects were shown to be due to a low molecular weight non-protein compound, neocarcinostatin chromophore (NCS-Chrom), which is tightly but non-covalently bound to the protein in the native antibiotic. The role of the protein component is merely to protect this otherwise very labile substance.[40e]

The structure of the active component **291**, established by Edo's group in 1985,[40f] turned out to be absolutely unprecedented (Scheme 4.89). In fact, the bicyclo[7.3.0]dodecadienediyne system present in **291** had never before been found in Nature or synthesized in a laboratory. Fortunately, several acyclic and

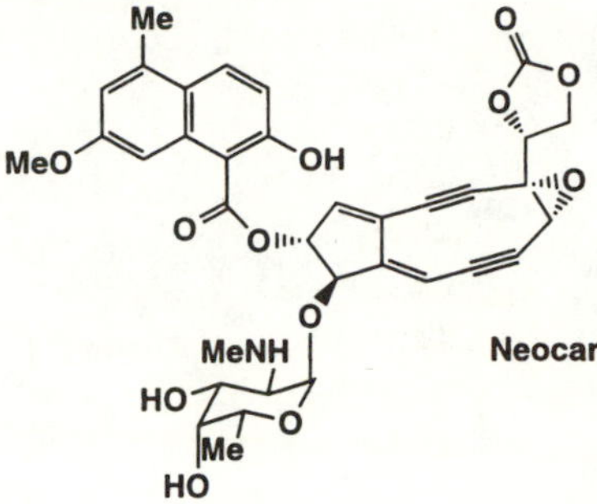

291
Neocarcinostatin chromophore

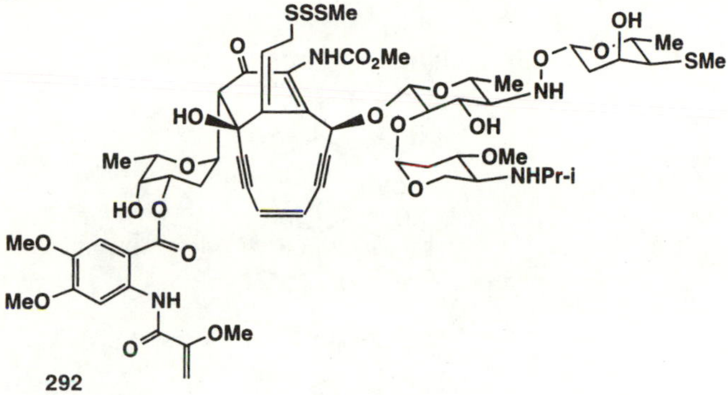

292
Esperamicin A_1

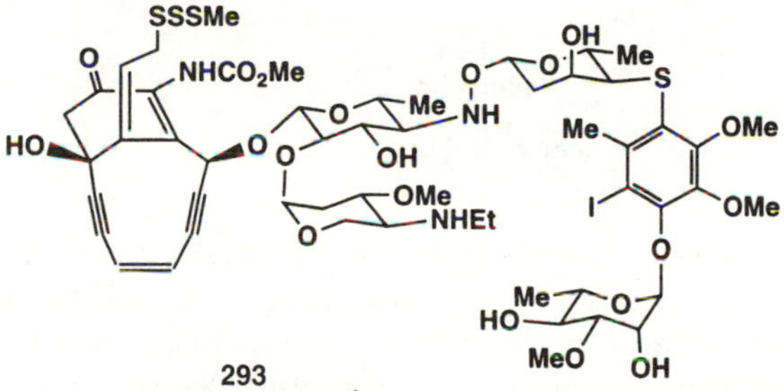

293
Calicheamicin γ^1_1

Scheme 4.89

monocyclic conjugated enediynes had already been prepared in the course of unrelated studies and experimental data on their reactivity provided some leading keys into the understanding of the chemical mechanism of activity of **291** (*vide infra*).

For almost 15 years, neocarcinostatin stood alone as an antibiotic with a unique pattern of DNA-damaging action. However, owing to an intensive search for other natural compounds with similar activity, several other antibiotics containing the enediyne fragment were found. Thus in 1987, almost simultaneously, esperamicin A_1 **292**[40g] and calicheamicin γ^1_1 **293**[40h] were

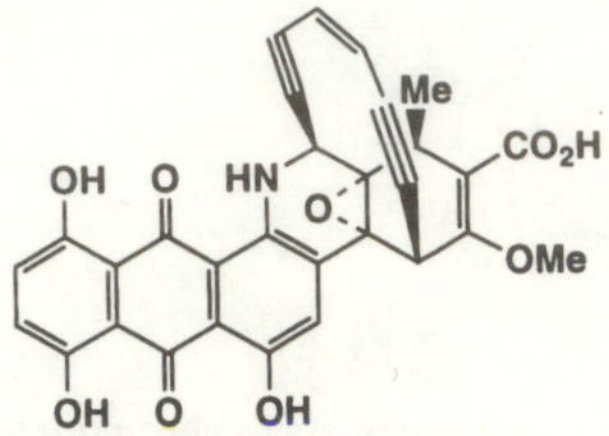

294
Dynemicin A

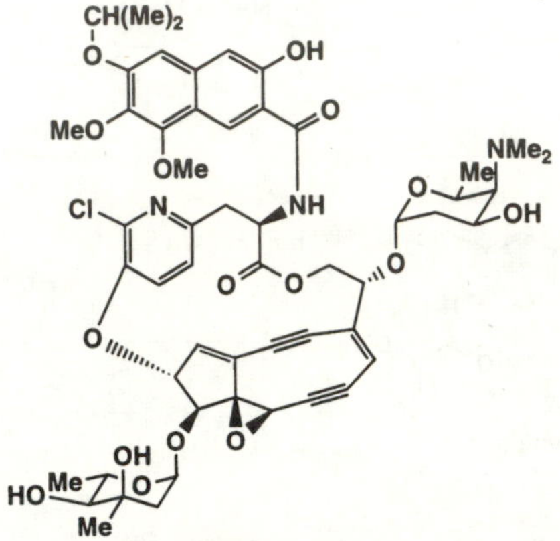

295
Kedarcidin

Scheme 4.89 (*continued*)

identified as antibiotics and isolated correspondingly, the former from the cultures of *Actinomadura verrucosospora* collected in Argentina, and the latter from the fermentation broth of *Micromonospora echinospora* sp. *calichensis*, a fungus from a soil sample collected in Texas. In 1989, dynamicin A **294** was identified in metabolites produced by *Micromonospora cherstina*, the mold cultured from soil samples from Gujarat State, India.[40i] Next came kedarcidin chromophore **295**, produced by a novel actinomycete strain.[40j] The diversity of the geographical areas and sources clearly attests to the wide (yet previously unsuspected!) occurrence of the enediyne family in Nature. One may safely claim that there are many more antibiotics of this type still awaiting discovery.

The common and most significant feature of this new class of antibiotic is an extraordinary activity as anti-tumor agents. Thus calicheamicin γ_1^1 **293** was approximately 4000-fold more active than the most potent and clinically useful antibiotic adriamicin. Owing to high toxicity, **293** cannot be directly used in medicine. However, conjugates of **293** derivatives with tumor-selective monoclonal antibodies revealed promising properties with high activity toward tumor cells in combination with low toxicity.[40b,h] Esperamicin A_1 **292** exhibits powerful activity against a number of murine tumor models at an extemely low injection dose (approximately 100 ng kg^{-1}). This compound is currently in phase II clinical trials.[40b,g]

As was established by extensive experiments *in vitro* and *in vivo*, DNA is a biological target for these antibiotics and, depending on the nature of the damaging agent, both single- or double-strand cleavage may be observed.[40b] Thus DNA interaction with **291** usually results in single-strand cuts, while double-strands cuts occur most typically with **293**. Dynamicin A **294** exhibits both modes of damaging action. This pattern can also be affected by changing the sugar appendage. For example, the intact esperamicin A_1 **292** induces single-strands scissions. Its analogs, produced by the partial hydrolysis of carbohydrate residues, acquire the capacity for double-strand cleavage of DNA.[40k]

Enediyne antibiotics act as site selective cleavers. For example, the tetramer tracts TCCT and CTCT were shown to be principal cleavage sites of DNA treated by **293**.[40l] According to current views, a molecule of the antibiotic intercalates into the minor groove between adjacent base pairs of the double helical DNA. Selectivity of site recognition and efficiency of the binding are determined by interactions of DNA with the carbohydrate and/or aromatic residues of the cleaving agent.[40b,e,l] For example, NMR data revealed that, in solution, **293** adopts an extended and highly organized conformation especially well suited for binding at a DNA minor groove. Contacts with the oligosaccharide domain of **293** are crucial for the selectivity and efficiency of this binding. Destruction of DNA can be also achieved under the action of calicheamicinone, the aglycon of **293**. In this case, a much higher concentration of the agent was required and the reaction was shown to proceed non-selectively and primarily as a single-strand cleavage. On the other hand, it was also discovered that the synthetic oligosaccharide, identical to the carbohydrate portion of **293**, binds to the same sites as the parent natural compound and thus may completely block the DNA cleavage caused by this antibiotic.[40m]

The presence of the enediyne moiety constitutes the most notable structural peculiarity of the antibiotics **291–295**. Cleavage of DNA strands under the action of these agents was shown to occur as the result of several sequential reactions. The major event in this sequence is connected with the transformations of the enediyne system.

An understanding of the chemistry of these transformations resulted from purely academic and seemingly (at that time!) irrelevent studies by Bergman's group in the early 1970s.[41a] At that time, a plethora of experimental evidence was accumulated showing that several aromatic substitution reactions proceed *via* the intermediate 1,2-dehydrobenzene, which was isolated in a matrix at 8 K. The isomeric 1,3- and 1,4-dehydrobenzenes represented equally challenging targets, especially because a biradical structure seemed to be more preferable for these species. According to Bergman's suggestion, the thermal isomerization of *cis*-diethynylalkenes may represent a viable pathway for the generation of 1,4-dehydrobenzene (Scheme 4.90). Experimental data provided evidence to substantiate the validity of this suggestion. In fact, it was found that heating enediyne **296** in a hydrocarbon solvent at 200 °C results in the formation of

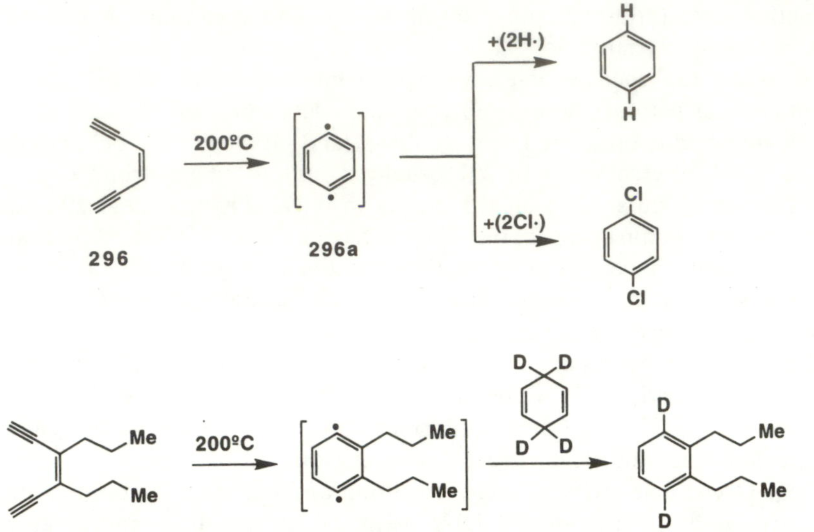

Scheme 4.90

benzene. At the same time, 1,4-dichlorobenzene was formed as the only product for the reaction carried out in CCl_4.[41b] If a similar cyclization of enediyne **297** was carried out in the presence of 2,2,5,5-tetradeutero-1,4-cyclohexadiene, the 1,4-dideuterated benzene derivative **298** was formed.

These observations, in conjunction with detailed kinetic studies,[41c] implied that, in the course of the thermal rearrangement of *cis*-diethynylalkenes, 1,4-dehydrobenzene intermediates are formed (*e.g.* **296a**, **297a**). The quenching of these biradical species with radical scavengers produced aromatic compounds like **298**. By a strange coincidence, these results were published right at the moment when they were crucially needed, not by theoreticians, but by those who were involved in the studies of the above-mentioned enediyne antibiotics.

The presence of the enediyne core and the radical-like pattern of DNA damage caused by these agents led to the recognition of Bergman's cycloaromatization reaction as a key lead toward understanding the chemistry involved in DNA interactions with enediyne antibiotics. The chemical soundeness of this suggestion was amply substantiated by the solid evidence provided in studies of various aspects of the reactivity of these agents in conjunction with experimental data on the pattern of their reactions with native DNA or synthetical oligonucleotides. Owing to these efforts, a fairly consistent description of the chemical events leading eventually to the cleavage of the DNA molecule was formulated for all antibiotics of the enediyne family, as is exemplified in Scheme 4.91 for the case of calicheamicin γ_1^I **293**.[41d]

In the absence of nucleophiles, **293** is a moderately stable compound. The cascade shown is initiated by the attack of the nucleophile (for example, glutathione[40b,l]) on the trisulfide moiety, which leads to the transient formation

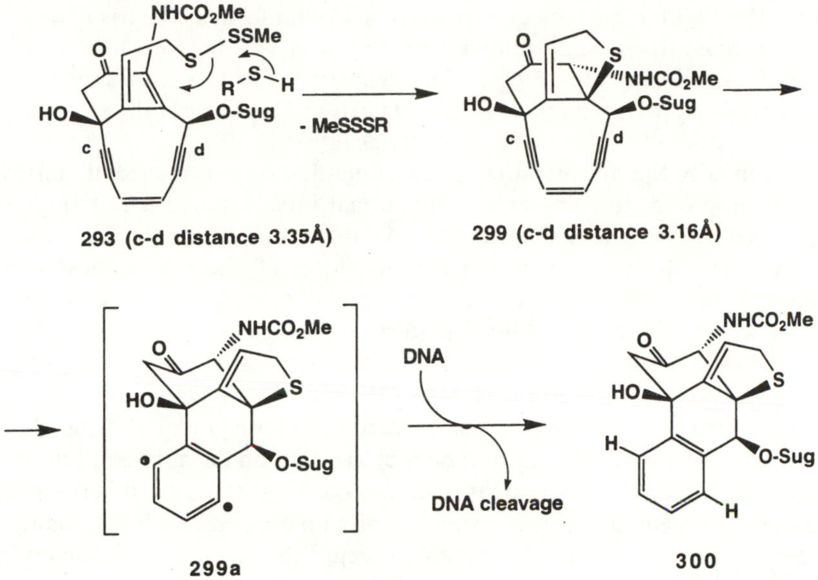

293 (c-d distance 3.35Å)

299 (c-d distance 3.16Å)

299a

300

Scheme 4.91

of allylthiol. Allylthiolate readily adds across the conjugated double bond of the enone fragment to give the Michael adduct **299**. According to molecular modeling calculations, the distance between the terminal atoms of enediyne system (**cd** distance) is significantly shortened as a result of this conversion. Dihydrothiophene **299** was identified as an intermediate stable at −67 °C for several hours. Upon warming up to −10 °C it undergoes spontaneous Bergman cyclization, presumably *via* benzenoid 1,4-biradical intermediate **299a**.[41d] This highly reactive species is capable of the ready abstraction of hydrogen atoms from available donors (in the absence of DNA, CD_2Cl_2 when used as a solvent was shown to act as a hydrogen donor). As was established experimentally in the study of DNA interaction with **293**, one hydrogen atom is abstracted from the C-5′ position of deoxycytidine from one strand and another from a ribose portion in the opposing strand. Thus both strands are affected and susceptible to cleavage. As a final result of the **293** conversion, the stable and non-active tetracyclic compound **300** is formed. In fact this compound was also isolated from natural sources. Its co-occurence with **293** serves as an important lead for the understanding of the chemical background of the observed biological activity of **293**.

An important conclusion might be drawn from the general outline of the main features of the interaction of **293** with DNA as given above. The complex molecule of this antibiotic is composed of several readily recognized fragments, responsible for separate parts of the overall job of DNA cleavage. The tetrasaccharide domain was identified as a delivery and binding system for **293**. The conjugated enediyne moiety embedded into the 10-membered ring represents a sort of 'warhead', capable of effecting DNA cleavage. Its 'triggering device', the allylthiolate moiety, is present in a latent form as the trisulfide and is thus protected from the spontaneous initiation of the 'explosion of the warhead', say during the process of its delivery to the target. A similar set of functionally different structural fragments can be identified in the structures of other antibiotics of the enediyne family (*vide infra*).

Undoubtedly the above rather picturesque description represents an oversimplification and, to some extent, functional interplay of various fragments might also occur. Nevertheless, the identification of the main principles employed by Mother Nature for the invention of these 'weapons' greatly facilitated the pursuits targeted at the rational design of artificially created analogs with a similar or modified pattern of biological activity. The ultimate goal of these studies was to create artificial mimics of the aforementioned antibiotics capable of exhibiting similar DNA-damaging activity and useful as anticancer drugs. In chemical terms it meant that these potential drugs should: (i) bear a moiety amenable to Bergman cyclization under mild conditions; (ii) contain the fragment which might operate as a trigger to start this reaction at the proper moment; and (iii) be provided with an appropriate functionality for the attachment of additional residues to secure the efficiency of delivery and formation of the site-selective drug–DNA intercalate. Initial studies in this area were aimed at the elucidation of structural prerequisites which determine the ease and efficency of the Bergman cyclization. In other words, this part of a general problem can be referred to as the design of the 'warhead'.

As was mentioned above, the classical Bergman cyclization of conjugated enediynes proceeds at elevated temperatures (160–200 °C), while a similar transformation within the structural framework of enediyne antibiotics occurs readily at 37 °C.[41d] Obvious considerations prompted the suggestion that the proximity of the terminal sp centers, as estimated by the critical **cd** distance, is the main factor governing the ease of the cyclization. This simple consideration led to the design of several monocyclic compounds containing the enediyne moiety, as shown in Scheme 4.92.[42a]

It was found that, among the monocyclic analogs, only 10-membered enediynes like **300–302**[42a–d] undergo cycloaromatization with an ease comparable to that of the antibiotics **292** or **293**. Molecular modeling calculations revealed that, for these and similar compounds, **cd** varies within the limits 3.20–3.40 Å[42a] (*cf.* the values 3.35 and 3.16 Å calculated for the conversion of **292**, Scheme 4.91). This distance is increased to 3.61 Å in the 11-membered enediyne **303** and, in accordance with the predictions, this compound is thermally quite stable at ambient temperature. The same refers to the 12-membered lactone **304** as well as to the 12- to 18-membered analogs of **303**.[42b]

Further elaboration of models involved the design of water-soluble derivatives like diol **305**, suitable for tests as DNA-affecting agents. It was rewarding to discover that the conversion **305** $\rightarrow$ **306** can be carried out in water in the presence of DNA. As a result, a temperature- and time-dependent cleavage of double-stranded DNA was observed.[42b] Hence diol **305**, the simplicity of its structure notwithstanding, can be properly considered as the synthetic mimic of the calicheamicin/esperamicin class of antibiotics. Most remarkably, it took less

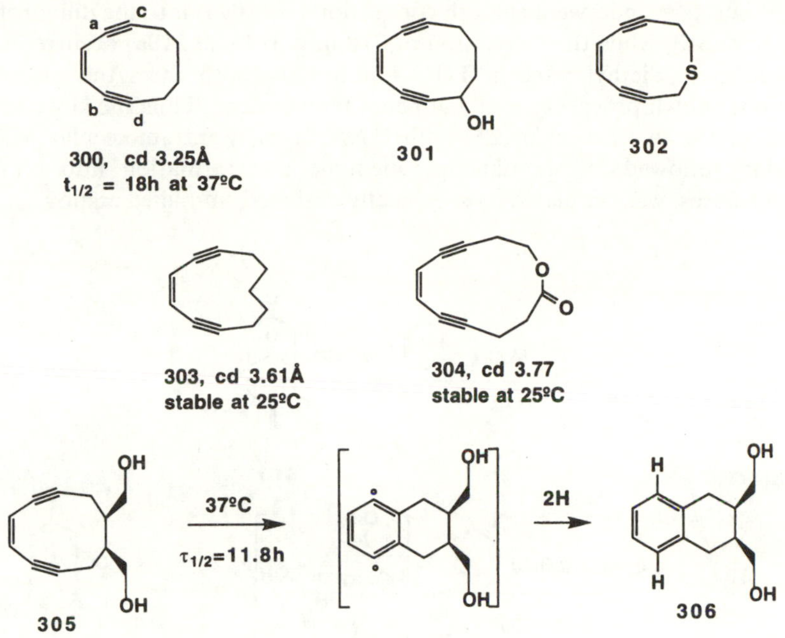

300, cd 3.25Å
$t_{1/2}$ = 18h at 37°C

301

302

303, cd 3.61Å
stable at 25°C

304, cd 3.77
stable at 25°C

305 → 37°C, $\tau_{1/2}$=11.8h → [ ] → 2H → **306**

Scheme 4.92

than one year after the isolation and structure elucidation of **292** to design and prepare this efficient mimic of the natural compound.

Studies by Danishefsky's group aimed at the elaboration of a general synthetic strategy for the total synthesis of the aglycon of **293** offered an opportunity to prepare a set of bicyclic models like **307–310**,[42e–g] which are structurally close to the natural compounds. As was to be expected, these compounds were prone to undergo Bergman cyclization (Scheme 4.93). More important was the observation that in this framework the rate of cycloaromatization can be easily controlled by changes in the geometry of the system brought about by routine transformations reminiscent of the triggering of **293** cyclization *via* its conversion into **299**. Thus the conversion of diketodiol **307** into **311** required prolonged heating, while tetrol **307a** (derived *in situ* from **307**) underwent spontaneous aromatization at 25 °C to give product **312**.[42e] In fact the authors were unable even to isolate this tetrol. In line with these observations, the incubation of DNA with **307** treated with $NaBH_4$ led to the noticeable cleavage of DNA. As had been established earlier, Bergman cyclization of antibiotics **292** or **293** is triggered by an intramolecular Michael addition (see Scheme 4.91). Attempts were made to mimic this process by an intermolecular Michael addition of a nucleophile to **308**.[42f] Rather surprisingly, standard nucleophiles like thiolate, cyanide, or cuprate failed to interact with this compound. On the other hand, intramolecular Michael addition within this framework was shown to be a feasible option for substrates equipped with a properly positioned nucleophilic group like **309** or **310**.[42f,g] Under mildly acidic conditions, **309** underwent smooth conversion (*via* **309a**) into the dihydrofuran derivative **313**, while the conversion of **310** into **314** (*via* **310a**) occurred under the action of diethylamine in THF solution. In both cases the cascade of reactions shown proceeded easily at room temperature. Thus the key chemical events in the interaction of **293** with DNA, namely intramolecular Michael addition followed by spontaneous enediyne transformation into biradical intermediates, was simulated by specifically designed simplified analogs.

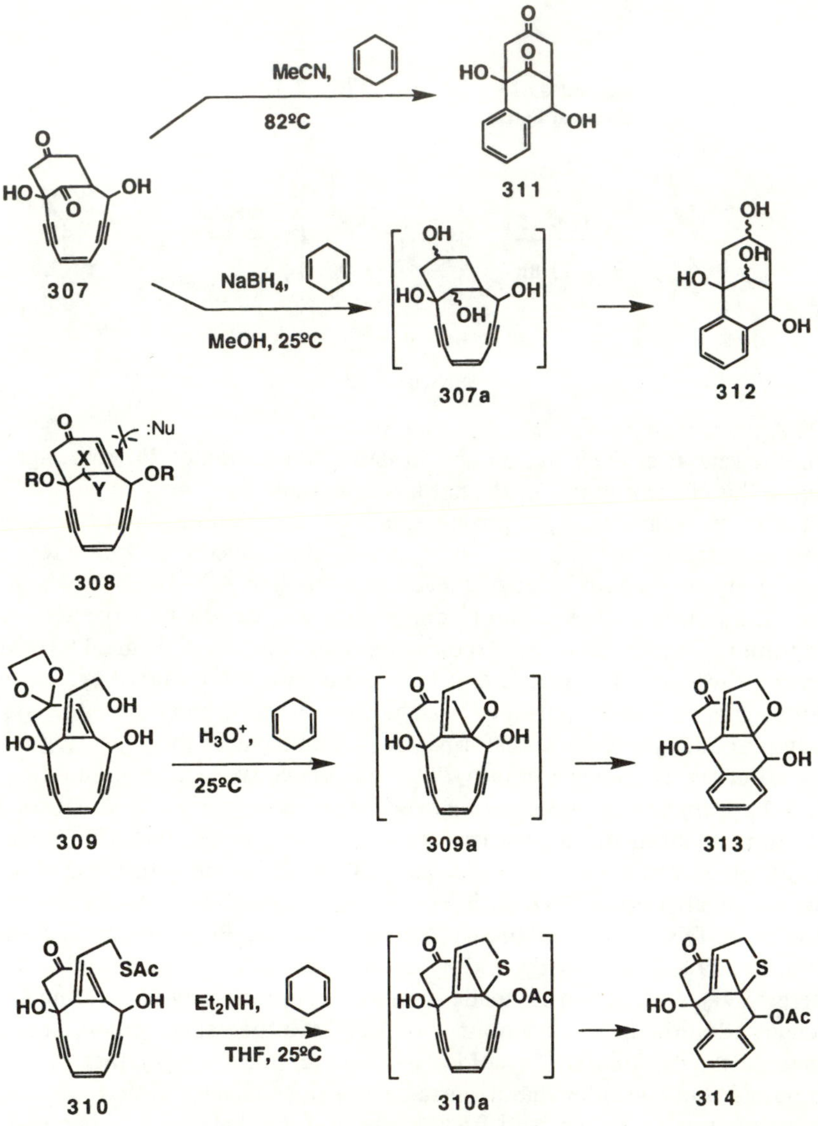

MeCN, 82ºC

NaBH$_4$, MeOH, 25ºC

307 311 307a 312 308

H_3O^+, 25ºC

309 309a 313

Et_2NH, THF, 25ºC

310 310a 314

Scheme 4.93

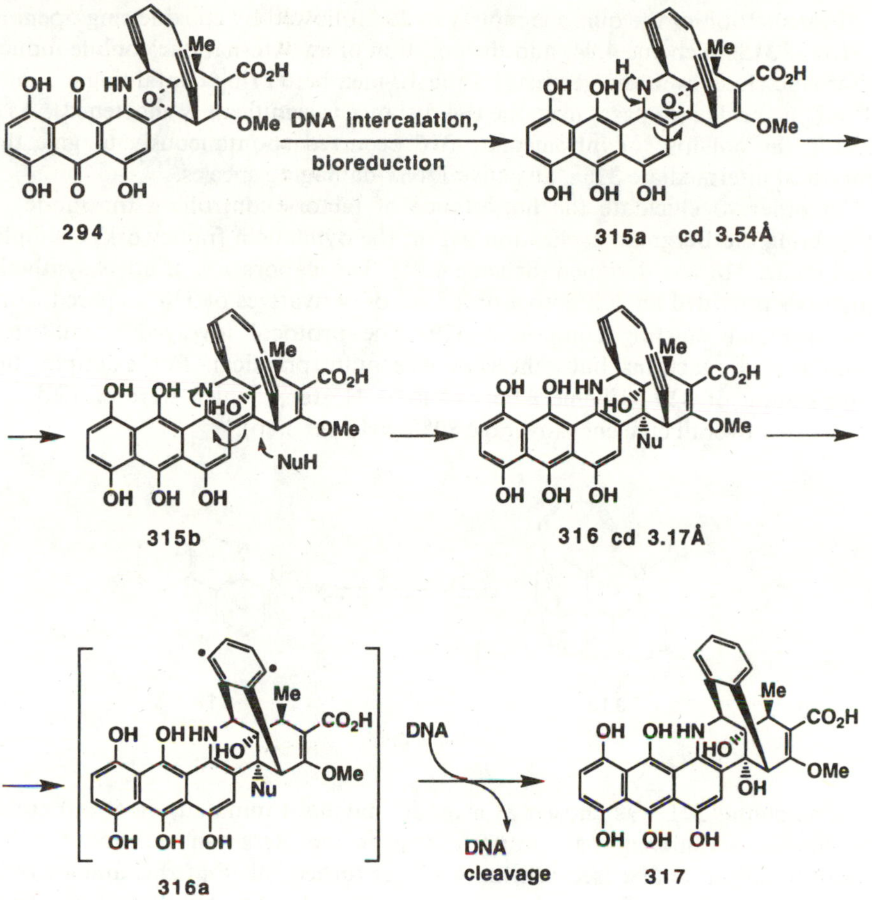

Scheme 4.94

Design of triggering devices is of special importance for the elaboration of drugs which must be safely delivered to the target and then activated under certain conditions. This problem was investigated especially thoroughly in Nicolaou's studies of a model system structurally related to dynamicin A **294**.[40b] The latter antibiotic exhibits potent inhibitory activity against various tumor cell lines and was found to prolong significantly the life span of mice inoculated with P388 leukemia and B16 melanoma cells. The molecule of **294** is a structural hybrid of two chemotypes of antitumor agents, enediyne and anthraquinone. In accordance with generally held suggestions, the antraquinone moiety present in **294** is responsible for the intercalation and specific binding of the agent in the minor groove of the DNA molecule. The DNA cleavage mode of **294**[42h] is fairly different from that of **292** or **293**, where the selectivity of the binding is determined by the presence of the oligosaccharide fragment.

Bioreduction of the quinone moiety in **294** followed by epoxide ring opening (**315a** → **315b**, Scheme 4.94) and the addition of an external nucleophile induce dramatic changes in the geomety of the 10-membered ring containing a 3-ene-1,5-diyne bridge (in particular, the **c–d** distance is significantly shortened). As a result, the collapse of intermediate **316** occurred spontaneously to give the biradical intermediate **316a**, an active DNA-damaging species.

In order to elucidate the importance of factors controlling the mode of triggering the Bergman cyclization within the dynamicin framework, a simplified model **318** was designed (Scheme 4.95). The elaboration of novel synthetic methods provided an access to a broad set of derivatives of **318** prepared from the available starting compound **319**. The protocol involved a multistep sequence of reactions but otherwise was quite practical. For example, the preparation of **320** (Scheme 4.96) required 11 steps, which were carried out with good overall efficiency (average 80% yield per step).[42i]

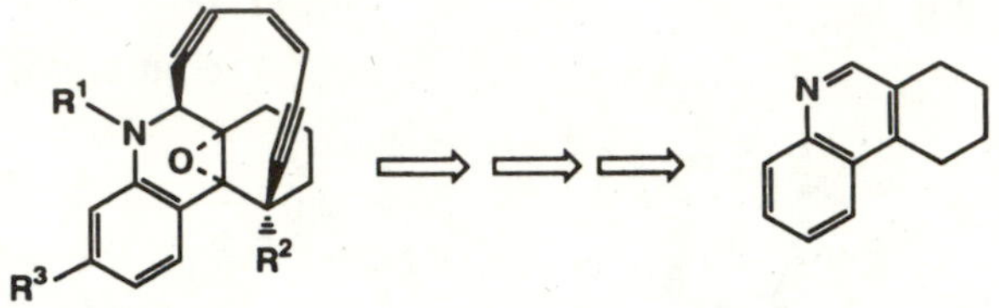

R1 N O R2 R3 **318** ⟹ ⟹ ⟹ N **319**

Scheme 4.95

Compound **321** was chosen as a model having a minimal set of structural fragments considered mandatory to trigger the Bergman reaction for the natural antibiotic **294** (see Scheme 4.94). It turned out that this amine could not be isolated in a free state owing to its ease of cycloaromatization. Thus reductive cleavage of the carbamoyl protecting group in **320** produced **321** (as ascertained by the spectral data), but transformation of the latter into diol **322** followed by Bergman cyclization to lead to **323** was shown to occur readily even upon storage in EtOH solution at 25 °C (Scheme 4.96). The introduction of a methoxy group at the bridgehead position noticeably increased the stability of the whole system. The conversion of **324** into **325** required moderate heating in the presence of acid.[42j]

The model **326a** was designed with the purpose of providing a sort of 'safety bolt' to the triggering system. In fact, the liability of this compound to undergo Bergman cyclization was greatly reduced, obviously due to the introduction of an electron withdrawing group at the nitrogen center. Yet under mildly basic conditions this group could be easily removed to give a transient **321**. In this case the cycloaromatization of the latter might be readily triggered even under the action of such weak acids as phenol or thiophenol to give the final products **327a** or **327b**.[42j] Structure **326a** was additionally modified by the installation of an ethylene glycol residue as a tethering group at C-2 to ensure the possibility of the future incorporation of other desirable moieties. The derivative **326b**

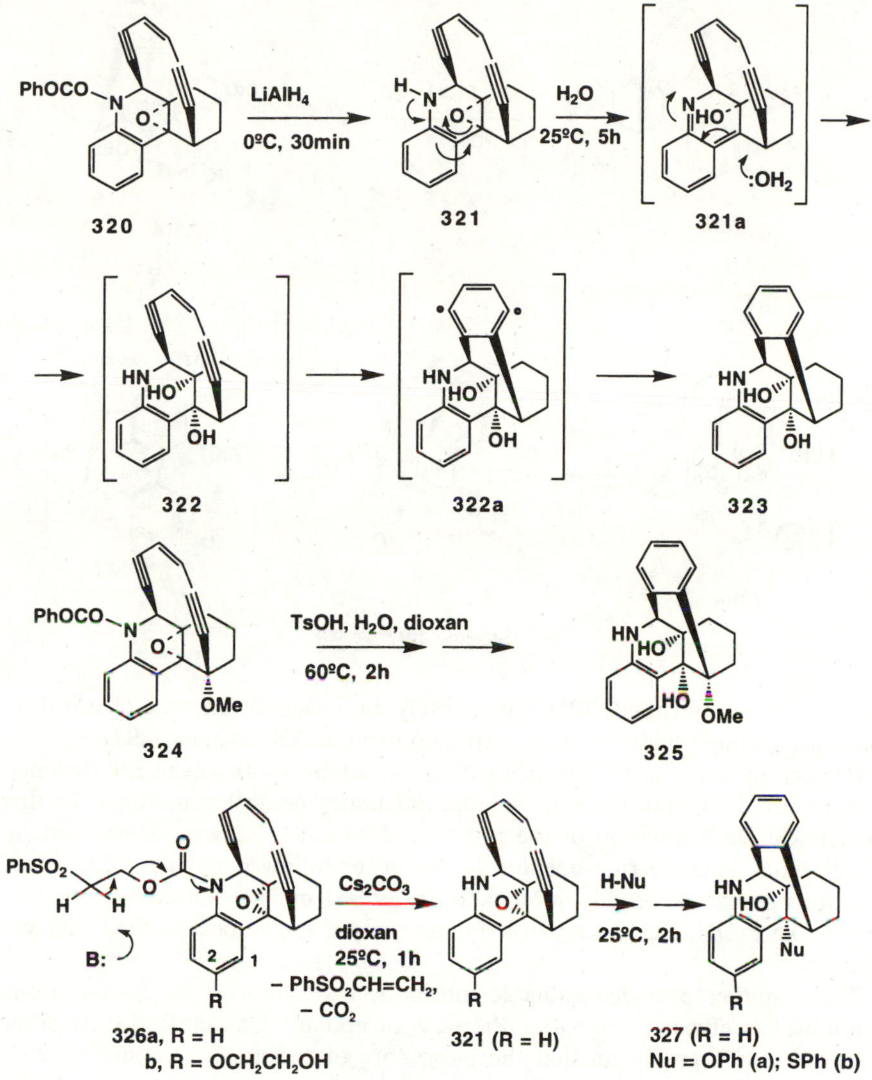

Scheme 4.96

displayed high activity as a DNA-cleaving agent and turned out to be among the most potent cytotoxic agents in tests against a variety of tumor cell lines.

An entirely different mode of triggering has been achieved for the series of models like **328** or **329** containing the protected hydroxyl group at C-3.[42k] It was anticipated that the presence of the hydroxyl group at this position should greatly facilitate epoxide ring opening and Bergman cyclization (see Schemes 4.97 and 4.98). In the model **328**, the phenolic hyroxyl was protected by a pivaloyl group. Removal of this group (with concomitant transesterification at the carbamoyl substituent) proceeded easily under mildly basic conditions. The

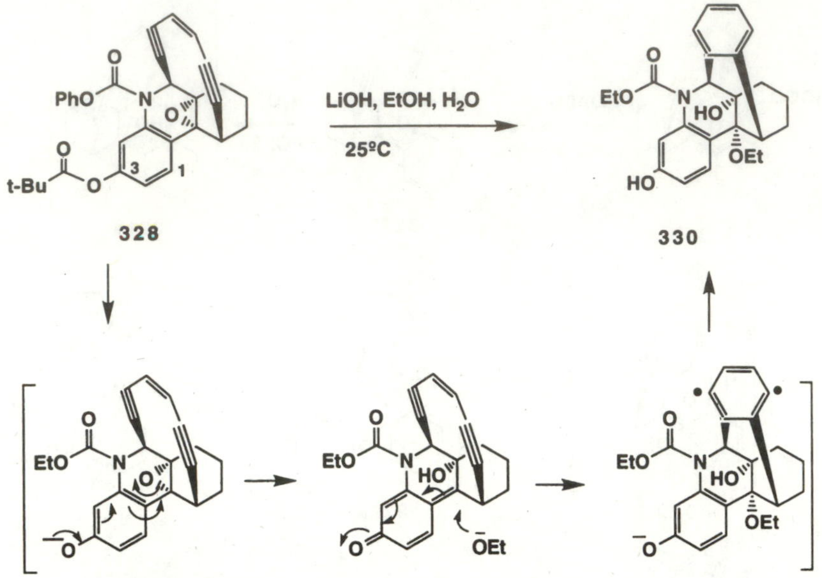

Scheme 4.97

initially formed phenolate **328a** immediately underwent a sequence of transformations, leading eventually to the expected product **330** (Scheme 4.97).

Protection of the same hydroxyl with a photolabile group as in **329** (Scheme 4.98) ensured its removal upon irradiation under neutral conditions. In this experiment the formation of the phenolic derivative **331** was ascertained by NMR data. Epoxide ring opening in the latter followed by the usual set of reactions proceeded readily upon exposure of **331** in THF solution to nucleophiles like EtOH, EtSH, or *n*-$PrNH_2$ and yielded the respective final adducts **332a–c**.[42k]

These studies provided valuable infomation attesting to the vast options available for efficient control of the ease of epoxide ring cleavage. In other words, it has been shown that the basic core of the artificial mimic of **294**, composed of the acting enediyne fragment and appropriately positioned epoxide trigger, can be furnished with additional devices capable of modulating the sensitivity of this trigger toward activation under a multitude of conditions.

The ongoing studies in this area represent a spectacular example of function-oriented molecular design carried out in a truly rational way. In fact, as a result of the initial investigations of the mechanism of the DNA-damaging action of natural antibiotics, a fairly consistent description of the chemistry of the elementary acts that lead ultimately to the DNA strand cleavage was developed. Thanks to the acquired understanding of the chemistry involved, the general problem of the creation of artificial mimics with the same pattern of activity could be formulated in purely chemical terms as the design of structures bearing a definite set of properly positioned functional groups.

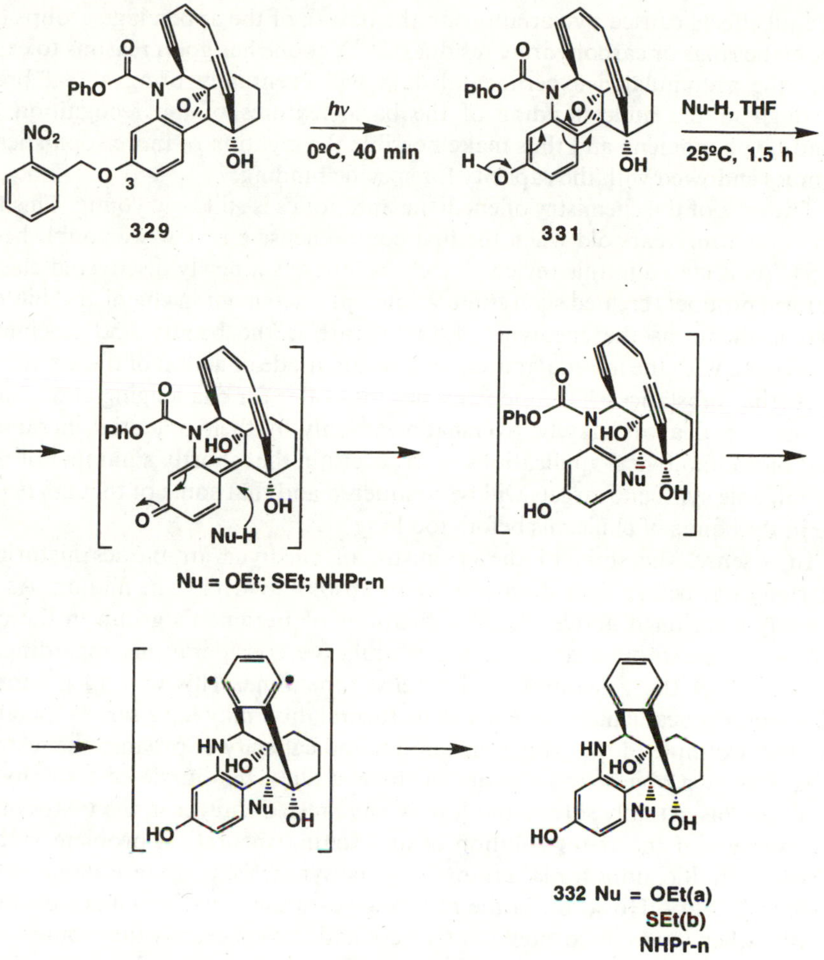

329

331

Nu = OEt; SEt; NHPr-n

332 Nu = OEt(a)
SEt(b)
NHPr-n

Scheme 4.98

It must be also emphasized that so far the achievements in this area refer mainly to the design of the simplified analogs capable of simulating the elementary acts of DNA cleavage by their natural counterparts. As was mentioned in the beginning of this section, the structures of all natural enediyne antibiotics also contain domains which constitute the parts of delivery and binding machinery. The functioning of this system is governed by much more complicated interactions between the participating molecules which are still difficult to interpret unambiguously in 'cause and effect' terms on the molecular level (*cf.* discussion of recognition phenomena in Section 4.8). Therefore it is a much more difficult task rationally to design structural fragments which should be attached to the molecule of enediyne analogs in order to ensure the required pattern of binding and/or selectivity of action. Thus far, achievements in this area are not very spectacular and are mainly due to empirical studies of the overall effects caused by variations in the nature of the appendage groups (like aromatic rings or carbohydrate residues).[40b] Yet one has good reasons to expect that the accumulated experimental data will eventually bring a real breakthrough in the understanding of the basic features of the recognition and binding phenomena and thus make possible the creation of more sophisticated mimics endowed with the capacity for specific binding.

The area of the chemistry of enediyne antibiotics is still very young. This field was only four years old when the first comprehensive review was published in 1991. As is stated in this review, 'rarely before has a newly discovered class of natural products created such stimulation and excitement in chemistry, biology, and medicine as the enediynes. Equally rare is the beauty and fascination associated with the molecular architecture and mode of action of these naturally occurring substances. The opportunities they offer for challenging and creative scientific endeavors may be overshadowed only by their potential therapeutic and biotechnological applications. ... It is certain that shortly a number of these formidable synthetic targets will be conquered and that some of them may even be in the hands of clinicians before too long'.[40b]

In a sense, the story of the chemistry of enediyne antibiotics historically started even before their discovery in an absolutely irrelevant finding. As was already mentioned above, the investigations of Bergman's group in the early 1970s[41a] were motivated by rather speculative considerations regarding the possibility of the generation of 1,4-dehydrobenzene. This was an interesting albeit purely academic problem and its formulation may have served merely as another example of the inbred inclination and capacity of organic chemistry to create its own objects. As a result of these studies, the puzzle of 1,4-dehydrobenzene was actually solved and it may have rested forever in the textbooks as an example of the clever solution of an exciting theoretical problem with no obvious ramifications for laboratory organic synthesis, to say nothing about its practical utility. However, in the next few years the situation changed dramatically, when it was discovered that Nature had chosen exactly that route for the generation of 1,4-biradicals as effective tools to cause severe damage in DNA. So it is not surprising to find references to the Bergman cyclization practically in all current publications dealing with the mechanism of action of natural antitumor antibiotics or attempts to create their mimics. It is also appropriate to mention in passing that the amazing rate of progress of synthetic efforts in this area was made possible owing to an array of novel and very efficient methods for the construction of enyne and enediyne fragments, which were elaborated previously, also as a result of purely academic studies. Thus, again and again we come across additional illustrations of the correctness of the prominent Russian chemist Alexander Nesmeyanov's paradoxical claim that 'nothing might be more practical than a good theory!'. It is truly unfortunate that these lessons seem to be so difficult to learn by those who are responsible for allocating money for the fundamental research and most typically would prefer to support studies having an immediate applied promise.

4.12 CONCLUDING REMARKS

It seems also obvious that molecular design can never be rigorously defined as purely structure or function oriented. This is an artificially introduced classification, somewhat convenient for the presentation of a huge and diversified amount of material in a more or less coherent way. In fact, it is easy to demonstrate that almost any structure of unusual design described in Part I is of immediate or at least potential interest for the function oriented design described in Part II of this chapter. Interestingly, the latter is not only based on the achievements of structure oriented design but, in turn, may provide the impetus toward the creation of novel structural types. The examples of both effects are in abundance and the reader would easily identify them in the preceding sections.

REFERENCES

[1] Eaton, P. E.; Cole, T. W., Jr. *J. Am. Chem. Soc.*, **1964**, *86*, 962; *ibid.*, **1964**, *86*, 3157.

[2] Ternansky, R. J.; Balogh, D. W.; Paquette, L. A. *J. Am. Chem. Soc.*, **1982**, *104*, 4503.

[3] (a) Maier, G.; Pfriem, S.; Schäfer, U.; Matush, R. *Angew. Chem., Int. Ed. Engl.*, **1978**, *17*, 520; (b) Maier, G.; Born, D. *Angew. Chem., Int. Ed. Engl.*, **1989**, *28*, 1050.

[4] (a) Reviews: Eaton, P. E. *Angew. Chem., Int. Ed., Engl.*, **1992**, *31*, 1421; Griffin, G. W.; Marchand, A. P. *Chem. Rev.*, **1989**, *89*, 997; (b) Martin, H.-D.; Urbanek, T.; Pföhler, P.; Walsh, R. *J. Chem. Soc., Chem. Commun.*, **1985**, 964; (c) Eaton, P. E.; Maggini, M. *J. Am. Chem. Soc.*, **1988**, *110*, 7230; see also: Lukin, K.; Eaton, P. E. *J. Am. Chem. Soc.*, **1995**, *117*, 7652; (d) Eaton, P. E.; Yang, C.-X.; Xiong, Y. *J. Am. Chem. Soc.*, **1990**, *112*, 3225; (e) Bingham, R. C.; Schleyer, P. *J. Am. Chem. Soc.*, **1971**, *93*, 3189; (f) Moriarty, R. M.; Tuladhar, S. M.; Penmasta, R.; Awasthi, A. K. *J. Am. Chem. Soc.*, **1990**, *112*, 3228; (g) Hrovat, D.A.; Borden, W. T. *J. Am. Chem. Soc.*, **1990**, *112*, 3227; (h) Eaton, P. E.; Galoppini, E.; Gilardi, R. *J. Am. Chem. Soc.*, **1994**, *116*, 7588; (i) Marchand, A. P. *Tetrahedron*, **1988**, *44*, 2377; (j) Eaton, P. E.; Xiong, Y.; Gilardi, R. *J. Am. Chem. Soc.*, **1993**, *115*, 10195.

[5] (a) See, for a review: Paquette, L. A. *Chem. Rev.*, **1989**, *89*, 1051; (b) Paquette, L. A.; Lagerwall, D. R.; King, J. L.; Niwayama, S.; Skerlj, R. *Tetrahedron Lett.*, **1991**, *32*, 5259; see also: (c) Olah, G. A.; Surya Prakash, G. K.; Kobayashi, T.; Paquette, L. A. *J. Am. Chem. Soc.*, **1988**, *110*, 1304; (d) Paquette, L. A.; Kobayashi, T.; Galucci, J. C. *J. Am. Chem. Soc.*, **1988**, *110*, 1305.

[6] (a) See, for a review: Prinzbach, H.; Fessner, W.-D. in *Organic Synthesis: Modern Trends*, Chizov, O., Ed., Blackwell, Oxford, **1987**, p. 23; (b) Melder, J.-P.; Pinkos, R.; Fritz, H.; Wörth, J.; Prinzbach, H. *J. Am. Chem. Soc.*, **1992**, *114*, 10213; (c) Pinkos, R.; Melder, J.-P.; Weber, K.; Hunkler, D.; Prinzbach, H. *J. Am. Chem. Soc.*, **1993**, *115*, 7173; (d) Fessner, W.-D.; Sedelmeier, G.; Spurr, P. R.; Rihs, G.; Prinzbach, H. *J. Am. Chem. Soc.*, **1987**, *109*, 4626.

[7] (a) See, for a review: Maier, G. *Angew. Chem., Int. Ed. Engl.*, **1988**, *27*, 309; for earlier data covering the tetrahedrane problem, see: Zefirov, N. S.; Koz'min, A. S.; Abramenkov, A. V. *Russ. Chem. Rev. (Engl. Transl.)*, **1978**, *47*, 163; (b) Seidl, E. T.; Shaefer, III, H. F. *J. Am. Chem. Soc.*, **1991**, *113*, 1915.

[8] (a) See, for a review: Fort, R. C., Jr. *Adamantane and Chemistry of Diamondoid Molecules*, Dekker, New York, **1976**; (b) Olah, G. A.; Surya Prakash, G. K.; Shih, J. G.; Krishnamurthy, V. V.; Mateesku, G. D.; Liang, G.; Sipos, G.; Buss, V.; Gund, T.

M.; Schleyer, P. v. R. *J. Am. Chem. Soc.*, **1985**, *107*, 2764; see also: Olah, G. A.; Lee, C. S.; Surya Prakash, G. K.; Moriarty, R. M.; Chander Rao, M. S. *J. Am. Chem. Soc.*, **1993**, *115*, 10728; (c) Stetter, H.; Krause, M. *Tetrahedron Lett.*, **1967**, 1841.

[9] (a) Olah, G. A.; Surya Prakash, G. K.; Fesner, W.-D.; Kobayashi, T.; Paquette, L. A. *J. Am. Chem. Soc.*, **1988**, *110*, 8599; (b) Surya Prakash, G. K.; Krishnamurthy, V. V.; Herges, R.; Bau, R.; Yuan, H.; Olah, G. A.; Fesner, W.-D.; Prinzbach, H. *J. Am. Chem. Soc.*, **1988**, *110*, 7764; (c) Bremer, M.; Schleyer, P. v. R.; Schötz, K.; Kausch, M.; Schindler, M. *Angew. Chem., Int. Ed. Engl.*, **1987**, *26*, 761; (d) see also the review: Saunders, M.; Jimênez-Vãzquez, H. A. *Chem. Rev.*, **1991**, *91*, 375.

[10] (a) Jones, D. E. H. *New Sci.*, **1966**, 3 November, p. 245; (b) Jones, D. E. H. *The Inventions of Daedalus*, Freeman, Oxford, **1982**.

[11] Fuller, R. B. *Inventions. The Patented Works of R. Buckminster Fuller*, St. Martin's Press, New York, **1983**.

[12] (a) Rohlfing, E. A.; Cox, D. M.; Kaldor, A. *J. Chem. Phys.*, **1984**, *81*, 3322; (b) Kroto, H. W.; Heath, J. R.; O'Brien, S. C.; Curl, R. F.; Smalley, R. E. *Nature*, **1985**, *318*, 162; see also: Curl, R. F.; Smalley, R. E. *Science*, **1988**, *242*, 1017; (c) Kroto, H. W. *Science*, **1988**, *242*, 1139; (d) Kroto, H. W. *Nature*, **1987**, *329*, 529; (e) Osawa, E. *Kagaku (Kyoto)*, **1970**, *25*, 854; (f) Bochvar, D. A.; Galpern, E. G. *Dokl. Akad. Nauk SSSR*, **1973**, *209*, 610.

[13] (a) Krätschmer, W.; Lamb, L. D.; Fostiropoulos, K.; Huffman, D. R. *Nature*, **1990**, *347*, 354; Krätschmer, W.; Fostiropoulos, K.; Huffman, D. R. *Chem. Phys. Lett.*, **1990**, *170*, 167; (b) Meijer, G.; Bethune, D. S. *J. Chem. Phys.*, **1990**, *93*, 7800; (c) Wragg, J. L.; Chamberlain, J. E.; White, H. W.; Krätschmer, W.; Huffman, D. R. *Nature*, **1990**, *348*, 623; (d) Haufler, R. E.; Conceicao, J.; Chibante, L. P. F.; Chai, Y.; Byrne, N. E.; Flanagan, S.; Haley, M. M.; O'Brien, S. C.; Pan, C.; Xiao, Z.; Billups, W. E.; Ciufolini, M. A.; Hauge, R. H.; Margrave, J. L.; Wilson, L. J.; Curl, R. F.; Smalley, R. E. *J. Phys. Chem.*, **1990**, *94*, 8634; (e) Ajie, H.; Alvarez, M. M.; Anz, S. J.; Beck, R. D.; Diederich, F.; Fostiropoulos, K.; Huffman, D. R.; Krätschmer, W.; Rubin, Y.; Schriver, K. E.; Sensharma, D.; Whetten, R. L. *J. Phys. Chem.*, **1990**, *94*, 8630; (f) Taylor, R.; Hare, J. P.; Abdul-Sada, A. K.; Kroto, H. W. *J. Chem. Soc., Chem. Commun.*, **1990**, 1423; (g) Becker, L.; Bada, J. L.; Winans, R. E.; Hunt, J. E.; Bunch, T. E.; French, B. M. *Science*, **1994**, *265*, 642; (h) Heymannn, D.; Chibante, L. P. F.; Brooks, R. R.; Wolbach, W. S.; Smalley, R. E. *Science*, **1994**, *265*, 645.

[14] Barth, W. E.; Lawton, R. G. *J. Am. Chem. Soc.*, **1971**, *93*, 1730.

[15] (a) Hawkins, J. M.; Meyer, A.; Lewis, T. A.; Loren, S.; Hollander, F. J. *Science*, **1991**, *252*, 312. (b) Fagan, P. J.; Calabrese, J. C.; Malone, B. *J. Am. Chem. Soc.*, **1991**, *113*, 9408; (c) Cox, D. M.; Behal, S.; Disko, M.; Gorun, S. M.; Greaney, M.; Hsu, C. S.; Kollin, E. B.; Millar, J.; Robbins, J.; Robbins, W.; Sherwood, R. D.; Tindall, P. *J. Am. Chem. Soc.*, **1991**, *113*, 2940; (d) Bausch, J. W.; Surya Prakash, G. K.; Olah, G. A; Tse, D. S.; Lorents D. C.; Bae, Y. K.; Malhotra, R. *J. Am Chem. Soc.*, **1991**,*113*, 3205. (e) Olah, G. A.; Bucsi, I.; Lambert, C.; Aniszfeld, R.; Trivedi, N. J.; Sensharma, D. K.; Surya Prakash, G. K. *J. Am Chem. Soc.*, **1991**, *113*, 9385; (f) Olah, G. A.; Bucsi, I.; Lambert, C.; Aniszfeld, R.; Trivedi, N. J.; Sensharma, D. K.; Surya Prakash, G. K. *J. Am. Chem. Soc.*, **1991**, *113*, 9387; (g) Rubin, Y.; Khan, S.; Freedberg, D. I.; Yeretzian, C. *J. Am. Chem. Soc.*, **1993**, *115*, 344; (h) Khan, S. I., Oliven, A. M.; Paddon-Row, M. N.; Rubin, Y. *J. Am . Chem. Soc.*, **1993**, *115*, 4919; (i) Prato, M.; Chan Li, Q.; Wudl, F.; Lucchini, V. *J. Am. Chem. Soc.*, **1993**, *115*, 1148; (j) Saunders, M.; Jimênez-Vásquez, H. A.; Bangerter, B. W; Cross, R. J.; Mroczkowski, S.; Freedberg, D. I.; Anet, F. A. L. *J. Am. Chem. Soc.*, **1994**, *116*, 3621; see, for examples of utilization: Smith III, A. B.; Strongin, R. M.; Brard, L.; Romanow, W. J.; Saunders, M.; Jimênez-Vásques, H. A.;

Cross, R. J. *J. Am. Chem. Soc.*, **1994**, *116*, 10831; (k) Haddon, R. C.; Hebard, A. F.; Rosseinsky, M. J.; Murphy, D. W.; Duclos, S. J.; Lyons, K. B.; Miller, B; Rosamilia, J. M.; Fleming, R. M.; Kortan, A. R.; Glarum, S. H.; Makhija, A. V.; Muller, A. J.; Eick, R. H.; Zahurak, S. M.; Tycko, R.; Dabbagh, G.; Thiel, F. A. *Nature*, **1991**, *350*, 320; (l) Hebard, A. F.; Rosseinsky, M. J.; Haddon, R. C.; Murphy, D. W.; Glarum, S. H.; Palstra, T. T. M.; Ramirez, A. P.; Kortan, A. R. *Nature*, **1991**, *350*, 600; (m) Holczer, K.; Klein, O.; Huang, S.-M.; Kaner, R. B.; Fu, K.-J.; Whetten, R. L.; Diederich, F. *Science*, **1991**, *252*, 1154; see also: *Phys. Chem. Lett.*, **1991**, *66*, 2830; (n) Cioslovsky, J. *J. Am. Chem. Soc.*, **1991**, *113*, 4139.

[16] (a) The story of the discovery of **59** is highlighted in: Stoddart, J. F. *Angew. Chem., Int. Ed. Engl.*, **1991**, *30*, 70; see also reviews: Kroto, H. W.; Allaf A. W.; Balm, S. P. *Chem. Rev.*, **1991**, *91*, 1213; Diederich, F.; Whetten, R. L. *Angew. Chem., Int. Ed. Engl.*, **1991**, *30*, 678; Sokolov, V. I. *Izv. Akad. Nauk SSSR, Ser. Khim.*, **1993**, 10; monographs: *The Fullerenes*, Kroto, H. W.; Fisher, J. E.; Cox, D. E., Eds., Pergamon, Oxford, **1993**; *Buckminsterfullerenes*, Billups, W. E.; Ciufolini, M. A., Eds., VCH, New York, **1993**; (b) Rabideau, P. W.; Abdourazak, A. H.; Folsom, H. E.; Marcinow, Z.; Sygula, A.; Sygula, R. *J. Am. Chem. Soc.*, **1994**, *116*, 7891; see also: Abdourazak, A. H.; Marcinow, Z.; Sygula, A.; Sygula, R.; Rabideau, P. W. *J. Am. Chem. Soc.*, **1995**, *117*, 6410.

[17] (a) See, for a comprehensive review: Tomalia, D. A.; Naylor, A.M.; Goddard III, W. A. *Angew. Chem., Int. Ed. Engl.*, **1990**, *29*, 138; (b) Padias, A. B.; Hall, Jr., H. K.; Tomalia, D. A.; McConnell, J. R. *J. Org. Chem.*, **1987**, *52*, 5305; (c) Bochkov, A. F.; Kalganov, B. E.; Chernetzky, V. N. *Izv. Akad. Nauk SSSR, Ser. Khim.*, **1989**, 2394; (d) Tomalia, D. A.; Hall, M.; Hedstrand, D. M. *J. Am. Chem. Soc.*, **1987**, *109*, 1601; (e) Bochkarev, M. N.; Silkin, V. B.; Mayorova, L. P.; Razuvaev, G. A.; Semchikov, Yu. D.; Scherstyanich, V. I. *Metalloorg. Khim.*, **1988**, *1*, 196; see also: Bochkarev, M. N.; Semchikov, Yu. D.; Silkin, V. B.; Scherstyanich, V. I.; Mayorova, L. P.; Razuvaev, G. A. *Vysokomol. Soedin.*, **1989**, *31*, 643; (f) Hawker, C. J.; Fréchet, J. M. J. *J. Am. Chem. Soc.*, **1990**, *112*, 7638; (g) Wooley, K. L.; Hawker, C. J; Fréchet, J. M. J. *J. Am. Chem. Soc.*, **1191**, *113*, 4252; see also the preparation of the fullerene-bound dendrimer in: Wooley, K. L.; Hawker, C. J.; Fréchet, J. M. J.; Wudl, F.; Srdanov, G.; Shi, S.; Li, C.; Kao, M. *J. Am. Chem. Soc.*, **1993**, *115*, 9836; (h) Newkome, G. R.; Baker, G. R.; Arai, S.; Saunders, M. J.; Russo, P. S.; Theriot, K. J.; Moorefield, C. N.; Rogers, L. E.; Miller, J. E.; Lieux, T. R.; Murray, M. E.; Phillips, B.; Pascal, L. *J. Am. Chem. Soc.*, **1990**, *112*, 8458; (i) Brabander-van der Berg, E. M. M.; Meijer, E. W. *Angew. Chem., Int. Ed. Engl.*, **1993**, *32*, 1308; see also: Wörner, C.; Mülhaupt, R. *Angew. Chem., Int. Ed. Engl.*, **1993**, *32*, 1306; (j) Farin, D.; Avnir, D. *Angew. Chem., Int. Ed. Engl.*, **1991**, *30*, 1379; (k) Turro, N. J.; Barton, J. K.; Tomalia, D. A. *Acc. Chem. Res.*, **1991**, *24*, 332; (l) Newkome, G. R.; Moorefield, C. N.; Baker, G. R.; Saunders, M. J.; Grossman, S. H. *Angew. Chem., Int. Ed. Engl.*, **1991**, *30*, 1178.

[18] (a) Frisch, H. L.; Wasserman, E. *J. Am. Chem. Soc.*, **1961**, *83*, 3789; (b) see, for a review: Shill, G. *Catenanes, Rotaxanes and Knots*, Academic Press, New York, **1971**; (c) Sokolov, V. I. *Russ. Chem. Rev. (Engl. Transl.)*, **1973**, *42*, 452; (d) for later data, see the review: Walba, D. M. *Tetrahedron*, **1985**, *41*, 3161; (e) Wasserman, E. *J. Am. Chem. Soc.*, **1960**, *82*, 4433; (f) Harrison, I. T. *J. Chem. Soc., Chem. Commun.*, **1972**, 231; (g) an aesthetical appeal of the studies in this area is impressively exposed in an essay: Hoffmann, R. *Am. Sci.*, **1988**, *76*, 604.

[19] (a) Ashton, P. R.; Goodnow, T. T.; Kaifer, A. E.; Reddington, M. V.; Slawin, A. M. Z.; Spencer, N.; Stoddart, J. F.; Vicent, C.; Williams, D. J. *Angew. Chem., Int. Ed. Engl.*, **1989**, *28*, 1396; (b) Ashton, P. R.; Brown, C. L.; Chrystal, E. J. T.; Goodnow, T.

T.; Kaifer, A. E.; Parry, K. P.; Philp, D.; Slawin, A. M. Z.; Spencer, N.; Stoddart, J. F.; Williams, D. *J. Chem. Soc., Chem. Commun.*, **1991**, 634; (c) Ashton, P. R.; Brown, C. L.; Chrystal, E. J. T.; Goodnow, T. T.; Kaifer, A. E.; Parry, K. P.; Slawin, A. M. Z.; Spencer, N.; Stoddart, J. F.; Williams, D. J. *Angew. Chem., Int. Ed. Engl.*, **1991**, *30*, 1039; see also: Ashton, P. R.; Brown, C. L.; Chrystal, E. J. T.; Parry, K. P.; Pietraszkiewicz, M.; Spencer, N.; Stoddart, J. F. *Angew. Chem., Int. Ed. Engl.*, **1991**, *30*, 1042; (d) Anelli, P. L.; Spencer, N.; Stoddart, J. F. *J. Am. Chem. Soc.*, **1991**, *113*, 5131; (e) Ashton, P. R.; Grognuz, M.; Slawin, A. M. Z.; Stoddart, J. F.; Williams, D. J. *Tetrahedron Lett.*, **1991**, *32*, 6235; (f) Anelli, P. L.; Ashton, P. R.; Spencer, N.; Slawin, A. M. Z.; Stoddart, J. F.; Williams, D. J. *Angew. Chem., Int. Ed. Engl.*, **1991**, *30*, 1036; (g) for a concise presentation of the ideology of the whole approach and its conceptual ramifications, see: Anelli, P. L.; Ashton, P. R.; Ballardini, R.; Balzani, V.; Delgado, M.; Gandolfi, M. T.; Goodnow, T. T.; Kaifer, A. E.; Philp, D.; Pietraszkiewicz, M.; Prodi, L.; Reddington, M. V.; Slawin, A. M. Z.; Spencer, N.; Stoddart, J. F.; Vicent, C.; Williams, D. J. *J. Am. Chem. Soc.*, **1992**, *114*, 193 and refs. cited therein; see also the later data in: Vögtle, F.; Müller, W. M.; Müller, U.; Bauer, M.; Rissanen, K. *Angew. Chem., Int. Ed. Engl.*, **1993**, *32*, 1295; Amabilino, D. A.; Ashton, P. R.; Tolley, M. S.; Stoddart, J. F.; Williams, D. J. *Angew. Chem., Int. Ed. Engl.*, **1993**, *32*, 1297; Ballaldini, R.; Balzani, V.; Gandolfi, M. T.; Prodi, L.; Venturi, M.; Philp, D.; Ricketts, H. G.; Stoddart, J. F. *Angew. Chem., Int. Ed. Engl.*, **1993**, *32*, 1301.

[20] (a) See, for reviews: Dietrich-Buchecker, C. O.; Sauvage, J.-P. *Chem. Rev.*, **1987**, *87*, 795; Sauvage, J.-P. *Acc. Chem. Res.*, **1990**, *23*, 319; (b) Dietrich-Buchecker, C. O.; Hemmert, C.; Khemiss, A.-K.; Sauvage, J.-P. *J. Am. Chem. Soc.*, **1990**, *112*, 8002; (c) Bitsch, F.; Dietrich-Buchecker, C. O.; Khêmiss, A.-K.; Sauvage, J.-P; Dorsselaer, A. V. *J. Am. Chem. Soc.*, **1991**, *113*, 4023; (d) Dietrich-Buchecker, C. O.; Sauvage, J.-P. *Angew. Chem., Int. Ed. Engl.*, **1989**, *28*, 189; (e) Dietrich-Buchecker, C. O.; Guilhem, J.; Pascard, C.; Sauvage, J.-P. *Angew. Chem., Int. Ed. Engl.*, **1990**, *29*, 1154; (f) in the subsequent studies the yield of **127** was increased to a respectable 24%, see: Dietrich-Buchecker, C. O.; Nierengarten, J.-F.; Sauvage, J.-P.; Armaroli, N.; Balzani, V.; De Cola, L. *J. Am. Chem. Soc.*, **1993**, *115*, 11237; (g) the methodology of the transition-metal templated synthesis of various helical complexes is highlighted by: Constable, E. C. *Angew. Chem., Int. Ed. Engl.*, **1991**, *30*, 1450.

[21] (a) Walba, D. M.; Richards, R. M.; Sherwood, S. P.; Haltiwanger, R. C. *J. Am. Chem. Soc.*, **1981**, *103*, 6213; (b) Walba, D. M.; Armstrong, III, J. D.; Perry, A. E.; Richards, R. M.; Homan, T. C.; Haltiwanger, R. C. *Tetrahedron*, **1986**, *42*, 1883; (c) Walba, D. M.; Simon, J.; Harary, F. *Tetrahedron Lett.*, **1988**, *29*, 731; (d) Mueller, J. E.; Du, S. M.; Seeman, N. C. *J. Am. Chem. Soc.*, **1991**, *113*, 6306; (e) a nice summary of the progress in this field and its biological implications is: Dietrich-Buchecker, C. O.; Sauvage, J.-P. in *Bioorganic Chemistry Frontiers*, Springer, Berlin, **1991**, vol. 2, pp. 197–246.

[22] (a) General aspects of the problem are explicitely discussed in the lecture: Keese, R. in *Organic Synthesis: Modern Trends*, 6th IUPAC Symposium, Chizov, O. S., Ed., Blackwell Scientific, Oxford, **1987**, p. 43; see also: Luef, W.; Keese, R., in *Advances in Strain in Organic Chemistry*, Halton, B., Ed., JAI Press, London, **1993**, vol. 3, p. 229; (b) for a review, see: Venepalli, B. R.; Agosta, W. C. *Chem. Rev.*, **1987**, *87*, 399; (c) Rao, V. B.; George, C. F.; Wolff, S.; Agosta, W. C. *J. Am. Chem. Soc.*, **1985**, *107*, 5732; (d) Kuck, D.; Bögge, H. *J. Am. Chem. Soc.*, **1986**, *108*, 8107; (e) Kuck, D.; Schuster, A. *Angew. Chem., Int. Ed. Engl.*, **1988**, *27*, 1192; (f) Kuck, D.; Schuster, A.; Krause, R. A. *J. Org. Chem.*, **1991**, *56*, 3472.

[23] (a) For a review, see: Wiberg, K. B. *Chem. Rev.*, **1989**, *89*, 975; (b) Wiberg, K. B.; Walker, F. H. *J. Am. Chem. Soc.*, **1982**, *104*, 5239; (c) Semmler, K.; Szeimies, G.; Belzner, J. *J. Am. Chem. Soc.*, **1985**, *107*, 6410; (d) Wiberg, K. B.; Waddell, S. T. *J. Am. Chem. Soc.*, **1990**, *112*, 2194; (e) Morf, J.; Szeimies, G. *Tetrahedron Lett.*, **1986**, *27*, 5363; (f) Pincock, R. E.; Torupka, E. J. *J. Am. Chem. Soc.*, **1969**, *91*, 4593; (g) Scott, W. B.; Pincock, R. E. *J. Am. Chem. Soc.*, **1973**, *95*, 2040.

[24] (a) General problems of double bond distortions are discussed in the review: Luef, W.; Keese, R. *Top. Stereochem.*, **1991**, *20*, 231; for the earlier data, see: Zefirov, N. S.; Sokolov, V. I. *Usp. Khim.*, **1967**, *36*, 243; (b) for a review on cyclopropene, see: Carter, F. L.; Frampton, V. L. *Chem. Rev.*, **1964**, *64*, 497; (c) Wiberg, K. B.; Bartley, W. K. *J. Am. Chem. Soc.*, **1960**, *82*, 6375; (d) Nesmeyanova, O. A.; Rudashevskaya, T. Y. *Izv. Akad. Nauk SSSR, Ser Khim.*, **1978**, 1562; Rudashevskaya, T. Y; Nesmeyanova, O. A. *Izv. Akad. Nauk SSSR, Ser Khim.*, **1979**, 669; see also: Lehmkuhl, H.; Mehler, K. *Liebigs Ann. Chem.*, **1978**, 1841; (e) Moiseenkov, A. M.; Cheskis, B. A.; Semenovski, A. V. *J. Chem. Soc., Chem. Commun.*, **1982**, 109.

[25] (a) Borden, W. T. *Chem. Rev.*, **1989**, *89*, 1095; (b) Szeimies, G.; Harnisch, J.; Baumgärtel, O. *J. Am. Chem. Soc.*, **1977**, *99*, 5183; see also: Szeimies-Seebach, U.; Szeimies, G. *J. Am. Chem. Soc.*, **1978**, *100*, 3966; Zoch, H.-G.; Shlüter, A.-D.; Szeimies, G. *Tetrahedron Lett.*, **1981**, *22*, 3835; (c) Casanova, J.; Bragin, J.; Cottrell, F. D. *J. Am. Chem. Soc.*, **1978**, *100*, 2264; see also: Wiberg, K. B.; Matturo, M . G.; Okarma, P. J.; Jason, M. E.; Dailey, W. P.; Burgmaier, G. J.; Bailey, W. F.; Warner, P. *Tetrahedron*, **1986**, *42*, 1895; (d) Wiberg, K. B.; Matturo, M. G.; Okarma, P. J.; Jason, M. E. *J. Am. Chem. Soc.*, **1984**, *106*, 2194; (e) Wiberg, K. B.; Adams, R. D.; Okarma, P. J.; Matturo, M. G.; Segmuller, B. *J. Am. Chem. Soc.*, **1984**, *106*, 2200; (f) Renzoni, G. E.; Yin, T.-K.; Borden, W. T. *J. Am. Chem. Soc.*, **1986**, *108*, 7121; (g) Radziszewski, J. G.; Yin, T.-K.; Miyake, F.; Renzoni, G. E.; Borden, W. T.; Michl, J. *J. Am. Chem. Soc.*, **1986**, *108*, 3544; see also: Renzoni, G. E.; Yin, T.-K.; Miyake, F.; Borden, W. T. *Tetrahedron*, **1986**, *42*, 1581; (h) Hrovat, D. A.; Miyake, F.; Trammell, G.; Gilbert, K. E.; Mitchell, J.; Clardy, J.; Borden, W. T. *J. Am. Chem. Soc.*, **1987**, *109*, 5524; (i) Ermer, O.; Bell, P.; Mason, S. A. *Angew. Chem., Int. Ed. Engl.*, **1989**, *28*, 1239; (j) Paquette, L. A.; Shen, C.-C. *J. Am. Chem. Soc.*, **1990**, *112*, 1159; (k) Paquette, L. A.; Kunzer, H.; Green, K. E. *J. Am. Chem. Soc.*, **1985**, *107*, 4788 and references therein.

[26] (a) Kane, V. V.; Wolf, A. D.; Jones, Jr., M. *J. Am. Chem. Soc.*, **1974**, *96*, 2643; (b) Tobe, Y.; Ueda, K.-I.; Kakiuchi, K.; Odaira, Y.; Kai, Y.; Kasai, N. *Tetrahedron*, **1986**, *42*, 1851; (c) Jenneskens, L. W.; De Kanter, F. J. J.; Turkenberg, L. A. M.; de Boer, H. J. R.; De Wolf, W. H.; Bickelhaupt, F. *Tetrahedron*, **1984**, *40*, 4401; see also: Jenneskens, L. W.; Klamer, J. C.; de Boer, H. J. R.; de Wolf, W. H.; Bickelhaupt, F. *Angew. Chem.*, **1984**, *96*, 236; (d) Jenneskens, L. W.; De Boer, H. J. R.; De Wolf, W. H.; BickelHaupt, F. *J. Am. Chem. Soc.*, **1990**, *112*, 8941; (e) for reviews, see: Bickelhaupt, F. *Pure Appl. Chem.*, **1990**, *62*, 373; Bickelhaupt, F.; De Wolf, W. H., in *Advances in Strain in Organic Chemistry*, Halton, B., Ed., JAI Press, London, **1993**, vol. 3, p. 185.

[27] (a) Boekelheide, V., in *Strategy and Tactics of Organic Synthesis*, Lindberg, C., Ed., Academic Press, New York, **1984**, vol. 1, ch. 1, p. 1; (b) Sekine, Y.; Boekelheide, V. *J. Am. Chem. Soc.*, **1981**, *103*, 1777; see also a review on superphanes: Gleiter, R.; Kratz, D. *Acc. Chem. Res.*, **1993**, *26*, 311; (c) Diercks, R.; Vollhardt, K. P. C. *J. Am. Chem. Soc.*, **1986**, *108*, 3150; (d) Mohler, D. L.; Vollhardt, K. P. C.; Wolff, S. *Angew. Chem., Int. Ed. Engl.*, **1990**, *29*, 1151.

[28] Olah, G.; Surya Prakash, G. K.; Williams, R. E.; Field, L. D.; Wade, K. *Hypercarbon Chemistry*, Wiley, New York, **1987**.

[29] (a) McMurry, J. E.; Lectka, T. *J. Am. Chem. Soc.*, **1990**, *112*, 869; (b) Kirchen, R. P.; Ranganayakulu, K.; Rauk, A.; Singh, B. P.; Sorensen, T. S. *J. Am. Chem. Soc.*, **1981**, *103*, 588; (c) McMurry, J. E. *Acc. Chem. Res.*, **1983**, *16*, 405; (d) McMurry, J. E.; Lectka, T.; Hodge, C. N. *J. Am. Chem. Soc.*, **1989**, *111*, 8867; see also: McMurry, J. E.; Hodge, C. N. *J. Am. Chem. Soc.*, **1984**, *106*, 6450; McMurry, J. E.; Lectka, T. *J. Am. Chem. Soc.*, **1993**, *115*, 10167.

[30] A number of exciting stories showing the role of serendipity in great chemical discoveries can be found in an excellent book: Roberts, R. M. *Serendipity. Accidental Discoveries in Science*, Wiley, New York, **1989**.

[31] (a) Alder, R. W.; Bowman, P. S.; Steele, W. R. S.; Winterman, D. R. *J. Chem. Soc., Chem. Commun.*, **1968**, 723; (b) Hünig, S.; Kiessel, M. *Chem. Ber.*, **1958**, *91*, 380; (c) Diem, M. J.; Burrow, D. F.; Fry, J. L. *J. Org. Chem.*, **1977**, *42*, 1801; (d) Hoffmann, H. M. R.; Tsushima, T. *J. Am. Chem. Soc.*, **1977**, *99*, 6008.

[32] (a) Pedersen, C. J. *J. Am. Chem. Soc.*, **1967**, *89*, 2495; (b) Pedersen, C. J. *Angew. Chem., Int. Ed. Engl.*, **1988**, *27*, 1021; (c) Pedersen, C. J. *Aldrichim. Acta*, **1971**, 6 (1), 1; (d) Pedersen, C. J. *J. Am. Chem. Soc.*, **1967**, *89*, 7017; for an extensive coverage of the area, see reviews in: (e) Hiraoka, M. *Crown Compounds: Their Characteristics and Applications*, Elsevier, Amsterdam, **1982**; (f) *Host Guest Complex Chemistry: Macrocycles; Synthesis, Structures, Applications*, Vogtle, F.; Weber, E., Eds., Springer, Berlin, **1985**; (g) *Supramolecular Chemie. Eine Einfuhrung*, Vögtle, F., Ed., Teubner, Stuttgart, **1989**; (h) Maeda, M.; Ouchi, M.; Kimura, K.; Shono, T. *Chem. Lett.*, **1981**, 1573; (i) Shinkai, S.; Koreishi, H.; Ueda, K.; Arimura, T.; Manabe, O. *J. Am. Chem. Soc.*, **1987**, *109*, 6371.

[33] (a) Lehn, J.-M. *Angew. Chem., Int. Ed. Engl.*, **1988**, *27*, 89; (b) Dietrich, B.; Lehn, J.-M.; Sauvage, J.-P. *Tetrahedron Lett.*, **1969**, 2885; (c) Dietrich, B.; Lehn, J.-M.; Sauvage, J.-P. *Tetrahedron Lett.*, **1969**, 2889; (d) Graf, E.; Kintzinger, J.-P.; Lehn, J.-M.; LeMoigne, J. *J. Am. Chem. Soc.*, **1982**, *104*, 1672; (e) Graf, E.; Lehn, J.-M.; *J. Am. Chem. Soc.*, **1976**, *98*, 6403; (f) see data in the comprehensive treatise: Ovchinnikov, Y. A.; Ivanov, V. T.; Shkrob, A. M. *Membrane Active Complexones*, Elsevier, Amsterdam, **1974**; for a concise review, see also: ch. 2 in ref. 32(f); (g) Hosseini, M. W.; Lehn, J.-M.; *J. Am. Chem. Soc.*, **1982**, *104*, 3525; (h) Pascard, C.; Riche, C; Cesario, M.; Kotzyba-Hibert, F.; Lehn, J.-M. *J. Chem. Soc., Chem. Commun.*, **1982**, 557.

[34] (a) For a general discussion of the problem, see: Diederich, F. *Angew. Chem., Int. Ed. Engl.*, **1988**, *27*, 362; (b) Diederich, F.; Dick, K. *J. Am. Chem. Soc.*, **1984**, *106*, 8024; (c) Carcanague, D. R.; Diederich, F. *Angew. Chem., Int. Ed. Engl.*, **1990**, *29*, 769.

[35] (a) Sasaki, H.; Ueno, A.; Osa, T. *Chem. Lett.*, **1986**, 1785; (b) Shinkai, S.; Ogawa, T.; Kusano, Y.; Manabe, O. *Chem. Lett.*, **1980**, 283; (c) Shinkai, S.; Minami, T.; Kusano, Y.; Manabe, O. *J. Am. Chem. Soc.*, **1983**, *105*, 1851; (d) Shinkai, S.; Inuzuka, K.; Miyazaki, O.; Manabe, O. *J. Am. Chem. Soc.*, **1985**, *107*, 3950; (e) Beer, P. D. *Chem. Soc. Rev.*, **1989**, *18*, 409; (f) Gagnaire, G.; Gellon, G.; Pierre, J.-L. *Tetrahedron Lett.*, **1988**, *29*, 933; (g) Rebeck, J., Jr. *Acc. Chem. Res.*, **1984**, *17*, 258.

[36] (a) For a general discussion of pertinent data, see: Dugas, H.; Penney, C. *Bioorganic Chemistry. A Chemical Approach to Enzyme Action*, Springer, New York, **1981**; (b) Menger, F. M. *Acc. Chem. Res.*, **1993**, *26*, 206 and references cited therein; (c) Houk, K. N.; Tucker, J. A.; Dorigo, A. E. *Acc. Chem. Res.*, **1990**, *23*, 107; see also discussion in: Menger, F. M.; Sherrod, M. J. *J. Am. Chem. Soc.*, **1990**, *112*, 8071; (d) Menger, F. M. *Acc. Chem. Res.*, **1985**, *18*, 128; (e) Menger, F. M.; Ladika, M. *J. Am. Chem. Soc.*, **1988**, *110*, 6794.

[37] (a) Deno, N. C.; Fishblein, R.; Pierson, C. *J. Am. Chem. Soc.*, **1970**, *92*, 1451; (b) Eden, Ch.; Shaked, Z. *Isr. J. Chem.*, **1975**, *13*, 1; (c) Barton, D. H. R.; Beaton, J. M. *J. Am. Chem. Soc.*, **1960**, *82*, 2641; (d) Barton, D. H. R.; Beaton, J. M.; Geller, L. E.; Pechet, M. M. *J. Am. Chem. Soc.*, **1961**, *83*, 4076; (e) Barton, D. H. R.; Basu, N. K.; Day, M. J.; Hesse, R. H.; Pechet, M. M.; Starrat, A. N. *J. Chem. Soc., Perkin Trans. I*, **1975**, 2243; see, for the story of these studies: Barton, D. H. R., in *Frontiers in Bioorganic Chemistry and Molecular Biology*, Ovchinnikov, Y. A.; Kolosov, M. N., Eds., Elsevier–North Holland Biomedical Press, Amsterdam, **1979**, p. 21; (f) Breslow, R. *Acc. Chem. Res.*, **1980**, *13*, 170; (g) Maitra, U.; Breslow, R. *Tetrahedron Lett.*, **1986**, *27*, 3087; (h) Lee, E.; Lee, H. H.; Chang, H. K.; Lim, D. Y. *Tetrahedron Lett.*, **1988**, 29, 339; (i) Stuk, T. L.; Grieco, P. A.; Marsh, M. M. *J. Org. Chem.*, **1991**, *56*, 2957; (j) Wiedenfeld, D.; Breslow, R. *J. Am. Chem. Soc.*, **1991**, *113*, 8977; (k) recent advances in the enantiospecific synthesis, including the utilization of chiral catalysts are highlighted in a review: Seebach, D. *Angew. Chem., Int. Ed. Engl.*, **1990**, *29*, 1320; (l) as representative examples of the elaboration of efficient catalysts for enantioselective reactions, see: Sharpless epoxidation of allylic alcohols: Hanson, R. M., Sharpless, K. B. *J. Org. Chem.*, **1986**, *51*, 1922; Corey, E. J. *J. Org. Chem.*, **1990**, *55*, 1693; carbene C–H insertion or cyclopropanation: for a review, see: Doyle, M. P. *Recl. Trav. Chim. Pays-Bas*, **1991**, *110*, 305; *e.g.* Doyle, M. P.; Dyatkin, A. B.; Roos, G. H. P.; Carias, F.; Pierson, D. A.; van Basten, A.; Müller, P.; Polleux, P. *J. Am. Chem. Soc.*, **1994**, *116*, 4507 and references cited therein; (m) Corey, E. J.; Bakshi, R. K.; Shibata, S. *J. Am. Chem. Soc.*, **1987**, *109*, 5551; see also: Corey, E. J.; Bakshi, R. K.; Shibata, S.; Chen, C.-P.; Singh, V. K. *J. Am. Chem. Soc.*, **1987**, *109*, 7925; (n) Corey, E. J. *Angew. Chem., Int. Ed. Engl.*, **1991**, *30*, 1455; see also: Corey, E. J. *Pure Appl. Chem.*, **1990**, *62*, 1209; (o) Corey, E. J.; Cimprich, K. A. *J. Am. Chem. Soc.*, **1994**, *116*, 3151; (p) Corey, E. J.; Imwinkelried, R.; Pikul, S.; Xiang, Y. B. *J. Am. Chem. Soc.*, **1989**, *111*, 5493; (q) Corey, E. J.; DaSilva Jarden, P.; Virgil, S.; Yuen, P.-W.; Connel, R. D. *J. Am. Chem. Soc.*, **1989**, *111*, 9243.

[38] (a) Cram, D. J. *Angew. Chem., Int. Ed. Engl.*, **1988**, *27*, 1009; (b) Cram, D. J.; Cram, J. M. *Acc. Chem. Res.*, **1978**, *11*, 8; (c) Trueblood, K. N.; Knobler, C. B.; Maverick, E.; Helgeson, R. C.; Brown, S. B.; Cram, D. J. *J. Am. Chem. Soc.*, **1981**, *103*, 5594; (d) Cram, D. J.; Kaneda, T.; Helgeson, R. C.; Brown, S. B.; Knobler, C. B.; Maverick, E.; Trueblood, K. N. *J. Am. Chem. Soc.*, **1985**, *107*, 3645; (e) Cram, D. J.; Carmack, R. A.; Helgeson, R. C.; *J. Am. Chem. Soc.*, **1988**, *110*, 571; (f) Moran, J. R.; Karbach, S.; Cram, D. J.; *J. Am. Chem. Soc.*, **1982**, *104*, 5826; (g) Cram, D. J.; Karbach, S.; Kim, Y. H.; Baczynskyj, L.; Kalleymeyn, G. W. *J. Am. Chem. Soc.*, **1985**, *107*, 2575; Cram, D. J.; Karbach, S.; Kim, Y. H.; Baczynskyj, L.; Marti, K; Sampson, R. M.; Kalleymeyn, G. W. *J. Am. Chem. Soc.*, **1988**, *110*, 2554; (h) Bryant, J. A.; Blanda, M. T.; Vincenti, M.; Cram, D. J. *J. Am. Chem. Soc.*, **1991**, *113*, 2167; (i) Sherman, J. C.; Knobler, C. B.; Cram, D. J. *J. Am. Chem. Soc.*, **1991**, *113*, 2194; (j) Chapman, R. G.; Chopra, N.; Cochien, E. D.; Sherman, J. C. *J. Am. Chem. Soc.*, **1994**, *116*, 369; (k) Chapman, R. G.; Sherman, J. C. *J. Am. Chem. Soc.*, **1995**, *117*, 9081; (l) Cram, D. J.; Tanner, M. E.; Thomas, R. *Angew. Chem., Int. Ed. Engl.*, **1991**, *30*, 1024; (m) Quan, M. L. C.; Cram, D. J. *J. Am. Chem. Soc.*, **1991**, *113*, 2754; (n) see, for the highlights of further development: Robbins, T. A.; Cram, D. J. *J. Am. Chem. Soc.*, **1993**, *115*, 12199; Robbins, T. A.; Knobler, C. B.; Bellew, D. R.; Cram, D. J. *J. Am. Chem. Soc.*, **1994**, *116*, 111; (o) Lehn, J. M. *Angew. Chem., Int. Ed. Engl.*, **1990**, *29*, 1304.

[39] (a) See for the references, a review: Braun, M. *Organic Synthesis Highlights*, Mülzer, J.; Altenbach, H.-J., Braun, M.; Kröhn, K.; Reissig, H.-U. Eds., VCH, Weinheim, **1991**, p. 309; (b) Lynch, J. E.; Volante, R. P.; Wattley, R. V.; Shinkai, I. *Tetrahedron Lett.*,

1987, *28*, 1385; (c) for a concise review, see: Cross, P. E.; Dickinson, R. P. *Chem. Br.*, **1991**, *27*, 911; (d) Bundy, G. L. *Tetrahedron Lett.*, **1975**, *24*, 1957; (e) Nicolaou, K. C., *et al.*, *Proc. Natl. Acad. Sci. USA*, **1979**, *76*, 2566; (f) Hiraku, S., *et al.*, *Jpn. J. Pharmacol.*, **1986**, *41*, 393; (g) Friedman, S. H.; DeCamp, D. L.; Sijbesma, R. P.; Srdanov, G.; Wudl, F.; Kenyon, G. L. *J. Am. Chem. Soc.*, **1993**, *115*, 6506; (h) Sijbesma, R.; Srdanov, G.; Wudl, F.; Castoro, J. A.; Wilkins, C.; Friedman, S. H.; DeCamp, D. L.; Kenyon, G. L. *J. Am. Chem. Soc.*, **1993**, *115*, 6510.

[40] (a) Initial advances in this field are highlighted in: Baum, R. M. *Chem. Eng. News.*, **1991**, May 6, 31; (b) see, for a review: Nicolaou, K. C.; Dai, W.-M. *Angew. Chem., Int. Ed. Engl.*, **1991**, *30*, 1387; (c) Ishida, N.; Miyazaki, K.; Kumagai, K.; Rikimary, M. *J. Antibiot.*, **1965**, *18*, 68; (d) Kumagai, K.; Ono, Y.; Nishikawa, T.; Ishida, N. *J. Antibiot.*, **1966**, *19*, 50; (e) see a detailed discussion of the mechanism of its action in the review: Goldgerg, I. H. *Acc. Chem. Res.*, **1991**, *24*, 191; (f) Edo, K.; Mizugaki, M.; Koido, Y.; Seto, H.; Furihata, K.; Otake, N.; Ishida, N. *Tetrahedron Lett.*, **1985**, *26*, 331; (g) Golik, J.; Clardy, J.; Dubay, G.; Groenewold, G.; Kawaguchi, H.; Konishi, M.; Krishnan, B.; Ohkuma, H.; Saitoh, K.-i.; Doyle, T. W. *J. Am. Chem. Soc.* **1987**, *109*, 3461; Golik, J.; Dubay, G.; Groenewold, G.; Kawaguchi, H.; Konishi, M.; Krishnan, B.; Ohkuma, H.; Saitoh, K.-i.; Doyle, T. W. *J. Am. Chem. Soc.* **1987**, *109*, 3462; (h) Lee, M. D.; Dunne, T. S.; Siegel, M. M.; Chang, C. C.; Morton, G. O.; Borders, D. B. *J. Am. Chem. Soc.*, **1987**, *109*, 3464; Lee, M. D.; Dunne, T. S.; Chang, C. C.; Ellestad, G. A.; Siegel, M. M.; Morton, G. O.; McGahren, W. J.; Borders, D. B. *J. Am. Chem. Soc.*, **1987**, *109*, 3466; (i) Konishi, M.; Ohkuma, H.; Tsuno, T.; Oki, T.; Van Duyne, G. D.; Clardy, J. *J. Am. Chem. Soc.*, **1990**, *112*, 3715; (j) Leet, J. E.; Schroeder, D. R.; Langley, D. R.; Colson, K. L.; Huang, S.; Klohr, S. E.; Lee, M. S.; Golic, J.; Hofstead, S. J.; Doyle, T. W.; Matson, J. A. *J. Am. Chem. Soc.*, **1993**, *115*, 8432; (k) Christner, D. F.; Frank, B. L.; Kozarich, J. W.; Stubbe, J.; Golik, J.; Doyle, T. W.; Rosenberg, I. E.; Krishnan, B. *J. Am. Chem. Soc.*, **1992**, *114*, 8763; (l) Lee, M. D.; Ellestad, G. A.; Borders, D. B. *Acc. Chem. Res.*, **1991**, *24*, 235; (m) for references, see the review: Halcomb, R. L. *Proc. Natl. Acad. Sci. USA*, **1994**, *91*, 9197.

[41] (a) Bergman, R. G. *Acc. Chem. Res.*, **1973**, *6*, 25; (b) Jones, R. R.; Bergman, R. G. *J. Am. Chem. Soc.*, **1972**, *94*, 660; (c) Lockhart, T. P.; Comita, P. B.; Bergman, R. G. *J. Am. Chem. Soc.*, **1981**, *103*, 4082; (d) De Voss, J. J.; Hangeland, J. J.; Townsend, C. A. *J. Am. Chem. Soc.*, **1990**, *112*, 4554.

[42] (a) Nicolaou, K. C.; Zuccarello, G.; Ogawa, Y.; Schweiger, E. J.; Kumazava, T. *J. Am. Chem. Soc.*, **1988**, *110*, 4866; (b) Nicolaou, K. C.; Ogawa, Y.; Zuccarello, G.; Kataoka, H. *J. Am. Chem. Soc.*, **1988**, *110*, 7247; (c) Crévisy, C.; Beau, J.-M. *Tetrahedron Lett.*, **1991**, *32*, 3171; (d) Sakai, Y.; Nishiwaki, E.; Shishido, K.; Shibuya, M.; Kido, M. *Tetrahedron Lett.*, **1991**, *32*, 4363; (e) Mantlo, N. B.; Danishefsky, S. J. *J. Org. Chem.*, **1989**, *54*, 2781; (f) Haseltine, J. N.; Danishefsky, S. J. *J. Org. Chem.*, **1990**, *55*, 2576; (g) Haseltine, J. N.; Danishefsky, S. J. *J. Am. Chem. Soc.*, **1989**, *111*, 7638; (h) Sugiura, Y.; Shiraki, T.; Konishi, M.; Oki, T. *Proc. Natl. Acad. Sci. USA*, **1990**, *87*, 3831; (i) Nicolaou, K. C.; Smith, A. L.; Wendeborn, S. V.; Hwang, C.-K. *J. Am. Chem. Soc.*, **1991**, *113*, 3106; (j) Nicolaou, K. C.; Maligres, P.; Suzuki, T.; Wendeborn, S. V.; Dai, W.-M.; Chadha, R. K. *J. Am. Chem. Soc.*, **1992**, *114*, 8890; (k) Nicolaou, K. C.; Dai, W.-M. *J. Am. Chem. Soc.*, **1992**, *114*, 8908; see also: Nicolaou, K. C.; Dai, W.-M.; Hong, Y. P.; Tsay, S.-C.; Baldridge, K. K.; Siegel, J. S. *J. Am. Chem. Soc.*, **1993**, *115*, 7944.

CHAPTER 5

Instead of Conclusion

5.1 A LITTLE BIT MORE ABOUT THE ROLE OF SYNTHESIS AND ITS RELATIONSHIP TO GENERAL ORGANIC CHEMISTRY

In the beginning of this book we discussed the specific goals of an organic synthesis in some detail. There is, however, still another function of a synthesis, consideration of which in the initial pages would have been premature. Now we are in a better position to treat this topic.

The whole manifold of organic compounds can be represented as a sort of multidimensional hyperspace pierced by numerous coordinates of classical systematics such as homologous series, types of functionality, series of structural isomers, and so on. Any one of these coordinates reflects some particular feature of the structural characteristics present in the organic molecule and hence any coordinate could be used to disclose the closeness (or distinction) between the compounds. Organic synthesis brings into this hyperspace an additional and important dimension based on synthetic relationships. If one takes into account this dimension, which reflects the possibilities of ready interconversions among different molecular structures, then apparently unrelated compounds or even whole classes of compounds (areas of our hyperspace) may turn out to be close neighbors. The proximity thus disclosed does not represent just an arbitrarily chosen artifact but reflects some intrinsic and meaningful connections. It is organic synthesis which integrates the separate parts of organic chemistry into a unified and closely bound system. Let us consider this claim more specifically.

Conventional texts on organic chemistry are usually divided into chapters corresponding to compound types, chemical reactions, synthetic methods, *etc.* From the viewpoint of synthesis, this traditional classification should be supplemented by consideration of the specific aspects of synthetic relationships *i.e.* into what a given compound can be converted and from what it can be obtained, in other words, its place in the solution of synthetic tasks. For example, in accordance with conventional classification, alkenes, alkynes, cyclopropanes, and oxiranes fall into substantially different and rather distant 'taxons' of organic chemistry systematics. These classes traditionally are treated as independent and only remotely related areas of organic chemistry. However, the demands of organic synthesis make it imperative to consider also the close genetic relationship between these types of compounds from the point of view of their potential synthetic equivalency.

The very nature of modern organic synthesis illustrates its truly powerful unifying and integrating function. One of the most striking features of the modern art of solving synthetic problems is an extremely wide application of the wealth of the accomplishments of organic chemistry procured from all branches of this science.

'Bridges' connecting rather dissimilar types of organic structures are at present being established with surprising ease. The following is just one example of such a 'bridge', constructed recently over an apparent 'abyss' of structural discrepancy. From the point of view of classical organic chemistry, carbohydrates, owing to the peculiarities of their structure and chemical behavior, have been set aside as a separate realm from the majority of other areas of organic chemistry and seem to have very little in common with, say, hydrocarbon chemistry. Yet it was recognized that carbohydrates can be used as an abundant pool of chiral precursors for the synthesis of enantiomerically pure compounds. Hence numerous, and at times ingenious, synthetic pathways have been elaborated for the conversion of carbohydrates into chiral compounds belonging to such diverse classes as aliphatic hydrocarbons or polycyclic heterocycles.

In the 1950s the choirmaster of organic synthesis, Robert Woodward, declared, 'that the time is over for the narrow professionals which are highly specialized, say in the field of carbohydrates or alkaloids, since modern synthetics are able to work successfully in any field of organic chemistry ...'. When issued, this claim was in obvious contrast to the dominating tendency for narrow specialization and hence was perceived by many organic chemists as belonging to the area of wishful thinking. Later developments confirmed fully the correctness of Woodward's prophecy and nowadays it is a true imperative for professional synthetic chemists to acquire a command in all areas of organic chemistry. Such conventional names as, for example, 'the laboratory of the chemistry of carbocyclic (or heterocyclic, aromatic, transition metal, *etc.*) compounds', became meaningless and archaic.

The new mindset of contemporary organic chemists is actually based on a *systems analysis* approach which implies a thorough and generalized examination of the phenomena in relation to fairly different areas of organic chemistry and science in general. In a sense, organic synthesis is emerging now as a sort of unifying 'meta-science'. Without a proper understanding of the connections within this meta-science, an organic chemist is liable to lose perception of the true value of his/her investigations and may overlook the most promising ramifications of the results procured by investigations in a particular area of organic chemistry.

It is instructive to bring up yet another feature of modern organic synthesis. Outstanding multiple-step syntheses of the 1930s and 1940s certainly caused true excitement among the chemical community. At the same time, however, these almost ecstatic feelings were often mixed with some perplexity concerning the justification of the tremendous efforts expended to achieve this or that goal. In fact, aside from meeting the challenge of preparation of the most complicated organic compounds, the net output of these 'marathon runs' for organic chemistry as a whole quite often was not very significant, insomuch as these syntheses represented more often than not the result of skillful manipulation of known preparative methods.

Now, as earlier, the successful synthesis of a complex molecule is still considered *per se* to be an outstanding achievement. However, much greater applause is gained by those total syntheses which are based on a conceptually non-trivial insight that ensured the accomplishment of the goals with a maximum economy of time and energy. Realization of such an approach usually requires the elaboration of novel methodology for carrying out a previously unknown conversion at a key step(s) of the strategic plan and thus tremendous efforts may be spent to achieve this purpose. However, here the final success means something more than finding an effective pathway for the synthesis of a selected target molecule. It also enriches greatly the entire arsenal of synthetic methods and stimulates the application of discovered approaches in the design of other complex and formally unrelated syntheses. Thus we may eyewitness the gradual disappearance of the traditional borderline which divided the chemists studying the reactivity pattern of organic compounds from those who preferred to devote their efforts to total synthesis.

5.2 ORGANIC CHEMISTRY AS A FUNDAMENTAL AND RIGOROUS SCIENCE

The words 'fundamental' (or 'pure') science sometimes are interpreted merely as an antonym to the term 'applied' science. In fact the term 'fundamental' science implies that such a science is capable of procuring some basic knowledge which can in turn provide an understanding of a higher order of complexity phenomena. We take it for granted that there is no need to argue that fundamental science as the source of our understanding of the world has its own worth, not related to any immediate practical value.

It is generally recognized that the most fundamental natural science is physics. Nevertheless, for a large number of natural and applied sciences, physics turns out to be 'too fundamental'. For example, an understanding of quantum mechanics as a fundamental discipline for both chemistry and biology is unlikely to be of any practical help to the clinical physician or agronomist for the solution of their immediate problems. The gap between Planck's constant and the problems of the given patient or a wheat field is painfully too wide. Fortunately, the existing system of knowledge has evolved in such a manner that its fairly distant parts are connected *via* intermediate links — areas of science — which serve as a means of adapting the more fundamental knowledge to the particular problem at hand. It is easy to trace the hierarchy of 'fundamentality', in which every level feeds its main accomplishments to the next 'floor' and, in turn, rests on more general and fundamental laws that were provided by the previous 'floor'. What can be said about the position of organic chemistry in such a hierarchy?

Human beings exist in a world of organic compounds and represent an essential part of this world. Organic compounds and their transformations constitute the material basis for the functioning of all known forms of life on our planet. Therefore without an understanding of the basic principles of organic chemistry one cannot rationally investigate the essence of biological phenomena. At the same time, biological sciences are fundamental in relation to such important applied sciences as agriculture or medicine. Hence the development of the latter areas depends on the achievements of biology, which in its turn is connected, *via* molecular biology, biochemistry, and the chemistry of natural compounds, to organic chemistry.

It is even easier to trace down rather short chains connecting organic chemistry as a fundamental science with applied areas such as technology, processing and utilization of organic raw materials, chemistry and technology of polymers, drugs, dyes, cleaning materials, perfume materials, *etc.* In reality, the very appearance of these areas of applied activity has been fully indebted to the progress of the academic investigations of organic chemists and any accomplishment in these applied sciences was always prefaced by certain achievements in pure organic chemistry.

In other words, organic chemistry holds a position that provides a theoretical basis for both pure and applied activities in any field dealing with organic compounds. Therefore the development of organic chemistry, and, in particular, organic synthesis as one of the most important components of this science, has a distinct value in its own right and does not require any justification referring to the immediate usefulness of its results.

What serves as the basis of organic chemistry? The answer seems to be obvious: it is quantum chemistry which is based, in turn, on quantum mechanics. However, sometimes these obvious considerations may lead one to rather unjustified generalizations. The typical way of reasoning in these cases runs as follows. In the final analysis there is nothing in any molecule except atomic nuclei and electrons and their behavior could be exhaustively described with the help of the Schrödinger equation. Hence, for the complete description of any chemical phenomenon it is necessary and sufficient to formulate and solve the corresponding equations, taking into account the position and mutual interaction of all nuclei and electrons of the system. From this point of view, chemistry cannot be considered as an independent science pursuing its own objectives. Hence the very existence of this science can be justified only up to the moment when a sufficiently sophisticated mathematical apparatus of quantum chemistry is developed and supplemented by the availability of adequate computing facilities. This way of reasoning ('reductionism'), while being logically correct (in a formal sense), is in fact open to serious criticism, to say the least.

First of all, it is highly doubtful in general that systems of higher complexity can be exhaustively described and understood in terms elaborated for the description of simpler systems. This is, in essence, a purely philosophical issue:

whether or not the reductionism can be accepted as a reliable way of reasoning and understanding in science. A high level philosophical discussion of this problem is hardly appropriate here, yet we can suggest some considerations based on a rather common practice of organic chemistry experimentation.

Any organic chemist has to deal daily with scores of organic compounds, and likewise must solve on a daily basis the tasks related to the expected chemical behavior of these compounds under highly variable circumstances (temperature, media, reaction partners, catalysts, *etc.*). He/she must obtain answers to these problems promptly and with an acceptable level of reliability. Again it might be argued that, in principle, quantum chemistry is capable of providing quantitatively exact answers to any question of this type. Yet in present practice, organic chemists are usually quite successful in solving these tasks without the help of quantum chemistry, using traditional approaches based mainly on the purely qualitative concepts which, nevertheless, enable them to recognize immediately the basic peculiarities of the chemical phenomenon being studied. In other words, organic chemistry still preserves the looks of a 'kingdom of the crawling empiricism', where reasoning by analogy is the most popular and, we may add, a very efficient tool for problem solving.

It is also fairly obvious that the 'territory' of this kingdom is now contracted owing to the development of computational methods and more powerful computers. Yet there are no truly compelling reasons for the claims that that the appearance of powerful computers of, say, the 5th, 6th, and *n*th generations will put an end to this empiricism and thus transform organic chemistry into a fully quantified and hence 'exact' science. In our opinion, these claims are not justified first of all because of the intrinsic propensity of organic chemistry to create permanently its own objects of exploration. Correspondingly, the diversity and complexity of organic chemistry problems tend to expand without any limits and with an increasing rate. One can hardly expect that a truly rigorous and complete quantitative description might be an achievable goal for these sorts of tasks, however powerful are the calculating facilities employed.

What then, can organic chemistry as a science draw out from quantum chemistry? In the search for the answer it is useful to look at the already accumulated experience of the interactions in these closely related areas of chemical science. In the last decades there have evolved various methods for the non-empirical and semi-empirical calculations of structure and reactivity of organic molecules based on quantum mechanics. In numerous cases these calculations turned out to be of extreme usefulness in obtaining quantitative information such as the charge distribution in a molecule, the reaction indices of alternate reaction centers, the energy of stabilization for various structures, the plausible shape of potential energy surfaces for chemical transformations, *etc.* This list seems to include almost all parameters that are needed for the explanation and prediction of the reactivity of a compound, that is, for solving the main chemical task. Yet there are several intrinsic defaults that impose rather severe limitations on the scope of the reliability of this approach.

First of all, such calculations can be carried out with sufficient accuracy only for rather simple structures and most frequently the results obtained cannot be accurately extrapolated even to related (but more complicated) systems. Secondly, these calculations refer to ideal situations such as the behavior of an isolated molecule. The validity of these results, strictly speaking, is thus limited to gas phase reactions. For these reasons, quantum mechanical calculations have not yet become daily working instruments in chemical practice and it is hardly to be expected that this approach might ever become a universal tool for the solution of chemical problems. At the same time, there are plenty of examples of entirely different and more fruitful ways for the application of quantum chemistry to organic chemistry.

Throughout this book we often referred to the effectiveness of simple qualitative concepts routinely used in organic chemistry. The majority of these concepts originated as the result of a rationalization of voluminous empirical data. With the elaboration of quantum chemistry methods it became possible to provide a rigorous formulation based on a non-empirical basis for a number of the major concepts of organic chemistry. Representative examples are served by Hückel's concept of aromaticity and the Woodward–Hoffmann rules of conservation of orbital symmetry. We rush to assert that the ramifications of these achievements for the development of organic chemistry are more significant than the contributions of all the other results of calculation methods taken together. Their very strength lies in the simplicity of the formulated concept, which can be widely employed not only for the consistent interpretation of a huge database of factual material, but also for predicting new phenomena with a high degree of accuracy. It is especially noteworthy that these concepts are ultimately based on solid theoretical grounds and hence are incomparably more valuable than the previously employed 'rule-of-thumb' generalizations. As was stated in the benchmark publication of Hoffmann and Woodward, 'we have relied on the most basic ideas of molecular orbital theory — the concepts of symmetry, overlap, interactions, bonding, and the nodal structure of the wave functions. The lack of numbers in our discussion is not a weakness — it is its greatest strength. ... But an argument from first principles or symmetry, of necessity qualitative, is in fact much stronger than the deceptively authoritative numerical result' (*Acc. Chem. Res.*, **1968**, *2*, 17). By no means are we trying to cast any doubts on the significance of the utilization of purely calculational methods for the solution of a number of specific tasks in organic chemistry. However, far more important and promising is the application of quantum chemistry for the elaboration of general concepts, such as those mentioned above.

We are now going a step further to present our conviction that organic chemistry *per se* is an exact science and its exactness is not at all determined by the extent to which it uses the calculation methods of quantum chemistry. It is expedient to discuss this claim more thoroughly, since among our colleagues (mathematicians, physicists, and even physical chemists) there still pervades a bias that organic chemistry cannot be correctly listed as an exact science, but rather as the sum of empirical knowledge that has not yet undergone a sound quantitative treatment.

It is certainly true that organic chemistry is almost devoid of the canonical entourage of an 'exact science' in the form of rigorous mathematical apparatus describing its basic phenomena. From this point of view, organic chemistry definitely does not fit into the most common understanding of how an 'exact science' should look. However, such a judgement is obviously superficial and oversimplified. As a matter of fact, organic chemistry operates with specific and rather complicated objects which are not necessarily subject to a quantitative description. We feel it justified to assert that organic chemistry (and probably chemistry in general) belongs to a category of science which is qualitative by its very nature and not because of the immaturity of its theory.

At the bottom line, the strictness of any science implies the exactness and reliability of predictions made on the basis of a few initial assumptions, regardless of whether the latter are formulated in the rigorous form of mathematical equations or bear a qualitative and heuristic character. By this criterion, organic chemistry, no doubt, should be considered as an exact science. To illustrate this claim we will confine ourselves to only a few examples.

The concept of steric effects, an extremely important component of the modern theory of organic chemistry, is based upon an oversimplified assumption which considers organic molecules in terms of ball-and-stick models. The latter seem to be absolutely inadequate or even basically incorrect from the point of view of quantum chemistry. Nevertheless, the imaginative and insightful application of this approach enabled organic chemists to start to develop conformational analysis as one of the most fruitful theories of modern organic chemistry, capable not only of explanation but also of prediction of chemical phenomena.

In a more refined and sophisticated way, this nearly mechanistic approach to the consideration of molecular structure emerged as a set of empirical force field or molecular mechanics methods that are widely used as a non-quantum mechanical way for computing the structure and properties of molecules as well as for transition-state modeling. These methods, while still being vulnerable to criticism because of the absence of unambiguous criteria to select the 'right' potential function and/or associated parameter sets, proved to be extremely helpful tools not only for the explanation of chemical phenomena but also for the prediction of the preferred reaction pathways and thus became instrumental in the planning of organic synthesis (see, for the pertinent data, the set of reviews in *Chem. Rev.*, **1993**, *93*, number 7).

It is also relevant to remember that the very idea of the rigorously defined, geometrically ordered positioning of the atoms forming an organic molecule, which had been formulated more than century ago as a purely qualitative concept, structural theory, proved to be of overwhelming importance for the development of organic chemistry to an extent comparable with the role of Darwin's evolution theory for biology.

An essentially qualitative character was likewise embedded in the theoretical representation of the mechanism of various organic reactions which implied more or less speculative suggestions about the structure of reactive intermediates. This, however, did not prevent the creation of new synthetic methods based on the assumption of the formation of these intermediates as discrete species and the consideration of their reactivity patterns.

There is also a logical fault in the assertion that organic chemistry is merely a descriptive science since it does not have at its disposal the mathematical apparatus to formulate its generalizations. As a matter of fact, there exists such an apparatus in organic chemistry but its shape is rather distinct from the conventional one used, for example, in physics or mathematics.

Consider, in a very general way, the major features of the mathematical apparatus as a universal logical method which enables one to arrive at a set of rigorous and general deductions from a few original premises. To achieve this goal it is necessary to have a system of reasoning, based on strict and well-defined rules for completing consecutive logical steps. The symbols utilized for the designation of parameters and operators are nothing more than merely the technical tools, which simplify the application of formal rules and the formulation of the final results. In all of this, obviously, the very shape of the employed symbols does not have any particular significance.

From this point of view, the structural symbols employed in organic chemistry can be correctly considered as the equivalents of mathematical symbols, specifically devised for the description of organic chemistry phenomena. Rules of operation with these symbols are explicitly formulated by the theory of organic chemistry. Numerous examples of the utilization of these symbols as tools to express the results of the chemical transformations can be found in any textbook on organic chemistry. Now, the course of reasoning in organic chemistry implies the manipulation with structural formulas, manipulations which are by no means arbitrary, but strictly formalized and subject to a well-defined logic. Thus organic chemistry does not use the traditional mathematical language primarily because it does not need them; it has been able to create instead its own language of symbols as well as the rules for the manipulation with these symbols to secure the fast and unamibiguous processing of input information. With this formalism and a few initial assumptions, an organic chemist can arrive at fairly general and reliable conclusions, valid for the prediction of yet unknown phenomena.

We can also add, in passing, that the formulation of principles of retrosynthetical analysis can be viewed as a further evolution of these tools in compliance with the emerging requirements of syntheses. The very fact that the conventional reasoning of organic chemistry can be successfully translated to machine language in a way as, for example, was done in the previously discussed LHASA program, may serve as an illustration to the mathematical rigor of the logic involved.

Finally, as was stated many times by Woodward and other outstanding organic chemists, the successful accomplishment of a total synthesis is the best proof of the rigor and exactness of organic chemistry, since the multistep synthesis of a structurally complicated compound may be considered a viable undertaking only under the condition of a high predictability of the results in the whole sequence of reactions involved even when applied to previously unknown substrates.

Thus, organic synthesis represents a huge and vigorously expanding area of scientific activity, the progress of which is fed by the successes of all fundamental chemical sciences, and the results of which enrich both organic chemistry and, in more general terms, the natural sciences.

Here is what we would like to add in conclusion. The reader, most likely, already has noticed that so often throughout this book we have resorted to a rather descriptive terminology as 'good' or 'bad' reactions (leaving group, disconnection), 'elegant' (method, synthesis, application), 'soft' or 'hard' (conditions, effect), *etc*. Such terminology seems to be more appropriate for art rather than scientific descriptions. Yet the utilization of these terms is by no means to be perceived as the authors' arbitrariness, or even less so as an indication of the non-strictness of organic chemistry. On the contrary, this is a mere manifestation of the intrinsic complexity of this science, reflection of the fact that all of organic synthesis rests on a delicate balance between contradictory extremes: generality *vs*. selectivity of reactions, high reactivity of intermediates *vs*. their stabilization, maximizing diversity *vs*. maximizing standardization, stable protecting *vs*. easily removable groups, *etc*. It is just this aspect that brings to the profession of the synthetic chemist a very special flavor of heuristics and intuitions, making organic synthesis reside on the edge between a grand science and a true art. The best examples of organic syntheses cannot help but stir up an admiration for the beauty and perfection of a discovered solution (here again the readers are addressed to Hoffmann's book *The Same and Not The Same*, mentioned in the Introduction).

We must not forget that at the bottom line of this art lies a rigorous analysis of the structure and chemical properties of the compounds used. It is exactly this well-developed apparatus of logical analysis and foresight that caused the 'breath' of modern syntheses at the highest level, and made it accessible not only to the singled-out members of a 'professional elite', but to a wide circle of qualified 'ordinary' professionals.

In the beginning of this century the outstanding Russian poet Osip Mandelstam wrote, looking at the exquisite shape of the Admiralty building in Saint Petersburg:

'. . . красота не прихоть полу-Бога,
а хищный глазомер простого столяра'.

('Камень', 1913, Санкт-Петербург)

These words, we believe, are equally applicable to the creations of organic synthesis.

'. . . beauty is no demigod caprice
But is caught by a simple carpenter's greedy eye'.

(Osip Mandelstam, 'Stone', Harville, Great Britain, 1991, translated by Robert Tracy)

Subject Index